# Symmetries in Physics

Andreas Wipf

# Symmetries in Physics

## Group and Representation Theory with Applications

 Springer

Andreas Wipf (iD)
Theoretisch-Physikalisches-Institut
Friedrich-Schiller-Universität Jena
Jena, Thüringen, Germany

ISBN 978-3-662-72674-7          ISBN 978-3-662-72675-4    (eBook)
https://doi.org/10.1007/978-3-662-72675-4

This book is a translation of the original German edition "Symmetrien in der Physik" by Andreas Wipf, published by Springer-Verlag GmbH, DE in 2023. The translation was done with the help of an artificial intelligence machine translation tool. A subsequent human revision was done primarily in terms of content, so that the book will read stylistically differently from a conventional translation. Springer Nature works continuously to further the development of tools for the production of books and on the related technologies to support the authors.

Translation from the German language edition: "Symmetrien in der Physik" by Andreas Wipf, © Der/die Herausgeber bzw. der/die Autor(en), exklusiv lizenziert an Springer-Verlag GmbH, DE, ein Teil von Springer Nature 2023. Published by Springer Berlin Heidelberg. All Rights Reserved.



Editorial Contact: Caroline Strunz
This Springer imprint is published by the registered company Springer-Verlag GmbH, DE, part of Springer Nature.
The registered company address is: Heidelberger Platz 3, 14197 Berlin, Germany

If disposing of this product, please recycle the paper.

# Preface

In both the sciences and the arts, symmetries and symmetry considerations play a prominent role. In particular, modern physics, equipped with the tools of group and representation theory, has placed symmetry considerations at the center of interest. Many advances in understanding and in simplifying and classifying phenomena, and the insights and knowledge that come with them, testify to the effectiveness of the symmetry principle. Symmetries and symmetry breakings are, for example, indispensable in the physics of condensed matter or particle physics. Without symmetry considerations, the efforts to unify the forces are unthinkable.

What exactly is meant by symmetries and how are they used? This book attempts to answer these questions. It is based on the content of my regular lectures on symmetries in physics at the Friedrich Schiller University Jena. For a better understanding of the material presented in this book, questions, short tasks, and examples highlighted in gray have been included in the text. Each chapter contains references to accompanying literature and at the end of the chapter some further exercises.

The presentation is at the level of an elective lecture in theoretical physics for advanced students in the bachelor's program or students in the first year of the master's program. However, the material presented goes beyond what can be discussed during one semester.

The content of the book can be divided into three parts. The first part on group theory discusses the mathematical foundations that are indispensable or at least helpful when working with group theoretical techniques. In structuring the material into more than 20 chapters and selecting numerous examples interspersed in the text, I was guided by the diverse applications in physics—and here in particular quantum mechanics and quantum field theory. Thus, the first part also includes three chapters on space-time symmetries, point groups, and space groups. These also contain simple applications from classical mechanics, molecular physics, and crystallography.

The second part focuses on the theory of Lie groups and Lie algebras with their representations. We start with the important lemma of Schur and many of its consequences for irreducible representations and their characters, for example, the important orthogonality theorems. An alternative, and often preferred by physicists, way to the irreducible representations of a Lie group leads via its Lie algebra. With the tools of linear algebra, it is possible to classify all representations of a semi-simple Lie algebra, and each of these representations provides a representation of the corresponding Lie group (or its universal cover). Also in this part, problems and examples from physics are addressed, for example, the spherical harmonics, which represent the rotation group SO(3). Also addressed are the universal cover SU(2) of SO(3) as rotational symmetry in quantum mechanics and the Lie group SU(3) with its representations appearing in quantum chromodynamics. The Lie algebras of the Lorentz and Poincaré groups appearing in every relativistic theory are treated as well.

The last part of the book contains selected applications of group and representation theory to problems of non-relativistic and relativistic physics. Here, the focus is on systems and properties for which the insights already gained can be utilized. We start with the theory of angular momentum and the Wigner-Eckart theorem and end with the Virasoro algebra in conformally invariant quantum field theories. We discuss the Noether theorem for internal and space-time symmetries in classical and quantized field theories in great detail. Of course, in a book about symmetries in physics, the eminently important gauge theories must not be missing. Their principles and their occurrence in particle physics are the content of Chap. 20.

In many of my scientific works, symmetries of systems with finite or infinite many degrees of freedom (and their breaking) play a significant role. I would have liked to include further interesting results—for example, about Kac-Moody or $W$-algebras, supersymmetry, or dynamic symmetries–in this book. This did not happen due to lack of space and time. However, I would like to take this opportunity to thank some colleagues with whom I was able to share many insights about symmetries in physics. First and foremost, I thank Lochlain O'Raifeartaigh, to whom I owe many insights about symmetries. I would also like to express my sincere thanks to Norbert Straumann, my doctoral supervisor. I have benefited from working with Manuel Asorey, Janos Balog, Laszlo Feher, Peter Forgacs, Luis Inzunza, Mihkail Plyushchay, Ivo Sachs, Thomas Strobl, Izumi Tsutsui, Sebastian Uhlmann, and Christian Wozar, in which symmetries were at the center of our discussions and research.

Special thanks go to Marc Steinhauser, who helped with the creation of figures and especially to Michael Mandl for reading through the German version of the entire manuscript chapter by chapter multiple times and his helpful suggestions for improvement.

This book is dedicated with love to my wife Ingrid, who has accompanied me over the years—even though I occasionally "lived in a parallel universe"—as well as to our children Leonie, Severin, and Valentin.

Jena, Germany                                                        Andreas Wipf

**Competing Interests** The author has no competing interests to declare that are relevant to the content of this manuscript.

# Contents

# Abbreviations

| | |
|---|---|
| $\circ$ | Group multiplication |
| $G$, $|G|$ and $g, h, \ldots$ | Group, group order, and group elements |
| $\mathrm{Aut}(G)$ | Automorphism group of $G$ |
| $T(G)$ | Group table |
| $\ell_g, r_g$ | Left and right multiplication with $g$ |
| $<, \leq, \lhd, \unlhd$ | Subgroups and invariant subgroups (normal subgroup) Lie subalgebras and invariant Lie subalgebras (ideals) |
| $\mathrm{Ker}(\varphi)$ and $\mathrm{Im}(\varphi)$ | Kernel and image of a homomorphism $\varphi$ |
| $H_1 \times H_2$ and $H \ltimes N$ | Direct product and semi-direct product of groups |
| $D$ and $\mathcal{R}$ | Representation and regular representation of a group |
| $D_1 \oplus D_2$ and $D_1 \otimes D_2$ | Direct sum and direct product of representations |
| $\chi_D$ and $\chi_\lambda(h)$ | Character of $D$ and the representation with highest weight $\lambda$ |
| $\mathcal{M}(f)$ | Invariant averaging of the function $f : G \mapsto \mathbb{C}$ |
| $\mathrm{d}\mu$ and $\mathrm{d}\mu_{\mathrm{red}}$ | Haar measure and reduced Haar measure |
| $\mathrm{Ad}$ | Adjoint mapping $G \mapsto \mathrm{Aut}(G)$, $g \mapsto \mathrm{Ad}_g$, $\mathrm{Ad}_g(a) = gag^{-1}$ |
| | Adjoint representation $G \mapsto \mathrm{GL}(\mathfrak{g})$, $g \mapsto \mathrm{Ad}(g)$, $\mathrm{Ad}(g) X = gXg^{-1}$ |
| $\mathfrak{g}$, $T_e G$ and $X, Y, \ldots$ | Lie algebra and Lie algebra of $G$ as well as vectors in $\mathfrak{g}$ |
| $K(X, Y)$ and $[X, Y]$ | Killing form and Lie bracket |
| $K_{ij}$ and $f_{ab}{}^c$ | Cartan matrix and structure constants |
| $\mathcal{D}$ and $D_*$ | Representation of $\mathfrak{g}$ and of $T_e G$ induced by $D$ |
| $\mathrm{ad}$ | Adjoint representation $\mathfrak{g} \mapsto \mathrm{L}(\mathfrak{g})$, $X \mapsto \mathrm{ad}_X$, $\mathrm{ad}_X(Y) = [X, Y]$ |
| $\alpha$, $\gamma_i$ and $\delta$ | Roots, simple roots and Weyl vector |
| $\Phi$, $\Phi^+$ and $\Delta$ | Set of roots, positive roots and simple roots |
| $\mu$, $\mu_i$, $\lambda$ | Weight, simple weight and highest weight |
| $[n_1, \ldots, n_r]$ | Dynkin labels |
| $\mathcal{W} = N_G(T)/T$ | Weyl group of $G$ |
| $\sigma_\alpha$ and $w$ | Weyl reflection and elements of $\mathcal{W}$ |
| $\mathcal{W}_\Delta$ | Fundamental Weyl chamber |
| $P_S$ and $P_A$ | Symmetrizer and antisymmetrizer |
| $T_\lambda$, $Y_\lambda$ | Young diagram and Young symmetrizer |

| | |
|---|---|
| $\mathrm{Mat}(n, \mathbb{K})$ | $n \times n$ matrices over field $\mathbb{K}$ |
| $\mathrm{GL}(n, \mathbb{K})$ and $\mathrm{SL}(n, \mathbb{K})$ | General linear and special linear groups over field $\mathbb{K}$ |
| $\mathrm{O}(n)$ and $\mathrm{SO}(n)$ | Orthogonal and special orthogonal groups |
| $\mathrm{U}(n)$ and $\mathrm{SU}(n)$ | Unitary and special unitary groups |
| $\mathrm{O}(1, 3)$ and $\mathrm{O}(p, q)$ | Lorentz group and pseudo-orthogonal groups |
| $\mathcal{S}_M$ | Set of bijective mappings $M \mapsto M$ |
| $\mathcal{S}_n$ | Symmetric group, group of permutations of $n$ objects |
| $\mathcal{A}_n \lhd \mathcal{S}_n$ | Alternating group |
| $\mathcal{T}, \mathcal{O}, \mathcal{Y}$ | Three of the five Platonic solids |
| $\mathcal{C}_n$ and $\mathcal{D}_n$ | Cyclic and dihedral groups |
| $\mathcal{L}$ and $\mathcal{L}^*$ | Bravais lattice and reciprocal Bravais lattice |
| $\pi_n(M)$ | $n$'th homotopy group of the space $M$ |
| $T_p(M)$ | Tangent space, tangential at $p \in M$ |
| $F^*$ and $F_*$ | Pull-back and push-forward mappings |
| $\mathcal{V}_1 \oplus \mathcal{V}_2$ and $\mathcal{V}_1 \otimes \mathcal{V}_2$ | Direct sum and tensor product of two vector spaces |
| $\Gamma_a$, $\Gamma_R$ and $\Gamma_U$ | Representation of translations and rotations on Hilbert space |
| $\langle j_1 m_1 j_2 m_2 | j_1 j_2 j m \rangle$ | Clebsch-Gordan coefficients |
| $T_M^J$ | Spherical components of a tensor operator |
| $Q$ and $W_\mu$ | Runge-Lenz and Pauli-Lubanski vector |
| $M_{\mu\nu}$, $S_{\mu\nu}$ and $J_{\mu\nu}$ | Infinitesimal Lorentz transformations |
| $\gamma^\mu$ and $\mathcal{C}$ | Gamma matrices and charge conjugation matrix |
| $\psi$ and $\bar{\psi}$ | Dirac spinor and Dirac-conjugated spinor |
| $d_s = 2^{[d/2]}$ | Number of components of a Dirac spinor in $d$ dimensions |
| $D_\mu$, $A_\mu$ and $F_{\mu\nu}$ | Covariant derivative, vector potential and field strength |
| $\mathcal{L}$, $\mathcal{H}$, $S$ and $T_{\mu\nu}$ | Lagrangian density, energy density, action and energy-momentum tensor |
| $\{F, G\}$ and $[F, G]$ | Poisson bracket and commutator |
| $L_X$ | Lie derivative |
| $c$ and $\Delta$ | Central charge and conformal weight |

# Introduction

**1**

> *Symmetry is the idea by which man for centuries tried to comprehend and create order, beauty, and perfection.*
>
> —Hermann Weyl

Symmetries and their breaking are fundamental concepts that extend far beyond the field of physics. This becomes clear when the concept of symmetry is defined as broadly as possible. For example, the formulation,

> a transformation is called symmetry if before and after cannot be distinguished,

includes applications that go far beyond our field of study. The basis for the exploitation of symmetries is their mathematical formulation. This is provided by group theory, because symmetry operations are always elements of a group.

## 1.1 Symmetries and Groups

Symmetries play a prominent role in finding conserved quantities and selection rules in many subfields of physics. For example, in classical mechanics we learn that the symmetries of spacetime form a group, the Galilei group. The invariance of a physical system with respect to Galilei transformations implies conservation laws:

- Translations in time $\longrightarrow$ Energy conservation,
- Translations in space $\longrightarrow$ Momentum conservation,
- Rotations in space $\longrightarrow$ Angular momentum conservation.


A. Wipf, *Symmetries in Physics*, https://doi.org/10.1007/978-3-662-72675-4_1

This profound relationship between symmetries and the existence of conserved quantities also applies in quantum theory. In fact, the most important physical applications of group theory are found in quantum mechanics. Consider the hydrogen atom: Its geometric symmetries, the spatial rotations, lead to the classification of eigenstates according to the eigenvalues of the angular momentum. In particular, the ground state has the same rotational symmetry as the Hamiltonian operator $H$.

However, there are important exceptions to the rule that the ground state is invariant under the symmetries of $H$. An example is the ferromagnet. Contrary to $H$, the ground state, in which all spins are aligned, changes with a simultaneous reversal of all spins. We will return to this phenomenon of *spontaneous symmetry breaking*.

What is meant by the *symmetries of a molecule*? These are transformations between different arrangements of atoms in the molecule that are physically indistinguishable. For example, a permutation of the coordinates of identical atoms, an inversion of the particle coordinates at the center of mass, or a rotation of the molecule about a fixed axis passing through the center of mass. Symmetries can explain many properties of molecules such as their dipole moment and their spectroscopic transitions.

Symmetry considerations are not only important in atomic and molecular physics, but also in *solid state theory*. Here, the focus is on discrete space groups corresponding to the lattice-periodic structure of crystals. We will examine these groups in more detail in Chap. 7. In the description and classification of *elementary particles*, simple discrete symmetries as well as continuous symmetries play a crucial role. These are the Poincaré group and internal symmetry groups. For semi-simple compact Lie groups, there is a theory mainly developed by Cartan (1869–1951) and Weyl (1885–1955). This is the content of the second half of the book.

Group theory was initially a mathematical discipline of quite abstract nature. Its relevance for physics arises from the observation that symmetries form representations of a group. Group theory and representation theory have been a useful tool in solid state physics or nuclear and elementary particle physics for decades and will continue to be so in the future. For example, in modern elementary particle theory, symmetry considerations are used to explain as many properties of the "elementary building blocks" as possible, without understanding the underlying dynamics in detail.

## 1.2  Group Theory in Mathematics and Physics

Groups were used long before they were axiomatically described. For example, all symmetries of a body, such as a Platonic solid, automatically satisfy the group axioms. The regular polyhedra named after Plato (428–347 BC)

- Tetrahedron made of 4 (tetra) triangles,
- Hexahedron made of 6 (hexa) squares,

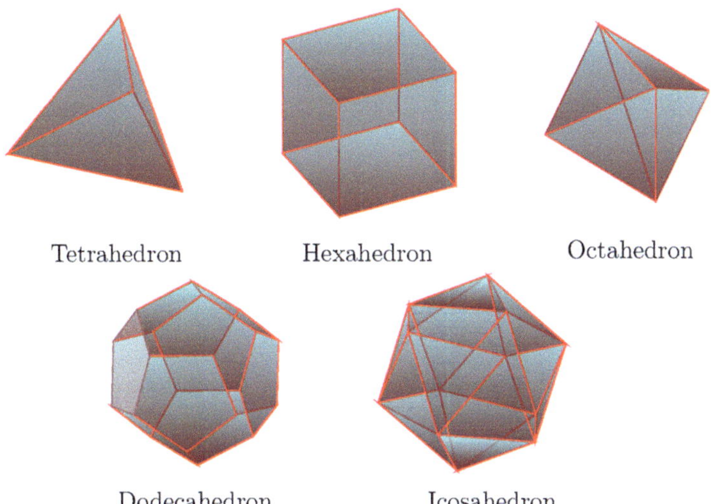

Tetrahedron            Hexahedron            Octahedron

Dodecahedron            Icosahedron

**Fig. 1.1**  The five Platonic solids

**Fig. 1.2**  Exterior angle $\chi$ and interior angle $\varphi$ in the regular hexagon

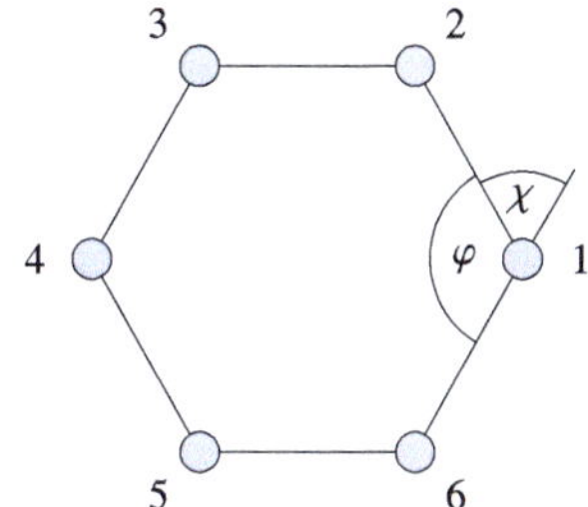

- Octahedron made of 8 (octa) triangles,
- Dodecahedron made of 12 (dodeka) pentagons,
- Icosahedron made of 20 (eikosi) triangles.

are shown in Fig. 1.1: Besides the five Platonic solids, there are no regular polyhedra. To see this, we consider the regular $n$-gon, e.g. the hexagon in Fig. 1.2. For the exterior angle $\chi$ of a corner, the relationship $2\pi = n\chi = n(\pi - \varphi)$ holds, or

$$\varphi = \left(1 - \frac{2}{n}\right) \cdot 180° = \begin{cases} 60° & 3 - gon, \\ 90° & 4 - gon, \\ 108° & 5 - gon, \\ 120° & 6 - gon . \end{cases}$$

In each spatial corner of a polyhedron, at least three polygons must meet. Since a regular polyhedron is convex, the sum of the angles of all triangles that meet at a corner of the polyhedron must always be less than 360°. Therefore, only 3, 4 or 5 regular triangles, 3 squares or 3 regular pentagons can meet at a corner. These five possible cases can be realized exactly by the given Platonic solids.

> **Platonic Groups**
> The rotations in space, which transform a Platonic solid into itself, form a discrete subgroup of the rotation group. We will analyze these Platonic groups in more detail in Sect. 6.1.1.

The foundations of modern group theory are largely due to Evariste Galois (1811–1832) and Niels Henrik Abel (1802–1829).

With the help of the concept of groups, they were able to prove that there can be no solution formula for equations of the fifth or higher degree, or that an angle cannot generally be divided into 3 equal parts with a compass and ruler.

The generation of mathematicians following Galois and Abel, Cayley (1821–1895), Dedekind (1831–1916), Kronecker (1823–1891) and Jordan (1838–1922) further developed group theory and brought it into the form we know today. The term *group* for such structures was first used in 1868 by Jordan, although he only explicitly required the axiom of closure with respect to the combination of two group elements. Since he studied symmetry groups, the other group properties were automatically fulfilled. In 1854, Cayley recognized the necessity of the associative law and the existence of the identity element. He defined the group operation using a table, which is now called a *Cayley table*. Two years later, Hamilton (1805–1865) gave a very space-saving method for representing a concrete group. Compared to the Cayley table for the icosahedron group with $60 \times 60$ entries, Hamilton only needed one line.

The first definition of a group with the axioms used today was made independently in 1882 by van Dyck (1856–1934) and Weber (1842–1913). Afterwards, a set $G$ of elements forms a group with respect to a binary operation $\circ : G \times G \to G$, often called multiplication, if the multiplication is closed and associative, an identity element exists and if every element has an inverse element. An important step in the development of group theory was the introduction of the concept of groups into geometry by Klein (1849–1925) and Lie (1842–1899).

In 1879, Killing (1847–1923) introduced the Lie algebras, the Cartan subalgebras and the Cartan matrices. We will discuss these structures in the second part of the book. The classification of semi-simple Lie algebras was completed in 1894 by Cartan, one of the outstanding mathematicians of his time. He studied the representations of semi-simple groups, and these studies were continued by Weyl. Weyl has greatly influenced the group and representation theory and successfully applied these theories to problems of quantum mechanics.

A similarly important development began with the determination of all 230 space groups by the Russian crystallographer Federov (1853–1919) in 1890, and independently by Schoenflies (1853–1928) and Barlow (1845–1934). Space groups are symmetry groups of three-dimensional point lattices, and each point lattice can be uniquely characterized by its symmetries. Periodic point lattices were early on considered as models for the structure of crystals made of atoms, although lattice structures were not experimentally confirmed until 1912 by Laue (1879–1960) and 1913 by Bragg (1862–1942). These works were continued by Coxeter (1907–2003), who in 1934 classified all spherical and Euclidean Coxeter groups now named after him. Independently of Dynkin (1924–2014), he discovered the *Dynkin diagrams* for the classification of semi-simple Lie algebras.

The group and representation theory of finite and continuous groups is still an active field of research. It would be a hopeless endeavor to fully present this beautiful and important theory in a textbook. Therefore, I will have to make a selection of topics on symmetries as well as group and representation theory that I find interesting. In some places, proofs will be completely omitted or only sketched. The focus is on methods and results that find application in physics. These are symmetries in atomic, molecular, solid state, and particle physics. After reading this book, the attentive reader should be able to use symmetry arguments in solving physical problems and successfully apply corresponding group-theoretical methods.

## 1.3    Literature, Software

The mathematical physicists H. Weyl and E. Wigner were responsible for the introduction of group theory as a mathematical method into quantum mechanics. In their books [1, 2] one can already find many applications of group theory in atomic physics with emphasis on permutations, rotations, and Lorentz group. Other recommended books are the two volumes by Cornwell [3], in which applications in particle physics and the role of Lie algebras receive more attention. An introduction to group theory from a more formal perspective can be found in the readable books [4–6].

Applications in the natural sciences or specifically in atomic, molecular and solid-state physics can be found in [7] or [8–12]. The books [13, 14] emphasize more the applications in quantum mechanics and elementary particle physics. Texts with a focus on finite groups as they occur in chemistry and physics are [15, 16]. The books [17, 18] come close in style and content to the present book.

If you want more facts about the history of group theory and the people involved in its development, I refer you to the websites of St. Andrews, https://mathshistory. st-andrews.ac.uk/HistTopics/Development_group_theory In the book, the freely available algebraic computer programs GAP and LiE http://www.gap-system.org/ Releases/index.html  http://wwwmathlabo.univ-poitiers.fr/~maavl/LiE/  are used. The first program is excellent for the analysis of finite groups, while the second can be useful in the investigation of Lie algebras.

# Elements of Group Theory

**2**

The Theory of Groups is a branch of mathematics in which one does something to something and then compares the result with the result obtained from doing the same thing to something else, or something else to the same thing.

—James R. Newman

We rely on the fact that the laws of nature will apply tomorrow just as they do today. This property is a symmetry property, an invariance of the laws with respect to translations in time. The same applies to translations of the coordinate origin or rotations of the coordinate system. The strength of gravity on Earth is different from that on the Moon, but the law of gravity is the same.

Such symmetries can be described by groups. The translations and rotations are elements of symmetry groups. However, groups may contain more abstract elements that either do not allow a geometric realization or whose geometric meaning is not obvious.

## 2.1  Groups and Group Tables

We will first define what a group is and then look at some well-known groups that will play a role in later chapters.

**Definition 2.1 (Group)** A group $(G, \circ)$ is a set $G$, for which an operation $\circ$ is defined on $G \times G$ with the following properties (group axioms):

1. Closure: $g_1, g_2 \in G \mapsto g_1 \circ g_2 \in G$,
2. Associative law: $g_1 \circ (g_2 \circ g_3) = (g_1 \circ g_2) \circ g_3$,
3. There exists a unit element $e \in G$ with $e \circ g = g \circ e = g$ for every element $g \in G$,



A. Wipf, *Symmetries in Physics*, https://doi.org/10.1007/978-3-662-72675-4_2

4. Every $g \in G$ has an inverse $g^{-1} \in G$ with $g \circ g^{-1} = g^{-1} \circ g = e$.

If only the first two properties apply, then it is a *semigroup*. A semigroup with a unit element is called a *monoid*.

From these properties, one can prove the uniqueness of the unit element $e$ and the inverse element $g^{-1}$ to $g$: If $e$ and $g$ are both unit elements, then

$$g \circ g = g \xRightarrow{g^{-1}\circ} g^{-1} \circ (g \circ g) = g^{-1} \circ g \xRightarrow{\text{assoc.}} g = e .$$

If $g'$ and $g''$ are inverse elements to $g$, then

$$g' \circ g = e = g \circ g'' \implies g' = g' \circ e = g' \circ (g \circ g'') = (g' \circ g) \circ g''$$
$$= e \circ g'' = g'' \implies g' = g''.$$

So there can only be one unit element and only one inverse element for each $g$.

The inverse element of a product is

$$(g \circ g')^{-1} = g'^{-1} \circ g^{-1}, \quad \text{because then} \quad (g \circ g') \circ (g \circ g')^{-1} . \tag{2.1}$$

Similarly, the inverse of a multiple product $g_1 \circ g_2 \circ \cdots \circ g_n$ is equal to the product of the inverse elements in reverse order,

$$(g_1 \circ g_2 \circ \cdots \circ g_n)^{-1} = g_n^{-1} \circ g_{n-1}^{-1} \circ \cdots \circ g_1^{-1} . \tag{2.2}$$

---

### Example: Examples of Groups

1. The set $\mathbb{Z}_2 = \{-1, 1\}$ with multiplication as operation $\circ$ defines the group $(\mathbb{Z}_2, \cdot)$. Multiplication is associative and each of the two group elements has an inverse element: itself.
2. The set $\{e, a, a^2, a^3, \ldots, a^{n-1}\}$, where e.g. $a \circ a = a^2$ and $a^n$ is identified with $e$, is the *cyclic group* $C_n$.
3. The integers with addition as operation form the group $\mathbb{Z}$.
4. The set of complex numbers of magnitude 1 with multiplication as operation form the continuous group U(1).

---

### Task

Show that U(1) is a group.

---

The number of elements of a group $G$ is called the *order* of $G$ and is denoted by $|G|$. If $|G| < \infty$, then $G$ is finite. The cyclic group $C_n$ is a finite group of order $n$. If the elements of a group are countable, then it is referred to as a *discrete group*.

For example, $\mathbb{Z}$ is a discrete group. On the other hand, U(1) is uncountable—it is a *continuous group*.

### 2.1.1  Group Table (Cayley Table)

For finite groups, the possible products of all group elements can be tabulated. The resulting *Cayley table* or *group table* is shown in Table 2.1:

All properties of the group can be read from the Cayley table. The existence of a unit element implies that one column matches the column on the far left and one row matches the header row. The column on the far left and the header row are obviously redundant and we can write group tables a bit more compactly:

$$T(G) = \begin{pmatrix} e & g_2 & g_3 & \cdots & g_n \\ g_2 & g_2 \circ g_2 & g_2 \circ g_3 & \cdots & g_2 \circ g_n \\ g_3 & g_3 \circ g_2 & g_3 \circ g_3 & \cdots & g_3 \circ g_n \\ \vdots & \vdots & \vdots & \ddots & \vdots \\ g_n & g_n \circ g_2 & g_n \circ g_3 & \cdots & g_n \circ g_n \end{pmatrix}, \quad \text{or} \quad T(G)_{ij} = g_i \circ g_j. \tag{2.3}$$

For $g \circ g'$ we also write $\ell_g(g')$ to indicate that it is the left multiplication with $g$. The existence of a unique inverse implies, that each row (column) contains the unit element once. Because $g \circ g^{-1} = g^{-1} \circ g = e$ it appears symmetrical to the main diagonal.

In each row (column) each element occurs exactly once, because from $g \circ g_p = g \circ g_q$ it follows after multiplication by $g^{-1}$ that $g_p = g_q$. The group elements of the row to $g$ are a *permutation* $\ell_g(g_p) \equiv g \circ g_p$ of the elements in the first row, and in particular $\ell_g(e) = \ell_e(g) = g$:

$$T(G) = \begin{pmatrix} \ell_e(e) & \ell_e(g_2) & \ell_e(g_3) & \ldots & \ell_e(g_n) \\ \ell_{g_2}(e) & \ell_{g_2}(g_2) & \ell_{g_2}(g_3) & \ldots & \ell_{g_2}(g_n) \\ \ell_{g_3}(e) & \ell_{g_3}(g_2) & \ell_{g_3}(g_3) & \ldots & \ell_{g_3}(g_n) \\ \vdots & \vdots & \vdots & \ldots & \vdots \\ \ell_{g_n}(e) & \ell_{g_n}(g_2) & \ell_{g_n}(g_3) & \ldots & \ell_{g_n}(g_n) \end{pmatrix}. \tag{2.4}$$

**Table 2.1** Each entry in the group table is the result of multiplying the element on the far left of the row with the element at the top of the column

| $G$ | $e$ | $g_2$ | $g_3$ | $\cdots$ | $g_n$ |
|---|---|---|---|---|---|
| $e$ | $e$ | $g_2$ | $g_3$ | $\cdots$ | $g_n$ |
| $g_2$ | $g_2$ | $g_2 \circ g_2$ | $g_2 \circ g_3$ | $\cdots$ | $g_2 \circ g_n$ |
| $g_3$ | $g_3$ | $g_3 \circ g_2$ | $g_3 \circ g_3$ | $\cdots$ | $g_3 \circ g_n$ |
| $\vdots$ | $\vdots$ | $\vdots$ | $\vdots$ | $\ddots$ | $\vdots$ |
| $g_n$ | $g_n$ | $g_n \circ g_2$ | $g_n \circ g_3$ | $\cdots$ | $g_n \circ g_n$ |

Similarly, the group elements of each column are a permutation of the elements of the first column. As a simple example, we consider the group table of the cyclic group $C_4$.

**Example: Group Table of the Cyclic Group $C_4$**

The table of this group is symmetric with respect to the main diagonal:

$$T(C_4) = \begin{pmatrix} e & a & a^2 & a^3 \\ a & a^2 & a^3 & e \\ a^2 & a^3 & e & a \\ a^3 & e & a & a^2 \end{pmatrix} .$$

For a group with a symmetric table, the operation $\circ$ is commutative, i.e. $g \circ g' = g' \circ g$ for all $g, g' \in G$. Such groups are called abelian. ◄

## 2.2  Matrix Groups

An important class of groups are the *matrix groups*. Let's first recall what a matrix with elements in a number field is:

**Definition 2.2 (Matrix)** An $m \times n$ matrix with elements in a number field $\mathbb{K}$ is a mapping

$$A : \{1, 2, \ldots, m\} \times \{1, 2, \ldots, n\} \mapsto \mathbb{K}, \quad (i, j) \mapsto a_{ij} .$$

A matrix is usually written as a family,

$$A = \left(a_{ij}\right)_{\substack{1 \le i \le m \\ 1 \le j \le n}} = \begin{pmatrix} a_{11} & a_{12} & \ldots & a_{1n} \\ a_{21} & a_{22} & \ldots & a_{2n} \\ \vdots & \vdots & \ldots & \vdots \\ a_{m1} & a_{m2} & \ldots & a_{mn} \end{pmatrix},$$

where $a_{ij}$ denotes the element in the $i$-th row and $j$-th column of $A$.

The set of $n \times n$ matrices with matrix elements in the number field $\mathbb{K}$ are denoted by $\mathrm{Mat}(n, \mathbb{K})$. They can be added and multiplied. In general, they do not form a group with respect to matrix multiplication, since the inverse of a matrix only exists if its determinant is not equal to zero. More generally, matrix groups with matrix elements in a ring can be defined, see Appendix 2.4.

### 2.2.1 The Groups GL($n$, $\mathbb{K}$) and SL($n$, $\mathbb{K}$)

The set of invertible $n \times n$−matrices

$$\mathrm{GL}(n, \mathbb{K}) = \{A \in \mathrm{Mat}(n, \mathbb{K}) \mid \det A \neq 0\} \tag{2.5}$$

forms a group with respect to matrix multiplication. The subset

$$\mathrm{SL}(n, \mathbb{K}) = \{A \in \mathrm{GL}(n, \mathbb{K}) \mid \det A = 1\} \tag{2.6}$$

is also a group. It is a subgroup of GL($n$, $\mathbb{K}$) (see Sect. 3.2).

- In most applications, $\mathbb{K} = \mathbb{R}$ or $\mathbb{C}$. The continuous groups $\mathrm{GL}(n, \mathbb{R})$ and $\mathrm{GL}(n, \mathbb{C})$ are called *general linear groups*. Their subgroups $\mathrm{SL}(n, \mathbb{R})$ and $\mathrm{SL}(n, \mathbb{C})$ are the *special linear groups*.
- If $\mathbb{K}$ is finite, then $\mathrm{GL}(n, \mathbb{K})$ is also finite. For example, the set $\mathbb{Z}_p$ with addition modulo a prime number $p$ is a finite number field, see Appendix 2.4 and Problem 2.2.

---

**Example: The Finite Matrix Group GL($2$, $\mathbb{Z}/2\mathbb{Z}$)**

The two-dimensional matrices in this group have elements in the number field $\mathbb{Z}_2 = \mathbb{Z}/2\mathbb{Z} \cong \{0, 1\}$ and the invertible ones have determinant 1 (in $\mathbb{Z}_2$ the elements $\pm 1$ are identified), so that $\mathrm{GL}(2, \mathbb{Z}/2\mathbb{Z}) \cong \mathrm{SL}(2, \mathbb{Z}/2\mathbb{Z})$. The group has the 6 elements

$$e = \begin{pmatrix} 1 & 0 \\ 0 & 1 \end{pmatrix}, \ a = \begin{pmatrix} 0 & 1 \\ 1 & 1 \end{pmatrix}, \ b = \begin{pmatrix} 1 & 1 \\ 1 & 0 \end{pmatrix}, \ c = \begin{pmatrix} 1 & 1 \\ 0 & 1 \end{pmatrix}, \ d = \begin{pmatrix} 1 & 0 \\ 1 & 1 \end{pmatrix}, \ f = \begin{pmatrix} 0 & 1 \\ 1 & 0 \end{pmatrix}.$$

The group table is asymmetric, i.e. the group is non-abelian:

$$\mathrm{T}(\mathrm{SL}(2, \mathbb{Z}/2\mathbb{Z})) = \left( \begin{array}{ccc|ccc} e & a & b & c & d & f \\ a & b & e & f & c & d \\ b & e & a & d & f & c \\ \hline c & d & f & e & a & b \\ d & f & c & b & e & a \\ f & c & d & a & b & e \end{array} \right). \tag{2.7}$$

The elements $\{e, a, b\}$ form an abelian subgroup. ◄

The elements of SL($2$, $\mathbb{Z}/2\mathbb{Z}$) can be defined and analyzed with the computer program GAP:

```
GAP> e:=Z(2)*[[1,0],[0,1]];c:=Z(2)*[[1,1],[0,1]];
GAP> a:=Z(2)*[[0,1],[1,1]];d:=Z(2)*[[1,0],[1,1]];
GAP> b:=Z(2)*[[1,1],[1,0]];f:=Z(2)*[[0,1],[1,0]];
GAP> G:=Group(e,a,b,c,d,f);G1:=Group(e,a,b);
GAP> IsSubgroup(G,G1);IsSimple(SL(2,2));IsNormal(G,G1);
```

The last line asks whether $G_1$ is a subgroup of $G$ (true), whether $G$ is simple
(false) and whether $G_1$ is a normal subgroup in $G$ (true). Normal subgroups and
simple groups are introduced in Sect. 3.2.3.

## 2.3   Dihedral Groups

The elements of the dihedral group $\mathcal{D}_n$ are the symmetry transformations of the
regular $n$-polygon in the plane, i.e. all linear transformations that map the polygon
onto itself. The group $\mathcal{D}_n$ contains $2n$ elements, namely $n$ rotations and $n$ reflections.

### 2.3.1   Symmetries of the Equilateral Triangle

As a simple example, we consider the symmetry transformations of the equilateral
triangle in the plane. These are the 3 rotations of the triangle around its center of
gravity by multiples of $2\pi/3$ and the 3 reflections $\sigma_v$, see Fig. 2.1.

The group operation is the composition of transformations. The rotation by $2\pi/3$
is denoted by $c_3$ and the reflection on the line passing through the $i$-th vertex by
$\sigma_v^{(i)}$. The symmetry transformations act on the vertices of the triangle. For example,

$$c_3 : (1, 2, 3) \longmapsto (3, 1, 2),$$

$$\sigma_v^{(1)} : (1, 2, 3) \longmapsto (1, 3, 2),$$

$$\sigma_v^{(2)} : (1, 2, 3) \longmapsto (3, 2, 1),$$

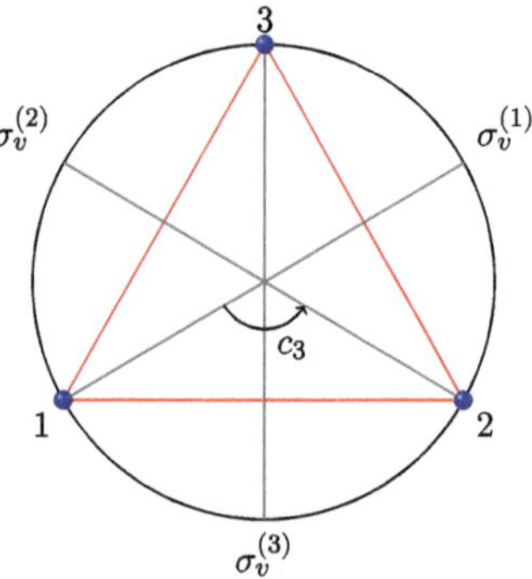


**Fig. 2.1** Symmetries of the
equilateral triangle

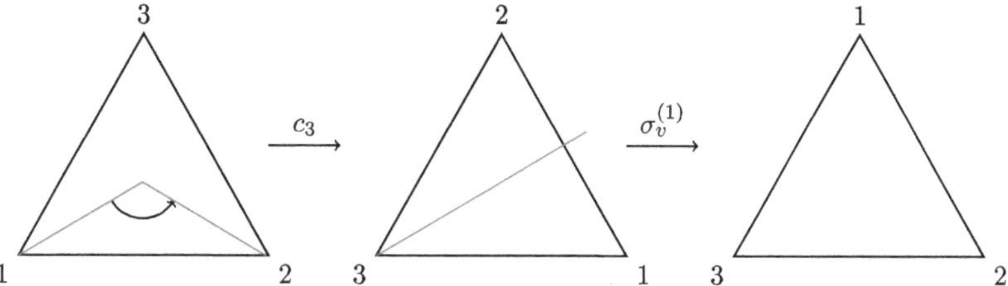

**Fig. 2.2** The product is the sequential execution of symmetry operations

as can be seen from Fig. 2.1. The rotation $c_3$ is to be read as follows: in the corner where the 1 was before the rotation, the 3 is afterwards, where the 2 was is afterwards the 1 and where the 3 was is afterwards the 2. If we carry out these symmetry transformations one after the other, we obtain for example

$$(1, 2, 3) \xmapsto{c_3} (3, 1, 2) \xmapsto{\sigma_v^{(1)}} (3, 2, 1), \quad \text{i.e.} \quad \sigma_v^{(1)} \circ c_3 = \sigma_v^{(2)},$$

The operations are shown in Fig. 2.2. With similar considerations for the remaining products, one finds the group table for $\mathcal{D}_3$:

$$\mathrm{T}(\mathcal{D}_3) = \left( \begin{array}{ccc|ccc} e & c_3 & c_3^2 & \sigma_v^{(1)} & \sigma_v^{(2)} & \sigma_v^{(3)} \\ c_3 & c_3^2 & e & \sigma_v^{(3)} & \sigma_v^{(1)} & \sigma_v^{(2)} \\ c_3^2 & e & c_3 & \sigma_v^{(2)} & \sigma_v^{(3)} & \sigma_v^{(1)} \\ \hline \sigma_v^{(1)} & \sigma_v^{(2)} & \sigma_v^{(3)} & e & c_3 & c_3^2 \\ \sigma_v^{(2)} & \sigma_v^{(3)} & \sigma_v^{(1)} & c_3^2 & e & c_3 \\ \sigma_v^{(3)} & \sigma_v^{(1)} & \sigma_v^{(2)} & c_3 & c_3^2 & e \end{array} \right). \tag{2.8}$$

**Question**

Why are there only rotations in the blocks top left and bottom right and only reflections top right and bottom left?

The symmetry transformations form the *smallest non-abelian group* $\mathcal{D}_3$ of order 6. Note that the top left block contains the group table of $\mathcal{C}_3$. The group table of $\mathcal{D}_3$ is identical to that of GL$(2, \mathbb{Z}/2\mathbb{Z})$ in (2.7).

## 2.4   Appendix A: Matrices with Elements in a Ring

The matrix elements $a_{ij}$ do not need to be elements of a number field. It is sufficient that they are elements of a ring. The axioms of a ring are modeled after the calculation rules for integers:

**Definition 2.3 (Ring)** A set $R$ with two binary operations, the addition $+ : R \times R \to R$ and the multiplication, $\cdot : R \times R \to R$ is a ring, if it forms a commutative group with respect to addition and a semigroup with identity element with respect to multiplication,[1] and both operations are connected via the distributive laws

$$(a + b) \cdot c = a \cdot c + b \cdot c \quad , \quad c \cdot (a + b) = c \cdot a + c \cdot b$$

The neutral element with respect to addition is called zero, $a + 0 = a$. If the multiplication is commutative, then $R$ is a commutative ring.

In a ring, therefore, addition, subtraction and multiplication can be performed. A commutative ring is a number field, if every element in $R \backslash \{0\}$ is invertible.[2] The *division* in a number field is then defined as follows:

$$a/b := ab^{-1}, \qquad b \neq 0.$$

---

**Example: Rings**

- $\mathbb{Q}, \mathbb{R}, \mathbb{C}$ are number fields, i.e. special rings,
- $\mathbb{Z}$ is a commutative ring,
- the even numbers in $\mathbb{Z}$ form a commutative ring, but without 1,
- $\mathbb{Z}/n\mathbb{Z}$ is a commutative ring.

◄

We want to prove the last statement: For this we define the *residue classes modulo n*. Two integers $m$ and $m'$ are in the same class if $m = m' + kn$ with $k \in \mathbb{Z}$ holds. There are obviously $n$ classes, and each class has a representative $m$ in the set $\{0, 1, \ldots, n - 1\}$. We denote this class with $[m]$. On the set of classes we define an addition by the addition of representatives,

$$\underbrace{[m]}_{\text{Class of } m} + \underbrace{[m']}_{\text{Class of } m'} = \underbrace{[m + m']}_{\text{Class of } m+m'}$$

This is well-defined, since $[m + m']$ does not depend on the representatives of the classes. In this ring, for example, it holds

$$\underbrace{1 + 1 + \cdots + 1}_{n} = 0.$$

---

[1] One can also find the definition without the assumption of the existence of the identity element.

[2] You can also find the definition without the assumption that the multiplication is commutative.

The residue classes mod $n$ with this addition form the abelian group $(\mathbb{Z}/n\mathbb{Z}, +)$ of order $n$. The rule

$$[m_1] \cdot [m_2] = [m_1 \cdot m_2]$$

defines an additional multiplication on the classes and $\mathbb{Z}/n\mathbb{Z}$ becomes a commutative ring. However, only all elements $\neq 0$ have an inverse if $n$ is a prime number.

**Question**

Why is the commutative ring $\mathbb{Z}/n\mathbb{Z}$ a field for a prime number $n$?

**Definition 2.4 (Matrix)**  An $m \times n$ matrix with elements in a commutative ring $R$ (or an $m \times n$ matrix over $R$) is a mapping

$$A : \{1, 2, \ldots, m\} \times \{1, 2, \ldots, n\} \mapsto R, \quad (i, j) \mapsto a_{ij} \, .$$

We also write $A = (a_{ij})$.

The set of $m \times n$ matrices with elements in a commutative ring $R$ is denoted by $\mathrm{Mat}(m, n, R)$. Matrices with elements in $\mathbb{Z}$ or $\mathbb{Q}$ are briefly called integer or rational matrices. Matrices can be added and can be multiplied by scalars:

**Definition 2.5**  Let $A, B \in Mat(m, n, R)$ and $\alpha \in R$. Then $A + B := \left(a_{ij} + b_{ij}\right)$ is called the sum of $A$ and $B$, and $\alpha \cdot A = \left(\alpha a_{ij}\right)$ the $\alpha-$fold scalar multiple of $A$.

In the following, we write $\alpha A$ instead of $\alpha \cdot A$. Two $n \times n$ matrices with elements in a ring $R$ can be multiplied,

$$A \cdot B = C = (c_{ij}), \qquad c_{ij} = \sum_k a_{ik} b_{kj} \, .$$

**Theorem 2.1 (Matrix Ring)**  *The set* $\mathrm{Mat}(n,R)$ *of* $n \times n$ *matrices is a ring with respect to the addition and multiplication of matrices. If* $n \geq 2$ *and* $R$ *has at least* 2 *elements, then* $\mathrm{Mat}(n,R)$ *is non-commutative.*

Here the question arises, when a matrix in $\mathrm{Mat}(n, R)$ is invertible. The following holds

**Theorem 2.2 (Invertible Matrices)**  *Let* $R$ *be a commutative ring. A matrix* $A \in$ $\mathrm{Mat}(n, R)$ *is invertible if and only if* $\det(A) \in R$ *is invertible.*

For the proof we recall the

**Theorem 2.3 (Cayley-Hamilton)** *Let $R$ be a commutative ring, $A \in \mathrm{Mat}(n, R)$ and $\chi_A(\lambda) = \det(\lambda - A) = \lambda^n + a_{n-1}\lambda^{n-1} + a_{n-2}\lambda^{n-2} + \cdots + a_0$ the characteristic polynomial of $A$. Then*

$$\chi_A(A) = A^n + a_{n-1}A^{n-1} + a_{n-2}A^{n-2} + \cdots + a_1 A + a_0 \cdot \mathbb{1} = 0. \qquad (A.1)$$

The coefficients $a_{n-1} = -\mathrm{Tr}(A), \ldots, a_0 = (-)^n \det A$ are polynomials in the matrix elements and are therefore in $R$.

---

**Question**

Where in the proof is it assumed that $R$ is a commutative ring?

With the formula of Cayley and Hamilton we can now prove that for invertible $\det A$ the matrix $A$ is invertible. By assumption, $\det A$ is invertible and we can multiply in

$$A \cdot \left(A^{n-1} - \mathrm{Tr}A \, A^{n-2} + \cdots + a_1 \mathbb{1}\right) = (-)^{n-1} \det A \, \mathbb{1},$$

with the inverse of $\det A$. We obtain

$$AA^{-1} = \mathbb{1}, \qquad A^{-1} = (-)^{n-1}\left(A^{n-1} - \mathrm{Tr}A \, A^{n-2} + \cdots + a_1 \mathbb{1}\right)(\det A)^{-1},$$

which shows that the inverse matrix to $A$ exists, when $\det A$ is invertible in $R$.

---

**Example: The Inverse of a $2 \times 2$ Matrix**

For an arbitrary $2 \times 2$ matrix $A = (a_{ij})$ the characteristic polynomial has the form

$$\chi_A(\lambda) = \lambda^2 - \mathrm{Tr}A\,\lambda + \det A.$$

The Cayley-Hamilton theorem implies for such matrices

$$A^{-1} = \frac{1}{\det A}\left((\mathrm{Tr}A)\mathbb{1} - A\right) = \frac{1}{\det A}\begin{pmatrix} a_{22} & -a_{12} \\ -a_{21} & a_{11} \end{pmatrix}. \qquad (A.2)$$

◄

For $3 \times 3$ matrices one can proceed similarly.

**Example: The Inverse of a $3 \times 3$ Matrix**

The inverse of an arbitrary $3 \times 3$ matrix $A = (a_{ij})$ is

$$A^{-1} = \frac{1}{\det A}\left(A^2 - (\operatorname{Tr} A)A + \Delta \mathbb{1}\right), \quad \text{with}$$

$$\Delta = a_{11}a_{22} - a_{12}a_{21} + a_{11}a_{33} - a_{13}a_{31} + a_{22}a_{33} - a_{23}a_{32}. \tag{A.3}$$

◄

The set of invertible matrices

$$\mathrm{GL}(n, R) = \{A \in \operatorname{Mat}(n, R)\,|\, \det A \text{ is invertible}\}$$

thus forms a group with respect to matrix multiplication. It is called the general linear group $\mathrm{GL}(n, R)$. The subset of matrices with determinant 1 defines the special linear group $\mathrm{SL}(n, R)$ with matrix elements in the ring $R$.

## 2.5   Appendix B: Presentation of a Group

An important role is played by generating sets in the group. If $S$ is a generating set then every element of $G$ can be written as a product of powers of generators, i.e. of elements in $S$. A minimal set of generators, the number of which is referred to as the *rank* $r$, is called the basis of the group. An elegant method to present groups compactly is to specify generating elements and relations between them. For example, the dihedral group $\mathcal{D}_3$ has the generators $c_3$ and $\sigma_v$ with the relations

$$c_3^3 = \sigma_v^2 = \mathbb{1}, \quad \sigma_v \circ c_3 \circ \sigma_v^{-1} = c_3^{-1}.$$

This simple method is not always intuitive. The properties of a group may not be apparent from looking at one of its presentation. It is often not even clear whether the presented group is finite. It is also not always clear from two presentations whether they describe isomorphic groups. Nevertheless, presentations play an important role in the classification of groups.

We return to $\mathcal{D}_3$ with the *presentation* (we set $c_3 = a$ and $\sigma_v = b$)

$$\mathcal{D}_3 = \left\{a, b \,|\, a^3 = b^2 = e, \; b \circ a \circ b^{-1} = a^{-1} \iff b \circ a = a^2 \circ b\right\} \tag{B.1}$$

The group generated by $a$ and $b$ contains all products of $a$ and $b$, the so-called words of the alphabet consisting of $a$ and $b$, for example

$$a, \; ab, \; ab^2, \; abab^2a^3, \ldots,$$

where $ab$ stands for $a \circ b$. We may use the relations (B.1) to simplify the words. The last relation allows us to move all $a$ to the left and all $b$ to the right in a word. If we then use the relations $a^3 = b^2 = e$, every word can be reduced to one of the following words:

$$\{e, a, a^2, b, ab, a^2 b\} \, .$$

Using the relations in (B.1), we obtain the multiplication table of $\mathcal{D}_3$ in (2.8):

$$\mathrm{T}(\mathcal{D}_3) = \left( \begin{array}{ccc|ccc} e & a & a^2 & b & ab & a^2 b \\ a & a^2 & e & ab & a^2 b & b \\ a^2 & e & a & a^2 b & b & ab \\ \hline b & a^2 b & ab & e & a^2 & a \\ ab & b & a^2 b & a & e & a^2 \\ a^2 b & ab & b & a^2 & a & e \end{array} \right) \, . \tag{B.2}$$

A similar presentation can be found for all dihedral groups $\mathcal{D}_n$. These are generated by the rotation $c_n$ with the angle $2\pi/n$ and the reflection $\sigma$ along a symmetry axis. A possible presentation is then (with $c_n = a$ and $\sigma = b$):

$$\mathcal{D}_n = \{a, b \mid a^n = b^2 = e, ba = a^{n-1} b\} \, . \tag{B.3}$$

These relations imply that every word from the alphabet with the letters $a, b$ can be transformed into one of the following words:

$$\{e, a, a^2, \ldots, a^{n-1}, b, ab, a^2 b, \ldots, a^{n-1} b\} \, .$$

The group table can now be calculated using the relations (B.3).

With GAP, one can inquire about the properties of a presentation. For example, the following defines the dihedral group $\mathcal{D}_6$, which is called up in GAP with DihedralGroup(12):

```
GAP> f:=FreeGroup("a","b");
GAP> f1:=f/[f.1^6,f.2^2,f.2*f.1*f.2*f.1];
GAP> Order(f1);f2:=DihedralGroup(12);Order(f2);
```

A group can have different presentations, which differ in the number of generators and/or relations, see http://brauer.maths.qmul.ac.uk/Atlas/v3/.

## 2.6    Exercises for Chap. 2

**Problem 2.1 (Group Axioms)** Prove that the following axioms already define a group $G$ with operation $\circ$:

1. Closure: $g_1, g_2 \in G \Rightarrow g_1 \circ g_2 \in G$,
2. Associativity: $g_1, g_2, g_3 \in G \Rightarrow g_1 \circ (g_2 \circ g_3) = (g_1 \circ g_2) \circ g_3$,
3. Existence of a left-neutral element: $\exists\, e \in G : \forall g \in G,\ e \circ g = g$,
4. Existence of a left-inverse element: $\forall g \in G\ \exists g^{-1} \in G : g^{-1} \circ g = e$.

**Problem 2.2 (The Fields $\mathbb{Z}/n\mathbb{Z}$)**  Show that for a prime number $n$, the commutative ring $\mathbb{Z}/n\mathbb{Z}$ is a field.

**Problem 2.3 (Symmetries of the Square)**  The elements of the dihedral group $\mathcal{D}_4$ are the symmetry transformations $\{e, c_4, c_2, c_4^3, \sigma_v, \sigma_v', \sigma_d, \sigma_d'\}$ of the equilateral square, where $\sigma_v, \sigma_v'$ denote the reflections on the two medians and $\sigma_d, \sigma_d'$ denote the reflections on the two diagonals.

1. The symmetry transformations act on the vertices. For example, $c_4$ : $(1, 2, 3, 4) \mapsto (4, 1, 2, 3)$. What are the remaining symmetry transformations?
2. Convince yourself that the group table has the following form:

$$
\mathrm{T}(\mathcal{D}_4) = \left(
\begin{array}{cccc|cccc}
e & c_4 & c_2 & c_4^3 & \sigma_v & \sigma_v' & \sigma_d & \sigma_d' \\
c_4 & c_2 & c_4^3 & e & \sigma_d' & \sigma_d & \sigma_v & \sigma_v' \\
c_2 & c_4^3 & e & c_4 & \sigma_v' & \sigma_v & \sigma_d' & \sigma_d \\
c_4^3 & e & c_4 & c_2 & \sigma_d & \sigma_d' & \sigma_v' & \sigma_v \\
\hline
\sigma_v & \sigma_d & \sigma_v' & \sigma_d' & e & c_2 & c_4 & c_4^3 \\
\sigma_v' & \sigma_d' & \sigma_v & \sigma_d & c_2 & e & c_4^3 & c_4 \\
\sigma_d & \sigma_v' & \sigma_d' & \sigma_v & c_4^3 & c_4 & e & c_2 \\
\sigma_d' & \sigma_v & \sigma_d & \sigma_v' & c_4 & c_4^3 & c_2 & e
\end{array}
\right) .
$$

**Problem 2.4 (Cayley Table)**  We want to become a bit more familiar with group tables here:

1. Complete the following schemes to group tables with elements $e, a, b, c$:

$$
\begin{array}{cccc}
e & a & b & c \\
a & e & \cdot & \cdot \\
b & \cdot & e & \cdot \\
c & \cdot & \cdot & \cdot
\end{array}
\qquad\qquad
\begin{array}{cccc}
e & a & b & c \\
a & c & \cdot & \cdot \\
b & \cdot & e & \cdot \\
c & \cdot & \cdot & \cdot
\end{array}
\qquad\qquad
\begin{array}{cccc}
e & a & b & c \\
a & c & \cdot & \cdot \\
b & \cdot & c & \cdot \\
c & \cdot & \cdot & \cdot
\end{array}
$$

2. Are the groups that belong to the first and last table isomorphic (do they have identical group tables except for renaming of the group elements)?
3. Which of the groups is the "Four-group of Klein".
4. Determine the group table of the group generated by the elements $a$ and $b$, which satisfy the relations $a^2 = b^3 = (ab)^2 = e$. Do you recognize the group?

**Problem 2.5 (Presentations)** Which two groups are defined by the following presentations

$$\{a^2 = b^2 = (ab)^3 = e\} \quad \text{and} \quad \{a^5 = b^2 = e, \; bab^{-1} = a^{-1}\} \; ?$$

# Homomorphisms, Subgroups and Classes **3**

> *Structures are the weapons of mathematicians.*
>
> —Nicolas Bourbaki

The structures of groups become more transparent through their division into equivalence classes. In this chapter, we discuss coset classes (cosets) and conjugacy classes. The latter are important for the theory of representations. We begin the chapter with structure-preserving mappings $\varphi : G \mapsto G'$. These lead to the concept of isomorphic groups, which are indistinguishable as abstract groups. Many groups appear as subgroups of larger groups. Important examples are normal subgroups or the center of a group. Finally, we define and discuss the semi-direct product of two groups. This often appears in geometry and physics.

**Notation**
In the following, we will write $gg'$ for the group multiplication of $g$ and $g'$ instead of $g \circ g'$.

## 3.1 Homomorphisms and Isomorphic Groups

The group tables of $\mathcal{D}_3$ in Sect. 2.3 and $\mathrm{SL}(2, \mathbb{Z}/2\mathbb{Z})$ in Sect. 2.2 are identical, if we make the following identification:

$$\left( e, c_3, c_3^2, \sigma_v^{(1)}, \sigma_v^{(2)}, \sigma_v^{(3)} \right) \longleftrightarrow \left( e, a, b, c, d, f \right).$$

© The Author(s), under exclusive license to Springer-Verlag GmbH, DE, part of Springer Nature 2026
A. Wipf, *Symmetries in Physics*, https://doi.org/10.1007/978-3-662-72675-4_3

The groups are the same except for a renaming of the group elements, i.e., they are isomorphic. More generally, we define homomorphic and isomorphic groups as follows:

**Definition 3.1 (Homomorphism, Isomorphism)** A mapping $\varphi : G \mapsto G'$ between two groups with the property $\varphi(g_1 g_2) = \varphi(g_1)\varphi(g_2)$ for all $g_1, g_2 \in G$ is called group homomorphism. A bijective homomorphism is called an isomorphism. If an isomorphism exists, then $G$ and $G'$ are called isomorphic, $G \cong G'$.

If $G = G'$, a homomorphism is called *endomorphism* and an isomorphism is called *automorphism*.

$\mathbf{Hom}(G, G')$
The set of homomorphisms $G \to G'$ is denoted by $\mathrm{Hom}(G, G')$.

**Theorem 3.1** *Let $G$ and $G'$ be groups with identity elements $e$ and $e'$ and $\varphi \in Hom(G, G')$. Then $\varphi(e) = e'$ and $\varphi(g^{-1}) = \left[\varphi(g)\right]^{-1}$ for all $g \in G$.*

***Proof*** These properties are easy to prove:

$$\varphi(g) = \varphi(ge) = \varphi(g)\varphi(e) \implies \varphi(e) = e',$$

$$e' = \varphi(gg^{-1}) = \varphi(g)\varphi(g^{-1}) \implies \varphi(g^{-1}) = \left[\varphi(g)\right]^{-1}. \tag{3.1}$$

$\square$

**Lemma 3.1 (Automorphism Group)**   *The set of automorphisms*

$$\mathrm{Aut}(G) = \{\varphi : G \mapsto G \,|\, \varphi \text{ is an isomorphism}\}$$

*forms a group and is called the automorphism group of $G$.*

The group operation is the composition of mappings, the identity element is the identical mapping $\varphi : g \to g$ and the inverse element to $\varphi$ is the inverse mapping.

For each group element $g$, there exists the conjugation with $g$,

$$\mathrm{Ad}_g : G \longmapsto G, \qquad \mathrm{Ad}_g(g') = gg'g^{-1}, \tag{3.2}$$

called adjoint mapping. Conjugations are special automorphisms called *inner automorphisms*.

---

**Task**

Convince yourself that $\mathrm{Ad}_g$ is an automorphism.

---

## 3.2  Subgroups

A *subgroup* of $G$ is a subset $H \subset G$, which is closed under the multiplication in $G$ and forms a group:

**Lemma 3.2** *Let $G$ be a group and $H \subset G$. Then $H$ is a subgroup of $G$ if, for all $a, b \in H$, $ab \in H$ and $a^{-1} \in H$.*

---

**Question**

How do the remaining axioms of a group follow from this?

A group always has the improper subgroups $G$ and $\{e\}$.

If $H$ is a subgroup of $G$, then we write $H \leq G$. If it is it a proper subgroup, then we write $H < G$. A proper subgroup $H < G$ is called *maximal* if there is no $H' < G$ with $H \subset H'$.

---

**Example: Subgroups**

- A subgroup of $\mathbb{C}^* = \mathbb{C}\backslash\{0\}$ with multiplication as operation is

$$\mathrm{U}(1) = \{z \in \mathbb{C}\,|\,|z| = 1\} < \mathbb{C}^* . \tag{3.3}$$

- $\mathbb{Z}$ contains the subgroup $(m\mathbb{Z}, +)$ of integer multiples of $m \in \mathbb{Z}$.
- The dihedral group $\mathcal{D}_3$ contains the cyclic group $\mathcal{C}_3$ as a subgroup. It also contains three times the subgroup $\mathcal{C}_2$, whose elements are each the identity and a reflection.

◄

### 3.2.1  Kernel and Image of a Homomorphism

Every homomorphism $G \mapsto G'$ leads to special subgroups of $G$ and $G'$.

**Definition 3.2 (Kernel, Image)**  The kernel of a homomorphism $\varphi$ is the set

$$\mathrm{Ker}(\varphi) = \{g \in G\,|\,\varphi(g) = e'\} \subset G , \tag{3.4}$$

where $e'$ denotes the identity element in $G'$. The image of the homomorphism is the set

$$\mathrm{Im}(\varphi) = \{\varphi(g) | g \in G\}. \tag{3.5}$$

Each $\varphi \in \mathrm{Hom}(G, G')$ now defines a subgroup of $G$ and of $G'$:

**Lemma 3.3** *For each $\varphi \in Hom(G, G')$, $\mathrm{Im}(\varphi) \leq G'$ and $\mathrm{Ker}\,(\varphi) \leq G$.*

To prove the first property, we have to show that $\mathrm{Im}(\varphi)$ is closed and contains the inverse element of each element:

$$\varphi(g_1)\varphi(g_2) = \varphi(g_1 g_2) \in \mathrm{Im}(\varphi) \quad , \quad \varphi(g)^{-1} = \varphi(g^{-1}) \in \mathrm{Im}(\varphi).$$

To prove the second property let $g_1$, $g_2$ and $g$ be in the kernel of the homomorphism $\varphi$. Then

$$\varphi(g_1 g_2) = \underbrace{\varphi(g_1)}_{e'} \underbrace{\varphi(g_2)}_{e'} = e', \qquad \varphi(gg^{-1}) = e' = \underbrace{\varphi(g)}_{e'} \varphi(g^{-1}) \Rightarrow \varphi(g^{-1}) = e',$$

which proves that $\mathrm{Ker}(\varphi)$ is a subgroup of $G$.

---

### Example: Powers of Group Elements in an Abelian Group

Let $G$ be an abelian group. Then for every integer $m$ the exponential mapping

$$\varphi_m : \quad G \longmapsto G, \qquad \varphi_m(g) = g^m$$

is a homomorphism. For $G = (\mathbb{Z}, +)$, $\mathrm{Ker}(\varphi_m) = \{0\}$ and $\mathrm{Im}(\varphi_m) = m\mathbb{Z}$.  ◄

---

### Example: Determinant Mapping

Because $\det(AB) = \det(A)\det(B)$, the determinant mapping $\det \colon \mathrm{GL}(n, \mathbb{K}) \mapsto (\mathbb{K}^*, \cdot)$ is a surjective group homomorphism into the multiplicative group $\mathbb{K}^* = \mathbb{K} \setminus \{0\}$ of invertible elements in $\mathbb{K}$. Its kernel is $\mathrm{SL}(n, \mathbb{K})$. From Lemma 3.3 it follows directly ◄

**Corollary 3.1** *If $\varphi \in Hom(G, G')$ and $H \leq G$, then $\varphi(H) \leq \varphi(G)$.*

### 3.2.2  Cyclic Subgroups

For each element $g \in G$ there is the *cyclic subgroup generated by g*

$$(g) = \left\{ g^k \,|\, k \in \mathbb{Z} \right\} \leq G, \quad \text{where} \quad g^0 = e \quad \text{and} \quad g^{-k} := (g^k)^{-1}.$$

The order of the subgroup $(g) \leq G$ is called the order of the group element $g$ in $G$. If $(g) = G$, then $G$ is a cyclic group and thus abelian.

---

**Example: There Can Be Several Generating Elements**

The cyclic group $\mathbb{Z}$ has the generating elements $1$ and $-1$. So there can be more than one generating element.  ◄

---

**Task**

Convince yourself that $\mathcal{D}_4$ and $\mathcal{D}_5$ have the cyclic subgroups $\{\mathcal{C}_4, \mathcal{C}_2, \mathcal{C}_2, \mathcal{C}_2, \mathcal{C}_2\}$ and $\{\mathcal{C}_5, \mathcal{C}_2, \mathcal{C}_2, \mathcal{C}_2, \mathcal{C}_2, \mathcal{C}_2\}$ respectively. What are the cyclic subgroups of $\mathcal{D}_p$ with prime number $p$?

### Finite Cyclic Groups

Cyclic groups can have finite or infinite order. If the order is *finite*, then there are numbers $k > l \in \mathbb{Z}$ for which $g^k = g^l$ or $g^{k-l} = e$. Therefore, there is an $n > 0$ with $g^n = e$. Let $n$ be the smallest positive number with this property. The cyclic group then consists of the elements

$$(g) = \{e, g, g^2, \ldots, g^{n-1} \,|\, g^n = e\}. \tag{3.6}$$

It then follows that cyclic groups of the same order are isomorphic:

**Lemma 3.4**  *A cyclic group of order $n < \infty$ is isomorphic to $\mathbb{Z}_n \equiv \mathbb{Z}/n\mathbb{Z}$.*

### Infinite Cyclic Groups

If the order of a cyclic group is *infinite*, all $g^k$ are different, and the mapping $\mathbb{Z} \ni k \mapsto g^k \in (g)$ is an isomorphism.

**Lemma 3.5**  *An infinite cyclic group is isomorphic to $(\mathbb{Z}, +)$.*

### 3.2.3  Properties of Important Subgroups

Now we will discuss some common subgroups. For each subgroup $H \leq G$ and each fixed element $g \in G$ we obtain a new subgroup by conjugating the elements in $H$ with $g$:

**Lemma 3.6 (Conjugate Subgroups)** *Let $H \leq G$. Then*

$$\mathrm{Ad}_g(H) \equiv gHg^{-1} = \{ghg^{-1}|h \in H\} \qquad (g \in G \text{ fixed}) \qquad (3.7)$$

*is a subgroup of G isomorphic to H. It is called the subgroup conjugate to H.*

**Proof** The conjugation $\mathrm{Ad}_g$ is an (inner) automorphism, and thus $\mathrm{Ad}_g(H) \leq G$. Since $\mathrm{Ad}_g$ is bijective, $\mathrm{Ad}_g(H) \cong H$.                                    □

Of importance are *invariant* subgroups. These are special self-conjugate subgroups:

**Definition 3.3 (Normal Subgroup)** A subgroup $N$ of $G$ is called a normal subgroup of $G$ (written $N \lhd G$ or $N \unlhd G$), if

$$\mathrm{Ad}_g(N) = gNg^{-1} = N \quad \text{for all} \quad g \in G. \qquad (3.8)$$

A group always has two *trivial normal subgroups*, namely the identity element and itself. We now define two families of groups that are important for the classification of groups:

**Definition 3.4 (Simple Groups)** A group is called simple if it contains no (non-trivial) normal subgroup; it is called semi-simple if it contains no (non-trivial) abelian normal subgroup.

Every simple group is semi-simple and an abelian group with proper subgroups is not semi-simple.

> **Classes of Simple Groups**
> All simple groups can be divided into a few classes:
>
> - $\mathcal{C}_p$ with $p$ prime number,
> - alternating groups $\mathcal{A}_n$, with $n \geq 5$ (see Sect. 4.2),
> - Lie type groups (Chevalley or twisted Chevalley and Tits groups),
> - 26 sporadic groups.

Among the sporadic groups is also "the monster", a finite group of rotations in 196 883-dimensional space. The order of this monster is a 54-digit number. Burnside conjectured that every finite simple group of non-primary order must have an even order or that the order of every simple non-Abelian group is an even number. The conjecture was proved in 1962 by W. Feit and J.G. Thompson.

We prove another result, according to which every group homomorphism defines a normal subgroup:

**Lemma 3.7**  *Let $\varphi \in \mathrm{Hom}(G, G')$. Then $Ker(\varphi) \trianglelefteq G$ is a normal subgroup of $G$*

***Proof***  We have already proven that $\mathrm{Ker}(\varphi) \leq G$. Let $n \in \mathrm{Ker}(\varphi)$ and $g \in G$ be arbitrary. Then $gng^{-1}$ is also in the kernel of $\varphi$:

$$\varphi\big(gng^{-1}\big) = \varphi(g)\varphi(n)\varphi(g^{-1}) = \varphi(g)\, e'\, [\varphi(g)]^{-1} = e' \implies gng^{-1} \in \mathrm{Ker}(\varphi) \,.$$

$\square$

The following also holds

**Lemma 3.8**  *Let $\varphi \in \mathrm{Hom}(G, G')$ and $N \trianglelefteq G$ be a normal subgroup. Then the image $\varphi(N)$ is a normal subgroup in the image $\varphi(G) \leq G$.*

This follows directly from

$$\varphi(g)\varphi(N)\varphi(g)^{-1} = \varphi(gNg^{-1}) = \varphi(N) \,.$$

If every element of a subgroup of $G$ commutes with all elements of $G$, then it is an abelian normal subgroup. The largest subgroup with this property is called the center:

**Definition 3.5 (Center)**  The center of a group $G$

$$Z = \{z \in G \,|\, zg = gz \quad \forall g \in G\} \trianglelefteq G \tag{3.9}$$

is a non-empty normal subgroup of $G$.

***Proof***  The set $Z$ contains $e$ and is therefore not empty. It is a subgroup of $G$,

$$z, z' \in Z \implies zz'g = zgz' = gzz' \quad \text{and} \quad gz^{-1} = z^{-1}(zg)z^{-1}$$
$$= z^{-1}(gz)z^{-1} = z^{-1}g \,.$$

The center is a very special normal subgroup: not only the set $Z$, but each individual element in $Z$ is invariant under all conjugations. An abelian group is equal to its center. For a strongly non-abelian group on the other hand, $Z = \{e\}$. $\quad\square$

## 3.3    Cosets and Factor Groups

If Peter has the same exam grade as Paul, then Paul has the same grade as Peter. If Maria has the same grade as Paul, then she also has the same grade as Peter. This is an example of an *equivalence relation*, which in general should have the following property:

**Definition 3.6 (Equivalence Relation)**    An equivalence relation $\sim$ on a set $M$ with elements $a, b, c, \ldots$ is a relation with the following properties:

- reflexive: $a \sim a$,
- symmetric: $a \sim b \Leftrightarrow b \sim a$,
- transitive: $a \sim b, b \sim c \Rightarrow a \sim c$.

A set of equivalent objects is referred to as an *equivalence class* or simply as a *class* of $M$. In our case, all students with the same grade form a class. Classifications are useful for subdividing sets and revealing structures.

**Task**

Convince yourself that the set $M$ decomposes into *disjoint* equivalence classes, $M = K_1 \cup K_2 \cup \ldots$ with $K_i \cap K_j = \emptyset$ for $i \neq j$.

In groups, there are two important types of divisions into classes, the *conjugacy classes* and the *cosets*. We start with the cosets.

**Definition 3.7**  Let $H \leq G$ be an arbitrary subgroup of $G$. We define the equivalence relation

$$a \overset{\cdot}{\sim} b \quad \text{if} \quad b \in aH = \{ah | h \in H\}.$$

This defines an equivalence relation

$$a = ae, \quad a = bh \Leftrightarrow b = ah^{-1}, \quad a = bh, \ b = ch' \Rightarrow a$$

$$= (ch')h \equiv ch'' \quad \text{with} \quad h'' = h'h.$$

For each subgroup $H \leq G$, the group can now be divided into disjoint equivalence classes, so-called coset classes or cosets for short. For this, one forms for each $a \in G$ the left coset $aH$ of $a$. Obviously, $aH = bH$ exactly when $a \overset{\cdot}{\sim} b$. Now we decompose the group into disjoint left cosets

$$G = eH \cup aH \cup a'H \cup \ldots.$$

The number of elements $|aH|$ of $aH$ is equal to the order $|H|$ of $H$. We conclude

**Theorem 3.2 (Lagrange)**   *G is finite and $H \leq G$. Then the index $j = |G|/|H|$ of the subgroup H in the group G is a natural number.*

From the theorem follows immediately

**Lemma 3.9 (Groups with Prime Order)**  *A finite group of prime order has no proper subgroup and is therefore simple.*

Instead of left, one can define right cosets $Ha$. They also allow a division of the group. In abelian groups, right and left cosets are equal. Note: Cosets are not subgroups—with the exception of $H$ itself.

**Lemma 3.10 (Factor or Quotient Group)**  *Let $N \trianglelefteq G$ be a normal subgroup. Then the cosets $\{gN | g \in G\}$ with the operation*

$$(gN) \cdot (g'N) = gNg'N = (gg')N$$

*form a group, the so-called factor group or quotient group.*

As shown in Fig. 3.1, the product of the classes $gN$ and $g'N$ is the coset in which $gg'$ lies,

$$(gN)(g'N) \overset{\text{associative}}{=} gNg'N \overset{\text{normal subgroup}}{=} gg'N \,.$$

***Proof*** The multiplication is well-defined, because the classes are independent of their representative. The coset $eN = N$ forms the identity element, since $eN \cdot gN = gN$, and the inverse of $gN$ is $g^{-1}N$. $\qquad\qquad\qquad\qquad\qquad\qquad\qquad\qquad\square$

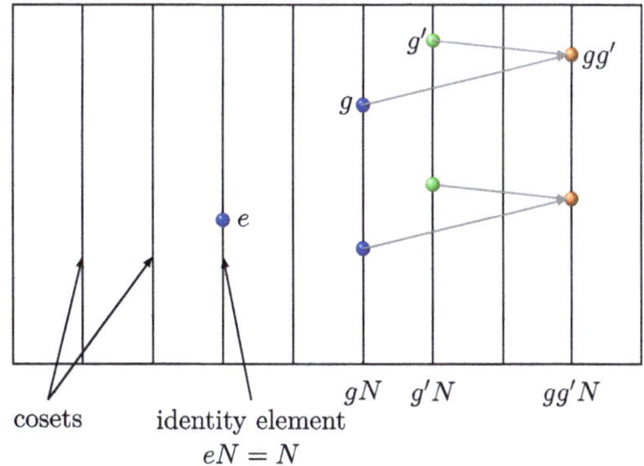


**Fig. 3.1**   The group is partitioned into disjoint coset classes (cosets). The cosets modulo a normal subgroup form a group themselves

---

**Example: Examples of Factor Groups**

- The following applies: $G/e = G$ and $G/G = e$.
- $m\mathbb{Z}$ is normal subgroup of $\mathbb{Z}$ and $\mathbb{Z}/m\mathbb{Z} = \mathbb{Z}_m$ ($\mathbb{Z}$ modulo $m$).

◀

The kernel of a homomorphism is a normal subgroup and the following result applies

**Theorem 3.3 (First Isomorphism Theorem)** *Let $\varphi \in \mathrm{Hom}(G, G')$ be a group homomorphism. Then the mapping*

$$\psi : \ G/\mathrm{Ker}(\varphi) \longmapsto \varphi(G) \leq G', \quad g\,\mathrm{Ker}(\varphi) \mapsto \varphi(g)\,. \tag{3.10}$$

*is an isomorphism. If $\varphi$ is surjective, then $G/\mathrm{Ker}(\varphi) \cong G'$.*

More precisely: If $\pi$ is the surjective homomorphism (also called projection) that assigns to a group element $g$ its left coset $g\,\mathrm{Ker}(\varphi)$, then the isomorphism is given by

$$\psi\,(\pi(g)) = \psi\,(g\,\mathrm{Ker}(\varphi)) = \varphi(g) \quad \text{or} \quad \psi \circ \pi = \varphi\,,$$

and this is shown in Fig. 3.2.

---

**Task**

Show that $\psi$ is a bijective homomorphism.

From the isomorphism theorem it follows in particular:

**Fig. 3.2** The homomorphism $\varphi$ is the composition of the projection $\pi$ and the isomorphism $\psi$

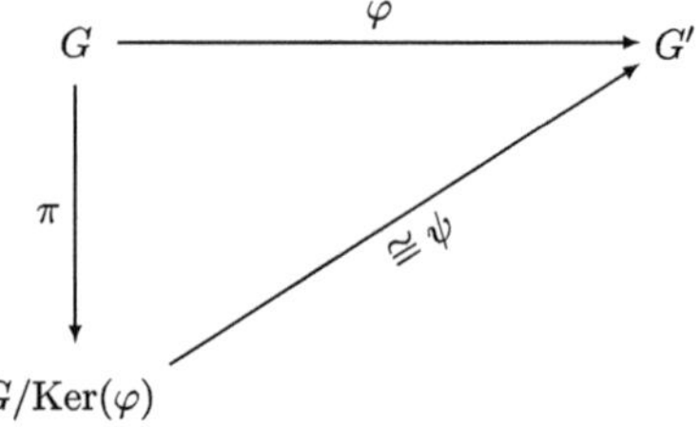

**Homomorphisms and Normal Subgroups**
All possible homomorphisms of $G$ are determined by $G$ alone by considering the projections $G \mapsto G/N$ for all normal subgroups $N \trianglelefteq G$.

In a non-abelian group, two elements do not generally commute, $gg' \neq g'g$, and their commutator $[g, g'] \equiv gg'g^{-1}g'^{-1}$ is not equal to the identity element.

**Lemma 3.11 (Commutator Group)** *The commutators $\{[a, b] \equiv aba^{-1}b^{-1} | a, b \in G\}$ generate a subgroup, the so-called commutator group of $G$.*

The set of commutators does not generally form a subgroup. Only the set generated by the commutators $[G, G]$—this contains all possible multiple products of commutators—is closed. Since $[a, a] = e$, it contains the identity element. The inverse of a commutator is again a commutator

$$[a, b]^{-1} = [b, a],$$

and this also applies to multiple products of commutators. Thus, $[G, G] \leq G$. The order of the commutator group is a measure of how much $G$ is non-abelian.

**Lemma 3.12** *The commutator group $[G, G]$ is a normal subgroup of $G$.*

**Proof** The conjugation of a commutator is again a commutator:

$$g(aba^{-1}b^{-1})g^{-1} = \underbrace{(gag^{-1})}_{a'}\underbrace{(gbg^{-1})}_{b'}(gag^{-1})^{-1}(gbg^{-1})^{-1}$$

$$= a'b'a'^{-1}b'^{-1} \in [G, G].$$

Any element in $[G, G]$ has the form $c_1 c_2 \cdots c_k$, with commutators $c_i$. For this general element

$$g(c_1 c_2 \cdots c_k)g^{-1} = \underbrace{(gc_1 g^{-1})}_{c_1' \in [G,G]}(gc_2 g^{-1}) \cdots (gc_k g^{-1}) \in [G, G]$$

is also in the commutator group. $\qquad\square$

For every group, we can form the factor group $G/N$ with $N = [G, G]$. It holds

**Lemma 3.13** *The factor group $G/[G, G]$ is abelian.*

***Proof***

$$gNg'Ng^{-1}Ng'^{-1}N = \underbrace{[g, g']}_{\in N} N = N(= e \text{ of the factor group}).$$

□

With GAP, one can determine the commutator subgroup of a group. For example, one defines the symmetric group $\mathcal{S}_3$ with the generators $(1, 2)$ and $(1, 2, 3)$ as follows (the symmetric group $\mathcal{S}_n$, also called permutation group, contains all bijective self-mappings of a set of $n$ elements and is analyzed in Chap. 4):

```
GAP> s3:=Group((1,2),(1,2,3));
```

One gets the answer `Group([(1,2),(1,2,3)])`. Now we define the commutator subgroup:

```
GAP> a3:=DerivedSubgroup(s3);
```

The program responds with `Group([(1,2,3)])`. It has the order 3, as one can check with `Order(a3);`. It is the alternating group $\mathcal{A}_3$ of even permutations of three elements. One can find $[\mathcal{S}_3, \mathcal{S}_3]$ also with the command

```
GAP> CommutatorSubgroup(s3,s3);
```

One can verify that the factor group $\mathcal{S}_8/[\mathcal{S}_8, \mathcal{S}_8]$ has two elements and is abelian. First, we calculate the factor group:

```
GAP> s8:=SymmetricGroup(8);;
GAP> a8:=CommutatorSubgroup(s8);;
GAP> g:=FactorGroup(s8,a8);;
```

Instead of the command "CommutatorSubgroup", we can also use "DerivedSubgroup". Now we determine the index of the commutator group in $\mathcal{S}_8$ and check whether the factor group is abelian:

```
GAP> Index(s8,a8);IsAbelian(g);
```

## 3.4    Conjugacy Classes

We introduce another useful equivalence relation on groups. Similar elements are thereby conjugate to each other and classes of similar elements form the conjugacy classes. These are different from the cosets studied above.

**Definition 3.8 (Conjugate Elements)** Two elements $a, b \in G$ are called conjugate to each other, $a \sim b$, if there is an element $g \in G$ such that

$$\mathrm{Ad}_g(a) = gag^{-1} = b .$$

This relation is reflexive, symmetric, and transitive. Transitivity follows from:

$$b = gag^{-1} \sim a, \quad c = \tilde{g}b\tilde{g}^{-1} \sim b \implies c = \tilde{g}(gag^{-1})\tilde{g}^{-1} = (\tilde{g}g)a(\tilde{g}g)^{-1} \sim a .$$

From these properties, it immediately follows that we can decompose any group into disjoint conjugacy classes $K_i$. The conjugacy class of an element $a$,

$$K_a = \{gag^{-1} | g \in G\}, \tag{3.11}$$

always contains the element $a$. Every center element (and in particular the identity element) forms a class by itself. Members of the same conjugacy class cannot be distinguished by using only the group structure, and therefore share many properties.

Let the number of elements of $K_i$ be $n(K_i)$. Then for a finite group

$$\sum_i n(K_i) = |G| . \tag{3.12}$$

The following theorem, which is useful in the representation theory of finite groups, holds

**Theorem 3.4** *All elements of a conjugacy class have the same order. For each conjugacy class $K_i$ of a finite group $G$, $n(K_i)$ is a divisor of $|G|$.*

***Proof*** The first statement follows from

$$a^n = e, \quad b = gag^{-1} \implies b^n = (gag^{-1})(gag^{-1}) \cdots (gag^{-1}) = ga^n g^{-1} = e.$$

$\square$

To prove the second statement we consider special subgroups in $G$:

**Definition 3.9 (Normalizer)** For each element $a \in G$ we define the normalizer $N_a$ according to

$$N_a = \{g \in G | gag^{-1} = a\} \leq G . \tag{3.13}$$

In other words: under the conjugation with $g \in N_a$ the element $a$ remains invariant. Obviously, $N_a$ is a subgroup of $G$ with $\{e, a, a^{-1}\} \in N_a$. According to the Lagrange's theorem, for a finite group the $|N_a|$ are divisors of $|G|$. We have

$$g'ag'^{-1} = gag^{-1} \iff g^{-1}g' \in N_a \iff g' \in gN_a .$$

This implies

$$g'ag'^{-1} \neq gag^{-1} \Longleftrightarrow g' \notin gN_a.$$

**Conjugacy Class $K_a$ and Normalizer $N_a$**
The number of elements in the conjugacy class $K_a$ is equal to the number of cosets of the normalizer $N_a$ or equal to the index $j(N_a)$ of $N_a$ in $G$,

$$n(K_a) = \frac{|G|}{|N_a|} \Longleftrightarrow \frac{|G|}{n(K_a)} = |N_a|. \tag{3.14}$$

The last equation means in particular that the number of elements of each conjugacy class divides the group order $|G|$.

**Lemma 3.14** *If $|G|$ is a prime number, then $G$ is abelian.*

**Proof** If the number of group elements is a prime number, then according to the above theorem each conjugacy class has only one element. Thus, each element forms a class by itself, i.e. it is $gag^{-1} = a$ or $ag = ga$ for two arbitrary group elements $g$ and $a$. Then the group is abelian.                                  □

Each *subgroup $H \leq G$* has its own normalizer:

**Definition 3.10 (Normalizer of a Subgroup $H \leq G$)**

$$N_G(H) = \left\{ g \in G | gHg^{-1} = H \right\}.$$

$N_G(H)$ is a subgroup of $G$, which contains the subgroup $H$. It is the largest subgroup of $G$, in which $H$ is a normal subgroup.

## 3.5  Direct and Semi-Direct Product of Groups

From two groups, a new group can be formed:

**Definition 3.11 (Direct Product)** The direct product $G = H_1 \times H_2$ of two groups $H_1$ and $H_2$ consists of the elements $g = (h_1, h_2)$ with $h_i \in H_i$ and the group multiplication

$$gg' = (h_1, h_2)(h'_1, h'_2) \equiv (h_1 h'_1, h_2 h'_2).$$

So the elements from $H_1$ are multiplied separately from those from $H_2$.

**Example: Klein Four-group**

This is the direct product $\mathbb{Z}_2 \times \mathbb{Z}_2$ and has the symmetric group table

| $\mathbb{Z}/2\mathbb{Z} \times \mathbb{Z}/2\mathbb{Z}$ | $(0,0)$ | $(1,0)$ | $(0,1)$ | $(1,1)$ |
|---|---|---|---|---|
| $(0,0)$ | $(0,0)$ | $(1,0)$ | $(0,1)$ | $(1,1)$ |
| $(1,0)$ | $(1,0)$ | $(0,0)$ | $(1,1)$ | $(0,1)$ |
| $(0,1)$ | $(0,1)$ | $(1,1)$ | $(0,0)$ | $(1,0)$ |
| $(1,1)$ | $(1,1)$ | $(0,1)$ | $(1,0)$ | $(0,0)$ |

It is the smallest non-cyclic group. It is named after Felix Klein, who referred to this group as the "four-group" in the solution of equations.  ◄

Both $H_1$ and $H_2$ are normal subgroups of the direct product, where $G/H_1 = H_2$ and $G/H_2 = H_1$. In general, the reverse does not hold for factor groups, i.e., $G \neq N \times (G/N)$.

Another construction often encountered in physics is the semi-direct product:

**Definition 3.12** Let $H$ and $N$ be groups and $\varphi : H \mapsto \text{Aut}(N)$ a group homomorphism. On $G := H \times N$, one can define an operation $*$ as follows:

$$(h_1, n_1) * (h_2, n_2) := \big(h_1 h_2, n_1 \varphi(h_1)(n_2)\big), \tag{3.15}$$

where the operations in $H$ and $N$ are used respectively.

**Lemma 3.15 (Semi-direct Product)**  *With this operation, $H \times N$ becomes a group. It is denoted by $H \ltimes N$ and is called the semi-direct product of $H$ with $N$.*

**Proof**  The identity element in the semi-direct product is $(e_N, e_H)$, as one can easily see,

$$(e_H, e_N) * (h, n) = \big(e_H h, e_N \underbrace{\varphi(e_H)(n)}_{n}\big) = (h, n).$$

The element inverse to $(h, n)$ is $\big(h^{-1}, [\varphi(h^{-1})(n)]^{-1}\big)$, because

$$\big(h^{-1}, [\varphi(h^{-1})(n)]^{-1}\big) * (h, n) = \big(h^{-1}h, [\varphi(h^{-1})(n)]^{-1}\varphi(h^{-1})(n)\big) = (e_H, e_N).$$

$\square$

The following holds

**Lemma 3.16**  *$H \ltimes N$ contains $H \times \{e_N\}$ as a subgroup and $\{e_H\} \times N$ as a normal subgroup.*

> **Task**
>
> Prove these properties of the semi-direct product $H \ltimes N$.

> **Example: Space-time Symmetries**
>
> - The Euclidean group $E_3$ is the semi-direct product of the group O(3) of spatial rotations and the abelian group of translations $\mathbb{R}^3$ (see Sect. 5.3). The homomorphism $\varphi : \mathrm{O}(3) \mapsto \mathrm{Aut}(\mathbb{R}^3)$ is $\varphi(R)(a) = Ra$ for $R \in \mathrm{O}(3)$ and $a \in \mathbb{R}^3$.
> - The symmetry group of non-relativistic physics—the Galilei group—is a semi-direct product with the group homomorphism (5.39).
> - The symmetry group of relativistic physics—the Poincaré group—is the semi-direct product of Lorentz transformations and translations in space and time, see Sect. 5.5.
>
> ◀

## 3.6  Exercises for Chap. 3

**Problem 3.1 (Isomorphic Groups)**  Which of the following groups are isomorphic and what are the isomorphisms?

1. the complex numbers $\{1, i, -1, -i\}$ with multiplication,
2. the integers $\{2, 4, 6, 8\}$ with respect to multiplication modulo 10,
3. the permutations

$$
\begin{pmatrix} 1 & 2 & 3 & 4 \\ 1 & 2 & 3 & 4 \end{pmatrix}, \quad
\begin{pmatrix} 1 & 2 & 3 & 4 \\ 2 & 1 & 3 & 4 \end{pmatrix}, \quad
\begin{pmatrix} 1 & 2 & 3 & 4 \\ 1 & 2 & 4 & 3 \end{pmatrix}, \quad
\begin{pmatrix} 1 & 2 & 3 & 4 \\ 2 & 1 & 4 & 3 \end{pmatrix};
$$

4. the permutations

$$
\begin{pmatrix} 1 & 2 & 3 & 4 \\ 1 & 2 & 3 & 4 \end{pmatrix}, \quad
\begin{pmatrix} 1 & 2 & 3 & 4 \\ 2 & 3 & 4 & 1 \end{pmatrix}, \quad
\begin{pmatrix} 1 & 2 & 3 & 4 \\ 4 & 1 & 2 & 3 \end{pmatrix}, \quad
\begin{pmatrix} 1 & 2 & 3 & 4 \\ 3 & 4 & 1 & 2 \end{pmatrix}.
$$

**Problem 3.2 (Automorphisms of the Klein Four-group)** Try to determine the automorphism group Aut(Four-group).
Hint: An automorphism $\varphi$ must preserve the orders of the group elements.

**Problem 3.3 (Center of a Group)** Show that the center defined in Sect. 3.2 is an abelian subgroup of $G$.

**Problem 3.4 (Conjugate Subgroups)** Let $H$ be a subgroup of $G$. Show that for every $g \in G$, the set $gHg^{-1}$ is also a subgroup of $G$ that is isomorphic to $H$.

**Problem 3.5 (Dihedral Groups as Semi-direct Products)** Show that the dihedral group $\mathcal{D}_n$ is a semi-direct product, $\mathcal{D}_n = \mathbb{Z}_2 \ltimes \mathbb{Z}_n$.

**Problem 3.6 (Normal Subgroups)** Let $N$ be a subgroup of $G$. Show that the following statements are equivalent:

- $N \trianglelefteq G$ is a normal subgroup, i.e., $gNg^{-1} = N$ for all $g \in G$.
- $gN = Ng$ for all $g \in G$.
- $gNg^{-1} \subseteq N$ for all $g \in G$.

**Problem 3.7 (Normalizer)** Let $H$ be a subgroup of the group $G$. Show that the normalizer of $H$,

$$N_G(H) = \{g \in G \,|\, gHg^{-1} = H\},$$

forms a subgroup of $G$.

**Problem 3.8 (Groups Whose Order Is a Prime Number)** Prove the statement that a finite group, whose order $p$ is a prime number, is isomorphic to a cyclic group.
Note: Consider the subgroup generated by an element $g \neq e$ and recall Lagrange's theorem.

**Problem 3.9 (Semi-direct Product)** Prove Lemma 3.16, namely that $\{e_H\} \times N$ is a normal subgroup and $H \times \{e_N\}$ is a subgroup of $H \ltimes N$.

# Finite Groups

**4**



Any possible arrangement of $n$ different elements is called a permutation of these elements. Permutations (without repetition) can also be interpreted as functions: they are bijective self-mappings that transform a fixed standard arrangement of elements (for example, the header of the group table) into another arrangement. Permutations appear in many branches of mathematics and sciences. In computer science, for example, they are used to analyze sorting algorithms [19], in physics and chemistry to describe many-particle systems [20] and in biology in the study of RNA sequences.

The multiplication of two permutations is the composition of the corresponding mappings. With this multiplication, the $n!$ permutations of $n$ elements form the symmetric group $\mathcal{S}_n$—often also called the group of permutations. From a group theoretical point of view, this group is as complicated as possible: according to a theorem by Cayley, every finite group $G$ is a subgroup of the symmetric group $\mathcal{S}_{|G|}$. An easy-to-read introduction to the theory of finite groups can be found in [21–23] and a detailed treatment of permutation groups in [24].

## 4.1 Subgroups of the Symmetric Group $\mathcal{S}_n$

In the group table of a finite group $G$ with header $g_1, g_2, \ldots, g_n$ an element $g$ in the left column is assigned the row with the permuted elements,

$$\{gg_1, \ldots, gg_n\} = \{\ell_g(g_1), \ldots, \ell_g(g_n)\}. \tag{4.1}$$

Each group element $g$ is thus assigned the left multiplication $\ell_g$

$$\ell : G \longmapsto \mathcal{S}_n, \qquad g \longmapsto \ell_g, \tag{4.2}$$

which permutes the group elements. The mapping $\ell$ defines a *group homomorphism* from $G$ into the group of all permutations, as one can easily verify:

$$(\ell_g \ell_{g'})(g'') = (\ell_g)(g'g'') = gg'g'' = \ell_{gg'}(g'') \implies \ell_g \ell_{g'} = \ell_{gg'}. \tag{4.3}$$

It has the trivial kernel $\{e\}$, because the rows of the group table are different. According to the isomorphism theorem $G$ is isomorphic to the subgroup $\mathrm{Image}(\ell) = \{\ell_g | g \in G\}$ of $\mathcal{S}_n$. It follows the

**Theorem 4.1 (Cayley)** *Every finite group $G$ is isomorphic to a subgroup of the symmetric group $\mathcal{S}_{|G|}$.*

The theorem ensures that every abstract finite group is isomorphic to a concrete group of permutations. This property distinguishes the symmetric groups among all finite groups. The Cayley theorem is important for the representation theory of finite groups and will therefore play a central role in Chap. 10.

## 4.2 Symmetric and Alternating Groups

We label the elements to be permuted—in many-particle theory these could be the coordinates of spinless distinguishable or identical particles—with the numbers $1, 2, 3, \ldots$. Then a permutation $\pi$ is given by the original order and the new order,

$$\pi : \{1, 2, 3, \ldots\} \mapsto \{\pi(1), \pi(2), \pi(3), \ldots\} \quad \text{or} \quad \pi = \begin{pmatrix} 1 & 2 & 3 & \ldots \\ \pi(1) & \pi(2) & \pi(3) & \ldots \end{pmatrix}. \tag{4.4}$$

We first consider the small symmetric group $\mathcal{S}_3$ with 6 elements. If we perform the permutation

$$a : (1, 2, 3) \rightarrow (2, 3, 1) \quad \text{or} \quad a = \begin{pmatrix} 1 & 2 & 3 \\ 2 & 3 & 1 \end{pmatrix}$$

twice in succession, then we obtain the permutation $b : (1, 2, 3) \mapsto (3, 1, 2)$, i.e. $a \circ a \equiv a^2 = b$.

### Example: The Symmetric Group $\mathcal{S}_3$

The 6 elements of the permutation group $\mathcal{S}_3$ are

$$e = \begin{pmatrix} 1\ 2\ 3 \\ 1\ 2\ 3 \end{pmatrix}, \qquad a = \begin{pmatrix} 1\ 2\ 3 \\ 2\ 3\ 1 \end{pmatrix}, \qquad b = a^2 = \begin{pmatrix} 1\ 2\ 3 \\ 3\ 1\ 2 \end{pmatrix},$$

$$c = \begin{pmatrix} 1\ 2\ 3 \\ 1\ 3\ 2 \end{pmatrix}, \quad d = ca = \begin{pmatrix} 1\ 2\ 3 \\ 3\ 2\ 1 \end{pmatrix} \quad f = cb = \begin{pmatrix} 1\ 2\ 3 \\ 2\ 1\ 3 \end{pmatrix}. \tag{4.5}$$

Here, $ca$ means to execute $a$ first and then $c$. The group is non-abelian. ◀

### Task

Compare the group tables of $\mathcal{S}_3$ and $\mathcal{D}_3$.

A permutation can also be represented by indicating only those positions that change with the specification of how they change. One then notes only the *closed subcycles*.

### Example: Cycle Representation of Elements from $\mathcal{S}_4$

The permutation $(1, 2, 3, 4) \rightarrow (1, 3, 4, 2)$ is denoted $(1)(2, 3, 4)$, in the sense of: 1 remains 1, 2 becomes 3, 3 becomes 4 and 4 becomes 2. If there are several cycles, they are separated by brackets. The element $(1, 2, 3, 4) \rightarrow (2, 1, 4, 3)$ is then written $(1, 2)(3, 4)$. In this way, one can easily calculate group multiplication, i.e., successive permutations. For example, $(2, 3, 4) \circ (2, 3, 4) = (2, 4, 3)$ or $(2, 3, 4) \circ (1, 2, 3) = (1, 3)(2, 4)$ (first right, then left!). ◀

The simplest permutations are the *transpositions*, which only swap two elements. A $k$-cycle and its inverse can be represented by the formula

$$(m_1, m_2, \ldots, m_k) = (m_1, m_k) \ldots (m_1, m_3)(m_1, m_2) \tag{4.6}$$

$$(m_1, m_2, \ldots, m_k)^{-1} = (m_1, m_k, \ldots, m_3, m_2) = (m_1, m_2)(m_1, m_3) \ldots (m_1, m_k)$$

as a product of $k - 1$ transpositions.. If there is an even number of transpositions, the permutation is called *even*, otherwise *odd*. Thus, $k$-cycles with even $k$ are odd permutations and $k$-cycles with odd $k$ are even permutations. $\mathcal{S}_n$ contains an equal number of even and odd permutations. The product of two even permutations is even and the inverse permutation of an even permutation is also even. Because the identity element is also even, we conclude:

**Alternating Groups**
The even permutations form a subgroup of $\mathcal{S}_n$, the so-called alternating group $\mathcal{A}_n$.

The product of an even and an odd permutation is odd, the product of two odd permutations is even and the inverse of an odd permutation is also odd. This means

$$\mathrm{Ad}_g(\mathcal{A}_n) = g\mathcal{A}_n g^{-1} \subset \mathcal{A}_n \quad \text{for all} \quad g \in \mathcal{S}_n ,$$

and proves the following result:

The alternating group $\mathcal{A}_n$ is a normal subgroup of $\mathcal{S}_n$, $\mathcal{A}_n \lhd \mathcal{S}_n$

$\mathcal{A}_n$ is the largest (proper) normal subgroup in $\mathcal{S}_n$ and the only normal subgroup for $n = 3$ and $n \geq 5$. In contrast, $\mathcal{S}_4$ has two normal subgroups: the alternating group $\mathcal{A}_4$ and the Klein four-group $\mathbb{Z}_2 \times \mathbb{Z}_2$. $\mathcal{A}_n$ is also the commutator subgroup of $\mathcal{S}_n$. The factor group is abelian,

$$\mathcal{S}_n/\mathcal{A}_n \cong \mathbb{Z}/2\mathbb{Z} = \mathbb{Z}_2 . \tag{4.7}$$

The cycles $(1, 2, \ldots, n)$ and $(1, 2)$ generate the group $\mathcal{S}_n$ and with GAP a symmetric group can be defined via its generators or directly:

```
GAP> s5:=Group((1,2,3,4,5),(1,2));
GAP> s5:=SymmetricGroup(5);
```

The generators are then extracted as follows:

```
GAP> GeneratorsOfGroup(s5);
```

We calculate the *normalizers* (in GAP called Stabilizer) of the following group elements in $\mathcal{S}_3$:

$$e, \quad a = (1, 2, 3), \quad b = (1, 3, 2), \quad c = (2, 3) \quad d = (1, 3) \quad \text{and} \quad f = (1, 2) .$$

The elements $a$ and $b$ define the subgroup $\mathcal{A}_3$. With GAP ($\mathcal{S}_3$ is defined):

```
GAP> na:=Stabilizer(s3,(1,2,3));
```

with the answer $\mathtt{Group([(1,2,3)]}$. Thus, the normalizer $N_a$ is the cyclic group $\mathcal{C}_3$ generated by $a$. Similarly,

```
GAP> nc:=Stabilizer(s3,(2,3));Order(nc);
```

returns 2 and accordingly, $N_c$ is the cyclic group $\mathcal{C}_2$ generated by $c$. Similar results are found for the remaining group elements:

$$N_e = \mathcal{S}_3, \quad N_a = N_b = \mathcal{C}_3, \quad N_c = \{e, c\}, \quad N_d = \{e, d\} \quad \text{and} \quad N_f = \{e, f\}. \tag{4.8}$$

Here one can explicitly see that the orders of the normalizers divide the order 6 of the group. The indices of the stabilizers are

$$j(N_e) = 1, \quad j(N_a) = j(N_b) = 2 \quad \text{and} \quad j(N_c) = j(N_d) = j(N_f) = 3\,,$$

and the conjugacy classes

$$K_e = e, \quad K_a = K_b = \{a, b\}, \quad K_c = K_d = K_f = \{c, d, f\}. \tag{4.9}$$

We see that for each group element, the number of elements of the conjugacy class $K_g$ is equal to the index of the stabilizer group $N_g$ in $G$, $|K_g| = j(N_g)$, in accordance with the general analysis in Sect. 3.4.

The permutation group $\mathcal{S}_5$ has 7 conjugacy classes. To the question

```
GAP> ConjugacyClasses(s5);
```

GAP responds $[()^G, (1, 2)^G, (1, 2)(3, 4)^G, (1, 2, 3)^G, (1, 2, 3)(4, 5)^G, (1, 2, 3, 4)^G, (1, 2, 3, 4, 5)^G]$.

The number of elements in each orbit can be determined as follows:

```
GAP> OrbitLength(s5,());
GAP> OrbitLength(s5,(1,2));
GAP> OrbitLength(s5,(1,2)(3,4));
GAP> OrbitLength(s5,(1,2,3));
GAP> OrbitLength(s5,(1,2,3)(4,5)));
GAP> OrbitLength(s5,(1,2,3,4));
GAP> OrbitLength(s5,(1,2,3,4,5));
```

The results are 1, 10, 15, 20, 20, 30 and 24. In accordance with the general result (3.12) we find

$$1 + 10 + 15 + 20 + 20 + 30 + 24 = 120 = 5! = |\mathcal{S}_5|\,.$$

### 4.2.1   Cycle Decomposition

When studying representations of a group, knowledge of its conjugacy classes is very helpful. We will now characterize the classes of $\mathcal{S}_n$. In the cycle notation for permutations, we arrange the cycles in decreasing order of length, for example

$$\mathcal{S}_6 \ni g = \begin{pmatrix} 1\ 2\ 3\ 4\ 5\ 6 \\ 4\ 2\ 5\ 6\ 3\ 1 \end{pmatrix} \sim (1, 4, 6)(3, 5)(2) \,.$$

Each permutation is a product of *disjoint* cycles and the decomposition of a permutation into cycles is unique, up to the order of cycles of the same length.

Two permutations are of the *same type* if cycles of length $\ell$ occur exactly the same number of times in both permutations. For example, the permutations

$$(1, 3, 8)(4, 5)(2, 6)(7) \quad \text{and} \quad (1, 4, 6)(3, 5)(7, 8)(2) \tag{4.10}$$

are of the same type. One can use Young tableaux (after Alfred Young) to represent permutations. A Young tableau is a collection of cells (usually symbolized by squares) arranged from top to bottom and left-aligned so that their number in each new row does not increase. Each row represents a cycle and the cycles are arranged from top to bottom in decreasing order of length. For example, the two permutations (4.10) have the representation

$$\begin{array}{|c|c|c|}
\hline
1 & 3 & 8 \\
\hline
4 & 5 \\
\cline{1-2}
2 & 6 \\
\cline{1-2}
7 \\
\cline{1-1}
\end{array}
\qquad
\begin{array}{|c|c|c|}
\hline
1 & 4 & 6 \\
\hline
3 & 5 \\
\cline{1-2}
7 & 8 \\
\cline{1-2}
2 \\
\cline{1-1}
\end{array}
\tag{4.11}$$

For each permutation, let $\nu_\ell$, $\ell = 1, 2, \ldots, n$, be the number of cycles of length $\ell$ in its decomposition. Since the total number of elements is equal to $n$, we have

$$\nu_1 + 2\nu_2 + \cdots + n\nu_n = n \,. \tag{4.12}$$

For the permutations in (4.10), $\nu = (\nu_1, \nu_2, \nu_3, \nu_4, \ldots, \nu_8) = (1, 2, 1, 0, \ldots, 0)$. In the Young tableau, $\nu_\ell$ is the number of rows with $\ell$ cells.

The number of different types of permutations in $\mathcal{S}_n$ is equal to the number of partitions $P(n)$ of the number $n$, i.e., the number of ways to write the natural number $n$ as a sum of positive natural numbers. For example,

$$4 = 3 + 1 = 2 + 2 = 2 + 1 + 1 = 1 + 1 + 1 + 1 \implies P(4) = 5 \,. \tag{4.13}$$

Euler devised an elegant method to calculate the numbers $P(n)$. For this, one considers the so-called $q$-series

$$(q)_\infty = \prod_{m=1}^{\infty} \left(1 - q^m\right) = \sum_{n=-\infty}^{\infty} (-1)^n q^{n(3n+1)/2}$$

$$= 1 - q - q^2 + q^5 + q^7 - q^{12} - q^{15} + q^{22} + q^{26} + \dots . \tag{4.14}$$

The exponents $0, 1, 2, 5, 7, 12, 15, 22, 26, 35 \dots$ are the so-called *pentagonal numbers*, and the sign of the $k$-th term is $(-)^{\lfloor (k+1)/2 \rfloor}$. Then the number of partitions $P(n)$ is given by the generating function

$$\frac{1}{(q)_\infty} = \sum_{n=1}^{\infty} P(n)q^n$$

$$= 1 + q + 2q^2 + 3q^3 + 5q^4 + 7q^5 + 11q^6 + 15q^7 + 22q^8 + \dots . \tag{4.15}$$

Another generating function is

$$\left( \frac{2t^{1/8}}{\vartheta_1'(0, \sqrt{t})} \right)^{1/3} = \sum_{n=0}^{\infty} P(n)t^n, \tag{4.16}$$

where $\vartheta_1'(0, x)$ is the derivative of the Jacobi theta function of the first kind. Hardy and Ramanujan found the asymptotic solution

$$P(n) \sim \frac{1}{4n\sqrt{3}} \; e^{\pi \sqrt{2n/3}} . \tag{4.17}$$

The number of partitions of a natural number can be determined with GAP as follows:

```
GAP> NrPartitions(5);NrPartitions(100);
```

The answers are 7 and 190 569 292. The partitions can be obtained with Partititons(5); with the answer [[1,1,1,1,1],[2,1,1,1],[2,2,1], [3,1,1],[3,2], [4,1],[5]].

### 4.2.2  Conjugacy Classes

Two permutations of the same type are conjugate to each other. As an illustrative example, we conjugate a particular $g \in \mathcal{S}_6$ with a transposition $a$:

$$g = (1, 2, 3)(4, 5)(6) \quad \text{and} \quad a = (2, 5) = a^{-1} .$$

We calculate the element conjugate to $g$,

$$aga^{-1} = a(1, 2, 4, 5, 3)(6) = (1, 5, 3)(2, 4)(6) \, .$$

The conjugate element is identical to $g$, except for the exchange of elements 2 and 5. The conjugation with $(2, 5)$ thus only swaps the two elements 2 and 5, without changing the type. If one conjugates with $(4, 5)$, then $g$ even remains fixed, since an exchange of these elements does not change $g$. These properties also apply in the general case. To prove this, one notes that disjoint cycles commute and, moreover

$$(m_a, m_b)(m_1, \ldots, m_a, \ldots, m_b, \ldots m_k)(m_a, m_b) = (m_1, \ldots, m_b, \ldots, m_a, \ldots m_k) \, ,$$

$$(m_1, \tilde{m}_1)(m_1, \ldots m_k)(\tilde{m}_1, \ldots \tilde{m}_{k'})(m_1, \tilde{m}_1) = (\tilde{m}_1, m_2, \ldots m_k)(m_1, \tilde{m}_2, \ldots, \tilde{m}_{k'}) \, .$$

In $\mathcal{S}_n$, too, each conjugation with a transposition only exchanges two elements without changing the type. Every permutation is a product of transpositions and therefore the type does not change even when conjugating with any permutation. Conversely, two permutations of the same type can be transformed into each other by swapping pairs of elements. This leads to the

**Theorem 4.2** *Two permutations $g$ and $g'$ of the symmetric group $\mathcal{S}_n$ are conjugate to each other, $g' = aga^{-1}$, if and only if they are of the same type.*

> **Conjugacy Classes of $\mathcal{S}_n$**
> All permutations of the same type form a conjugacy class. The number of conjugacy classes is equal to the number of partitions $P(n)$ of $n$.

For each partition and thus for each conjugacy class, there is a *Young diagram $T_v$*. This is a Young tableau *without entries* in the cells (in the representation theory of $\mathcal{S}_n$, one encounters such diagrams again). For example, the following diagrams are assigned to the 5 partitions of $\mathcal{S}_4$ in (4.13):

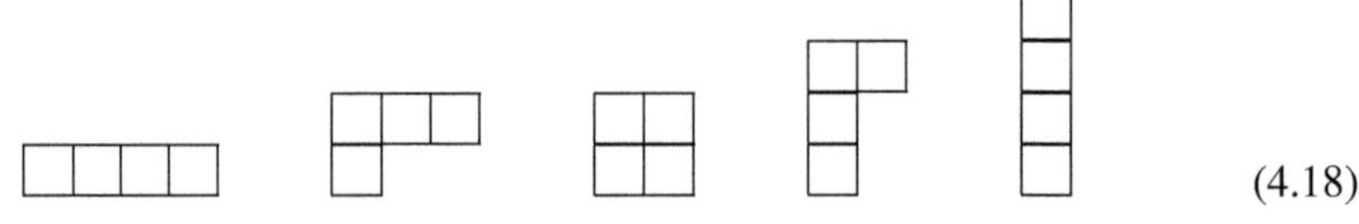

$$(4.18)$$

The second diagram represents, for example, the eight permutations with a 3-cycle and a 1-cycle,

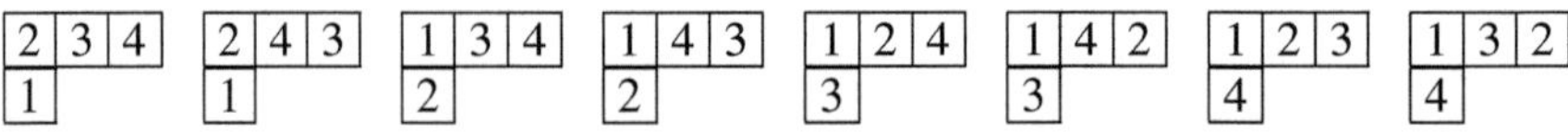

Without proof, we note the

**Theorem 4.3** *The number of permutations of type* $(v_1, v_2, \ldots, v_n)$ *is*

$$P(n, v) = n! \left( \prod_{\ell=1}^{n} v_\ell! \cdot \ell^{\,v_\ell} \right)^{-1} \implies \sum_{v} P(n, v) = n! \,. \tag{4.19}$$

$P(n, v)$ *is the number of elements of the conjugacy class determined by* $v_1, \ldots, v_n$.

The $P(5) = 7$ conjugacy classes of $\mathcal{S}_5$ are given in the following table.

**Example: Conjugacy Classes of the Symmetric Group $\mathcal{S}_5$**

| class | partition | | parity | $v_5$ | $v_4$ | $v_3$ | $v_2$ | $v_1$ | $P(n, v)$ |
|---|---|---|---|---|---|---|---|---|---|
| 1 | 5 | (.....) | even | 1 | 0 | 0 | 0 | 0 | 24 |
| 2 | 4 + 1 | (....)(.) | odd | 0 | 1 | 0 | 0 | 1 | 30 |
| 3 | 3 + 2 | (...)(..) | odd | 0 | 0 | 1 | 1 | 0 | 20 |
| 4 | 3 + 1 + 1 | (...)(.)(.) | even | 0 | 0 | 1 | 0 | 2 | 20 |
| 5 | 2 + 2 + 1 | (..)(..)(.) | even | 0 | 0 | 0 | 2 | 1 | 15 |
| 6 | 2 + 1 + 1 + 1 | (..)(.)(.)(.) | odd | 0 | 0 | 0 | 1 | 3 | 10 |
| 7 | 1 + 1 + 1 + 1 + 1 | (.)(.)(.)(.)(.) | even | 0 | 0 | 0 | 0 | 5 | 1 |

◀

The numbers in the last column add up to the group order $5! = 120$. A $k$-cycle is even for odd $k$ and odd for even $k$, see the discussion after (4.6).

According to Cayley's theorem, the cyclic group $\mathcal{C}_n$ is a subgroup of $\mathcal{S}_n$. It is generated by the cyclic permutation of $n$ elements,

$$g = (1, 2, \ldots, n) = \begin{pmatrix} 1 & 2 & \cdots & n \\ 2 & 3 & \cdots & 1 \end{pmatrix} .$$

Thus, $\mathcal{C}_n$ can be realized as cyclic permutations of $n$ elements or equivalently as rotations in the plane by multiples of $2\pi/n$.

## 4.3  Small Groups

Several finite groups have already been introduced and discussed. Table 4.1 contains (up to isomorphism) all groups up to order 15 and their most important properties. These properties can be obtained, for example, with the help of GAP after entering

**Table 4.1** List of all finite groups up to order 15. Here, a stands for abelian, na for non-abelian, c for cyclic, sol for solvable and $a^b$ is the abbreviation for $bab^{-1}$

| $|G|$ | Isomorphism type of G | Presentation | Isomorphic groups | Important properties |
|---|---|---|---|---|
| 2 | $\mathbb{Z}_2$ | | | c |
| 3 | $\mathbb{Z}_3$ | | | c |
| 4 | $\mathbb{Z}_2 \times \mathbb{Z}_2$ | | | a, Klein four-group |
| | $\mathbb{Z}_4$ | | | c |
| 5 | $\mathbb{Z}_5$ | | | c |
| 6 | $\mathbb{Z}_2 \times \mathbb{Z}_3$ | | $\mathbb{Z}_6$ | c |
| | $\mathcal{S}_3$ | $a^3 = b^2 = e, a^b = a^{-1}$ | $\mathcal{D}_3 \cong \mathbb{Z}_2 \ltimes \mathbb{Z}_3$ | dihedral group |
| | | | $\mathrm{SL}_2(2, \mathbb{Z}/2\mathbb{Z})$ | na, sol |
| 7 | $\mathbb{Z}_7$ | | | c |
| 8 | $\mathbb{Z}_2 \times \mathbb{Z}_2 \times \mathbb{Z}_2$ | | | a |
| | $\mathbb{Z}_2 \times \mathbb{Z}_4$ | | | a |
| | $\mathbb{Z}_8$ | | | c |
| | $\mathcal{D}_4$ | $a^4 = b^2 = e, a^b = a^{-1}$ | $\mathbb{Z}_2 \ltimes \mathbb{Z}_4$ | na, sol, dihedral group |
| | $\mathcal{Q}_8$ | $a^4 = e, b^2 = a^2, a^b = a^{-1}$ | $\mathbb{H}_2$ | na, sol, quaternion group |
| 9 | $\mathbb{Z}_3 \times \mathbb{Z}_3$ | | | a |
| | $\mathbb{Z}_9,$ | | | c |
| 10 | $\mathbb{Z}_2 \times \mathbb{Z}_5$ | | $\mathbb{Z}_{10}$ | c |
| | $\mathcal{D}_5$ | $a^5 = b^2 = e, a^b = a^{-1}$ | $\mathbb{Z}_2 \ltimes \mathbb{Z}_5$ | na, sol, dihedral group |
| 11 | $\mathbb{Z}_{11}$ | | | c |
| 12 | $\mathbb{Z}_2 \times \mathbb{Z}_2 \times \mathbb{Z}_3$ | | $\mathbb{Z}_2 \times \mathbb{Z}_6$ | a |
| | $\mathbb{Z}_4 \times \mathbb{Z}_3$ | | $\mathbb{Z}_{12}$ | c |
| | $\mathcal{A}_4$ | $a^3 = b^3 = (ab)^2 = e$ | $(\mathbb{Z}_2 \times \mathbb{Z}_3) \ltimes \mathbb{Z}_2$ | na, sol, alternating group |
| | $\mathcal{D}_6$ | $a^6 = b^2 = e, a^b = a^{-1}$ | $\mathbb{Z}_2 \ltimes \mathbb{Z}_6$ | na, sol, dihedral group |
| | $\mathbb{H}_3$ | $a^6 = e, b^2 = a^3, a^b = a^{-1}$ | $\mathbb{Z}_4 \ltimes \mathbb{Z}_3$ | na, sol, dicyclic group |
| 13 | $\mathbb{Z}_{13}$ | | | c |
| 14 | $\mathbb{Z}_2 \times \mathbb{Z}_7$ | | $\mathbb{Z}_{14}$ | c |
| | $\mathcal{D}_7$ | $a^7 = b^2 = e, a^b = a^{-1}$ | $\mathbb{Z}_2 \ltimes \mathbb{Z}_7$ | na, sol, dihedral group |
| 15 | $\mathbb{Z}_3 \times \mathbb{Z}_5$ | | $\mathbb{Z}_{15}$ | c |

```
GAP> SmallGroupsInformation(6);
```

and this leads to

```
There are 2 groups of order 6, 1 of type S3, 2 of type Z6.
```

The following remarks help to understand the list:

- For each prime number $p$ there is only the group $\mathcal{C}_p \cong \mathbb{Z}_p$.
- For each prime number $p$ there exist two groups of order $p^2$: $\mathbb{Z}_p \times \mathbb{Z}_p$ and $\mathbb{Z}_{p^2}$.
- For coprime $p$ and $q$ it holds: $\mathbb{Z}_p \times \mathbb{Z}_q \cong \mathbb{Z}_{pq}$.

- Currently, a list of groups of order $\leq 2000$ exists. This contains $49\,910\,529\,484$ groups [25].

In the list, the group of *quaternions* appears. This is generated by the matrices

$$a = \begin{pmatrix} 0 & i \\ i & 0 \end{pmatrix} \quad \text{and} \quad b = \begin{pmatrix} 0 & 1 \\ -1 & 0 \end{pmatrix} \tag{4.20}$$

They fulfill the relations

$$a^4 = \mathbb{1}, \quad a^2 = b^2 \quad \text{and} \quad bab^{-1} = a^{-1}. \tag{4.21}$$

The group has order 8 and consists of the elements

$$Q = \{\mathbb{1}, a, a^2, a^3, b, ab, a^2 b, a^3 b\}. \tag{4.22}$$

Further properties of the quaternion group are discussed in Problem 4.5. The quaternion group can be defined and then analyzed with GAP, e.g.

```
GAP>          f:=FreeGroup("a","b");
GAP>          f1:=f/[f.1\^4,f.2*f.1*f.2^(-1)*f.1,f.1^2*f.2^(-2)];
GAP>          Order(f1);Elements(f1);
```

## 4.4   Exercises for Chap. 4

**Problem 4.1 (Symmetric Groups $\mathcal{S}_3$ and $\mathcal{S}_4$)** Calculate the multiplication table for the symmetric groups $\mathcal{S}_3$ and $\mathcal{S}_4$.

**Problem 4.2 (Generators of the Symmetric Groups)** Show that $\pi = (1, 2)$ and $\pi' = (1, 2, \ldots, n)$ generate the symmetric group $\mathcal{S}_n$.

**Problem 4.3 (Symmetric and Alternating Groups)** Let $\mathcal{S}_n$ be the symmetric group and $\mathcal{A}_n$ its alternating subgroup of even permutations. Show that $\mathcal{A}_n \triangleleft \mathcal{S}_n$ is a normal subgroup. What group is the quotient group $\mathcal{S}_n/\mathcal{A}_n$?

**Problem 4.4 (Conjugacy Classes of $\mathcal{S}_6$)** Consider the symmetric group $\mathcal{S}_6$ with 720 elements. How many conjugacy classes does it have? Determine the table of conjugacy classes (similar to $\mathcal{S}_5$ in Sect. 4.2.2).

**Problem 4.5 (Quaternion Group)** The *quaternion* group is generated by the matrices in (4.20).

1. Convince yourself that $Q$ in (4.22) contains all words of the presentation and determine the group table.
2. Find a subgroup of $Q$. Is it a normal subgroup?
3. The elements

$$\{\pm 1, \pm i, \pm j, \pm k\} \quad \text{with} \quad i^2 = j^2 = k^2 = ijk = -1 \tag{4.23}$$

form a basis of the quaternions,[1] similar to 1 and i for the complex numbers. The set of quaternions $\mathbb{H} = \{x_0 + x_1 i + x_2 j + x_3 k \,|\, x_a \in \mathbb{R}\}$ extends the range of real numbers. How can the basis (4.23) be represented by the generators of the quaternion group in (4.20)?

---

[1] These multiplication rules can be found on a memorial plaque at Broom Bridge in Dublin, where they were spontaneously carved into the stone by William Rowan Hamilton in October 1843.

# Space-Time Symmetries

5

> *If a system possesses a certain group of symmetry operations,*
> *then every physical observable of this system must also possess*
> *the same symmetry.*
>
> Principle of Neumann

Groups primarily appear as symmetry groups, which play a prominent role in *all* areas of physics. In classical mechanics and quantum mechanics, these are the rotations and translations in space, or more generally, the elements of the Galilean group. In relativistic physics, these are the Lorentz and Poincaré groups. The point groups of molecular physics are finite subgroups of the rotation group, and the crystal groups of solid state theory are discrete subgroups of the group of rigid motions.

## 5.1  Group Actions

In physics, symmetries are realized by their action on elements of a physical model. This set is denoted by $M$. The elements of $M$ can be material objects such as elementary particles, atoms, molecules, or crystals, or abstract quantities such as fields or state vectors in a Hilbert space. $M$ does not need to be a linear space.[1] For example, in Hamiltonian mechanics, $M$ is a symplectic manifold with a symplectic group action. In the following, let $G$ be an arbitrary group and $M \neq \emptyset$ an arbitrary set.

**Definition 5.1 (Group Action)**  A left group action is a function $\Phi : G \times M \mapsto M$ with the properties

---

[1] Group actions on linear spaces are called representations.



A. Wipf, *Symmetries in Physics*, https://doi.org/10.1007/978-3-662-72675-4_5

- $\Phi(e, p) = p$ for all $p \in M$,
- $\Phi\big(g_1, \Phi(g_2, p)\big) = \Phi\big(g_1 g_2, p\big)$ for all $g_1, g_2 \in G$ and $p \in M$.

The transformation group $G$ is said to act from the left on the $G$-set $M$.

**Notation**
One often writes $\Phi(g, p) = \Phi_g(p)$. Then $\Phi_e = id$ and $\Phi_g \circ \Phi_{g'} = \Phi_{gg'}$ hold for all $g, g'$.

Each mapping $\Phi_g : M \mapsto M$ has the inverse mapping $\Phi_{g^{-1}}$ and it follows that

**Lemma 5.1**  *For each $g \in G$, the mapping $\Phi_g : M \to M$ is bijective.*

The bijective mappings

$$\mathcal{S}_M = \{f : M \to M \mid f \text{ is bijective}\} \tag{5.1}$$

with the composition as operation form a group. The neutral element is the identity mapping and the inverse of a mapping is the inverse mapping. The property $\Phi_{gg'} = \Phi_g \circ \Phi_{g'}$ now means:

**Group Homomorphism** $g \to \Phi_g$
The mapping $g \to \Phi_g$ is a homomorphism $G \to \mathcal{S}_M$.

Conversely, each $\varphi \in \mathrm{Hom}(G, \mathcal{S}_M)$ defines a group action of $G$ on $M$ through $\Phi_g(p) = \varphi(g)(p)$.

Group actions generalize the homomorphisms of finite groups into the symmetric groups introduced in Chap. 3. For $M = G$, $\mathcal{S}_G$ is the symmetric group $\mathcal{S}_{|G|}$ and $g \mapsto \ell_g$ is that homomorphism which identifies $G$ as a subgroup of $\mathcal{S}_{|G|}$.

### 5.1.1  Special Types of Group Actions

Group actions can be faithful, free, or transitive.

**Definition 5.2 (Faithful, Free and Transitive Actions)**  A left action is called

- faithful, if $\Phi_g$ is the identity mapping only for $g = e$,
- free, if for each $p \in M$ only the mapping $\Phi_e$ leaves the point $p$ fixed,

- transitive, if for each pair $p, q \in M$ there exists a group element $g$ with $\Phi_g(p) = q$.

For a faithful action $\Phi_g$, the kernel of $\mathrm{Hom}(G, \mathcal{S}_M)$ contains only $e \in G$ and therefore $g \to \Phi_g$ is injective.

The orbit of a point in the $G$-set $M$ appears in symmetry considerations. In general, one defines

**Definition 5.3 (Orbit)** The orbit (or path) of a point $p \in M$ under the action of $G$ is the set

$$B(p) = \left\{ \Phi_g(p) \mid g \in G \right\} \subset M. \tag{5.2}$$

The orbits form equivalence classes with respect to the following equivalence relation: $p$ is similar to $q$ if there is a group element $g$ such that $q = \Phi_g(p)$.

A group $G$ acts transitively on $M$ if and only if $B(p) = M$ for all $p \in M$. If it does not act freely, then there exist points in $M$, which are mapped onto themselves by all $\Phi_g$. This leads to

**Definition 5.4 (Isotropy Group)** The isotropy group of $p \in M$ is

$$N_p = \{ g \in G \mid \Phi_g(p) = p \} \leq G. \tag{5.3}$$

It is the natural generalization of the normalizer $N_a$ of $a \in G$, see (3.13). As indicated by the symbol $\leq$, $N_p$ is a subgroup of $G$, since with $g, g'$ also $gg'$ and $g^{-1}$ lie in $G$,

$$\Phi_g(p) = \Phi_{g'}(p) = p \implies \Phi(g'g, p) = \Phi(g', \Phi(g, p)) = \Phi(g', p) = p.$$

The following holds

**Lemma 5.2** *If $G$ acts transitively on $M$, then all isotropy groups $N_p$ are conjugate to each other.*

To prove this, we consider two arbitrary points $p, q$ in $M$. By assumption, there is a group element $g'$ with $\Phi(g', p) = q$. Then it follows that

$$\Phi(g, p) = p \iff \Phi\left( g'gg'^{-1}, q \right) = \Phi(g'g, p) = \Phi(g', p) = q.$$

Therefore, the isotropy groups of $p$ and $q$ are conjugate,

$$N_q = g' N_p g'^{-1}. \tag{5.4}$$

This proves that two points $p, q$ in the same orbit have isomorphic isotropy groups.

**Example: Action of SL(2, $\mathbb{R}$) on the Complex Half-plane**

The continuous group SL(2, $\mathbb{R}$) acts on the upper complex half-plane $\mathcal{H} = \{z|\Im(z) > 0\}$ as follows:

$$\Phi_S(z) = \frac{az + b}{cz + d}, \quad S = \begin{pmatrix} a & b \\ c & d \end{pmatrix} \in \text{SL}(2, \mathbb{R}). \tag{5.5}$$

An explicit calculation yields,

$$\Im\big(\Phi_S(z)\big) = \frac{\Im(z)}{|cz + d|^2} \quad \text{and} \quad \Phi_{S'} \circ \Phi_S(z) = \Phi_{S'S}(z),$$

which means that $\Phi_S$ defines an action of SL(2, $\mathbb{R}$) on $\mathcal{H}$. However, this is not faithful, as $\pm \mathbb{1}$ turn into the identical mapping. If we identify the two elements $\pm S$ in SL(2, $\mathbb{R}$), which leads to the factor group PSL(2, $\mathbb{R}$) = SL(2, $\mathbb{R}$)/$\{\pm\mathbb{1}\}$, then $\Phi_S$ becomes a faithful action of PSL(2, $\mathbb{R}$) on $\mathcal{H}$. An arbitrary $z \in \mathcal{H}$ is mapped by

$$S = \frac{1}{\sqrt{\Im(z)}} \begin{pmatrix} 0 & \Im(z) \\ -1 & \Re(z) \end{pmatrix} \in \text{SL}(2, \mathbb{R})$$

to i, so that SL(2, $\mathbb{R}$) acts transitively on $\mathcal{H}$. The isotropy group $N_z$ of $z = i$ is the subgroup of proper rotations,

$$\text{SO}(2, \mathbb{R}) = \left\{ \begin{pmatrix} \cos\theta & -\sin\theta \\ \sin\theta & \cos\theta \end{pmatrix} \,\Big|\, \theta \in [0, 2\pi) \right\}.$$

◀

We have already seen earlier that the left multiplication and the adjoint action, given by

$$\ell_g(a) = ga \quad \text{and} \quad \text{Ad}_g(a) = gag^{-1}, \tag{5.6}$$

are special group actions, in which $G$ acts on itself. The left multiplication is free and transitive, and these properties were used in the proof of Cayley's theorem. The adjoint action is faithful if and only if $G$ has a trivial center. It is neither free, since $\text{Ad}_g(g) = g$, nor transitive, since $\text{Ad}_g(e) = e$ for all $g$. The orbit of $a$ under the adjoint action is the conjugacy class of $a$ in $G$.

## 5.2    Rotations in Space

The rotations in Euclidean space $\mathbb{R}^3$ define a continuous non-abelian group, which appears, for example, in the treatment of the rigid body in classical mechanics or the solution of the hydrogen atom in quantum mechanics. The Galilean group examined further below contains this group as a subgroup.

The treatment of rotations in $\mathbb{R}^n$ is not much more complicated than in $\mathbb{R}^3$, and therefore we consider here the more general case. We introduce a Cartesian basis $\{e_1, e_2, \ldots, e_n\}$ in $\mathbb{R}^n$, i.e. an oriented basis with $(e_i, e_j) = \delta_{ij}$. Under a rotation, $e_i$ transforms into

$$e_i' = \sum_{j=1}^{n} e_j R_{ji} \equiv e_j R_{ji} \,, \tag{5.7}$$

where we used the Einstein summation convention: we sum over index pairs. In rotations, neither the lengths of vectors nor the angles between vectors change. Therefore, the scalar product is invariant, i.e.

$$\delta_{ij} = (e_i, e_j) = (e_i', e_j') = \left( \sum_p e_p R_{pi}, \sum_q e_q R_{qj} \right) = \sum_p R_{pi} R_{pj} \,. \tag{5.8}$$

We combine the $n^2$ real numbers $R_{ij}$ into a real matrix,

$$(R_{ij}) = R = \begin{pmatrix} R_{11} & R_{12} & \ldots & R_{1n} \\ R_{21} & R_{22} & \ldots & R_{2n} \\ \vdots & \vdots & & \vdots \\ R_{n1} & R_{n2} & \ldots & R_{nn} \end{pmatrix} .$$

On the right side in (5.8) is the product of the transposed matrix of $R$, $(R^T)_{ij} = R_{ji}$, and the matrix $R$. Therefore, we can write this condition as follows,

$$(R^T R)_{ij} = \delta_{ij} \quad \text{or} \quad R^T R = \mathbb{1} \,.$$

We have thus the

**Theorem 5.1 (Rotation Matrices)** *The real matrix R of a rotation with respect to a Cartesian basis satisfies the condition*

$$R^T = R^{-1} \quad or \quad R^T R = R R^T = \mathbb{1} \,. \tag{5.9}$$

For a fixed basis $e_i$, we identify a rotation with "its matrix" $R$. The sequential execution of two rotations corresponds to the product of the associated matrices:

$$R_2 R_1 \sim \text{first the rotation } R_1 \text{ and then the rotation } R_2 .$$

Rotations are invertible linear transformations that leave the scalar product invariant. We conclude:

> **Orthogonal Groups**
> The rotations in $\mathbb{R}^n$ form a continuous group—the orthogonal group $O(n)$.

After choosing a Cartesian basis, rotations are characterized by a matrix $R$ in (5.9) and it follows

**Corollary 5.1** *The real $n \times n$ matrices with $R^T R = R R^T = \mathbb{1}$ form a group.*

This can also be explicitly checked: The unit matrix obviously satisfies (5.9). We still need to check whether the matrix product or the inverses of matrices from this set are again in this set:

$$(R_2 R_1)^T (R_2 R_1) = R_1^T \underbrace{R_2^T R_2}_{=\mathbb{1}} R_1 = R_1^T R_1 = \mathbb{1}, \quad (R^{-1})^T R^{-1} = (R R^T)^{-1} = \mathbb{1} .$$

One should distinguish between *active* and *passive transformations*. In an actively interpreted rotation, the position vector is mapped to a new position vector,

$$r = x_i \, e_i \mapsto r' = x_i \, e_i' \equiv x_i' e_i . \tag{5.10}$$

With a *fixed basis*, this corresponds to the transformation

$$x \mapsto x' = R x . \tag{5.11}$$

Alternatively, we can imagine that the *same vector* is viewed from different frames. In the rotated frame, the Cartesian basis is $e_i'$ and the vector has the transformed coordinates $x_i'$, so that

$$x_i \, e_i = x_i' \, e_i' . \tag{5.12}$$

In a passively interpreted rotation, the transformation (5.11) just compensates for the rotation (5.7) of the basis. Whether a transformation is to be interpreted passively or actively usually becomes clear from the context.

For a rotation matrix, $\det(R^T R) = 1$ and thus $\det R = \pm 1$. From

$$\det\left(\frac{\partial x_i'}{\partial x_j}\right) = \det R \tag{5.13}$$

it follows that rotations with $\det R = 1$ preserve the orientation and rotations with $\det R = -1$ reverse it.

> **Proper and Improper Rotations**
> The orientation-preserving rotations are called *proper*, the others *improper*. As the kernel of the determinant mapping, the proper rotations form the normal subgroup $SO(n) \lhd O(n)$. The group $SO(n)$ is called *special orthogonal group* .

**Theorem 5.2 (Euler)** *In odd dimensions, every proper rotation and in even dimensions every improper rotation has a one-dimensional subspace of fixed points. Specifically in 3 dimensions, every proper rotation has a rotation axis.*

We need to show that $Rn = n$ for a $n \neq 0$ is solvable. The line defined by $n$ is then the fixed point axis of $R$. A solution exists when $R$ has the eigenvalue 1 or when $\det(R - \mathbb{1})$ vanishes. Because

$$\det(R - \mathbb{1}) = \det(R - \mathbb{1})^T = \det(R^{-1} - \mathbb{1}) = \det\left[R^{-1}(\mathbb{1} - R)\right]$$

$$= \det R^{-1} \det(\mathbb{1} - R) = (-)^n \det R \det(R - \mathbb{1}) \tag{5.14}$$

$\det(R - \mathbb{1})$ vanishes for $(-1)^n \det R = -1$, i.e. for proper rotations and odd $n$ or for improper rotations and even $n$. In both cases, $R$ has the eigenvalue 1 and thus has a fixed axis. If we repeat the argument for the determinant of $R + \mathbb{1}$, then it follows that

**Lemma 5.3 (Rotoreflection)** *An improper rotation has the eigenvalue $-1$ in all dimensions.*

We now consider rotations in $\mathbb{R}^3$ in more detail: According to the just proven theorem by Euler every proper rotation has a rotation axis. Therefore, it can be transformed with a second proper rotation $S$ (which rotates the rotation axis into the 3-direction) into a rotation around the third axis, $SRS^{-1} = R(e_3, \varphi)$. We first consider such rotations around the 3-axis with the angle $\varphi$, see Fig. 5.1.

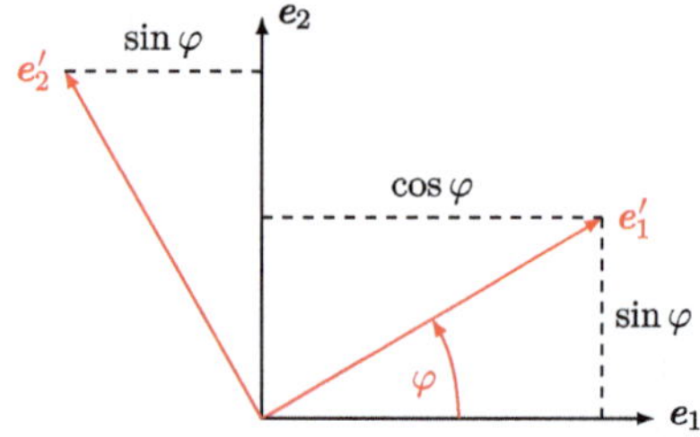



**Fig. 5.1** Rotation of the basis vectors around the 3-axis

For these rotations, $e_3' = e_3$ and, as can be seen from Fig. 5.1, the transformation formulas for the remaining basis vectors are

$$
\begin{aligned}
e_1' &= \phantom{-}e_1 \cos\varphi + e_2 \sin\varphi\,, \\
e_2' &= -e_1 \sin\varphi + e_2 \cos\varphi\,.
\end{aligned}
\tag{5.15}
$$

The corresponding rotation matrix has the form

$$
R(e_3, \varphi) = \begin{pmatrix} R_2(\varphi) & 0 \\ 0 & 1 \end{pmatrix}, \quad
R_2(\varphi) = \begin{pmatrix} \cos\varphi & -\sin\varphi \\ \sin\varphi & \cos\varphi \end{pmatrix}.
\tag{5.16}
$$

The transformation of the Cartesian coordinates is obtained from $x_i\, e_i' = x_i'\, e_i$ and reads

$$
x' = R(e_3, \varphi)x\,.
$$

Similarly, we find the following rotation matrices for rotations around the other two axes:

$$
R(e_1, \varphi) = \begin{pmatrix} 1 & 0 \\ 0 & R_2(\varphi) \end{pmatrix}, \quad
R(e_2, \varphi) = \begin{pmatrix} \cos\varphi & 0 & \sin\varphi \\ 0 & 1 & 0 \\ -\sin\varphi & 0 & \cos\varphi \end{pmatrix}.
\tag{5.17}
$$

---

**Canonical Form of Rotations in $\mathbb{R}^3$**

For every *proper rotation*, the Cartesian basis $\{e_i\}$ in $\mathbb{R}^3$ can be chosen such that the rotation matrix has the simple form (5.16).

---

In the *theory of the gyroscope*, an alternative parametrization of rotations using the Euler angles $\varphi$, $\vartheta$, $\psi$ is used. It is shown that every proper rotation is the product of three elementary rotations around the third and first axis,

$$
R(\varphi, \vartheta, \psi) = R(e_3, \varphi)R(e_1, \vartheta)R(e_3, \psi)\,.
\tag{5.18}
$$

## 5.3    The Euclidean Groups $E_n$

The Euclidean group describes the motion of rigid bodies in Euclidean space $\mathbb{R}^3$ or more generally in $\mathbb{R}^n$. Rigid motions are affine transformations that leave the distance between two points in $\mathbb{R}^n$ invariant. We choose a *Cartesian basis* $e_i$ and an origin $\mathcal{O}$, so that every point $P$ is uniquely characterized by a vector $\overline{OP} = r = \sum x_i\, e_i$ and thus by its coordinate $n$-tuple $x$. The distance between two points $P$ and $Q$ with coordinates $x$ and $y$ is

$$d(P, Q) = \|x - y\|, \quad \text{with} \quad \|x\|^2 = \sum_1^n x_i^2 \,. \tag{5.19}$$

In the case of a rigid motion, $x$ and $y$ transform into $Rx + a$ and $Ry + a$, and the distance between the points does not change if $R$ is a rotation.

---

**Euclidean Group $E_n$**
A rigid motion consists of rotations and translations,

$$(R, a) : x \mapsto x' = Rx + a, \quad a \in \mathbb{R}^n, \quad R \in \mathrm{O}(n)\,. \tag{5.20}$$

---

Rigid motions with $a \neq 0$ have no fixed point. Since rigid motions are defined by an invariance property, they form a group called *Euclidean group*, denoted by $E_n$. Let now

$$(R', a') : x' \mapsto x'' = R'x' + a' \tag{5.21}$$

be another motion. Then the composition of the rigid motions (5.20) and (5.21) is given by

$$(R', a') \circ (R, a) : x \mapsto x'' = R'(Rx + a) + a' = R'Rx + R'a + a' \,.$$

---

**Composition Law for the Euclidean Group**
The composition of two rigid motions in Euclidean space $\mathbb{R}^n$ is given by

$$(R', a') \circ (R, a) = (R'R, R'a + a')\,. \tag{5.22}$$

---

Accordingly, the inverse to $(R, a)$ is

$$(R, a)^{-1} = \left(R^{-1}, -R^{-1}a\right) \in E_n \,. \tag{5.23}$$

If we describe a motion $(R, a)$ relative to a frame with an origin shifted by $a_1$ and a Cartesian basis rotated by $R_1 \in SO(n)$, then it has the form

$$(R', a') = (R_1, a_1)(R, a)(R_1, a_1)^{-1} = \left( R_1 R R_1^{-1}, a_1 + R_1 a - R_1 R R_1^{-1} a_1 \right) .$$

$$(5.24)$$

The transition between the frames is thus described by a conjugation. In each conjugacy class, there are particularly simple representatives. These describe the motion in an adapted Cartesian system. These representatives, also known as normal forms, are discussed in Appendix 5.6 to this chapter.

The Euclidean group $E_n$ is the *semi-direct product* (see Sect. 3.5) of the group of rotations and the translations, i.e. $E_n = O(n) \ltimes \mathbb{R}^n$. The homomorphism $\varphi : O(n) \mapsto \mathrm{Aut}(\mathbb{R}^n)$ is given by $\varphi(R)(a) = Ra$ for $R \in O(n)$ and $a \in \mathbb{R}^n$. In particular, it follows that

$$(R, a)(\mathbb{1}, a')(R, a)^{-1} = (\mathbb{1}, Ra') ,$$

which means that the subgroup of translations is a normal subgroup.

Another subgroup is that of the *proper* rigid motions

$$E_n^+ = \left\{ (R, a) | a \in \mathbb{R}^n, R \in SO(n) \right\} .$$

$$(5.25)$$

If we decompose the Euclidean group $E_n$ according to its subgroup $E_n^+$ into right cosets, we obtain two classes, the class $E_n^+$ of proper rigid motions and the class of *improper rigid motions*

$$E_n^- = \left\{ (R, a) | a \in \mathbb{R}^n, R \in O(n), \det R = -1 \right\} .$$

$$(5.26)$$

They are also left cosets, so that $E_n^+ < E_n$ is a normal subgroup, called special Euclidean group. We prove the useful

**Theorem 5.3 (Fixed Points)** *Every finite subgroup $G$ of the Euclidean group $E_n$ has a fixed point. This applies in particular to all point groups.*

***Proof*** Let $x \in \mathbb{R}^n$ be the coordinates of an arbitrary point in Euclidean space and

$$B(x) = \{ x_i = R_i x + a_i | (R_i, a_i) \in G \}$$

the *orbit* of this point under the action of $G < E_n$. For a finite subgroup, the orbit consists of finitely many points. The center of mass of the orbit, given by

$$\bar{x} = \frac{1}{n} \sum_{i=1}^{|G|} x_i ,$$

$$(5.27)$$

is obviously a fixed point for each group element $(R, a) \in G$, because this only permutes the elements of the orbit, so that $R\bar{x} + a = \bar{x}$ for all all $(R, a) \in G$.   □

## 5.4   The Galilean Group

Space-time models, in which the structure of space and time is independent of the existing matter, are the *Galilei-Newtonian model* with an absolute time and the *Einstein-Poincaré model* with relative time. The first model is a limiting case of the second. In both models, there are distinguished reference frames. Such a frame is almost ideal or inertial if the Galilean law of inertia holds sufficiently accurately for a sufficiently force-free body, i.e., if the body remains in its state of rest or uniform linear motion. For three point masses moving on non-parallel lines, this seems to be an empty statement, but for any further force-free motion, it provides an operational definition of inertial frames (also called inertial systems or inertial reference frames).

---

**Example: Realization of an Inertial Frame**

Within a box freely falling to the earth or in a spaceship flying without propulsion far away from celestial bodies, one has a (local) inertial frame to a good approximation.   ◄

In this section, we limit ourselves to the non-relativistic limit. Two arbitrary inertial frames $I$ and $I'$ may differ in that their origins $\mathcal{O}$ and $\mathcal{O}'$

- are shifted against each other by a *spatial translation* $a \in \mathbb{R}^3$ and/or
- move at a constant velocity $u \in \mathbb{R}^3$ relative to each other.

An equivalent formulation is

**Transformation Between Inertial Frames**

If a particle in the inertial frame $I$ has the position vector $r(t) = r(0) + v\,t$, then in the inertial frame $I'$ it has the position vector $r'(t) = (r(0) + a) + (v + u)\,t$.

We have only changed the inertial frame to describe the motion of the particle and not the motion itself (passive view). We choose in each inertial frame a Cartesian basis, and identify a vector $r$ with its 3-tuple $x$ in $r = x_i\,e_i$. The set of *spatial translations*[2]

$$x \mapsto x' = x + a \tag{5.28}$$

---

[2] We choose the same symbol $a$ for the displacement vector and its 3-tuple.

are elements of the above-discussed 3-parameter Abelian translation group. Similarly, the set of *special Galilean transformations*

$$\boldsymbol{x} \mapsto \boldsymbol{x}' = \boldsymbol{x} + \boldsymbol{u}\,t \qquad\qquad (5.29)$$

forms a 3-parameter Abelian group. The time origins in the inertial frames can be different, and the transition between the frames occurs with a *time shift*,

$$t \mapsto t' = t + \tau\,. \qquad\qquad (5.30)$$

These shifts form a 1-parameter Abelian group. We now assume that the frames $I$ and $I'$ have the same origin, that there is no time shift and that they have no relative velocity, so that $\boldsymbol{a}$, $\tau$ and $\boldsymbol{u}$ are zero. Then the *Cartesian* bases in $I$ and $I'$ as in (5.7) can still be rotated against each other,

$$\boldsymbol{e}_i' = \boldsymbol{e}_j\,R_{ji}\,. \qquad\qquad (5.31)$$

The corresponding rotation of the 3-tuple is

$$\boldsymbol{x} \mapsto \boldsymbol{x}' = R\,\boldsymbol{x}, \qquad R^T R = \mathbb{1}\,. \qquad\qquad (5.32)$$

Let's summarize:

---

**Galilean Transformations**

Let $(t, \boldsymbol{x})$ and $(t', \boldsymbol{x}')$ be the coordinate of a *fixed event* with respect to two inertial frames with origins $\mathcal{O}$, $\mathcal{O}'$ and Cartesian bases $\boldsymbol{e}_i$, $\boldsymbol{e}_i'$. Then the following transformations between the two coordinate systems are possible:

| type of transformation | time coordinate | space coordinates |
|---|---|---|
| shifting the origin in space | $t' = t$ | $\boldsymbol{x}' = \boldsymbol{x} + \boldsymbol{a}$ |
| shifting the time origin | $t' = t + \tau$ | $\boldsymbol{x}' = \boldsymbol{x}$ |
| rotation of the system | $t' = t$ | $\boldsymbol{x}' = R\boldsymbol{x}$ |
| special Galilean transformation | $t' = t$ | $\boldsymbol{x}' = \boldsymbol{x} + \boldsymbol{u}\,t$ |

$$(5.33)$$

The Galilean transformations form the Galilean group.

---

Time intervals and spatial distances of simultaneous events are independent of the inertial frame. They do not change under Galilean transformations.

Any group element is a composition of translations, rotations, and special Galilean transformations and has the form

$$t' = t + \tau \quad \text{and} \quad \boldsymbol{x}' = R\boldsymbol{x} + \boldsymbol{u}\,t + \boldsymbol{a}, \qquad \boldsymbol{u}, \boldsymbol{a} \in \mathbb{R}^3, \qquad R^T R = \mathbb{1}\,. \qquad (5.34)$$

The transformation is determined by the 10 parameters $(R, u, a, \tau)$, where the rotation matrix depends on three real parameters, for example, the Euler angles. If we switch from the reference frame $I$ to another inertial frame $I'$ with (5.34) and subsequently with the transformation $(R', u', a', \tau')$ from $I'$ to $I''$, then the following composite Galilean transformation results for the transition from $I \rightarrow I''$:

**Composition of Galilean Transformations**

$$(R', u', a', \tau')(R, u, a, \tau) = \left(R'R, u' + R'u, a' + R'a + u'\tau, \tau' + \tau\right).$$

$$(5.35)$$

We can assign $5 \times 5$ matrices to the Galilean transformations

$$\begin{pmatrix} x' \\ t' \\ 1 \end{pmatrix} = \begin{pmatrix} R & u & a \\ 0 & 1 & \tau \\ 0 & 0 & 1 \end{pmatrix} \begin{pmatrix} x \\ t \\ 1 \end{pmatrix} .$$

$$(5.36)$$

The identity element corresponds to the matrix $\mathbb{1}_5$ and to the inverse transformation the matrix

$$\begin{pmatrix} R^{-1} & -R^{-1}u & R^{-1}(u\tau - a) \\ 0 & 1 & -\tau \\ 0 & 0 & 1 \end{pmatrix} .$$

$$(5.37)$$

This representation shows that the Galilean group can be considered as a subgroup of $\mathrm{GL}(5, \mathbb{R})$. Under a conjugation with $(\tau_1, a_1, u_1, R_1)$, a Galilean transformation transforms into

$$R' = R_1 R R_1^{-1} ,$$

$$u' = R_1 u + u_1 - R' u_1 ,$$

$$a' = R'(\tau_1 u_1 - a_1) - R_1(\tau_1 u - a) + (\tau - \tau_1)u_1 + a_1 ,$$

$$(5.38)$$

$$\tau' = \tau .$$

The following normal subgroups are recognized:

- The translations in space and time $(\mathbb{1}, 0, a, \tau)$.
- The spatial translations and special Galilean transformations $(\mathbb{1}, u, a, 0)$.

**Galilean Group as a Semi-direct Product**
The Galilean group is the semi-direct product $E_3 \ltimes \mathbb{R}^4$ of the Euclidean group with elements $(R, u)$ and the translation group in time and space with elements $(a, \tau)$.

The action of $(R, u) \in E_3$ on $(a, \tau) \in \mathbb{R}^4$ is given by

$$\varphi(R, u)(a, \tau) = (Ra + \tau u, \tau).$$

(5.39)

**Task**

Prove the last statement. Note that $E_3$ contains the $(R, u)$ and not the $(R, a)$ as elements.

We have already seen that the factor $E_3$ itself is a semi-direct product, $E_3 = O(3) \ltimes \mathbb{R}^3$.

## 5.5   Lorentz and Poincaré Transformations

Let $M$ be the four-dimensional affine Minkowski spacetime, whose points $\{P, Q, \dots\}$ describe *events*. As reference frames, we choose inertial frames. These are equipped with families of identical and synchronized clocks, and in all inertial frames, the same units for lengths and times should be used.[3] An event $P$ is uniquely characterized by time and Cartesian coordinates, i.e., by specifying 4-standard coordinates (the concept of standard coordinates in an inertial frame is used in [26])

$$x = \left(x^0, x^1, x^2, x^3\right)^T = \begin{pmatrix} ct \\ x \end{pmatrix}.$$

(5.40)

We often write $x = (x^\mu)$ with $\mu = 0, 1, 2, 3$. We now consider an spherical wave that lights up at the spacetime point $y$—only very briefly and spatially localized, to come close to the idealization of a point in space and time. Spacetime points $x$, where the emitted spherical wave lights up, satisfy the condition

$$x^0 - y^0 = \pm |x - y|.$$

(5.41)

---

[3] For a discussion of scales and clocks in an inertial frame consult a textbook on special relativity theory [26–28].

The negative sign belongs to an incoming spherical wave, which arrives at $y$ and for which $x^0 - y^0$ is negative. It is advisable to use the covariant notation with 4-vectors. Then the condition (5.41) for the components $\xi^\mu$ of the difference vector $\xi = x - y$ reads

$$\xi^0 = \pm|\boldsymbol{\xi}| \iff (\xi^0)^2 - |\boldsymbol{\xi}|^2 = \xi^0\xi^0 - \xi^1\xi^1 - \xi^2\xi^2 - \xi^3\xi^3 = 0, \qquad (5.42)$$

and gives rise to the introduction of the following bilinear form of vectors,

$$(\xi, \zeta) = \xi^0\zeta^0 - \xi^1\zeta^1 - \xi^2\zeta^2 - \xi^3\zeta^3 . \qquad (5.43)$$

This can be written with the help of the metric tensor

$$\eta = \begin{pmatrix} 1 & 0 & 0 & 0 \\ 0 & -1 & 0 & 0 \\ 0 & 0 & -1 & 0 \\ 0 & 0 & 0 & -1 \end{pmatrix} = (\eta_{\mu\nu}) \quad \text{or} \quad \eta^{-1} = (\eta^{\mu\nu})$$

as follows:

$$(\xi, \zeta) = \sum_{\mu\nu} \eta_{\mu\nu}\xi^\mu\zeta^\nu = \xi^T\eta\zeta . \qquad (5.44)$$

In particular, the "relativistic distance square" between emission and observation event in (5.42) is equal to

$$d(x, y) \equiv (\xi, \xi), \quad \text{with} \quad \xi = x - y . \qquad (5.45)$$

Indices are lowered or raised with the covariant and contravariant components of the metric tensor, i.e., with $\eta_{\mu\nu}$ and $\eta^{\mu\nu}$. For example

$$\xi_\mu = \eta_{\mu\nu}\xi^\nu \quad \text{or} \quad \xi^\mu = \eta^{\mu\nu}\xi_\nu, \quad \text{so that} \quad (\xi, \zeta) = \xi^\mu\zeta_\mu = \xi_\mu\zeta^\mu .$$

For a fixed time difference $\xi^0$, the quadratic equation for $\boldsymbol{\xi}$ in (5.42) describes a spherical surface, whose radius $|\boldsymbol{\xi}|$ grows at the speed of light. In spacetime, these are surfaces of a double light cone, whose tips are at $\xi = 0$ or $x = y$. Figure 5.2 shows the situation in two spatial dimensions.

We now switch to another inertial frame $I'$, which is in constant uniform motion relative to the original inertial frame $I$. With no measurement can we distinguish between the two inertial frames. This is the essential content of Einstein's equivalence principle:

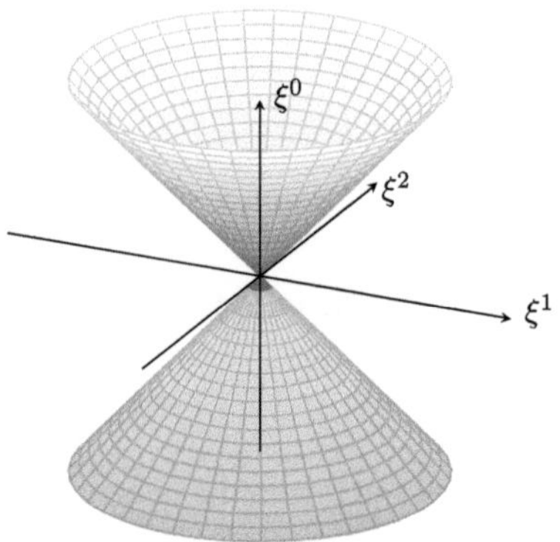



**Fig. 5.2** Causal structure of $M$: the double light cone with tips at the origin defines the light cone

---

**Equivalence Principle**
The laws of nature are the same in all inertial frames. In particular, the speed of light is the same in all inertial frames.

---

From this principle, the Lorentz transformations can be derived, as found in many textbooks on Special Relativity. The argument is roughly as follows: A point event is described in $I$ and $I'$ by standard coordinates $x$ and $x'$. The relationship between the coordinates has the form

$$x' = a + f(x), \quad \text{with} \quad f(0) = 0 \quad \text{and} \quad a \in \mathbb{R}^4 \text{ constant.}$$

Due to the homogeneity of spacetime, $x'' = x' - a$ are also standard coordinates in a third, shifted inertial frame $I''$. Then $x'' = f(x)$ with $f(0) = 0$. Now it is argued that the four functions $f^\mu$ must be linear. For example, in [26] this linearity is justified with the help of the first law of Newton and the homogeneity of space and time. Then $x'' = \Lambda x$ with a matrix $\Lambda$, or

$$x' = \Lambda x + a \quad \text{with} \quad \Lambda \in \mathrm{GL}(4, \mathbb{R}). \tag{5.46}$$

Let $x$ be the coordinates of the spherical wave emitted at $y$ in $I$. With respect to $I'$, the same wave is emitted at $y'$ and has the coordinates $x'$. In both inertial frames, the speed of light is the same, i.e.

$$0 = (\xi, \xi) = (\xi', \xi'), \qquad \xi = x - y, \quad \xi' = \Lambda \xi.$$

A sufficient and necessary condition for this is

$$\Lambda^T \eta \Lambda = \kappa(\Lambda)\eta \quad \text{with} \quad \kappa(\mathbb{1}) = 1 \quad \text{and} \quad \kappa(\Lambda) \geq 0.$$

If $\kappa \neq 1$, then we can always reach $\kappa = 1$ by a scale change $x' \mapsto \sqrt{\kappa}\, x'$. Therefore, we only need to consider matrices that satisfy the relation

$$\Lambda^T \eta \Lambda = \eta \iff \eta_{\alpha\beta}\, \Lambda^\alpha_{\ \mu} \Lambda^\beta_{\ \nu} = \eta_{\mu\nu} \tag{5.47}$$

For such transformations, the relativistic distance square of two events $x, y$ in Minkowski space is independent of the inertial frame, $(\xi', \xi') = (\xi, \xi)$. We summarize:

**Poincaré Group (Inhomogeneous Lorentz Group)**
The mappings (5.46) between inertial frames form the group named after Poincaré when $\Lambda$ satisfies the condition (5.47),

$$iO(1,3) = \left\{ (\Lambda, a)\, \big|\, a \in \mathbb{R}^4,\ \Lambda \in \mathrm{GL}(4, \mathbb{R}),\ \Lambda^T \eta \Lambda = \eta \right\}. \tag{5.48}$$

The notation $O(1,3)$ is analogous to the notation $O(n)$ for the rotations in $\mathbb{R}^n$, which leave the distance determined by the Euclidean metric $\mathrm{diag}(1, \dots, 1)$ invariant. Similarly, the Lorentz transformations leave the distance determined by the Lorentz metric $\mathrm{diag}(1, -1, -1, -1)$ invariant. The $i$ in front of $O(1, 3)$ in (5.48) stands for inhomogeneous.

The group multiplication in the Poincaré -Group is

$$(\Lambda_2, a_2)(\Lambda_1, a_1) = (\Lambda_2 \Lambda_1, a_2 + \Lambda_2 a_1), \tag{5.49}$$

and the inverse Poincaré -Transformation has the form

$$(\Lambda, a)^{-1} = (\Lambda^{-1}, -\Lambda^{-1}a). \tag{5.50}$$

**Task**

Show that a conjugation with $(\Lambda_1, a_1)$ is given by

$$(\Lambda_1, a_1)(\Lambda, a)(\Lambda_1, a_1)^{-1} = \left(\Lambda', a_1 + \Lambda_1 a - \Lambda' a_1\right), \quad \Lambda' = \Lambda_1 \Lambda \Lambda_1^{-1}, \tag{5.51}$$

and thus the translations $(a, \mathbb{1})$ define a normal subgroup.

**Lemma 5.4 (Poincaré Group)** *The Poincaré group $iO(1,3)$ is the semi-direct product of the subgroup $O(1,3)$ of Lorentz transformations, consisting of spatial* *rotations and Lorentz boosts,*

$$x' = \Lambda x, \quad \Lambda^T \eta \Lambda = \eta, \tag{5.52}$$

*and the abelian normal subgroup of translations in space and time,*

$$x' = x + a, \quad a \in \mathbb{R}^4. \tag{5.53}$$

### 5.5.1   Connected Components of the Lorentz Group

The difference vector $x - y$ connects the events $P$ and $Q$ in $M$ with coordinates $x$ and $y$. These events are spacelike, timelike, or lightlike separated if their (Lorentz-invariant) distance square is negative, positive, or zero. Under a Lorentz transformation, the difference vector transforms into $\Lambda(x - y)$. The question now is: what kind of geometric object does the orbit of the difference vector $B_{x-y} = \{\xi = \Lambda(x - y) | \Lambda \in O(1,3)\}$ describe under the action of the Lorentz group? Because of

$$(\xi, \xi) = (\xi^0)^2 - \boldsymbol{\xi}^2 = \text{const} \tag{5.54}$$

the image set $\{\xi^0, \boldsymbol{\xi}\}$ forms a hyperboloid. For a positive constant, it is a two-sheeted hyperboloid and for a negative constant a one-sheeted hyperboloid. For very large $|\xi^0|$ or $|\boldsymbol{\xi}|$ we can neglect the constant in the quadratic equation, so that the hyperboloid asymptotically approaches the light cone.

**Causal Structure**

1. For timelike separated events, $(\xi, \xi) > 0$ and the orbit $B_\xi$ is a two-sheeted hyperboloid, as shown in the left image in Fig. 5.3. One sheet lies in the forward light cone with positive $\xi^0$ and one in the backward light cone with negative $\xi^0$, and the two sheets form two disjoint sets. Each sheet intersects the $\xi^0$-axis, where $\boldsymbol{\xi}$ vanishes. In the corresponding inertial frame, $P$ and $Q$ are at the same point in space.
2. For spacelike separated events, $(\xi, \xi) < 0$ and the orbit $B_\xi$ describes a one-sheeted hyperboloid as shown in the right image in Fig. 5.3. On each one-sheeted hyperboloid there is a point with $\xi^0 = 0$. This means, there is an inertial frame, in which $P$ and $Q$ are simultaneous. Also, $\xi^0$ has both signs on each hyperboloid, and therefore the direction of time can change.
3. For lightlike separated events, $(\xi, \xi) = 0$ and $\xi$ lies on the forward or backward light cone. Then the orbit $B_\xi$ as shown in Fig. 5.2, describes one cone directed into the future and one into the past with a 45° opening angle.

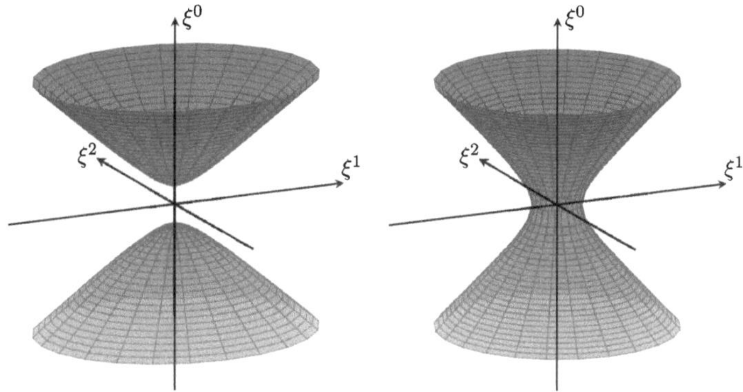

**Fig. 5.3** Orbits of the Lorentz group: on the left a two-sheeted hyperboloid for timelike $\xi$ and on the right a one-sheeted hyperboloid for spacelike $\xi$

## 5.5.2   Global Structure of the Lorentz Group

The 00-component of the matrix equation (5.47) implies $(\Lambda^0_{\ 0})^2 \geq 1$. For $\Lambda^0_{\ 0} \geq 1$ the forward light cone is mapped onto itself and the direction of time remains preserved. For $\Lambda^0_{\ 0} \leq -1$ the forward light cone is mapped with a time reversal into the backward light cone. Because of (5.47) also $\det \Lambda = \pm 1$. The signs of $\det(\Lambda)$ and $\Lambda^0_{\ 0}$ determine the *connected components* of the Lorentz group,

$$O(1,3) = O^{\uparrow}_{+}(1,3) \cup O^{\uparrow}_{-}(1,3) \cup O^{\downarrow}_{+}(1,3) \cup O^{\downarrow}_{-}(1,3) , \tag{5.55}$$

with the following meaning of the indices:

$$\pm : \ \det \Lambda = \pm 1 , \quad \uparrow : \text{no time reversal } (\Lambda^0_{\ 0} \geq 1) , \quad \downarrow : \text{time reversal } (\Lambda^0_{\ 0} \leq -1) .$$

Introducing the space inversion $P$, time reversal $T$ and spacetime inversion $PT$ (given by $P = -T = \eta$ and $PT = -\mathbb{1}_4$) every Lorentz transformation is then from

$$O(1,3)^{\uparrow}_{+} \cup P\, O^{\uparrow}_{+}(1,3) \cup PT\, O^{\uparrow}_{+}(1,3) \cup T\, O^{\uparrow}_{+}(1,3) . \tag{5.56}$$

The first coset class of proper (or special) orthochronous Lorentz transformations contains neither time reversal nor space inversion and forms a *normal subgroup*. The elements of the second class invert space and are called improper orthochronous, those of the third class reverse time and those of the last class invert space and time.

The normal subgroup of the special orthochronous Lorentz transformations is denoted by

$$SO^{\uparrow}(1,3) \equiv O(1,3)^{\uparrow}_{+} = \left\{ \Lambda \in O(1,3) \mid \det \Lambda = 1, \ \Lambda^0_{\ 0} \geq 1 \right\} . \tag{5.57}$$

The S in SO indicates that the transformations are special, $\det(\Lambda) = 1$.

The factor group of the Lorentz group by the group of special orthochronous Lorentz transformations is

$$O(1,3)/SO^{\uparrow}(1,3) = \{\mathbb{1}, P, T, PT\} \cong \text{Klein's four-group } \mathbb{Z}_2 \times \mathbb{Z}_2. \qquad (5.58)$$

Without proof, we finally note the four types of conjugacy classes in $SO^{\uparrow}(1,3)$:

**Conjugacy Classes of the Special Orthochronous Lorentz Group**

1. *Elliptical transformations* are conjugate to a spatial rotation with angle $\varphi$ around the third axis,

$$\Lambda_e = \begin{pmatrix} 1 & 0 \\ 0 & R(e_3, \varphi) \end{pmatrix}. \qquad (5.59)$$

2. *Hyperbolic transformations* are conjugate to a Lorentz boost with rapidity $\beta$ in the direction of the third axis,

$$\Lambda_h = \begin{pmatrix} \cosh\beta & 0 & 0 & \sinh\beta \\ 0 & 1 & 0 & 0 \\ 0 & 0 & 1 & 0 \\ \sinh\beta & 0 & 0 & \cosh\beta \end{pmatrix}. \qquad (5.60)$$

3. *Loxodromic transformations* are conjugate to the composition of the above two transformations, $\Lambda_l = \Lambda_e \Lambda_h$.
4. *Parabolic transformations* are conjugate to

$$\Lambda_p = \begin{pmatrix} 1 + \alpha^2/2 & \alpha & 0 & -\alpha^2/2 \\ \alpha & 1 & 0 & -\alpha \\ 0 & 0 & 1 & 0 \\ \alpha^2/2 & \alpha & 0 & 1 - \alpha^2/2 \end{pmatrix}. \qquad (5.61)$$

A parabolic transformation can be realized by a boost along the $x$-axis and a rotation around the $z$-axis. If the magnitudes of the rotation and the boost exactly "balance out", then we obtain a parabolic transformation.

With the Lorentz group, the Poincaré group also decomposes into four connected components,

$$iO(1,3) = iO^{\uparrow}_+(1,3) \cup iO^{\uparrow}_-(1,3) \cup iO^{\downarrow}_+(1,3) \cup iO^{\downarrow}_-(1,3). \qquad (5.62)$$

In Sect. 14.5 we will discuss the local properties of the Poincaré group. These are encoded in the Poincaré algebra—this is the Lie algebra of the Lie group $i$ SO(1, 3).

## 5.6   Appendix A: Normal Form of Rotations and Rigid Motions

The conjugacy classes of rotations and rigid motions in $\mathbb{R}^n$ have simple representatives. These are often referred to as "normal forms".

### 5.6.1   Normal Forms for Rotations in $\mathbb{R}^n$

By first diagonalizing the rotation matrices in the complex plane and then returning to a real basis, one proves:

- In even dimensions $n$, every *proper rotation* is conjugate to the normal form

$$N_n(\boldsymbol{\varphi}) \equiv D_n(\boldsymbol{\varphi}) = \begin{pmatrix} R_2(\varphi_1) & 0 & \ldots & 0 \\ 0 & R_2(\varphi_2) & \ldots & 0 \\ \vdots & & \ddots & \vdots \\ 0 & 0 & \ldots & R_2(\varphi_{n/2}) \end{pmatrix}, \tag{A.1}$$

and every *improper rotation* to the normal form

$$N_n(\boldsymbol{\varphi}) = \begin{pmatrix} D_{n-2}(\boldsymbol{\varphi}) & 0 & 0 \\ 0 & -1 & 0 \\ 0 & 0 & 1 \end{pmatrix}. \tag{A.2}$$

Here, $R_2(\varphi)$ denotes the rotation matrix in (5.16) and $D_{n-2}$ the proper rotation (A.1) in $n-2$ dimensions.
- In odd dimensions $n$, every *proper rotation* is conjugate to

$$N_n(\boldsymbol{\varphi}) = \begin{pmatrix} D_{n-1}(\boldsymbol{\varphi}) & 0 \\ 0 & 1 \end{pmatrix}, \tag{A.3}$$

and every *improper rotation* to

$$N_n(\boldsymbol{\varphi}) = \begin{pmatrix} D_{n-1}(\boldsymbol{\varphi}) & 0 \\ 0 & -1 \end{pmatrix}. \tag{A.4}$$

These normal forms of proper and improper rotations are in accordance with Euler's theorem in Sect. 5.2.

### 5.6.2   Normal Forms for Rigid Motions in $\mathbb{R}^n$

We can now apply the results obtained to the rigid motions in space, consisting of rotations and translations.

**Theorem 5.4 (Normal Form of Rigid Motions)** *For every motion $(R, a)$ in $\mathbb{R}^n$ there is a Cartesian basis $\{e_1, \ldots, e_n\}$ and an origin $\mathcal{O}$, with respect to which it takes one of the following normal forms:*

1. *If 1 is an eigenvalue of $R$, then the normal form is $x' = N_n(\varphi)x + a\,e_n$.*
2. *If 1 is not an eigenvalue of $R$, then the normal form is $x' = N_n(\varphi)x$.*

**Proof** Let $R_1$ be a rotation that transforms $R$ into its normal form $N_n$,

$$R_1 R R_1^{-1} = N_n \, . \tag{A.5}$$

If we also shift the origin by $a_1$, then according to (5.24) the transformed $a$ has the form

$$a' = R_1 a + (\mathbb{1} - N_n) a_1 \, . \tag{A.6}$$

Now we decompose $\mathbb{R}^n$ into two orthogonal subspaces,

$$\mathbb{R}^n = K \oplus K^{\perp} \quad \text{with} \quad K \equiv \text{Kernel}\,(\mathbb{1} - N_n) \, . \tag{A.7}$$

If 1 is not an eigenvalue of $R$ (and therefore not an eigenvalue of $N_n$) then $K = \emptyset$.
$\square$

The decomposition (A.7) is invariant under the action of $N_n$: The kernel $K$ is obviously invariant and its orthogonal complement is also, since for an arbitrary $b \in K^{\perp}$ we have

$$0 = (K, b) = (N_n K, N_n b) = (K, N_n b) \, .$$

We decompose both sides of the relation (A.6) into their $K$-component, denoted by $\|$, and their $K^{\perp}$-component, denoted by $\perp$,

$$a'_{\|} = (R_1 a)_{\|} \quad \text{and} \quad a'_{\perp} = (R_1 a)_{\perp} + (\mathbb{1} - N_n) a_{1\perp} \, . \tag{A.8}$$

On $K^{\perp}$, $\mathbb{1} - N_n$ is invertible and therefore there exists a $a_1$, for which $a'_{\perp}$ vanishes. After choosing such a $a_1$ we have $a' = a'_{\|}$. To proceed, we note that $R_1$ in (A.5) is not uniquely determined. Instead of $R_1$, we can also choose $SR_1$ with an $S$ that satisfies

$$SN_n S^{-1} = N_n \, . \tag{A.9}$$

Since the rotation $S$ commutes with $N_n$, it also respects the decomposition (A.7):

$$S : K \mapsto K \quad \text{and} \quad S : K^\perp \mapsto K^\perp \, .$$

If you replace $R_1$ in the first equation in (A.8) with $SR_1$, we get the following result,

$$a'_\parallel = S(R_1 a)_\parallel \, . \tag{A.10}$$

Now $S$ can be chosen so that $a'_\parallel$ points in any direction in $K$. Since $a'_\perp = 0$ was chosen, this proves the theorem. In particular, if 1 is not an eigenvalue of $R$, then $K = \emptyset$ and thus $a' = 0$.

**Types of Rigid Motions**

- A *proper rigid motion* in odd dimensions is a screw motion about the $e_n$-axis.
- An *improper rigid motion* in even dimensions is a glide reflection on the hyperplane perpendicular to $e_n$
- For proper rigid motions in even or improper rigid motions in odd dimensions, generically 1 is not an eigenvalue of $R$. Then it is a rotation in the first case and a rotoinversion (a combination of a rotation and inversion) in the second case. However, if at least one of the angles $\varphi_i$ in (A.1) is a multiple of $2\pi$, then in even dimensions a proper motion becomes a screw motion and in odd dimensions an improper motion becomes a glide reflection.

## 5.7 Exercises for Chap. 5

**Problem 5.1 (Group Actions from the Right)** Similar to the left group action, a right group action can be defined:

**Definition 5.5** A right group action is a function $\Phi_r : G \times M \mapsto M$ such that:

- $\Phi_r(e, p) = p$ for all $p \in M$,
- $\Phi_r(g_1, \Phi_r(g_2, p)) = \Phi_r(g_2 g_1, p)$ for all $g_1, g_2 \in G$ and $p \in M$.

Prove the following connection between left and right actions: If $(g, p) \to \Phi(g, p)$ is a left action, then $(g, p) \to \Phi_r(g, p) = \Phi(g^{-1}, p)$ is a right action.

Remark: It can be useful to identify a right action with the corresponding left action.

**Problem 5.2 (Action of $\mathrm{PSL}(2, \mathbb{Z})$ on the Poincaré -half-plane)** Consider the group action (5.5) on the Poincaré -half-plane $\mathcal{H}$, but now with $S$ from the discrete modular group $\mathrm{PSL}(2, \mathbb{Z}) = \mathrm{SL}(2, \mathbb{Z})/\{\pm\mathbb{1}\}$. Analyze the group action similarly as we did for $\mathrm{SL}(2, \mathbb{R})$ in Sect. 5.1. What is meant by a fundamental domain of a group action? The fundamental domain for the action of $\mathrm{PSL}(2, \mathbb{Z})$ on $\mathcal{H}$ is of interest in conformal field theory (a picture of the domain can be found in the Wikipedia article "Modular Group").

**Problem 5.3 (Proper Rotations)**  We consider the group $\mathrm{SO}(3)$ of proper rotations in Euclidean space $\mathbb{R}^3$ and the set $\mathrm{SO}(e)$ of rotations about the fixed axis, defined by the unit vector $e$. Show that $\mathrm{SO}(e)$ is a subgroup of $\mathrm{SO}(3)$ and for every pair of unit vectors $\mathrm{SO}(e)$ and $\mathrm{SO}(e')$ are conjugate to each other.

**Problem 5.4 (Galilean Group)**  A Galilean transformation (5.34) is determined by the 10 parameters $(R, u, a, \tau)$.

1. What is the inverse Galilean transformation to $(R, u, a, \tau)$?
2. Find the invariant subgroups of the Galilean group.
3. Can you write the Galilean group as a semi-direct product?

**Problem 5.5 (Lorentz Group)**  A four-dimensional matrix $\Lambda$ is called a Lorentz matrix, if

$$\Lambda^\top \eta \Lambda = \eta, \quad \text{where} \quad \eta = \mathrm{diag}(1, -1, -1, -1)$$

is the metric tensor. We denote the set of these matrices with $\mathrm{O}(1, 3)$.

1. Show that $\mathrm{O}(1, 3)$ is a subgroup of $\mathrm{GL}(4, \mathbb{R})$,
2. show that for $\Lambda \in \mathrm{O}(1, 3)$ the determinant is $\det \Lambda = \pm 1$,
3. prove that the set $\mathrm{O}_+(1, 3) = \{\Lambda \in \mathrm{O}(1, 3) \,|\, \det \Lambda = 1\}$ is a normal subgroup in $\mathrm{O}(1, 3)$. These are the proper (or special) Lorentz transformations.

**Problem 5.6 (Dimension of Symmetry Groups)**  How many real parameters does a Galilean transformation or a Poincaré  transformation depend on in a (hypothetical) spacetime with $n$ dimensions?

**Problem 5.7 (Intervals in Minkowski Space)**  Two events $P_1$ and $P_2$ are separated by a (1) space-like, (2) time-like and (3) light-like interval. Show that

1. an inertial frame exists in which the $P_i$ are simultaneous and that their time order can be reversed by a suitable choice of the frame,

2. an inertial frame exists in which the $P_i$ are at the same spatial point.
3. Determine the hypersurface in spacetime where $P_2$ can lie relative to $P_1$.

Hint: Choose $P_1$ as the origin of coordinates.

**Problem 5.8 (Time Dilation)** Derive the formula for time dilation using the Lorentz transformation. First, define two events for the reading of a clock moving in inertial frame $I$ and at rest in inertial frame $I'$. Note that the reading in $I'$ takes place at the same location and the one in $I$ at different locations.

# Point Groups

6

> *In my work, I have always tried to reconcile the true with the beautiful; but when I had to choose between one or the other, I usually choose the beautiful.*

—Hermann Weyl

Point groups are finite subgroups of the rotation group $O(3)$. For example, the symmetries of a Platonic solid or a molecule form point groups. The latter are briefly called *molecular symmetries*. Also, those symmetry transformations of an ideal crystal that leave a point fixed, form a point group. An understanding of point groups is essential, to deduce the symmetry of an atom or molecule from spectroscopic data, or conversely to predict their physical properties based on the known symmetries. One defines:

**Definition 6.1** A (finite) point group $G < O(3)$ is called chiral if it contains no reflections, otherwise it is called achiral.

We will characterize the point groups and show that there are only five types of chiral and nine types of achiral point groups. We start with the chiral groups and use the results in Sect. 5.2.

In the textbooks [9, 12, 29, 30] point groups are discussed and applied and in [31, 32] one can find further applications of the theory of point groups to atoms, molecules and solids.

## 6.1 Chiral and Achiral Point Groups

If the rotation of a molecule around an axis by the angle $\varphi_n = 2\pi/n$ leads to a configuration that is indistinguishable from the original one, the axis is called an $n$-fold symmetry axis $C_n$. The symmetry axis is characterized by a unit vector $e$

© The Author(s), under exclusive license to Springer-Verlag GmbH, DE, part of Springer Nature 2026
A. Wipf, *Symmetries in Physics*, https://doi.org/10.1007/978-3-662-72675-4_6

and the corresponding rotation is denoted by

$$c_n(e) = R(e, \varphi_n), \qquad \varphi_n = \frac{2\pi}{n}. \tag{6.1}$$

If the rotation is carried out $n$ times in succession, then one obtains the identity element, i.e. $c_n(e)$ has the order $n$. For a rotation about the $z$-axis, the rotation matrix has the form

$$c_n(e_3) = \begin{pmatrix} \cos\varphi_n & -\sin\varphi_n & 0 \\ \sin\varphi_n & \cos\varphi_n & 0 \\ 0 & 0 & 1 \end{pmatrix}, \tag{6.2}$$

as given in (5.16).

> **Rotations and Principal Axis**
> The principal axis of a molecule is the symmetry axis with maximal order $n$.
> If $e$ points in the direction of the principal axis, one writes $c_n$ instead of $c_n(e)$.
> A $m$-fold rotation about an axis perpendicular to $e$ is denoted by $c_{h,m}$.

### 6.1.1   Platonic Groups

The *proper* symmetry transformations of the five Platonic solids tetrahedron, cube, octahedron, icosahedron and dodecahedron form the Platonic groups.[1] To study them, we first count their number of vertices, edges and faces:

|  | V = No. of vertices | E = No. of edges | F = No. of faces |
| --- | --- | --- | --- |
| Tetrahedron | 4 | 6 | 4 |
| Cube | 8 | 12 | 6 |
| Octahedron | 6 | 12 | 8 |
| Icosahedron | 12 | 30 | 20 |
| Dodecahedron | 20 | 30 | 12 |

We will now determine the elements of the tetrahedral group $\mathcal{T}$, cube group $\mathcal{O}$ and dodecahedral group.

---

[1] A visit to the Wikipedia page on Platonic solids is worthwhile.

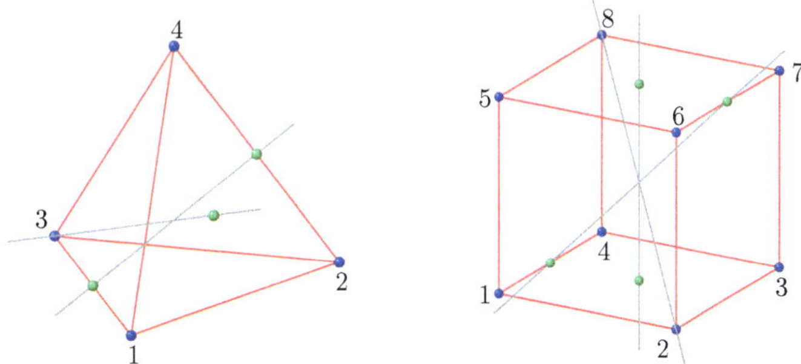

**Fig. 6.1** The symmetry transformations of tetrahedron and cube define the tetrahedral group $\mathcal{T}$ and cubic group $\mathcal{O}$

## Tetrahedral Group

The symmetry transformations of a *tetrahedron* are shown in Fig. 6.1. This solid has four threefold and three twofold rotation axes. The threefold axes connect the vertices with the centers of the opposite faces and the twofold axes run through the centers of opposite edges. The tetrahedral group $\mathcal{T}$ thus has the order

$$|\mathcal{T}| = 1 + V \times 2 + \frac{E}{2} \times 1 = 12\,, \tag{6.3}$$

and possesses the cyclic subgroups $\mathcal{C}_3$ and $\mathcal{C}_2$. It is isomorphic to the alternating group $A_4$.

## Cubic Group

The symmetries of a *cube* can be seen in Fig. 6.1. It has three fourfold rotation axes through the centers of opposite faces, six twofold rotation axes through the centers of opposite edges and four diagonals as fourfold rotation axes. The order of the *cubic group* is

$$|\mathcal{O}| = 1 + \frac{F}{2} \times 3 + \frac{K}{2} \times 1 + \frac{E}{2} \times 2 = 24\,. \tag{6.4}$$

**Relation to $\mathcal{S}_4$**

The cubic group is isomorphic to the *octahedron group* and both are isomorphic to $\mathcal{S}_4$.

This is no coincidence, as octahedron and cube are dual to each other: One can place an octahedron into a cube in such a way that the vertices of the octahedron are equal to the centers of the cube faces.

## Dodecahedron Group

The *dodecahedron* shown in Fig. 1.1 has six fivefold rotation axes through the centers of opposite faces, ten threefold axes through two opposite corners and fifteen twofold rotation axes through the centers of two opposite, parallel edges. Thus, the order of the dodecahedron group $\mathcal{Y}$ is

$$|\mathcal{Y}| = 1 + \frac{F}{2} \times 4 + \frac{E}{2} \times 2 + \frac{K}{2} \times 1 = 60 \,. \tag{6.5}$$

The icosahedron and the dodecahedron are dual to each other so that their symmetry groups are isomorphic. The group order is 60 and is equal to the order of $\mathcal{A}_5$ and indeed:

> **Relation to $\mathcal{A}_5$**
>
> The dodecahedron group (icosahedron group) is isomorphic to $\mathcal{A}_5$.

Here, the fivefold rotations in $\mathcal{Y}$ correspond to the 5-cycles in $\mathcal{A}_5$.

## Analysis of the Cubic Group

As generating elements of the cubic group $\mathcal{O}$ we choose the quarter rotation about the $z$-axis and the quarter rotation about the $x$-axis:

$$a = \begin{pmatrix} 0 & -1 & 0 \\ 1 & 0 & 0 \\ 0 & 0 & 1 \end{pmatrix} \quad \text{and} \quad b = \begin{pmatrix} 1 & 0 & 0 \\ 0 & 0 & -1 \\ 0 & 1 & 0 \end{pmatrix} \,.$$

First we define the group over its generators,

```
GAP>
    a:=[[0,-1,0],[1,0,0],[0,0,1]];;b:=[[1,0,0],[0,0,-1],[0,1,0]];
GAP> g:=Group(a,b);
```

and analyze it. The order is `Order(g)=24`. We further define the commutator subgroup and associated factor group $G/[G, G]$:

```
GAP> gk:=CommutatorSubgroup(g,g);
GAP> gf:=FactorGroup(g,gk);;Order(gf);
```

The order of `gk` is 12 and that of `gf` is 2. We can also ask for the order of the normalizers (isotropy groups) of the elements in $g$, e.g.

```
GAP> Order(Stabilizer(g,a));Order(Stabilizer(g,a*b));
```

and get the answers 4 and 3. The number of conjugacy classes of g is 5,

```
GAP> NrConjugacyClasses(g);
```

The representatives of the conjugacy classes are asked with

```
GAP> ConjugacyClasses(g);
```

with the result

$$\begin{pmatrix} 1 & 0 & 0 \\ 0 & 1 & 0 \\ 0 & 0 & 1 \end{pmatrix}, \quad \begin{pmatrix} -1 & 0 & 0 \\ 0 & -1 & 0 \\ 0 & 0 & 1 \end{pmatrix}, \quad \begin{pmatrix} -1 & 0 & 0 \\ 0 & 0 & -1 \\ 0 & -1 & 0 \end{pmatrix}, \quad \begin{pmatrix} 0 & -1 & 0 \\ 0 & 0 & -1 \\ 1 & 0 & 0 \end{pmatrix}, \quad \begin{pmatrix} 0 & -1 & 0 \\ 1 & 0 & 0 \\ 0 & 0 & 1 \end{pmatrix}.$$

The number of elements in the corresponding orbits are $1, 3, 6, 8$ and $6$. The cubic group has 8 maximal subgroups, all of which are non-abelian:

```
GAP> msg:=MaximalSubgroups(g);
```

Their orders are obtained with

```
GAP> Order(msg[1]);Order(msg[2]);Order(msg[3]);Order(msg[4]);
GAP> Order(msg[5]);Order(msg[6]);Order(msg[7]);Order(msg[8]);
```

They are $12, 8, 8, 8, 6, 6, 6, 6$. We can analyze these subgroups in more detail. In particular, we are interested in the generators of the subgroups:

```
GAP> GeneratorsOfGroup(msg[1]);GeneratorsOfGroup(msg[2]);
GAP> GeneratorsOfGroup(msg[5]);
```

The subgroup of order 12 is generated by the 3 matrices

$$\begin{pmatrix} 0 & -1 & 0 \\ 0 & 0 & -1 \\ 1 & 0 & 0 \end{pmatrix}_3, \quad \begin{pmatrix} 1 & 0 & 0 \\ 0 & -1 & 0 \\ 0 & 0 & -1 \end{pmatrix}_2, \quad \begin{pmatrix} -1 & 0 & 0 \\ 0 & 1 & 0 \\ 0 & 0 & -1 \end{pmatrix}_2.$$

The indices indicate the order of these elements and can be queried with

```
GAP> Order([[0,-1,0],[0,0,-1],[1,0,0]]);
```

The matrices correspond to a rotation with rotation angle $\varphi_3$ around the axis from the corner $(1, -1, 1)$ to the opposite corner and a rotation with $\varphi_2$ around the $x$ and $y$-axis. All elements of this subgroup are generated by

```
GAP> Orbits(g,msg[1]);
```

## 6.1.2  Classification of Chiral Point Groups

In addition to the Platonic groups $\mathcal{T}$, $\mathcal{O}$ and $\mathcal{Y}$ there are only two series of point groups:

**Theorem 6.1 (Chiral Point Groups)** *There are 5 classes of chiral point groups: the infinite series $\mathcal{C}_n$ and $\mathcal{D}_n$ and the three special point groups $\mathcal{T}$, $\mathcal{O}$ and $\mathcal{Y}$.*

The symmetry transformations of a pyramid over the regular $n$-gon form the cyclic group $\mathcal{C}_n$. It is generated by the $n$-fold rotation $c_n$ around the symmetry axis. The symmetry transformations of a perpendicular prism over the regular $n$-gon form the larger dihedral group $\mathcal{D}_n$. It contains in addition to the $n$-fold rotation $c_n$ around the principal axis $n$ further twofold rotations $c_{h,2}$ around $n$ axes perpendicular to the principal axis. The group is generated by $c_n$ and one of these twofold rotations. On the plane perpendicular to the principal axis, the twofold rotations act like reflections. This explains why the groups $\mathcal{D}_n$ introduced in Sect. 2.3 contain reflections in $\mathbb{R}^2$, in contrast to the proper symmetry transformations of prisms in $\mathbb{R}^3$.

The following table contains the chiral point groups and some of their properties. Indicated are the cyclic subgroups, non-abelian maximal subgroups, the group order $|G|$, the number of conjugacy classes $k(G)$ and isomorphic groups.

**Properties of the Chiral Point Groups**

| $G$ | cyclic subgroups | max. subgroups | $\|G\|$ | $k(G)$ | $\cong$ |
|---|---|---|---|---|---|
| $\mathcal{C}_n$ | $\mathcal{C}_m$, $m$ divides $n$ | | $n$ | $n$ | |
| $\mathcal{D}_{2n}$ | $\mathcal{C}_m$, $m$ divides $2n$ | $\mathcal{D}_m$, $m$ divides $2n$ | $4n$ | $n+3$ | |
| $\mathcal{D}_{2n+1}$ | $\mathcal{C}_m$, $m$ divides $2n+1$ | $\mathcal{D}_m$, $m$ divides $2n+1$ | $4n+2$ | $n+2$ | |
| $\mathcal{T}$ | $3\,\mathcal{C}_2$, $4\,\mathcal{C}_3$ | $\mathcal{D}_2$ | $12$ | $4$ | $\mathcal{A}_4$ |
| $\mathcal{O}$ | $6\,\mathcal{C}_2$, $4\,\mathcal{C}_3$, $3\,\mathcal{C}_4$ | $4\,\mathcal{D}_3$, $3\,\mathcal{D}_4$, $\mathcal{A}_4$ | $24$ | $5$ | $\mathcal{S}_4$ |
| $\mathcal{Y}$ | $15\,\mathcal{C}_2$, $10\,\mathcal{C}_3$, $6\,\mathcal{C}_5$ | $10\,\mathcal{D}_3$, $6\,\mathcal{D}_5$, $5\,\mathcal{A}_4$ | $60$ | $5$ | $\mathcal{A}_5$ |

Proof of Theorem 6.1 (see also [4]): To classify the chiral point groups we recall Euler's theorem, according to which every rotation is a rotation around an axis. This axis intersects the unit sphere $S^2 \subset \mathbb{R}^3$ at two points. We call these points on $S^2$ the poles of the group element.

**Definition 6.2** The pole set of $G$ is $P := \{x \in S^2 | Rx = x$ for a $R \in G \setminus \mathbb{1}\}$.

Figure 6.2 shows the poles lying on the equator of $S^2$ of the dihedral group $\mathcal{D}_3$, when the threefold principal axis is perpendicular to the plane of the paper. The (not

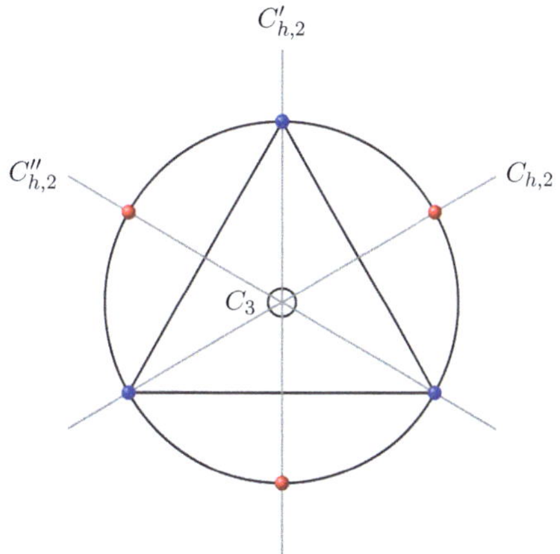


**Fig. 6.2** The prism over the equilateral triangle has a threefold rotation axis—this is the principal axis—and three twofold symmetry axes perpendicular to the principal axis. In the figure, one is looking in the direction of the principal axis. Shown are the poles on the equator of the unit sphere

shown here) two poles at the North and South pole of the unit sphere form an orbit, the three red points on the equator form an orbit and the three blue points on the equator also form an orbit.

The elements of the point group map poles to poles:

**Lemma 6.1** *The pole set $P$ is invariant under the action of the point group $G$, and poles in the same orbit have conjugate isotropy groups, $H_{x'} = R H_x R^{-1}$ if $x' = Rx$.*

For the proof, we consider a $x \in P$ with $Rx = x$ and an arbitrary second element $R'$ of the point group. It holds

$$\left( R' R R'^{-1} \right) \left( R'x \right) = R' Rx = R'x \,, \tag{6.6}$$

and thus $R'x$ is also in the pole set.

The mapping $\Phi(R, x) = Rx$ defines an action of the point group $G$ on the pole set $P$ and accordingly $\Phi_R(x) = Rx$ a group homomorphism $G \to \mathcal{S}_P$. Now we divide $P$ into *orbits* under the group action. Since $G$ acts transitively on each orbit, according to the results of Sect. 5.1.1, the isotropy groups of two poles in the same orbit are isomorphic. In fact, the order $m_x = |H_x|$ of the isotropy group to $x$ determines the number of elements in the orbit $B(x)$ of $x$:

**Theorem 6.2** *The orbit $B(x)$ contains $n/m_x$ poles, where $n = |G|$ and $m_x = |H_x|$.*

***Proof*** We consider a pole $x$ with isotropy group $H_x$ of order $m_x$. The point group is the union of the disjoint left cosets $R_1 H_x, \ldots, R_r H_x$ with $r = n/m_x$. All elements of a coset $R_i H_x$ map $x$ to the same point $R_i x$: $R_i H_x x = R_i x$. If $R_i$ and $R_j$ are in different cosets, then $R_i x \neq R_j x$, because from $R_i x = R_j x$ immediately

$R_i^{-1} R_j \in H_x$ would follow. Therefore, the orbit of $x$ contains the $n/m_x$ elements $\{R_1 x, \ldots, R_r x\}$.                                                                                                    $\square$

---

**Example: The Group $\mathcal{D}_3$**

North and South poles have the isotropy group $\mathcal{C}_3$ and all 6 poles on the equator have the isotropy group $\mathcal{C}_2$. Under $\mathcal{C}_3$ North and South poles are invariant and the three blue (red) poles on the equator in Fig. 6.2 form an invariant set. The twofold rotations with rotation axes $C_{h,2}$, $C'_{h,2}$ and $C''_{h,2}$ each exchange 2 elements in the 3 orbits. The orbit of the North pole contains $|\mathcal{D}_3|/|\mathcal{C}_3| = 2$ elements. The two orbits of the poles on the equator each contain $|\mathcal{D}_3|/|\mathcal{C}_2| = 3$ elements.  ◄

After these preparations, we can now classify all point groups. All poles in an orbit $B$ have the same order, which we denote by $m_B$. The $n-1$ *non-trivial* rotations in $G$ contain $m_B - 1$ rotations for each pole in $B$, according to Theorem 6.2 thus $\frac{1}{2}(m_B - 1)n/m_B$ rotations for the poles in the orbit $B$. For each non-trivial $R$ there are two poles, and this explains the factor $1/2$. Accordingly, the sum rule applies

$$n - 1 = \frac{1}{2} \sum_{\text{orbits } B} (m_B - 1) \frac{n}{m_B} \iff 2 - \frac{2}{n} = \sum_{\text{orbits}} \left(1 - \frac{1}{m_B}\right). \tag{6.7}$$

For each orbit, $m_B \geq 2$, and therefore each summand in the last sum is greater than or equal to $1/2$. The pole set can therefore only contain 1, 2 or 3 orbits.

- **One orbit**: In this case, $n(1 + m) = 2m$ would apply, which is impossible for $G \neq \{e\}$. So there are at least two orbits.
- **Two orbits:** These have $n/m_1$ and $n/m_2$ elements, respectively. Formula (6.7) then implies

$$2 = \frac{n}{m_1} + \frac{n}{m_2} . \tag{6.8}$$

Since $m_1$ and $m_2$ divide the group order $n$, only the solution $m_1 = m_2 = n$ exists. Accordingly, the two orbits each contain exactly one pole. Since the poles do not move under all rotations, $G \cong \mathcal{C}_n$ and the group is generated by a rotation $c_n$.

- **Three orbits:** Here, (6.7) leads to

$$1 + \frac{2}{n} = \frac{1}{m_1} + \frac{1}{m_2} + \frac{1}{m_3} . \tag{6.9}$$

If all $m_i \geq 3$, then the right side is never greater than one and the equation is never satisfied. Therefore, there exists at least one isotropy group with two elements. We choose $m_1 = 2$. It follows

$$\frac{1}{2} + \frac{2}{n} = \frac{1}{m_2} + \frac{1}{m_3} \iff (m_2 - 2)(m_3 - 2) = 4\left(1 - \frac{m_2 m_3}{n}\right) < 4 .$$

There are only a few possibilities to satisfy this inequality:

| $m_1$ | $n/m_1$ | $m_2$ | $n/m_2$ | $m_3$ | $n/m_3$ | $n$ | point group |
|---|---|---|---|---|---|---|---|
| 2 | $m$ | 2 | $m$ | $m$ | 2 | $2m$ | $\mathcal{D}_m$ |
| 2 | 6 | 3 | 4 | 3 | 4 | 12 | $\mathcal{T}$ |
| 2 | 12 | 3 | 8 | 4 | 6 | 24 | $\mathcal{O}$ |
| 2 | 30 | 3 | 20 | 5 | 12 | 60 | $\mathcal{Y}$ |

For each allowed combination of orders $m_1$, $m_2$, $m_3$ of the three isotropy groups, the identification of the point group is not difficult. For example, in the case $(m_1, m_2, m_3) = (2, 2, m)$, the twofold rotation axes must be perpendicular to the $m$-fold principal axis. For the group $\mathcal{Y}$, for example, there are $n/2m_1 = 15$ twofold, 10 threefold and 6 fivefold axes of rotation.

### 6.1.3   Achiral Point Groups

Achiral point groups are extensions of chiral point groups. We first consider reflections, which reverse the orientation, in more detail.

**Reflection on Planes**
Elementary are the reflections on the three planes through the origin and perpendicular to the base vectors

$$\sigma_x = \begin{pmatrix} -1 & 0 & 0 \\ 0 & 1 & 0 \\ 0 & 0 & 1 \end{pmatrix}, \quad \sigma_y = \begin{pmatrix} 1 & 0 & 0 \\ 0 & -1 & 0 \\ 0 & 0 & 1 \end{pmatrix}, \quad \sigma_z = \begin{pmatrix} 1 & 0 & 0 \\ 0 & 1 & 0 \\ 0 & 0 & -1 \end{pmatrix}. \tag{6.10}$$

The position of the mirror plane relative to the principal axis (which points in the direction $e$) is expressed in the notation, see Fig. 6.3:

**Horizontal, Vertical and Dihedral Reflections**

1. A reflection on the plane perpendicular to $e$ is denoted by $\sigma_h$.
2. A reflection on a plane containing $e$ is denoted by $\sigma_v$.
3. A reflection on a dihedral plane, which contains $e$ and bisects the angle between two adjacent and to $e$ orthogonal, twofold rotation axes, is denoted by $\sigma_d$.

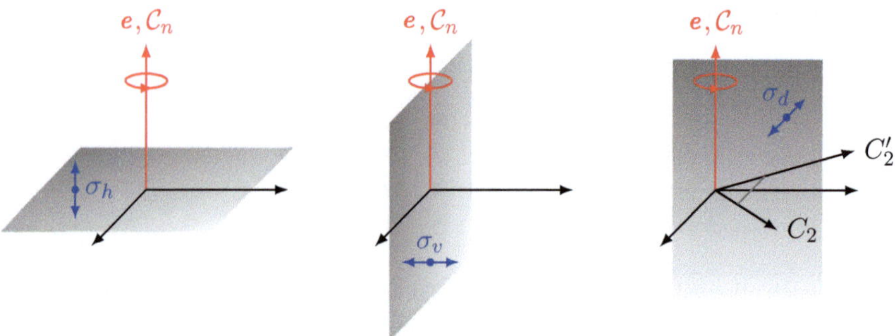

**Fig. 6.3** Notations for the three types of reflections on planes

If the $z$-axis is the principal axis, then $\sigma_h$ is the reflection on the $(x, y)$-plane,

$$\sigma_h = \sigma_z \quad \text{with} \quad \sigma_h c_n = c_n \sigma_h . \tag{6.11}$$

The reflection on the $(y, z)$-plane containing the principal axis has the form

$$\sigma_v = \sigma_x \quad \text{with} \quad \sigma_v c_n \neq c_n \sigma_v \quad \text{for} \quad n = 3, 4, \dots . \tag{6.12}$$

The product of two reflections $\sigma_1$, $\sigma_2$ is a rotation, whose rotation axis coincides with the line of intersection of the two mirror planes,

$$\sigma_1 \sigma_2 = c(\varphi) \implies \sigma_2 \sigma_1 = (\sigma_1 \sigma_2)^{-1} = c^{-1}(\varphi) = c(-\varphi) . \tag{6.13}$$

The rotation angle $\varphi$ is twice the angle enclosed by the two mirror planes. By left-multiplying the first relation with $\sigma_1$ we get

$$\sigma_2 = \sigma_1 c(\varphi) . \tag{6.14}$$

Thus, the product of a rotation and a reflection, whose plane contains the rotation axis, can be replaced by a reflection, which again contains the rotation axis. Because of (6.13), two reflections commute exactly when $c(\varphi) = c(-\varphi)$ holds, or when the rotation angle takes the values $0$ or $\pi$. In the first case $c = e$ and the reflections are identical, in the second case $c = c_2$ and they are reflections on mutually perpendicular mirror planes.

### Inversion and Rotoreflections

In addition to reflections on planes, there are reflections at a fixed point, the inversion center. Such an *inversion* has order 2. If the inversion center coincides with the origin, then $i = \sigma_x \sigma_y \sigma_z$. Another decomposition of the inversion contains a twofold rotation and a reflection on the plane perpendicular to the rotation axis,

$$i = c_2\sigma_h = \sigma_h c_2 \quad \text{or} \quad \sigma_h = ic_2 . \tag{6.15}$$

This means that a reflection on a plane is equal to a twofold rotation with axis perpendicular to the mirror plane and a subsequent inversion.

> **Rotoreflections**
> The product of an $n$-fold rotation $c_n$ with a reflection $\sigma_h$ on a plane perpendicular to the axis of rotation is called a rotation-reflection or rotoreflection and denoted by $s_n$.

If we choose the $z$-axis in the direction of the rotation axis, then the transformation matrix is

$$s_n = c_n\sigma_h = \begin{pmatrix} \cos\varphi_n & -\sin\varphi_n & 0 \\ \sin\varphi_n & \cos\varphi_n & 0 \\ 0 & 0 & -1 \end{pmatrix} . \tag{6.16}$$

The $n$-fold application of an $n$-fold rotoreflection yields

$$s_n^n = c_n^n\sigma_h^n = \sigma_h^n = \begin{cases} e & \text{for } n \text{ even,} \\ \sigma_h & \text{for } n \text{ odd} . \end{cases} \tag{6.17}$$

Thus, the $s_n$ have the order $n$ for even $n$ and the order $2n$ for odd $n$. Specifically for $n = 2$ we obtain the inversion, $s_2 = c_2\sigma_h = i$.

### Example: All Symmetries for a Prism Over the Regular $n$-gon

As previously discussed, the proper symmetry transformations form the dihedral group $\mathcal{D}_n$, cf. Fig. 6.4. If we also allow improper symmetries, then $\sigma_h$ is added. Because $s_n = c_n\sigma_h$, the rotation-reflections $s_n$ and the $n$ reflections $\sigma_v = c_{2,h}\sigma_h$ on planes, which contain the principal axis, also occur.  ◄

**Lemma 6.2**  *An achiral point group contains as many proper rotations as improper ones.*

If $G_0$ is the normal subgroup of the proper rotations in $G$ and $\sigma$ reverses the orientation, then the coset partition of $G$ is

$$G = G_0 \cup G_1, \qquad G_1 = \sigma \cdot G_0 . \tag{6.18}$$

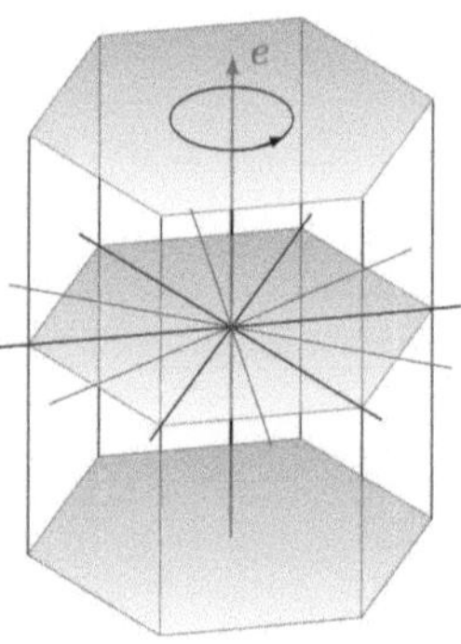

**Fig. 6.4** The symmetry transformations of a prism over an equilateral $n$-gon form the group $\mathcal{D}_{nh}$

This is particularly true for the group generated by the identity and the inversion

$$\mathcal{C}_i = \{e, i\} \cong \mathbb{Z}_2 \ . \tag{6.19}$$

The following theorem is useful for a classification of achiral point groups:

**Theorem 6.3** *An achiral point group, which does not contain the inversion, is isomorphic to a chiral point group.*

**Proof** Let $G$ be an achiral point group and $G = G_0 \cup G_1$ with $G_1 = \sigma \cdot G_0$ for an improper $\sigma \neq i$ the decomposition of $G$ according to the normal subgroup $G_0$. Then $G' = G_0 \cup iG_1$ is a chiral point group with $G_0 \cap iG_1 = \emptyset$. The mapping $\varphi : G \rightarrow G'$ with

$$\varphi(R_0) = R_0 \quad \text{for} \quad R_0 \in G_0 \quad \text{and} \quad \varphi(R_1) = i \cdot R_1 \quad \text{for} \quad R_1 \in G_1$$

is then an isomorphism.                                                                                  $\square$

---

**Task**

Convince yourself that the last statement is correct.

New perspectives will thus be brought by point groups, which contain the inversion $i$. Because the inversion commutes with all rotations, these point groups are a direct product

$$G = G_0 \times \mathcal{C}_i$$

of the normal subgroup $G_0$ with $\mathcal{C}_i$ in (6.19).

> **Rules for the Formation of Achiral Point Groups from Chiral Point Groups**
>
> 1. Form the direct product of a chiral point group with $C_i$.
> 2. Decompose the chiral point group $G$ into cosets with respect to a normal subgroup $G_0$ of index 2, $G = G_0 \cup gG_0$ with $g \notin G_0$ and transform it to $G' = G_0 \cup igG_0$.

With these rules, nine types of achiral point groups can be constructed:

**Theorem 6.4 (Achiral Point Groups)** *There are nine types of achiral point groups: five series* $C_{nh}, C_{nv}, D_{nh}, D_{nd}, S_{2n}$ *as well as four special point groups* $T_h, T_d, O_h, \mathcal{Y}_h$.

Some important properties of the achiral point groups are given in Table 6.1.

For odd $n$, $C_n$ has no normal subgroup with index 2 and the second formation rule is not applicable—therefore there is no achiral point group isomorphic to $C_n$ with odd $n$. The simple groups $T$ and $\mathcal{Y}$ have no (non-trivial) normal subgroups and the second formation rule is also not applicable. But it is applicable to $C_n$ with even $n$, to $D_n$ and to $O$. For example, the Platonic group $O \cong S_4$ has the normal subgroup $T \cong A_4$. If you decompose $O$ into two cosets $O = T \cup iG_1$, then you get the achiral point group according to $T_d = T \cup G_1 \cong O$. The groups $D_{nh}$ and $D_{nd}$ often appear in molecular physics. In contrast to $D_{nd}$, $D_{nh}$ contains the

**Table 6.1** Complete list of the achiral point groups, i.e., the finite subgroups of O(3). Groups isomorphic to $G \times C_i$ were constructed with the first formation rule, the other groups with the second formation rule. $|G|$ is the group order and $k(G)$ the number of conjugacy classes

| Group | | Normal subgroup | Isomorphic to | $|G|$ | $k(G)$ |
|---|---|---|---|---|---|
| $C_{nh}$ | $n$ even | $C_n$ | $C_n \times C_i$ | $2n$ | $2n$ |
| | $n$ odd | $C_n$ | $C_{2n}$ | $2n$ | $2n$ |
| $C_{nv}$ | $n$ even | $C_n$ | $D_n$ | $2n$ | $(n+6)/2$ |
| | $n$ odd | $C_n$ | $D_n$ | $2n$ | $(n+3)/2$ |
| $S_{2n}$ | | $C_n$ | $C_{2n}$ | $2n$ | $2n$ |
| $D_{nh}$ | $n$ even | $D_n$ | $D_n \times C_i$ | $4n$ | $n+6$ |
| | $n$ odd | $D_n$ | $D_{2n}$ | $4n$ | $n+3$ |
| $D_{nd}$ | $n$ even | $D_n$ | $D_{2n}$ | $4n$ | $n+3$ |
| | $n$ odd | $D_n$ | $D_n \times C_i$ | $4n$ | $n+3$ |
| $T_h$ | | $T$ | $T \times C_i$ | 24 | 8 |
| $T_d$ | | $T$ | $O$ | 24 | 5 |
| $O_h$ | | $O$ | $O \times C_i$ | 48 | 10 |
| $\mathcal{Y}_h$ | | $\mathcal{Y}$ | $\mathcal{Y} \times C_i$ | 120 | 10 |

reflection $\sigma_h$. Graphically well-prepared information on molecular symmetries in organic chemistry can be found at at http://csi.chemie.tu-darmstadt.de/ak/immel/misc/oc-scripts/symmetry.html

**Presentations of the Point Groups**

Finally, we give possible representations of all chiral point groups by means of presentations:

$$\mathcal{C}_n = \{c_n\}\,,$$

$$\mathcal{D}_n = \left\{c_n, c_{h,2}\middle| (c_n c_{h,2})^2 = e\right\}\,,$$

$$\mathcal{T} = \left\{c_3, c_2'\middle| (c_3 c_2')^3) = e\right\}\,,$$

$$\mathcal{O} = \left\{c_4, c_4'\middle| (c_4 c_4')^2 = e\right\}\,,$$

$$\mathcal{Y} = \left\{c_5, c_2'\middle| (c_5 c_2')^3 = e\right\}\,.$$

And similarly of all achiral point groups (recall that $[a, b] = aba^{-1}b^{-1}$):

$$\mathcal{C}_{nh} = \left\{c_n, \sigma_h\middle| [c_n, \sigma_h] = e\right\}\,,$$

$$\mathcal{C}_{nv} = \left\{c_n, \sigma_v\middle| (c_n \sigma_v)^2 = e\right\}\,,$$

$$\mathcal{D}_{nh} = \left\{c_n, c_{h,2}, i\middle| (c_n c_{h,2})^2 = [c_n, i] = [c_{h,2}, i] = e\right\}\,,$$

$$\mathcal{D}_{nd} = \left\{s_{2n}, \sigma_v\middle| (s_{2n} \sigma_v)^2 = e\right\}\,,$$

$$\mathcal{S}_{2n} = \{s_{2n}\}\,,$$

$$\mathcal{T}_h = \left\{c_3, c_2', i\middle| (c_3 c_2')^3 = [c_3, i] = [c_2', i] = e\right\}\,,$$

$$\mathcal{T}_d = \left\{c_3, c_2', \sigma_v\middle| (c_3 c_2')^3 = (c_3 \sigma_v)^2 = (c_2' \sigma_v)^2 = e\right\}\,,$$

$$\mathcal{O}_h = \left\{c_4, c_4', i\middle| (c_4 c_4')^2 = [c_4, i] = [c_4', i] = e\right\}\,,$$

$$\mathcal{Y}_h = \left\{c_5, c_2', i\middle| (c_5 c_2')^3 = [c_5, i] = [c_2', i] = e\right\}\,.$$

In *crystallography*, 32 of these point groups play a role.

### 6.1.4 Inertia Tensor of Symmetric Bodies

We consider a rigid body consisting of $N$ point masses with masses $m_i$ and choose the center of mass as the origin of the coordinate system. The inertia tensor of the rigid body is

$$\Theta_{ab} = \sum_i m_i \left( r_i^2 \delta_{ab} - x_{ia} x_{ib} \right) . \tag{6.20}$$

It transforms during a rotation $x \to x' = Rx$ like a second-order tensor,

$$\Theta'_{ab} = \sum_i m_i \left( r_i'^2 \delta_{ab} - x'_{ia} x'_{ib} \right) = R_{ac} R_{bd} \sum_i m_i \left( r_i^2 \delta_{cd} - x_{ic} x_{id} \right) ,$$

or in matrix notation

$$\Theta \mapsto \Theta' = R \Theta R^{-1} . \tag{6.21}$$

This transformation law also applies to bodies with continuous mass distribution.

**Symmetries of the Inertia Tensor**
If a rotation $R$ maps the rigid body onto itself, then it must hold

$$\Theta' = \Theta \quad \text{or} \quad \Theta = R \Theta R^{-1} . \tag{6.22}$$

The reverse is generally not true: not every symmetry of the inertia tensor is a symmetry transformation of the body.

The set of symmetry transformations forms the *symmetry group* of the body. This is a subgroup of the symmetries of the inertia tensor. A full sphere with homogeneous mass density is invariant under all rotations. Here, the inertia tensor and the sphere have the same symmetry group O(3). A cube with identical mass points at the vertices is only invariant under the cubic group $\mathcal{O}$ and its achiral extension $\mathcal{O}_h$. But the inertia tensor still has the symmetry group O(3).

**Task**

Convince yourself that a cube with edge length $L$ and identical point masses $m$ at the corners has the inertia tensor $\Theta_{ab} = 4mL^2 \delta_{ab}$ with respect to its center of mass. This shows that the cube is a spherical spinning top.

To examine the symmetry group of a general body, we select the main axes of the inertia tensor as coordinate directions. If the principal moments of inertia in $\Theta = \mathrm{diag}(A, B, C)$ are different, then only diagonal $R$ commute with $\Theta$. Since the matrices must be orthogonal we conclude

**Symmetries of the Asymmetric Top**
The symmetry group of an asymmetric top is a subgroup of the Abelian group

$$\mathcal{C} = \{e, c_2(e_1), c_2(e_2), c_2(e_3), \sigma_x, \sigma_y, \sigma_z, i\}. \tag{6.23}$$

The inertia tensor of a symmetric top $\Theta = \mathrm{diag}(A, A, C)$ with $C \neq A$ commutes instead with all rotations, which leave the symmetry axis defined by $e_3$ fixed. The symmetry group of a symmetric top is therefore a subgroup of $\mathcal{C}_s = \{R \in \mathrm{O}(3) | Re_3 = \pm e_3\}$. The chiral finite subgroups of $\mathcal{C}_s$ are the groups $\mathcal{C}_n$ or $\mathcal{D}_n$. More generally, the groups $\mathcal{C}_n, \mathcal{C}_{nh}, \mathcal{C}_{nv}, \mathcal{D}_n, \mathcal{D}_{nh}, S_n$ or $\mathcal{D}_{nd}$ can occur as symmetry groups of a symmetric top. Conversely, it holds

**Lemma 6.3** *If a symmetry group of a top contains a $\mathcal{C}_n$ with $n \geq 3$ as a subgroup, then it is a symmetric top.*

For the proof, consider the condition $R\Theta = \Theta R$ for a symmetric matrix $\Theta$ and a rotation $R$ with angle $\varphi$ about the third axis. If $\sin \varphi$ is not zero, then one finds $\Theta = \mathrm{diag}(A, A, C)$. If the symmetry group contains a second, at least threefold rotation $c_n(e)$ about another axis $e \neq e_3$, then it is a spherical top with $\Theta = A\mathbb{1}$.

**Corollary 6.1** *The Platonic solids with equal masses at the vertices are spherical tops with inertia tensor $\Theta = A\mathbb{1}$ (see Problem 6.7)*

## 6.2    Molecular Symmetries

We neglect the "intrinsic motions" of electrons and atomic nuclei and base our symmetry considerations on the molecule's nuclear framework. This is a point mass system of finitely many rigidly connected atoms. For example, the nitrogen atom and the three hydrogen atoms of the *ammonia molecule* $NH_3$ form the four vertices of a tetrahedron. On the other hand, Platonic hydrocarbons are saturated hydrocarbons, whose carbon atoms form the vertices of a Platonic solid. The synthesis of cuban (cube) was first achieved in 1964 and that of dodecahedrane (dodecahedron) in 1982. In this way, symmetry investigations on molecules often lead to the study of standard geometric figures in stereometry. The symmetries of some molecules are shown in Fig. 6.5.

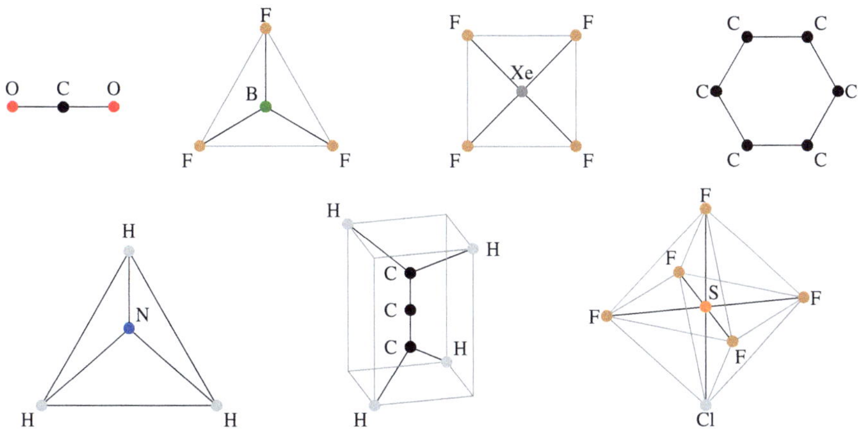

**Fig. 6.5** The symmetries of carbon dioxide ($CO_2$), boron trifluoride ($BF_3$), xenon tetrafluoride ($XeF_4$), benzene ($C_6H_6$), ammonia ($NH_3$), allene ($C_3H_4$) and sulfur chloro pentafluoride ($SF_5Cl$) are the symmetry transformations of the line, the equilateral triangle, square, regular hexagon, tetrahedron, cuboid and octahedron

**Fig. 6.6** The position of the atoms in the Allen molecule $C_3H_4$

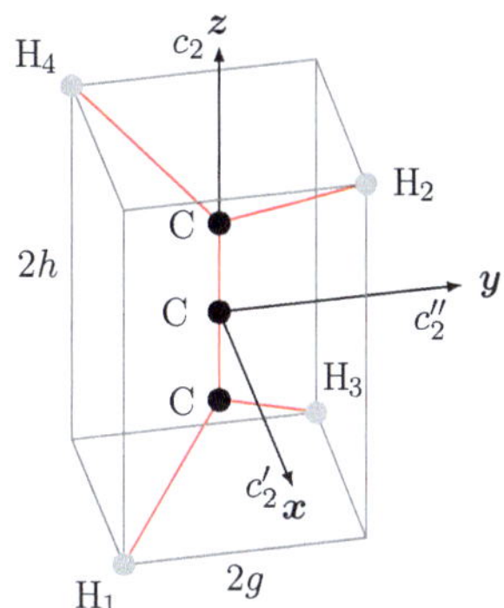

## 6.2.1  The Point Mass System Allene $C_3H_4$

In a Cartesian coordinate system, we consider the cuboid in axis-parallel center position, as sketched in Fig. 6.6. Its height is $2h$ and its square base has the edge length $2g < 2h$. As shown in the figure, the nuclear structure of an allene molecule can be inscribed in the cuboid: The four H atomic nuclei are distributed at the corners

$$H_1 : (g, -g, -h), \quad H_2 : (g, g, h), \quad H_3 : (-g, g, -h), \quad H_4 : (-g, -g, h)$$

and the three carbon nuclei are located at the points

$$(0, 0, -d), \quad (0, 0, 0), \quad (0, 0, d), \qquad 0 < d < h$$

of the $z$-axis.

If we rotate by $180^0$ around the $z$-axis, the nuclear framework comes back to itself: The three C atomic nuclei remain fixed, $H_1$ and $H_3$ as well as $H_2$ and $H_4$ exchange their places. The molecule also allows a rotation of $180^0$ around the $x$- and $y$-axis. *Rotational symmetries* are consistently denoted by the same letter $\mathcal{C}$ in the notation of Schoenflies. The $x$, $y$ and $z$-axes are twofold rotation axes and the corresponding symmetries are denoted by $c_2'$, $c_2''$ and $c_2$. We can see

$$c_2'(H_1) = H_2, \quad c_2''(H_3) = H_2, \quad c_2(H_1) = H_3 \quad \text{etc.}$$

Apart from $\mathcal{C}_2, \mathcal{C}_2'$ and $\mathcal{C}_2''$, the Allen molecule has no other rotational symmetries. However, if we rotate by $90^0$ counterclockwise around the $z$-axis and then reflect in the $x, y$-plane, $H_1$ takes the place of $H_2$, $H_2$ that of $H_3$, $H_3$ that of $H_4$, $H_4$ that of $H_1$ and the lower and upper C atoms exchanges places. The middle C atom remains fixed and its location is a *fixed point* of the symmetry operation. This *rotoreflection*, consisting of a fourfold rotation and a subsequent reflection in the plane perpendicular to the axis of rotation, is denoted by $s_4$. The molecule allows neither the sole rotation nor the reflection that make up $s_4$, but only their composition. The molecule also allows the rotoreflection $s_4^3$, which is created by a rotation around the $z$-axis by $270^0$ and subsequent reflection in the $x-y$-plane. Allen does not have any further rotoreflections. Rotoreflections are consistently denoted by the letter $s$ according to Schoenflies. Specifically, we write $s_n$ if the rotation-reflection has order $n$. For Allen, $n = 4$ and the fourfold vertical rotoreflection axis is the principal axis.

From Fig. 6.6 it can be seen that Allen allows the *reflections* denoted by $\sigma_d$ and $\sigma_d'$ of the planes determined by the principal axis and $H_1$ or the principal axis and $H_2$. For example

$$\sigma_d(H_1) = H_1, \quad \sigma_d(H_2) = H_4, \quad \sigma_d'(H_2) = H_2, \quad \sigma_d'(H_1) = H_3 \,.$$

*Reflection symmetries* are denoted by $\sigma$ in the Schoenflies notation, to distinguish several reflections, also with $\sigma$, $\sigma'$ etc. The position of the mirror axis to the principal axis is possibly expressed by an index on $\sigma$. The vertical mirror planes in the Allen molecule are of this kind. Apart from $\sigma_d$ and $\sigma_d'$, Allen has no other reflection symmetries.

---

### Example: Symmetries of the Allen Molecule

The symmetry group of the Allen molecule is

$$\mathcal{D}_{2d} = \{e, s_4, s_4^2, s_4^3, c_2', c_2'', \sigma_d, \sigma_d'\} \,.$$

Because $s_4^2 = c_2$, only $s_4^2$ is noted as an element. The symmetry group $\mathcal{D}_{2d}$ has order 8. It is an achiral extension of the dihedral group $\mathcal{D}_2$ (see Table in Sect. 6.1.3). ◄

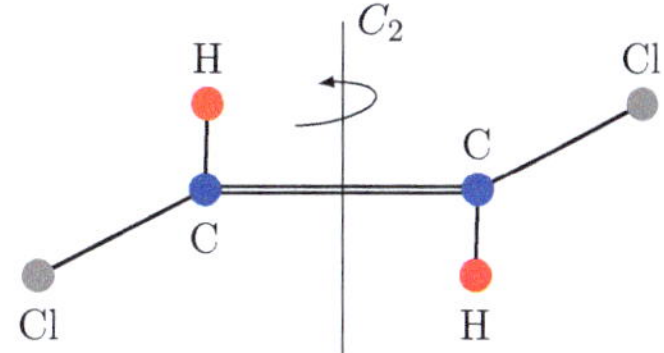

**Fig. 6.7** The nuclear framework of Dichloroethylene



## 6.2.2  Trans-Dichloroethylene

The nuclear framework of the planar molecule Trans-Dichloroethylene $H_2C_2Cl_2$ is shown in Fig. 6.7. It serves to illustrate the symmetry $C_{2h}$. The rotation of the molecule by $180°$ around the axis perpendicular to the molecular plane and the inversion at the center of mass of the molecule are symmetry transformations.

**Example: Symmetry Group of Trans-Dichloroethylene**

The symmetry group of the $H_2C_2Cl_2$ molecule is an achiral extension $C_{2h} = \{\mathbb{1}, c_2, \sigma_h, i\}$ of the group $C_2$ (see Table in Sect. 6.1.3). ◄

## 6.3  Exercises for Chap. 6

**Problem 6.1 (Point Groups in 2 Dimensions)** In two dimensions, a point group is a finite subgroup of O(2). Classify all point groups in two dimensions.

**Problem 6.2 (Examples of Molecular Symmetries)** Convince yourself that $H_2O$ has the symmetry $C_{2v}$, trans-1,2-dichloroethane has the symmetry $C_{2h}$ and benzene has the symmetry $D_{6h}$.

You can find many more examples of molecules with their symmetries in the book [32].

**Problem 6.3 (Group Table of $C_{3v}$)** Determine the group table of the point group $C_{3v}$ with the elements $e, c_3, c_3^2, \sigma_v^a, \sigma_v^b, \sigma_v^c$ and convince yourself of $C_{3v} \cong D_3$.

**Problem 6.4 (Conjugacy Classes of $D_{3h}$)** Prove that $D_{3h}$ has the following six conjugacy classes:

$$K_1 = e, \quad K_2 = c_3, c_3^2, \quad K_3 = \sigma_v^a, \sigma_v^b, \sigma_v^c, \quad K_4 = \sigma_h, \quad K_5 = s_3, s_3',$$

$$K_6 = c_2^a, c_2^b, c_2^c.$$

**Problem 6.5 (Symmetries of the $CH_4$ Molecule)** Consider the core nuclear framework of the (hypothetical) molecule $CH_4$, in which the four H atoms are at the corners of the square $(\pm a, 0, 0)$ and $(0, \pm a, 0)$, and the C atom at the point

**Fig. 6.8** In a tetrahedron, each node is connected to all other nodes

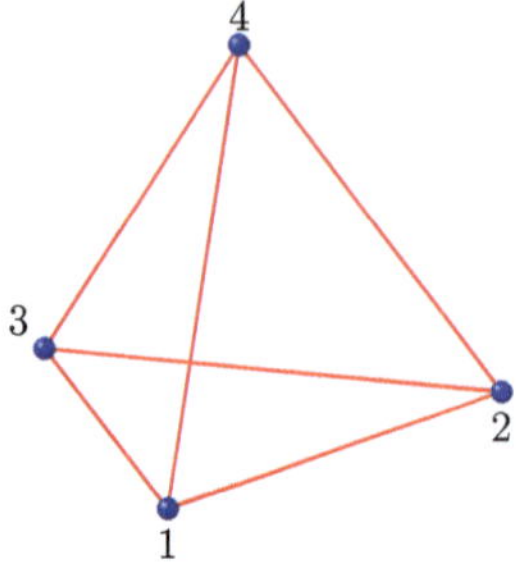

$(0, 0, z)$, with $z < a$. What is the symmetry group of the molecule. What is the extended symmetry in the planar limit case $z \to 0$?

**Problem 6.6 (Incidence Matrix and Symmetry Transformations)** An incidence matrix (sometimes also adjacency matrix) of an undirected graph is a matrix that stores which nodes of the graph are connected by an edge.

We can view the core structure of a molecule or also of a Platonic solid as a graph and assign it an incidence matrix $A$. Here, $a_{ij} = 1$ if $i$ and $j$ are connected, otherwise $a_{ij} = 0$. For example, the tetrahedron in the Fig. 6.8 is assigned the incidence matrix

$$A = \begin{pmatrix} 0 & 1 & 1 & 1 \\ 1 & 0 & 1 & 1 \\ 1 & 1 & 0 & 1 \\ 1 & 1 & 1 & 0 \end{pmatrix},$$

A symmetry transformation of the body is a permutation of the vertices, which does not change the neighborhood relations. This permutation is an inner automorphism

$$A \mapsto S^{-1} A S,$$

where the orthogonal permutation matrix $S$ describes the permutation. For example, in GAP the permutation $(1, 2, 3)(4)$ is assigned the following permutation matrix:

$$S = \begin{pmatrix} 0 & 1 & 0 & 0 \\ 0 & 0 & 1 & 0 \\ 1 & 0 & 0 & 0 \\ 0 & 0 & 0 & 1 \end{pmatrix}.$$

The element $S_{ij}$ is exactly 1 if $i$ goes into $j$, otherwise 0. The permutation represented by $S$ is a symmetry transformation when $S^{-1} A S = A$ holds. The following GAP program calculates the symmetries of the graph with incidence matrix $A$. It is taken into account that $S^{-1} = S^T$.

```
GAP> A:=[[0,1,1,1],[1,0,1,1],[1,1,0,1],[1,1,1,0]];
GAP> n:=4;i:=0;H:=[];
GAP> t:=SymmetricGroup(n);
GAP> tt:=Elements(t);
GAP> for a in tt do
GAP> S:=PermutationMat(a,n);
GAP> St:=TransposedMat(S);
GAP> y:=St*A*S;
GAP> if y=A then AddSet(H,a);fi;
GAP> od;
GAP> G:=Group(H);
```

1. The incidence matrix for a prism over the equilateral triangle is

$$
A = \begin{pmatrix} 0 & 1 & 1 \\ 1 & 0 & 1 \\ 1 & 1 & 0 \end{pmatrix}
$$

   Modify the above GAP program accordingly and analyze the symmetry group of
   the prism.
2. Determine an incidence matrix for the cube and analyze the associated symmetry
   group.

**Problem 6.7 (Inertia Tensors of the Platonic Solids)**  Determine the inertia tensors of the Platonic solids with respect to their center of mass, if equal masses are
located at the corners of the body.

# Space Groups and Crystals 7

*The chief forms of beauty are order and symmetry and
definiteness, which the mathematical sciences demonstrate in a
special degree.*

—Aristotle

An *ideal crystal* is a solid whose microscopic atomic structure is spatially periodic.[1]
This implies a periodicity of the densities of atomic nuclei and electrons, possibly
also the spin density of the electrons. Figure 7.1 shows the crystal lattice of $NaCl$.

An ideal crystal has as its symmetry group a discrete subgroup of the Euclidean
group—this is called the space group of the crystal. In this chapter, we classify
those combinations of rotations, reflections, and translations that transform a space
lattice into itself. In addition to three-dimensional ones, two-dimensional crystals,
which are crystalline materials, consisting of only one layer of atoms or molecules,
have been known for a long time. Due to their unusual properties, they are the
subject of numerous current research projects. In 2004, Andre Geim and Konstantin
Novoselov succeeded in producing the first two-dimensional crystals from carbon
atoms (graphene). The symmetry groups of two-dimensional crystals are called
wallpaper groups. Space groups and wallpaper groups and their role in crystal
physics are treated in many textbooks, see for example [12]. An illustrative journey
through shapes, patterns, symmetries, and structures in crystallography can be found
in [33].

---

[1] A real crystal differs from an ideal one by its finite extent and structural defects.



A. Wipf, *Symmetries in Physics*, https://doi.org/10.1007/978-3-662-72675-4_7

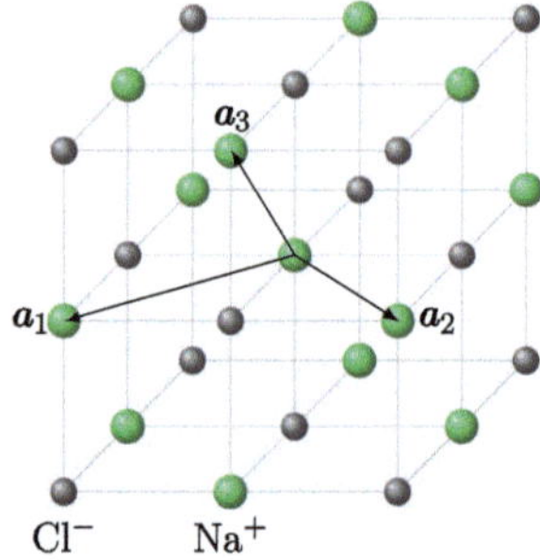



**Fig. 7.1** The building blocks of the NaCl lattice are the Na$^+$ and Cl$^-$ ions: These each form a cubic face-centered lattice and show the sketched nested arrangements, which arise from the translation by $\frac{1}{2}(a_1 + a_2)$. The unit cell is spanned by the vectors $a_i$. The symmetries translations, rotations, inversions, reflections and combinations of these can be read from the figure

**Fig. 7.2** A primitive cell of the lattice with basis vectors $a_1$, $a_2$ and $a_3$

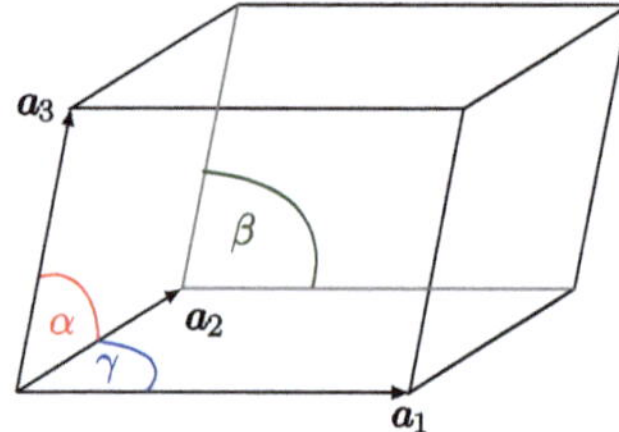



## 7.1  Lattice Vectors and Primitive Cells

The construction of a crystal lattice $\mathcal{L}$ is achieved by the periodic repetition of the so-called *primitive cell*, which divides the entire space into congruent cells. The primitive cell—it is the smallest structural unit, which contains the full lattice symmetry—forms a parallelepiped, which is spanned by primitive translations $a_i$ after choosing an origin. This is sketched in Fig. 7.2. Each corner point of a parallelepiped in $d$ dimensions is thus determined by the $d$ integer coefficients $n_1, \ldots, n_d$ in the expansion

$$a = \sum_{i=1}^{d} n_i \, a_i, \quad n_i \in \mathbb{Z} \tag{7.1}$$

Each lattice point $a$ is common to all $2^d$ neighboring parallelepipeds and each primitive cell contains exactly one lattice point—in Fig. 7.2 one can assign the lattice point at the bottom left to the shown cell and the 7 other lattice points to the neighboring cells. The choice of the origin and the $a_i$ and thus the shape of the cell are not unique. Figure 7.3 shows three primitive bases of a two-dimensional square lattice.

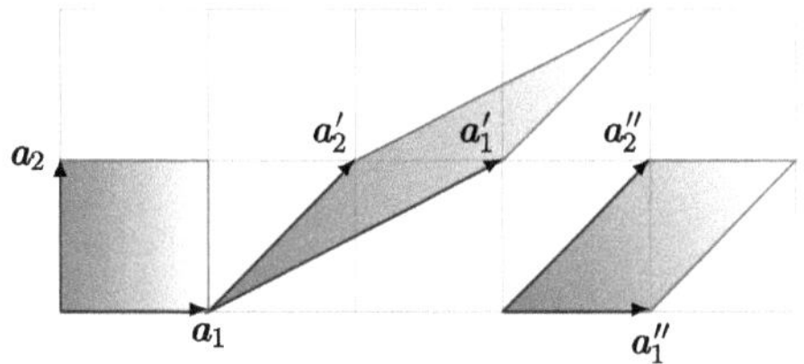

**Fig. 7.3** The basis of a primitive cell is not unique



**Lemma 7.1 (Basis Change)** *If the $a_1, \ldots, a_d$ form a basis of primitive translations of a $d$-dimensional lattice, then the vectors*

$$a_i' = \sum_{j=1}^{d} a_j G_{ji}, \quad i = 1, 2, \ldots, d, \qquad G_{ji} \in \mathbb{Z} \tag{7.2}$$

*also form a basis, if*

$$|\det G| = 1, \qquad G = (G_{ij}) \in GL(d, \mathbb{Z}). \tag{7.3}$$

***Proof*** By definition, all integer linear combinations of the $a_i$ transform the lattice into itself. We still need to check whether the volume $V_e$ of the parallelepiped spanned by the $a_i$, given by

$$V_e^2 = \det(A), \quad A_{ij} = (a_i, a_j),$$

is equal to the volume of the parallelepiped spanned by the $a_i'$. Because

$$A_{ij}' = \sum_{p,q} G_{pi} G_{qj} A_{pq} \quad \text{or} \quad A' = G^T A G$$

it immediately follows that $V_e' = |\det G| V_e$, which leads exactly to condition (7.3). $\qquad\square$

---

**Wigner-Seitz Cell**

There is a rule to define a unique primitive cell called *Wigner-Seitz cell*. The center of the Wigner-Seitz cell is chosen to be a lattice point and the cell contains those points that are closer to the selected lattice point than to all other lattice points.

---

A Wigner-Seitz cell can be constructed by choosing any lattice point and drawing connecting lines to its nearest neighbours. In a second step one constructs the

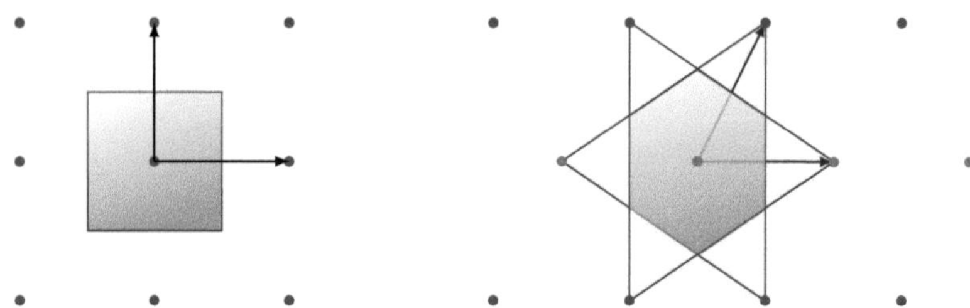

**Fig. 7.4** The Wigner-Seitz cells for the square and the rectangular centered Bravais lattice

perpendicular bisector planes of the connecting lines. The region enclosed by the planes is the Wigner-Seitz cell, see Fig. 7.4.

It turns out that the Wigner-Seitz cells in two dimensions are bounded by up to three pairs of parallel lines, while in three dimensions up to seven pairs of parallel planes are needed. The Wigner-Seitz cell contains all the information about the symmetry of the lattice. However, the symmetry of a crystal lattice is often easier to recognize when choosing a (generally non-primitive) unit cell that contains more than one lattice point. More on this later.

### 7.1.1 The Reciprocal Lattice

In X-ray, electron, and neutron diffraction on crystals one is inevitably led to the concept of the *reciprocal lattice*. To introduce it, we consider an arbitrary lattice-periodic function,

$$f(r + a) = f(r), \quad a \in \mathcal{L}. \tag{7.4}$$

It can be represented as a Fourier series with coefficients $\tilde{f}(k)$,

$$f(r) = \sum_{k} \tilde{f}(k)\, e^{ik \cdot r}. \tag{7.5}$$

For the Fourier coefficients, the inverse formula applies

$$\tilde{f}(k) = \frac{1}{V_e} \int_{EZ} f(r)\, e^{-ik \cdot r} d^3 r, \tag{7.6}$$

where the integration is over the primitive cell defined by the basis vectors with volume $V_e$. For the function $f$ to be lattice-periodic, it must hold

$$e^{ik \cdot (r+a)} = e^{ik \cdot r} \quad \text{or} \quad e^{ik \cdot a} = 1$$

for all lattice vectors $\boldsymbol{a}$. But since every lattice vector is an integer linear combination of the basis vectors, it is sufficient to demand this condition for the basis vectors:

$$e^{i\boldsymbol{k}\cdot\boldsymbol{a}_i} = 1 \quad \text{or} \quad \boldsymbol{k}\cdot\boldsymbol{a}_i \in 2\pi\,\mathbb{Z}. \tag{7.7}$$

Like the set $\mathcal{L}$ of lattice vectors, the allowed $\boldsymbol{k}$-vectors define a discrete additive group, which we denote by $\mathcal{L}^*$ and which defines the *reciprocal lattice*. For $\mathcal{L}^*$ there is again a basis $\{\boldsymbol{a}_i^*\}$, so that every $\boldsymbol{k}$-vector is an integer linear combination of the $\boldsymbol{a}_i^*$.

**The Reciprocal Lattice**
The lattice reciprocal to $\mathcal{L}$ is

$$\mathcal{L}^* = \left\{ \boldsymbol{k}\,\middle|\,\boldsymbol{k} = \sum_{i=1}^{d} k_i\,\boldsymbol{a}_i^*,\; k_i \in \mathbb{Z} \right\},$$

where the basis $\{\boldsymbol{a}_i^*\}$ of the reciprocal lattice is determined by $\boldsymbol{a}_i^*\cdot\boldsymbol{a}_j = 2\pi\,\delta_{ij}$.

This is the convention commonly used in solid state physics. In crystallography, on the other hand, the normalization $\boldsymbol{a}_i^* \cdot \boldsymbol{a}_j = \delta_{ij}$ is chosen for the basis vectors of the reciprocal lattice.

In three dimensions, the explicit solution of the system of equations for the $\boldsymbol{a}_i^*$ is

$$\boldsymbol{a}_1^* = \frac{2\pi}{V_e}\,\boldsymbol{a}_2 \times \boldsymbol{a}_3, \quad \boldsymbol{a}_2^* = \frac{2\pi}{V_e}\,\boldsymbol{a}_3 \times \boldsymbol{a}_1, \quad \boldsymbol{a}_3^* = \frac{2\pi}{V_e}\,\boldsymbol{a}_1 \times \boldsymbol{a}_2. \tag{7.8}$$

The volume of the primitive cell of $\mathcal{L}^*$ is $V_e^* = (2\pi)^d / V_e$. The Wigner-Seitz cell of the reciprocal lattice with the center at $\boldsymbol{k} = \boldsymbol{0}$ is the first Brillouin zone for the lattice.

### 7.1.2  Crystal Systems in the Plane

As the first primitive basis vector $\boldsymbol{a}_1$, one chooses the shortest vector in $\mathcal{L}$ (or one of the shortest vectors, if there are several of them), and as the second basis vector $\boldsymbol{a}_2$, a shortest vector in $\mathcal{L}$ that is not parallel to $\boldsymbol{a}_1$. The distinction of the crystal systems is done with the help of the lengths $a_i = |\boldsymbol{a}_i|$ and the angle $\varphi$ between $\boldsymbol{a}_1$ and $\boldsymbol{a}_2$.

**Fig. 7.5** The 4 crystal
systems in 2 dimensions.
Shown are primitive basis
vectors

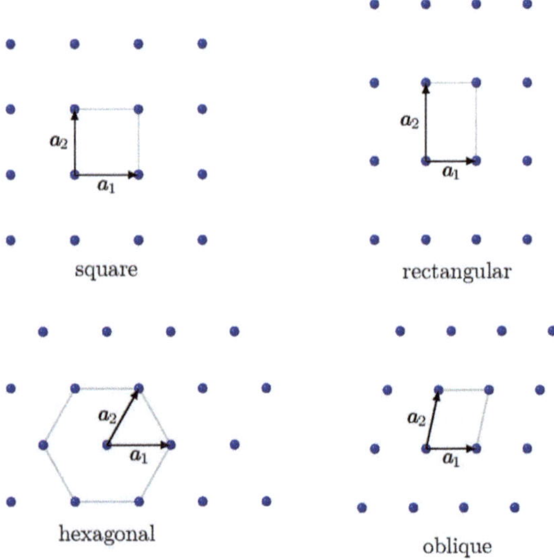



We may assume that $\varphi \in (0, \pi/2]$ (otherwise replace $a_2$ with $-a_2$). There are the 4 crystal systems listed in (7.9)

$$
\begin{aligned}
&1.\ \text{square system:} && a_1 = a_2,\ \varphi = \pi/2\,,\\
&2.\ \text{rectangular system:} && a_1 < a_2,\ \varphi = \pi/2\,,\\
&3.\ \text{hexagonal system:} && a_1 = a_2,\ \varphi = \pi/3\,,\\
&4.\ \text{oblique system:} && a_1 \neq a_2,\ \varphi < \pi/2\,,
\end{aligned}
\tag{7.9}
$$

and these are shown in Fig. 7.5.

### 7.1.3  Crystal Systems in Space

The different crystal systems in space are distinguished similarly to those in the plane by the ratios of the lengths $a_i$ of the basis vectors and their angles. The naming of the angles is as in Fig. 7.2. There are 7 different crystal systems: triclinic, monoclinic, orthorhombic, tetragonal, trigonal, hexagonal and cubic, and these are listed in (7.10).

$$
\begin{aligned}
&1.\ \text{triclinic system:} && a_1 \neq a_2 \neq a_3,\ \alpha \neq \beta \neq \gamma \neq 90\,,\\
&2.\ \text{monoclinic system:} && a_1 \neq a_2 \neq a_3,\ \alpha = \gamma = 90,\ \beta \neq 90\,,\\
&3.\ \text{orthorhombic system:} && a_1 \neq a_2 \neq a_3,\ \alpha = \beta = \gamma = 90\,,\\
&4.\ \text{tetragonal system:} && a_1 = a_2 \neq a_3,\ \alpha = \beta = \gamma = 90\,,\\
&5.\ \text{trigonal system:} && a_1 = a_2 = a_3,\ \alpha = \beta = \gamma \neq 90\,,\\
&6.\ \text{hexagonal system:} && a_1 = a_2 \neq a_3,\ \alpha = \beta = 90,\ \gamma = 120\,,\\
&7.\ \text{cubic system:} && a_1 = a_2 = a_3,\ \alpha = \beta = \gamma = 90\,.
\end{aligned}
\tag{7.10}
$$

## 7.2   Primitive Cells, Unit Cells and Bravais Lattices

If one wants to use axis systems adapted to the elements of a space group, one must also allow for non-primitive *unit cells* containing multiple lattice points. These then also contain atoms with non-integer coordinates, and their volume is a multiple of the volume of the primitive cell. This leads from the crystal systems to the Bravais lattices, which have the advantage that they allow the maximum possible symmetry of a crystal to be immediately recognized. The property that *unit cells* of Bravais lattices are not always primitive cells is accepted, because in general the recognition of symmetries is enormously helpful. Therefore, Bravais lattices are often used.

The points $a = m_1 a_1 + \ldots$ of a Bravais lattice $\mathcal{L}$ are obtained, when one acts on the origin in Euclidean space with all translations of the translation group $\{a\}$. Therefore, after defining a zero point, the translation group can be identified with the Bravais lattice $\mathcal{L}$.

---

**Example: Centered Rectangular Bravais Lattice**

For the special angle with $\cos\varphi = a_1/2a_2$ in the oblique crystal system in Fig. 7.6 there is an increased symmetry. This becomes obvious, when instead of the primitive cell spanned by the $a_i$ we use the rectangular unit cell with double area spanned by $a_1' = a_1$ and $a_2' = 2a_2 - a_1$. Then one recognizes that the centered rectangular system has all symmetry properties of the rectangular system. It is therefore considered as its own Bravais lattice within the rectangular crystal system. Thus, there are 5 Bravais lattices in the plane. ◀

### 7.2.1   Bravais Lattices in Space

In three-dimensional space, a unit cell of a Bravais lattice is spanned by the $a_i'$ and can contain several atoms. For $r$ atoms in the unit cell, their locations are determined

**Fig. 7.6** Primitive and unit cells

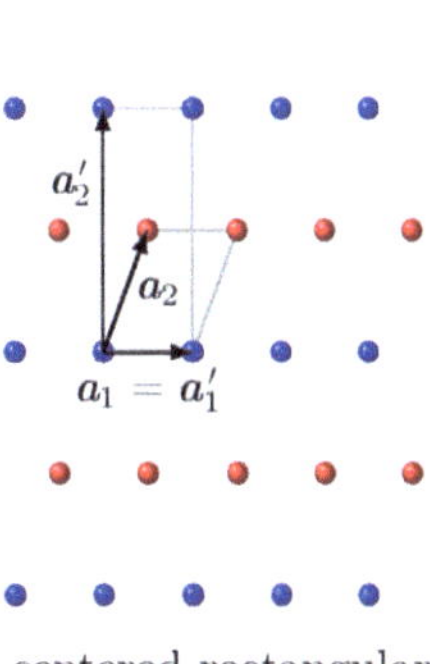


centered rectangular

by tuples $s_1, \ldots, s_r$,

$$r_{m,k} = \sum m_i \, a_i' + s_k = \sum_i (m_i + s_{k,i}) a_i', \qquad k = 1, \ldots, r, \qquad (7.11)$$

where in the last step we expanded the $s_k$ according to the basis $a_i'$.

There are 3 Bravais lattices of the cubic crystal system. One is the primitive lattice with the maximum symmetry, and two others are non-primitive with several atoms in the unit cell.

### Example: Simple Cubic (sc) Lattice

For this primitive system, the unit cell is simultaneously a primitive cell, thus contains one atom. The lattice has the cubic subgroup of the rotation group as symmetry group (more on this later). After choosing a basis, directions in the Bravais lattice can be specified by the *Miller indices*. For example, $\langle 1\bar{2}3 \rangle$ means the direction $a_1 - 2a_2 + 3a_3$. This is shown in Fig. 7.7 for two directions. ◀

### Example: Body-centered Cubic (bcc) Lattice

For the bcc lattice, the $a_i$ in Fig. 7.8 span a primitive cell. Instead, one chooses the unit cell with basis

$$a_1' = a_1, \quad a_2' = a_2, \quad a_3' = 2a_3 - a_1 - a_2 \,.$$

The cell has the shape of a cube and the cubic symmetry of the lattice is obvious. The unit cell contains two points at $s_1 = 0$ and $s_2 = a_3 = \frac{1}{2}(a_1' + a_2' + a_3')$. It has twice the volume of the primitive cell. ◀

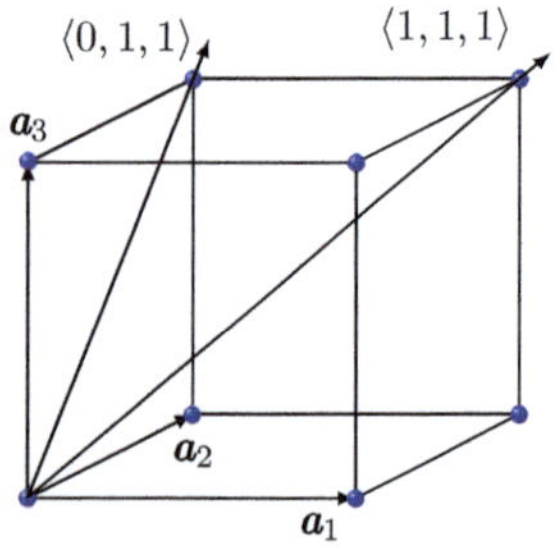

**Fig. 7.7** Primitive and unit cells

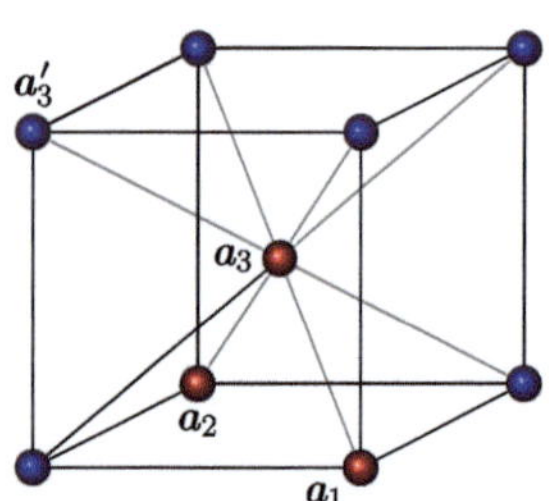


**Fig. 7.8** Primitive and unit cells

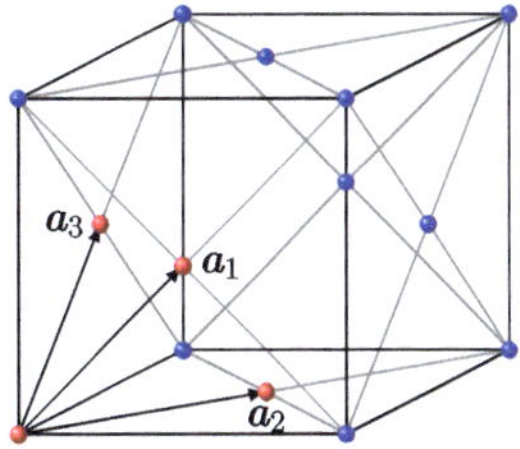

**Fig. 7.9** Primitive and unit cells

## Example: Face-centered Cubic (fcc) Lattice

If one chooses for the fcc lattice the primitive cell spanned by the $a_i$ in Fig. 7.9, then the cubic symmetry is hardly visible. If, on the other hand, one chooses the unit cell spanned by the equal-length and orthogonal vectors

$$a_1' = a\,e_1 = a_1 + a_2 - a_3,$$

$$a_2' = a\,e_2 = a_2 + a_3 - a_1,$$

$$a_3' = a\,e_3 = a_3 + a_1 - a_2.$$

then it is recognizable. The unit cell contains 4 atoms, whose locations are given according to (7.11). It has four times the volume of the primitive cell. ◀

The following table contains the three cubic Bravais lattices—the simple-cubic (sc), the body-centered cubic (bcc), and the face-centered cubic (fcc) lattice. The size of the unit cell is given in the last column. The positions of the $r$ atoms in the unit cell are determined by specifying the tuples $(s_1, s_2, s_3)$ in the expansion $s = \sum s_i\, a_i'$.

$$
\begin{array}{c|c|llll|c}
lattice & r & s_1 & s_2 & s_3 & s_4 & V_e \\
\hline
sc & 1 & (000) & & & & a^3 \\
bcc & 2 & (000) & (\tfrac{1}{2}\tfrac{1}{2}\tfrac{1}{2}) & & & a^3/2 \\
fcc & 4 & (000) & (0\tfrac{1}{2}\tfrac{1}{2}) & (\tfrac{1}{2}0\tfrac{1}{2}) & (\tfrac{1}{2}\tfrac{1}{2}0) & a^3/4
\end{array}
\tag{7.12}
$$

The sc Bravais lattice of the cubic crystal system with equal-length and perpendicular basis vectors has the highest-symmetry point group $\mathcal{O}_h$ as symmetry group.

## Example: The Orthorhombic System

As another example, we consider the orthorhombic crystal system with three unequal-length, perpendicular axes,

$$a_1 \neq a_2 \neq a_3 \quad , \quad \alpha = \beta = \gamma = 90^0.$$

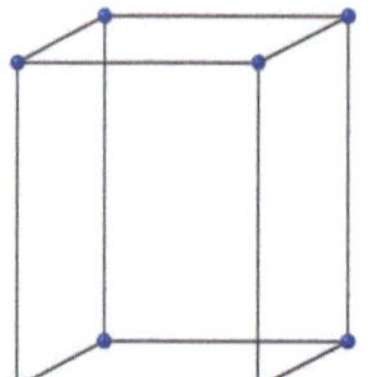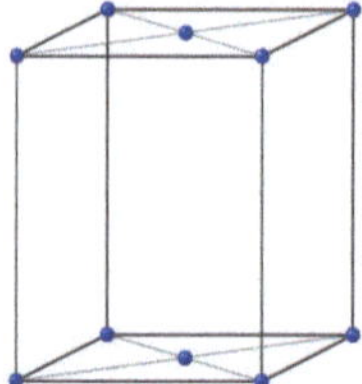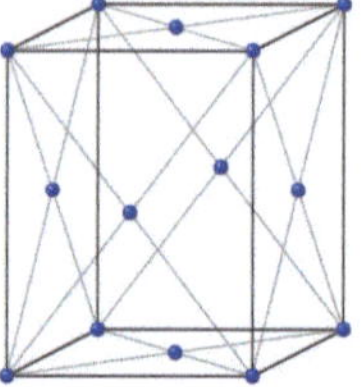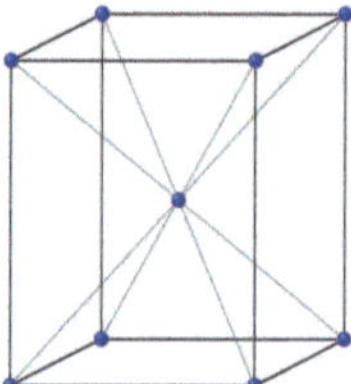

**Fig. 7.10**  There are 4 Bravais lattices in the orthorombic system

**Table 7.1**  The 7 crystal systems and 14 Bravais lattices. Only the orthorhombic system appears as a simple (P), base-centered (C), face-centered (F) and body-centered (I) Bravais lattice

| System | P | C | I | F | # |
|---|---|---|---|---|---|
| Triclinic | 1 | 0 | 0 | 0 | 1 |
| Monoclinic | 1 | 1 | 0 | 0 | 2 |
| Orthorhombic | 1 | 1 | 1 | 1 | 4 |
| Tetragonal | 1 | 0 | 1 | 0 | 2 |
| Trigonal | 1 | 0 | 0 | 0 | 1 |
| Hexagonal | 1 | 0 | 0 | 0 | 1 |
| Cubic | 1 | 0 | 1 | 1 | 3 |

A crystal belongs to this system if at least 2 twofold rotation axes or at least 2 symmetry planes are present. The four Bravais lattices shown in Fig. 7.10 exist. Note that a base-face-centered orthorhombic lattice also occurs.  ◄

Following the international Table of Crystallography, we denote primitive (or simple) lattices with P, face-centered with F, body-centered with I, and base-face-centered with C (instead of s, fc, bc, and bac). In this way, one finds the 14 Bravais lattices listed in Table 7.1. The triclinic, trigonal, and hexagonal systems each exist only as primitive lattices. To describe all possible structures of three-dimensional crystals, one needs 7 primitive and 7 centered Bravais lattices.

## 7.3  Space and Wallpaper Groups of Crystals

After the previous focus on translations, we now turn to all symmetries of a crystal in $d = 2$ or $d = 3$ dimensions. These symmetry transformations are those motions in Euclidean space, which map the crystal onto itself. They define a discrete subgroup of the Euclidean group $E_d$ introduced in Sect. 5.3.

**Definition 7.1 (Wallpaper and Space Group)**  The symmetry transformations of a crystal form a discrete subgroup $\mathcal{R} < E_d$ of the Euclidean group with multiplication (5.22). For two-dimensional crystals it is called wallpaper group (or space group in 2 dimensios) and for three-dimensional crystals space group.

### 7.3.1   Normal Forms of Rigid Motions

In Sect. 5.6 we dealt with the normal forms of rigid motions in Euclidean space, see Theorem 5.4. Every rigid motion in space is conjugate to one of these normal forms. To analyze the space groups in 2 and 3 dimensions, we recall these normal forms in these dimensions.

**Normal Forms in d=2**

In two dimensions, the normal forms given in Theorem 5.4 simplify because two proper rotations always commute and an improper rotation is conjugate to $\sigma_3$. In addition, every improper rotation has the eigenvalue 1, in contrast to the proper rotations, among which only $R = \mathbb{1}$ has the eigenvalue 1. It follows that

**Lemma 7.2** *For every rigid motion in* $\mathbb{R}^2$ *there is a Cartesian basis and an origin, with respect to which it takes one of the following normal forms:*

$$(\mathbb{1}, a\,e_1) \ \text{for } R = \mathbb{1}, \qquad (R, \mathbf{0}) \ \text{for } \det(R) = 1, \ R \neq \mathbb{1},$$

$$(\sigma_3, \mathbf{c}_{\parallel}) \ \text{for } \det(R) = -1 \,.$$

*with* $\sigma_3 \mathbf{c}_{\parallel} = \mathbf{c}_{\parallel}$.

For $\mathbf{c}_{\parallel} = \mathbf{0}$ the last motion is a reflection, otherwise a glide reflection with the $x$-axis as symmetry element. Every symmetry operation is thus a pure translation, a pure rotation, a reflection or a glide reflection.

**Normal Forms in d=3**

In three dimensions, every proper rotation has the eigenvalue 1, but an improper rotation only if $\text{Tr}\,R = 1$ is. It follows

**Lemma 7.3** *For every rigid motion in* $\mathbb{R}^3$ *there is a Cartesian basis and an origin, with respect to which it takes one of the following normal forms. For*

1. $\det(R) = 1$ *a screw rotation about the third axis:* $x'_{\perp} = R_2(\varphi)x_{\perp}$ *and* $z' = z + a$,
2. $\det(R) = -1$ *and* $\text{Tr}\,R < 1$ *a rotoinversion:* $x'_{\perp} = R_2(\varphi)x_{\perp}$ *and* $z' = -z$,
3. $\det(R) = -1$ *and* $\text{Tr}(R) = 1$ *a glide reflection:* $x'_{\perp} = x_{\perp} + a\,e_2$ *and* $z' = -z$.

*We chose the 3-axis and x-y-plane as symmetry elements and defined* $x_{\perp} = (x, y)^T$.

With the help of Lemmas (7.2) and (7.3) one proves the following corollary for proper rotations:

**Corollary 7.1** *Let $R \neq \mathbb{1}$ be a proper rotation with $R^n = \mathbb{1}$. Then*

$$\left(\mathbb{1} + R + \cdots + R^{n-1}\right)a = \begin{cases} 0, & d = 2, \\ n\,a_{\parallel}, & d = 3, \end{cases} \tag{7.13}$$

*where $a_{\parallel}$ is the component of $a$ in the direction of the rotation axis.*

The proof of these facts is the content of Problem 7.2.



Why is $R^2 = \mathbb{1}$ for arbitrary improper rotations in 2 dimensions, just as for improper rotations with $\operatorname{Tr} R = 1$ in 3 dimensions?

The counterpart of Corollary 7.1 for improper rotations (rotoinversion) is:

**Corollary 7.2** *For every improper rotation in 2 dimensions or in 3 dimensions with $\operatorname{Tr} R = 1$ we have $R^2 = \mathbb{1}$ and*

$$\left(\mathbb{1} + R\right)a = 2a_{\parallel}\,. \tag{7.14}$$

*Here, $a_{\parallel}$ is the projection of $a$ onto the mirror line or the mirror plane. If $R$ is a improper rotation in 3 dimensions with $\operatorname{Tr} R \neq 1$ and $R^n = \mathbb{1}$, then $(\mathbb{1} + R + \ldots R^{n-1})a = 0$.*

### 7.3.2   Wallpaper and Space Groups and Associated Point Groups

After these preparations, we now analyze wallpaper and space groups $\mathcal{R}$, consisting of those rigid motions $(R, a)$, which map the crystal onto itself, in detail. Because the group multiplication is inherited from the Euclidean group, the pure translations

$$T = \{(\mathbb{1}, a) \in \mathcal{R}\} \equiv (\mathbb{1}, \mathcal{L}) \tag{7.15}$$

form a *normal subgroup* $T \lhd \mathcal{R}$. The vectors in $\mathcal{L}$ are integer linear combinations of $d$ linearly independent basis vectors $a_i$, and define (after choosing the origin $0$) the Bravais lattice of the crystal. The cosets modulo this normal subgroup are (see Sect. 3.3)

$$(\mathbb{1}, \mathcal{L}), \ \ (R_2, c_2 + \mathcal{L}), \ \ (R_3, c_3 + \mathcal{L}), \ \ \ldots, (R_r, c_r + \mathcal{L})\,, \tag{7.16}$$

and as representatives of the classes we can choose the $(R_k, c_k) \in \mathcal{R}$. The countably infinite many rigid motions in a class thus differ only in their translational part. The conjugation of a translation with an arbitrary element $(a, R) \in \mathcal{R}$ yields

$$(R, a)(\mathbb{1}, b)(R, a)^{-1} = (\mathbb{1}, Rb) . \tag{7.17}$$

and we conclude:

**Lemma 7.4 (The Point Group Acts on $\mathcal{L}$)** *The rotations $R$ in $(R, a)$ map the Bravais lattice $\mathcal{L}$ onto itself, $R : \mathcal{L} \mapsto \mathcal{L}$.*

We denote basis vectors of the *unit cells* with $a_i$, and no longer with $a_i'$, as we will mainly study Bravais lattices in the following. For the vectors $c_k$ in the coset division (7.16) the following holds

**Lemma 7.5** *The $c_k$ can be chosen from the Wigner-Seitz cell with $0$ in the center and are then unique.*

**Proof** The set $c_k + \mathcal{L}$ always contains an element in the distinguished Wigner-Seitz cell, so that $c_k$ can be chosen from this set. If $c_k$ were not unique, then there would be a $c_k' \neq c_k$ in the same Wigner-Seitz cell with $(R_k, c_k')$ in $\mathcal{R}$. This would mean

$$(R_k, c_k)(R_k, c_k')^{-1} = (\mathbb{1}, c_k - c_k') \in \mathcal{R} \implies c_k - c_k' \in \mathcal{L} .$$

By assumption, $c_k$ and $c_k'$ are in the same Wigner-Seitz cell and therefore $c_k = c_k'$.
$\square$

In addition to the translations, the point groups of wallpaper and space groups play an important role. However, such a point group need not be the subgroup $\{(R, 0) \in \mathcal{R}\}$ of pure rotations, as one might suspect:

**Definition 7.2 (Point Group $\mathcal{R}_0$)** The point group $\mathcal{R}_0$ of a wallpaper or space group $\mathcal{R}$ contains all elements of $\mathcal{R}$, where the translation is set to zero,

$$\mathcal{R}_0 = \{(R, 0)|\, (R, a) \in \mathcal{R}\} \tag{7.18}$$

---

**Question**

Why does $\mathcal{R}_0$ not need to be a subgroup of $\mathcal{R}$?

The answer is that $\mathcal{R}$ may contain composite symmetry operations $(R, a)$—these are screw rotations or glide reflections—for which $(R, 0)$ is not in $\mathcal{R}$, but by definition is an element in $\mathcal{R}_0$. Accordingly, we distinguish between symmorphic and non-symmorphic wallpaper and space groups:

**Definition 7.3 (Symmorphic Wallpaper and Space Groups)**   $\mathcal{R}$ is called symmorphic if every symmetry operation can be represented as a product of a translation $(\mathbb{1}, a) \in \mathcal{R}$ and a rotation $(R, 0) \in \mathcal{R}_0$

In a symmorphic space group the point group of the crystal $\mathcal{R}_0$ is a subgroup of $\mathcal{R}$ and accordingly $\mathcal{R}$ is the semi-direct product of the normal subgroup of pure translations with the point group, $\mathcal{R} = T \rtimes \mathcal{R}_0$ . If a space group has neither glide reflections nor screw rotations, then it is symmorphic.

**Lemma 7.6**  *If $\mathcal{R}$ is a space group with translation group $T$, then $\mathcal{R}_0 \cong \mathcal{R}/T$.*

**Proof**  The mapping $\varphi : (R, a) \mapsto (R, 0)$ is, as one can easily verify, a surjective homomorphism from $\mathcal{R}$ to $\mathcal{R}_0$ with the trivial kernel $\{(\mathbb{1}, a)\} = T$. This proves the lemma.                                                                                      $\square$

**Lemma 7.7**  *Let $\mathcal{R}$ be a space group with translation group $T$. Let furthermore for $n \in \mathbb{N}$*

$$\mathcal{R}_n = \left\{ (R, a) \in \mathcal{R} | (R, a)(R_1, a_1)^n = (R_1, a_1)^n (R, a) \text{ for all } (R_1, a_1) \in \mathcal{R} \right\}. \tag{7.19}$$

*If $n$ is a multiple of the order of the point group $\mathcal{R}_0$, then $\mathcal{R}_n = T$. If in addition $\varphi : \mathcal{R} \mapsto \mathcal{R}'$ is an isomorphism of space groups with translation groups $T$ and $T'$, then $\varphi(T) = T'$:*

**Proof**  If $r$ is the order of the cyclic subgroup generated by $(R_1, 0)$ in $\mathcal{R}_0$, then $(R_1, a_1)^r = (\mathbb{1}, \dots)$ is a translation in $T$. By Lagrange's theorem, $n_0 = |\mathcal{R}_0|$ is a multiple of $r$ and thus $(R_1, a_1)^{n_0}$ is also a translation. Because $T$ is abelian, all translations then commute with all $(R_1, a_1)^{n_0}$ and thus lie in $\mathcal{R}_{n_0}$. This statement also holds for all multiples of $n_0$. Conversely, if we conjugate the $n$th power of any shift $(\mathbb{1}, b) \in T$ with $(R, a) \in \mathcal{R}_n$, then it follows

$$(R, a)(\mathbb{1}, b)^n (R, a)^{-1} = (R, a)(\mathbb{1}, nb)(R, a)^{-1} = (\mathbb{1}, nRb) \stackrel{(7.19)}{=} (\mathbb{1}, nb), \tag{7.20}$$

which implies $Rb = b$ for all $b \in \mathcal{L}$ and thus $(R, a) = (\mathbb{1}, a) \in T$. This then proves $\mathcal{R}_n = T$.                                                                                      $\square$

---

**Task**

Prove that for an isomorphism $\varphi : \mathcal{R} \mapsto \mathcal{R}'$ holds: $\varphi(\mathcal{R}_n) = \mathcal{R}'_n$ for all $n = 1, 2, \dots$.

Now let $n_0$ and $n'_0$ be the orders of the point groups $\mathcal{R}_0$ and $\mathcal{R}'_0$. Since $n = n_0 n'_0$ is a multiple of $n_0$ and of $n'_0$, we conclude, according to the first part of the lemma, that $T = \mathcal{R}_n$ and $T' = \mathcal{R}'_n$. Because $\varphi(\mathcal{R}_n) = \mathcal{R}'_n$ then also $\varphi(T) = T'$, which was to be shown.

**Lemma 7.8 (Isomorphic $\mathcal{R}$ have Isomorphic $\mathcal{R}_0$)** *If $\mathcal{R}$ and $\mathcal{R}'$ are isomorphic, then their point groups $\mathcal{R}_0$ and $\mathcal{R}'_0$ are isomorphic.*

***Proof*** For an isomorphism $\varphi : \mathcal{R} \mapsto \mathcal{R}'$ according to Lemma 7.7 also $\varphi(T) = T'$. Thus, $\varphi$ induces an isomorphism between $\mathcal{R}/T$ and $\mathcal{R}'/T'$. Because these factor groups according to Lemma 7.6 are isomorphic to the point groups, the isomorphism of $\mathcal{R}_0$ and $\mathcal{R}'_0$ follows. $\qquad\qquad\square$

To check whether two wallpaper or space groups $\mathcal{R}$ and $\mathcal{R}'$ with point groups $\mathcal{R}_0$ and $\mathcal{R}'_0$ are isomorphic, the following lemma is useful.

**Lemma 7.9** *Let $\varphi : \mathcal{R} \mapsto \mathcal{R}'$ be an isomorphism and $a_i$ and $a'_i$ integral bases of the translation lattices $\mathcal{L}$ and $\mathcal{L}'$. Then the restriction of $\varphi$ to $T$ is given by a linear mapping $\varphi((\mathbb{1}, a)) = (\mathbb{1}, Ga)$. For integral bases, the matrix $(G_{ji})$ in $Ga_i = \sum_j a'_j G_{ji}$ is in $GL(d, \mathbb{Z})$. The induced isomorphism $\bar{\varphi} : \mathcal{R}_0 \mapsto \mathcal{R}'_0$ is the conjugation with $G$.*

An integral basis $a_i$ has the property that for all $a \in \mathcal{L}$ the coefficients $m_i$ in the expansion $a = \sum m_i a_i$ are integers.

***Proof*** The restriction of the isomorphism to $T$ defines a linear isomorphism $\mathcal{L} \mapsto \mathcal{L}'$. $\qquad\qquad\square$

---



Why does the isomorphism property of $\varphi$ imply the linearity of the restriction?

If $a_i$ and $a'_i$ are integral bases of $\mathcal{L}$ and $\mathcal{L}'$, then it must hold

$$\varphi : \; a_i \mapsto \sum_j a'_j \, G_{ji} \tag{7.21}$$

with integer coefficients $G_{ji}$. Because the restriction of $\varphi$ is linear, an arbitrary vector $a = \sum n_i a_i \in \mathcal{L}$ is mapped to the vector $Ga \equiv \sum n'_i a'_i \in \mathcal{L}'$ with $n'_i = \sum_j G_{ij} n_j$. If one repeats the argument for the inverse isomorphism $\varphi^{-1}$, then one sees that $G^{-1}$ also has integer matrix elements and therefore lies in $GL(d, \mathbb{Z})$. According to Theorem 2.2, the invertible elements in $GL(d, \mathbb{Z})$ have the determinant $\pm 1$.

To prove the second statement in the lemma, we conjugate a translation $(\mathbb{1}, b) \in T$ with an arbitrary element $(R, a)$ of the wallpaper or space group, which is mapped by $\varphi$ into $(R', a')$. Because $\varphi : (\mathbb{1}, b) = (\mathbb{1}, Gb)$ is the image of the conjugated element

$$\varphi\big((R, a)(\mathbb{1}, b)(R, a)^{-1}\big) = (R', a')(\mathbb{1}, Gb)(R', a')^{-1}. \qquad (7.22)$$

The left side is $\varphi((\mathbb{1}, Rb))$ and the right side is $(\mathbb{1}, R'Gb)$. With the already proven first statement in the lemma, it then follows

$$(\mathbb{1}, R'Gb) = \varphi((\mathbb{1}, Rb)) = (\mathbb{1}, GRb). \qquad (7.23)$$

The relation $R'Gb = GRb$ means, however, that the induced mapping $\mathcal{R}_0 \mapsto \mathcal{R}'_0$ by $\varphi$ is a conjugation with $G$, $R' = GRG^{-1}$.

Point groups are finite subgroups of the orthogonal groups O(2) or O(3). The orders of the cyclic subgroups generated by individual rotations must be less than or equal to the order $n_0$ of $\mathcal{R}_0$. Indeed, the orders are significantly more restricted:

**Theorem 7.1 (Point Groups of Crystals)** *Proper rotations that transform a space lattice $\mathcal{L}$ into itself only contain rotation axes of order 2, 3, 4 and 6.*

In 1984, an unusual quasicrystal was found that behaves in many experiments like a crystal with a fivefold symmetry.

*Proof* We consider an $n$−fold rotation $c_n(e)$ of the lattice around the rotation axis $e$. There are always lattice vectors that are perpendicular to $e$: If $a$ is a lattice vector, which is not parallel to $e$, then the lattice vector $c_n(e)a - a$ is not equal to zero and perpendicular to $e$:

$$(e, c_n(e)a - a) = (c_n(e)e, c_n(e)a) - (e, a) = 0.$$

We consider the shortest lattice vector $a$ perpendicular to $e$. Any lattice vector $a'$ parallel to $a$ must be an integer multiple of $a$. If $a' = \lambda a$ with a non-integer $\lambda > 0$, then one could construct the vector $a' - [\lambda]a$ parallel to $a$, which is shorter than $a$, contradicting our assumption. In particular, for a lattice vector perpendicular to an $n$-fold rotation axis, it holds

$$c_n(e)a + c_n^{-1}(e)a = 2\cos\varphi_n\, a, \qquad \varphi_n = \frac{2\pi}{n},$$

and this is sketched in Fig. 7.11. The lattice vector must be an integer multiple $m$ of $a$, and since the cosine function takes values between $-1$ and $1$, we have

$$-2 \leq 2\cos\varphi_n = m \leq 2, \qquad m \in \mathbb{Z}.$$

$\square$

**Fig. 7.11** On the allowed rotations of a three-dimensional lattice

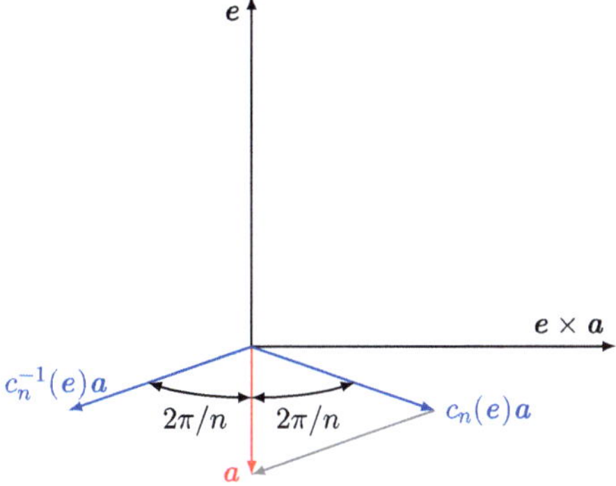


The allowed values for $m$ with corresponding rotation angle and order of the rotation axis are listed in the following table:

| $m$ | 2 | 1 | 0 | $-1$ | $-2$ |
|---|---|---|---|---|---|
| rotation angle | 0 | $\pi/3$ | $\pi/2$ | $2\pi/3$ | $\pi$ |
| order | 1 | 6 | 4 | 3 | 2 |

**Rotoinversions**

Exactly the same arguments also apply to improper rotations, and therefore point groups of crystals may only contain rotoinversions $s_n$ with $n = 1, 2, 3, 4$ and 6. These results apply in two and three dimensions.

After the restriction on the order of rotations and rotoinversions in the crystal, we can further characterize the unique vectors $c_k$ in (7.15) located in a fixed Wigner-Seitz cell.

**Lemma 7.10** *If $(R, c)$ is a screw operation in 3 dimensions, then the projection $c_\parallel$ of $c$ onto the screw axis satisfies the quantization condition $n c_\parallel \in \mathcal{L}$ with $n \in \{2, 3, 4, 6\}$. If $(R, c)$ is a glide reflection with $\mathrm{Tr}\, R = 1$, then the projection onto the mirror plane satisfies the condition $2c_\parallel \in \mathcal{L}$. If $(R, c)$ is a glide reflection in 2 dimensions, then the projection onto the mirror line satisfies the condition $2c_\parallel \in \mathcal{L}$. In 2 dimensions there are no screw operations.*

**Proof** If $\{R, c + \mathcal{L}\}$ with $R \neq \mathbb{1}$ is one of the equivalence classes in (7.15) then $R^n = \mathbb{1}$ for a $n \in \{1, 2, 3, 4, 6\}$ and correspondingly

$$(R, c)^n = (\mathbb{1}, c + Rc + \ldots R^{n-1}c) \implies (\mathbb{1} + R + \cdots + R^{n-1})c \in \mathcal{L}. \quad (7.24)$$

After Corollary 7.1 for a proper $R$ in 3 dimensions $n c_\parallel \in \mathcal{L}$. After Lemma 7.3 $c_\parallel$ cannot be removed by a change of the Cartesian basis and the origin. For this intrinsic part it holds $n c_\parallel \in \mathcal{L}$. Thus, for a screw operation $(R, c)$ in 3 dimensions $n c_\parallel$ with $n \in \{2, 3, 4, 6\}$ is a vector of the Bravais lattice $\mathcal{L}$. If, on the other hand, $R$ is any rotoinversion in 2 dimensions or a rotoinversion with $\mathrm{Tr}\, R = 1$ in 3 dimensions, then $(R, c)$ is a glide reflection with $2 c_\parallel$ in $\mathcal{L}$. $\qquad\square$

## 7.4    Two-Dimensional Crystals

According to Theorem 7.1 (and the corresponding result for improper rotations), the point groups of wallpaper groups are isomorphic to one of the following 10 groups:

$$
\begin{aligned}
\text{proper:} \quad & \mathcal{C}_1 \;\; \mathcal{C}_2 \;\; \mathcal{C}_3 \;\; \mathcal{C}_4 \;\; \mathcal{C}_6 \,, \\
\text{improper:} \quad & \mathcal{C}_{1v} \;\; \mathcal{C}_{2v} \;\; \mathcal{C}_{3v} \;\; \mathcal{C}_{4v} \;\; \mathcal{C}_{6v} \,.
\end{aligned}
\tag{7.25}
$$

In two dimensions, the $\mathcal{C}_{nv}$ are isomorphic to the dihedral groups $\mathcal{D}_n$, which were discussed in detail in Sect. 2.3. Therefore, in the literature, the improper point groups are also referred to as $\mathcal{D}_1$, $\mathcal{D}_2$, $\mathcal{D}_3$, $\mathcal{D}_4$ and $\mathcal{D}_6$. But with regard to the comparison with point groups in 3-dimensions, where the dihedral groups are proper, the notation in (7.25) is more appropriate. Next we construct the lattices with point groups $\mathcal{R}_0$ isomorphic to $\mathcal{C}_{1v}$ and $\mathcal{C}_{2v}$ explicitly (see also [34,35]). Both point groups contain a reflection $(\sigma, \mathbf{0})$.

### 7.4.1    Construction of Lattices with Point Groups $\mathcal{C}_{1v}$

We choose a vector $a \in \mathcal{L}$, which is not parallel to the reflection line. Then, according to Lemma 7.4, $\sigma a \in \mathcal{L}$. Since $a \pm \sigma a \in \mathcal{L}$ are parallel and perpendicular to the reflection line (see Fig. 7.12), $\mathcal{L}$ always contains non-vanishing vectors parallel and perpendicular to the reflection line. Let $a_1$ and $a_2$ be vectors of minimal length parallel and perpendicular to the reflection line, respectively. Then every vector parallel or perpendicular to the reflection line is an integer multiple of $a_1$ or $a_2$, respectively, so that $a + \sigma a = m_1 a_1$ and $a - \sigma a = m_2 a_2$ with $m_i \in \mathbb{Z}$.

Adding the two equation leads to $a = \frac{1}{2}(m_1 a_1 + m_2 a_2)$. If for each $a$ the $m_i$ are even numbers, then the $a_i$ form a basis of lattice. But if $m_1$ or $m_2$ is odd, then

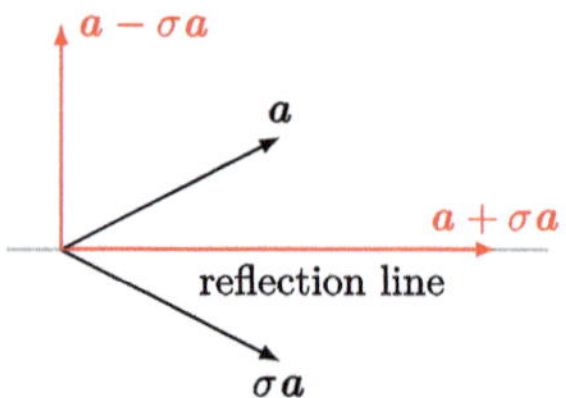


**Fig. 7.12** Construction of a lattice with point group $\mathcal{C}_{1v}$

both must be odd. Otherwise, $\frac{1}{2}a_1$ or $\frac{1}{2}a_2$ would be in $\mathcal{L}$, which is not possible by assumption. In the second case of odd $m_1$ and $m_2$ we set

$$c_1 = \frac{1}{2}(a_1 + a_2) \quad \text{and} \quad c_2 = \frac{1}{2}(a_1 - a_2) = \sigma\, c_1 \,.$$

Expressed through the new basis we have

$$a = m_1' c_1 + m_2' c_2 \quad m_1' = \frac{m_1 + m_2}{2} \in \mathbb{Z}, \quad m_2' = \frac{m_1 - m_2}{2} \in \mathbb{Z} \,.$$

This shows that the $c_i$ form an integral basis of $\mathcal{L}$. We conclude that

- either there are basis vectors $a_1 \perp a_2$, where $a_1$ points in the direction of the axis of reflection,
- or there are basis vectors $c_1$, $c_2$ of equal length, and the reflection swaps the two.

In the first case, it is a rectangular lattice and in the second case, it is a rhombic or rectangular centered lattice, as can be seen in the Fig. 7.13.

Under the reflection, $(a_1, a_2)$ is mapped to $(a_1, -a_2)$ and $(c_1, c_2)$ to $(c_2, c_1)$. Thus, in the rectangular and rectangular-centered lattice the following matrices are assigned to $\sigma$:

$$\sigma \mapsto R = \sigma_3, \quad \text{or} \quad \sigma \mapsto R' = \sigma_1, \tag{7.26}$$

where $\sigma_i$ are the Pauli matrices. If the wallpaper groups of the two lattices were isomorphic, then according to Lemma 7.9 a $G \in \mathrm{GL}(2, \mathbb{Z})$ would exist with

$$G\sigma_3 = \sigma_1 G \implies G = \begin{pmatrix} a & -b \\ a & b \end{pmatrix} \,.$$

The determinant $\det G = 2ab$ cannot be $\pm 1$, because $a, b$ must be integers. Therefore, the two wallpaper groups are not isomorphic. The wallpaper group for the rectangular-centered lattice contains glide reflections, e.g. the glide reflection $(\sigma_1, c_1)$ along the reflection line shifted parallel by $\frac{1}{2}c_1$. For the rectangular lattice, there are no glide reflection lines away from the reflection line.

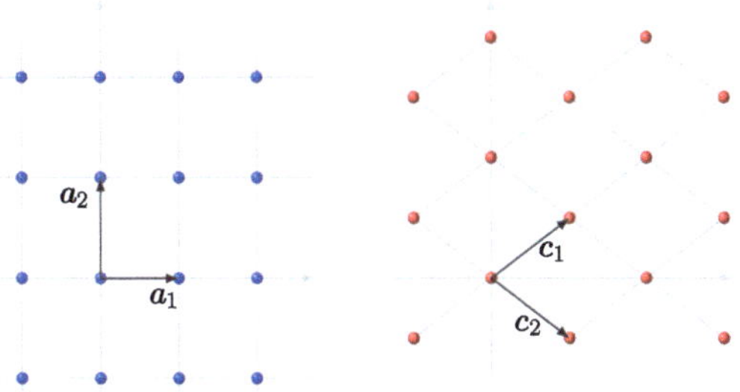


**Fig. 7.13** Construction of lattice with point group $\mathcal{C}_{1v}$

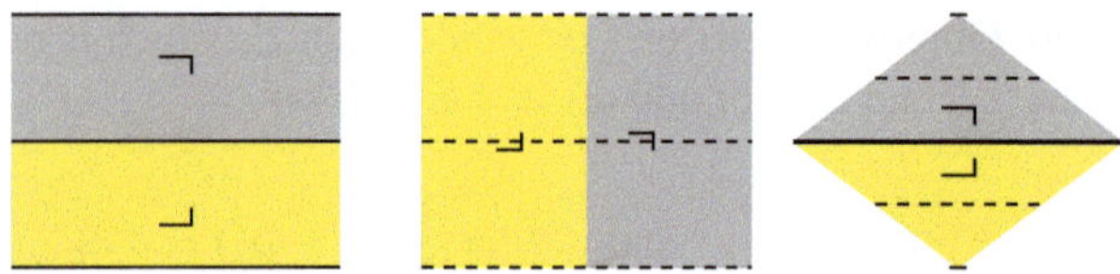

**Fig. 7.14** Wallpaper diagrams for the groups pm (left), pg (middle) and cm (right)

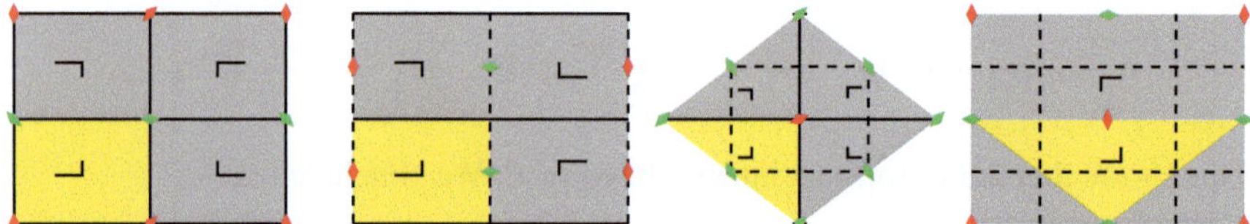

**Fig. 7.15** From left to right, the cells of the non-isomorphic wallpaper groups $pmm$, $pmg$, $cmm$ and $pgg$ are shown. Of these, the groups $pmm$ and $cmm$ are symmorphic

In Wikipedia, the 17 wallpaper groups are graphically represented,[2] where reflection lines are indicated with solid lines, glide reflection lines with dashed lines and centers of $n$-fold rotations with special symbols. The diagrams of the $pm$ and $pg$ groups for the rectangular and the diagram of the $cm$ group (right) for the rectangular-centered lattice have the shapes shown in Fig. 7.14. The $pg$ group with the middle diagram is non-symmorphic. The names of the wallpaper groups can be found, for example, in [12].

### 7.4.2    Construction of Lattices with Point Group $\mathcal{C}_{2v}$

The point group $\mathcal{C}_{2v}$ contains an additional rotation around $\pi$, and accordingly the point groups of the rectangular or rectangular-centered lattice are generated by

$$\{\sigma_3, -\sigma_0\} \quad \text{or} \quad \{\sigma_1, -\sigma_0\} \tag{7.27}$$

In the graphical representation, the twofold rotations are marked by the diamonds. The fundamental domain of the point group—this is the yellow area in Fig. 7.15—is only a quarter of the cell.

### 7.4.3    Point Groups and Bravais Lattices in 2 Dimensions

Similar analyses of the other proper and improper point groups in (7.25) then yield the assignments of the point groups to the Bravais lattices given in Table 7.2.

---

[2] See https://commons.wikimedia.org/wiki/Wallpaper_group_diagrams.

**Table 7.2** The ten crystallographic point groups in 2 dimensions

| Lattice | Oblique | Rectangular | Rect. centered | Square | Hexagonal |
|---|---|---|---|---|---|
| Point group | $\mathcal{C}_1, \mathcal{C}_2$ | $\mathcal{C}_{1v}, \mathcal{C}_{2v}$ | $\mathcal{C}_{1v}, \mathcal{C}_{2v}$ | $\mathcal{C}_4, \mathcal{C}_{4v}$ | $\mathcal{C}_3, \mathcal{C}_{3v}, \mathcal{C}_6, \mathcal{C}_{6v}$ |

**Table 7.3** The 32 crystalline point groups in three dimensions. The 10 red marked groups also occur in two dimensions

| point groups | crystalline point groups | | | | | Number |
|---|---|---|---|---|---|---|
| $\mathcal{C}_n$-series | $\mathcal{C}_1$ | $\mathcal{C}_2$ | $\mathcal{C}_3$ | $\mathcal{C}_4$ | $\mathcal{C}_6$ | 5 |
| $\mathcal{S}_n$-series | | $\mathcal{S}_2$ | | $\mathcal{S}_4$ | $\mathcal{S}_6$ | 3 |
| $\mathcal{D}_n$-series | | $\mathcal{D}_2$ | $\mathcal{D}_3$ | $\mathcal{D}_4$ | $\mathcal{D}_6$ | 4 |
| $\mathcal{C}_{nh}$-series | | $\mathcal{C}_{2h}$ | $\mathcal{C}_{3h}$ | $\mathcal{C}_{4h}$ | $\mathcal{C}_{6h}$ | 4 |
| $\mathcal{C}_{nv}$-series | $\mathcal{C}_{1v}$ | $\mathcal{C}_{2v}$ | $\mathcal{C}_{3v}$ | $\mathcal{C}_{4v}$ | $\mathcal{C}_{6v}$ | 5 |
| $\mathcal{D}_{nh}$-series | | $\mathcal{D}_{2h}$ | $\mathcal{D}_{3h}$ | $\mathcal{D}_{4h}$ | $\mathcal{D}_{6h}$ | 4 |
| $\mathcal{D}_{nd}$-series | | $\mathcal{D}_{2d}$ | $\mathcal{D}_{3d}$ | | | 2 |
| Platonic | $\mathcal{T}$ | $\mathcal{T}_h$ | $\mathcal{T}_d$ | $\mathcal{O}$ | $\mathcal{O}_h$ | 5 |

For the role of wallpaper groups in mathematics and art, and the question of how to apply symmetries to create impressive patterns, consult the book [36]. It also explains why there are 17 basic types of wallpaper. The graphical encoding of the wallpaper groups used in the above examples is also explained in more detail.

## 7.5 Three-Dimensional Crystals

Theorem 7.1 severely restricts the crystalline point groups, since only rotations and rotoinversion with 1, 2, 3, 4 and sixfold axes are allowed. According to the results in Sect. 6.1.2. only $\mathcal{C}_n$, $\mathcal{D}_n$, $\mathcal{T}$ and $\mathcal{O}$ are to be considered as proper point groups in three dimensions and $\mathcal{C}_{nh}$, $\mathcal{C}_{nv}$, $\mathcal{S}_{2n}$, $\mathcal{D}_{nh}$, $\mathcal{D}_{nd}$, $\mathcal{T}_h$, $\mathcal{T}_d$ and $\mathcal{O}_h$, as improper point groups. The allowed groups are listed in Table 7.3. In this table, the icosahedral groups $\mathcal{Y}_*$ (the star stands for $h, v, d$) and the dihedral groups $\mathcal{D}_{5*}$ do not appear because they have fivefold rotation axes. Also, the groups $\mathcal{D}_{4d}$ and $\mathcal{D}_{6d}$ are absent, as they contain the subgroups $\mathcal{S}_8$ and $\mathcal{S}_{12}$, respectively. Together with $\mathcal{S}_3 = \mathcal{C}_{3h}$, $\mathcal{D}_1 = \mathcal{C}_2$, $\mathcal{C}_{1h} = \mathcal{C}_{1v}$ and $\mathcal{D}_{1d} = \mathcal{C}_{2h}$ this explains the gaps in Table 7.3, which contains all 32 crystalline point groups in three dimensions. The allowed symmetries are all realized in real crystals. 10 of the crystalline point groups also occur in two dimensions and are written in red.

The subsequent Table 7.4 compares the number of crystal systems, Bravais lattices, point groups, and space groups in two and three dimensions. The number of space groups increases significantly with increasing dimension. For example, in four dimensions there are already 4895 "space groups". While in 2 dimensions the symmorphic wallpaper groups dominate, in 3 dimensions the non-symmorphic space groups prevail. The highest symmetry point group of a crystal system is called a holohedry and the corresponding crystal is called a holohedron. Its shape therefore has the highest possible number of crystal faces, and all possible symmetry elements

**Table 7.4**  Comparison of crystallographies in 2 and 3 dimensions

|         | Crystal-systems | Bravais-lattices | Point-groups | Space-groups | Symmorphic | Non-symmorphic |
|---------|-----------------|------------------|--------------|--------------|------------|----------------|
| $d = 2$ | 4               | 5                | 10           | 17           | 13         | 4              |
| $d = 3$ | 7               | 14               | 32           | 230          | 73         | 157            |

**Table 7.5**  How the 14 Bravais lattices, 32 crystalline point groups and 230 space groups are divided among the 7 crystal systems

| Systems      | Holohedry       | # Bravais lattices | # Point groups | Space group No.        |
|--------------|-----------------|--------------------|----------------|------------------------|
| Triclinic    | $\mathcal{S}_2$ | 1                  | 2              | $1, 2$                 |
| Monoclinic   | $\mathcal{C}_{2h}$ | 2               | 3              | $3, 4, \ldots, 15$     |
| Orthorhombic | $\mathcal{D}_{2h}$ | 4               | 3              | $16, 17, \ldots, 74$   |
| Tetragonal   | $\mathcal{D}_{4h}$ | 2               | 7              | $75, 76, \ldots, 142$  |
| Trigonal     | $\mathcal{D}_{3d}$ | 1               | 5              | $143, 144, \ldots, 167$ |
| Hexagonal    | $\mathcal{D}_{6h}$ | 1               | 7              | $168, 169, \ldots, 194$ |
| Cubic        | $\mathcal{O}_h$ | 3                  | 5              | $195, 196, \ldots, 230$ |

in its crystal system are present. In addition to the properties of the crystalline point groups derived above, the following facts are helpful in the classification of holohedries:

1. Every holohedry contains the inversion $i$, since with $a$ also $-a$ is a lattice vector.
2. If a system is invariant under $\mathcal{C}_n$ with $n > 2$, then it is also invariant under $\mathcal{C}_{nv}$.
3. Every 3, 4 or sixfold rotation axis contains a mirror plane through this axis.

**Point Groups of Crystals (Crystal Systems)**
If you scan the point groups according to these criteria, then exactly seven holohedries remain, and these are listed in the second column in Table 7.5.

Each of the seven crystal systems can contain several crystal classes, and each of the 32 classes is characterized by its crystallographic point group $\mathcal{R}_0$. For example, the cubic crystal system has the holohedry $\mathcal{O}_h$ and appears in the 5 classes $\mathcal{T}, \mathcal{T}_h, \mathcal{T}_d, \mathcal{O}$ and $\mathcal{O}_h$. The hexagonal system has the holohedry $\mathcal{D}_{6h}$ and contains the 7 classes $\mathcal{C}_{3h}, \mathcal{C}_6, \mathcal{C}_{6h}, \mathcal{C}_{6v}, \mathcal{D}_{3h}, \mathcal{D}_6$ and $\mathcal{D}_{6h}$. The 7 crystal systems with holohedry, number of Bravais lattices and point groups for each system and the assignment of the 230 space groups are listed in Table 7.5.[3]

---

[3] More detailed information can be found e.g. on the page http://de.wikipedia.org/wiki/Punktgruppe.

## 7.6 The 7 Crystal Systems

In this section, we look more closely at the 7 crystal systems and discuss their geometry, holohedries and point groups. We refer to the geometric relationships of the crystal systems or primitive Bravais lattices in (7.10).

**The Cubic Crystal System** $\mathcal{O}_h$ with three equal-length, perpendicular axes is the system with the highest symmetry. There are 3 Bravais lattices and 5 cubic crystal classes and these are listed in Table 7.6. The table contains for each class the number of rotation axes for each of the possible multiplicities and the number of mirror planes. Only the classes $\mathcal{T}_h$ and $\mathcal{O}_h$ contain the inversion. Examples of cubic systems are rock salt CaCl, fluorite $CaF_2$ and zinc blende ZnS.

**The Hexagonal Crystal System** $\mathcal{D}_{6h}$ has an axial cross with two equal-length axes intersecting at an angle of 120°, as well as a perpendicular axis that is longer or shorter. A crystal is hexagonal, if a sixfold rotation or rotation inversion axis exists. There is only one primitive hexagonal Bravais lattice, but there are 7 crystal classes, which are listed in Table 7.7. Hexagonal systems are found in mercury oxide, zinc, magnesium and graphite.

**The Trigonal Crystal System** $\mathcal{D}_{3d}$ has an axial cross with equal-length axes and equal angles. Only the primitive Bravais lattice is found in the 5 classes listed in Table 7.8—all contain a threefold rotation axis. Trigonal systems form calcite $CaCO_3$, quartz $SiO_2$ and sodium nitrate $NaNO_3$.

**The Tetragonal Crystal System** $\mathcal{D}_{4h}$ has an axial cross with two equal-length axes and one axis that is longer or shorter than the other two. A crystal is tetragonal if

**Table 7.6** The three cubic Bravais lattices P, I, B break down into five classes, each with four threefold rotation axes

| $G$ | $\mathcal{T}$ | $\mathcal{T}_h$ | $\mathcal{O}$ | $\mathcal{T}_d$ | $\mathcal{O}_h$ |
|---|---|---|---|---|---|
| $c_2$ | 3 | 4 | 6 | 0 | 6 |
| $c_3$ | 4 | 4 | 4 | 4 | 4 |
| $c_4$ | 0 | 0 | 3 | 3 | 3 |
| mirror planes | 0 | 3 | 0 | 6 | 9 |
| inversion center | 0 | 1 | 0 | 0 | 1 |

**Table 7.7** The hexagonal Bravais lattice P appears in seven crystal classes

| $G$ | $\mathcal{C}_6$ | $\mathcal{C}_{6h}$ | $\mathcal{D}_6$ | $\mathcal{D}_{6h}$ | $\mathcal{C}_{6v}$ | $\mathcal{D}_{3h}$ | $\mathcal{C}_{3h}$ |
|---|---|---|---|---|---|---|---|
| $c_2$ | 0 | 0 | 6 | 6 | 0 | 3 | 0 |
| $c_3$ | 0 | 0 | 0 | 0 | 0 | 0 | 1 |
| $c_6$ | 1 | 1 | 1 | 1 | 1 | 1 | 0 |
| mirror planes | 0 | 1 | 0 | 7 | 6 | 3 | 1 |
| inversion center | 0 | 1 | 0 | 1 | 0 | 0 | 0 |

**Table 7.8** The trigonal Bravais lattice P breaks down into five crystal classes

| $G$ | $\mathcal{C}_3$ | $\mathcal{C}_{3h}$ | $\mathcal{D}_3$ | $\mathcal{C}_{3v}$ | $\mathcal{D}_{3d}$ |
|---|---|---|---|---|---|
| $c_2$ | 0 | 0 | 3 | 0 | 3 |
| $c_3$ | 1 | 1 | 1 | 1 | 1 |
| mirror planes | 0 | 1 | 0 | 3 | 3 |
| inversion center | 0 | 0 | 0 | 0 | 1 |

**Table 7.9** The tetragonal Bravais lattices P and I break down into seven crystal classes

| $G$ | $\mathcal{C}_4$ | $\mathcal{S}_4$ | $\mathcal{C}_{4h}$ | $\mathcal{D}_4$ | $\mathcal{C}_{4v}$ | $\mathcal{D}_{2d}$ | $\mathcal{D}_{4h}$ |
|---|---|---|---|---|---|---|---|
| $c_2$ | 0 | 0 | 0 | 4 | 0 | 2 | 4 |
| $c_4$ or $s_4$ | 1 | 1 | 1 | 1 | 1 | 1 | 1 |
| mirror planes | 0 | 0 | 1 | 0 | 4 | 2 | 5 |
| inversion center | 0 | 0 | 0 | 0 | 0 | 0 | 1 |

**Table 7.10** Orthorhombic crystal classes have no three, four or sixfold rotation axes. The Bravais lattices P, C, I and F exist

| $G$ | $\mathcal{D}_2$ | $\mathcal{C}_{2v}$ | $\mathcal{D}_{2h}$ |
|---|---|---|---|
| $c_2$ | 3 | 1 | 3 |
| Mirror planes | 0 | 2 | 3 |
| Inversion center | 0 | 1 | 1 |

there is a single fourfold symmetry axis. There must not be a threefold axis. The seven classes are listed in Table 7.9. Tin dioxide $SnO_2$, titanium dioxide $TiO_2$ and lead tungstate $PbWO_4$ form tetragonal systems.

**The Orthorhombic Crystal System** $\mathcal{D}_{2h}$ has three unequal-length axes, perpendicular to each other. It has at least 3 perpendicular twofold rotation axes or mirror planes, but no other rotation axes. The classes of this system are listed in Table 7.10. It is realized in potassium nitrate $KNO_3$, barium sulfate $BaSO_4$ and potassium sulfate $K_2SO_4$.

**The Monoclinic Crystal System** $\mathcal{C}_{2h}$ has three unequal-length axes, two of which intersect at an oblique angle, while the third is perpendicular to these. There is only one twofold rotation axis and/or a plane of symmetry. The three classes are listed in Table 7.11. Examples are Gypsum $CaSO_4 \cdot 2H_2O$ and Borax $Na_2B_4O_7 \cdot 10H_2O$.

**The Triclinic Crystal System** $\mathcal{S}_2$ has 3 unequal axes, which all intersect at oblique angles. It has the smallest symmetry of all systems. A crystal is triclinic if neither symmetry planes nor rotation axes are present. The two classes are listed in Table 7.12. There is only the simple Bravais lattice. Examples of triclinic systems are copper sulfate $CuSO_4 \cdot 5H_2O$ and potassium dichromate $K_2Cr_2O_7$.

The Table 7.5 at the beginning of the section contains, in addition to the number of Bravais lattices and crystal classes in the last column, the assignment of the 230 crystalline space groups to the crystal systems. For example, the cubic

**Table 7.11** The monoclinic crystal classes are characterized by exactly one axis with symmetry $\neq 1$. There are the Bravais lattices P and C

| $G$ | $C_2$ | $C_s$ | $C_{2h}$ |
|---|---|---|---|
| $c_2$ | 1 | 0 | 1 |
| Mirror planes | 0 | 1 | 1 |
| Inversion center | 0 | 0 | 1 |

**Table 7.12** The two triclinic crystal classes

| $G$ | $C_1$ | $C_i$ |
|---|---|---|
| Inversion center | 0 | 1 |

system contains the space groups 195, 196, 197, ..., 230, thus in total 36 space groups. Hence there are 230 possibilities to generate space groups from allowed rotations, reflections, inversion and lattice translations. These 230 space groups were determined in 1891 by Arthur Schoenflies and Evgraf Fedorov. In crystallography, space groups are also called crystallographic groups or Fedorov groups.

## 7.7  Exercises for Chap. 7

**Problem 7.1 (Wigner-Seitz Cells of Two-dimensional Lattices)** Determine the Wigner-Seitz cells for the orthogonal, centered orthogonal and oblique Bravais lattices.

**Problem 7.2 (Proof of the Corollaries 7.1 and 7.2)** Prove the two corollaries at the beginning of the chapter.
Hint: The Lemmas 7.2 and 7.3 as well as the explanations in Sect. 5.6 should be helpful.

**Problem 7.3 (Lattice)** Let $e_1, e_2, \ldots, e_n$ be a basis of $\mathbb{R}^n$. The set

$$\left\{ \sum_{k=1}^{n} m_k\, e_k \,\middle|\, m_1, \ldots, m_k \in \mathbb{Z} \right\}$$

is called the $n$-dimensional lattice generated by the basis.

1. Show that an $n$-dimensional lattice is a subgroup of $(\mathbb{R}^n, +)$!
2. Show that every lattice in $\mathbb{R}^n$ is isomorphic to the $n-$fold direct product of $(\mathbb{Z}, +)$ with itself.

**Problem 7.4 (Two-dimensional Lattices)** We consider the point groups $\mathcal{R}_0$ of the two-dimensional lattices in Fig. 7.16. These contain all symmetry transformations that leave a lattice point fixed.

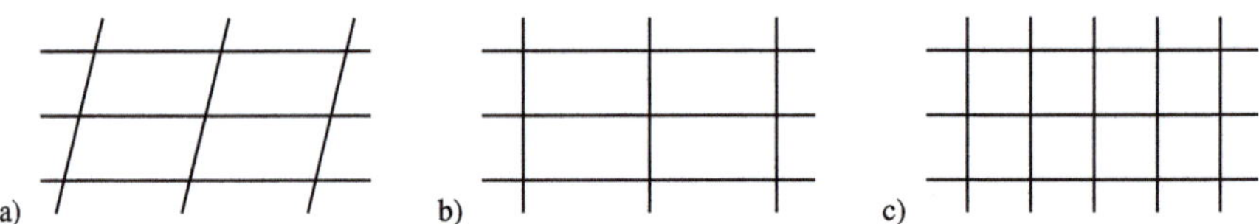

**Fig. 7.16** The oblique system in (**a**) rectangular system in (**b**) and quadratic system in (**c**) have different point symmetries

**Fig. 7.17** Primitive basis of the 2$d$ hexagonal lattice

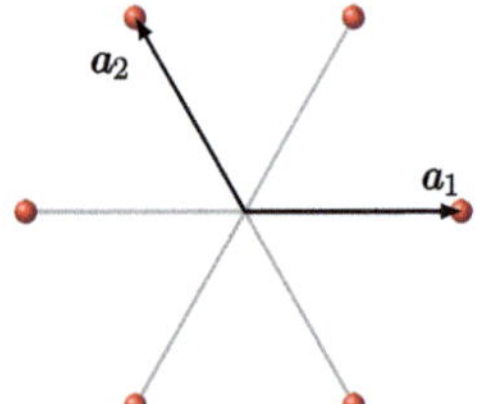

Justify the following statements:

1. For the left lattice, $\mathcal{R}_0$ contains only the inversion in addition to the identity.
2. For the middle lattice, $\mathcal{R}_0$ additionally contains two reflections on the vertical and horizontal through a lattice point, and thus four elements.
3. For the right square lattice, there are additionally two rotations by $\pi/2$ and $3\pi/2$ as well as two reflections on the diagonals. Thus, $\mathcal{R}_0$ has eight elements.

**Problem 7.5 (Point Groups Are Subgroups of GL($d$, $\mathbb{Z}$))** In this problem we characterize the elements of a point group $\mathcal{R}_0$.

1. Why does the action of $\mathcal{R}_0$ on $T$ define a group homomorphism $\mathcal{R}_0 \mapsto \mathrm{Aut}(T)$?
2. Why is $\mathrm{Aut}(T) \cong \mathrm{GL}(d, \mathbb{Z})$?
3. The vectors $a_1$, $a_2$ in Fig. 7.17 form a primitive basis for the two-dimensional hexagonal lattice. Confirm that the rotations in $\mathcal{C}_6$ with respect to this basis are generated by the matrix

$$c_6 = \begin{pmatrix} 1 & -1 \\ 1 & 0 \end{pmatrix} \quad \text{with} \quad c_6^6 = \mathbb{1} \tag{7.28}$$

Why do all rotation matrices have determinant 1.

**Problem 7.6 (Construction of Lattices with Given Point Groups)** We follow here [35] and choose a vector of minimal length $a_1 \in \mathcal{L}$ and rotate it by a rotation $c_3$ by 120° into a vector $a_2$, as shown in Fig. 7.17. The cyclic subgroup $\mathcal{C}_6$ is generated by $c_6$ in (7.28) and the subgroup $\mathcal{C}_3$ by

$$c_3 = c_6^2 = \begin{pmatrix} 0 & -1 \\ 1 & -1 \end{pmatrix}.$$

The 6 vectors that arise from repeatedly applying $c_6$ all have minimal length.

1. The point group $\mathcal{R}_0 = \mathcal{C}_{6v}$ contains in addition to the rotations generated by $c_6$ also 6 reflections. Show that the group is generated by

$$c_6 \quad \text{and} \quad \sigma = \sigma_1 c_3 \quad (\sigma_3 \text{ Pauli matrix}) \tag{7.29}$$

Define $\mathcal{C}_{6v}$ with GAP, for example according to

```
GAP> a:=[[1,-1],[1,0]];b:=[[1,-1],[0,-1]];
GAP> g:=Group(a,b);
GAP> g1:=Group(a);
```

and analyze the group (e.g. the order, subgroups, …).

2. The group $\mathcal{C}_{3v}$ can act on the basis in two different ways. First with rotations generated by $c_3$ and the reflections on the lines, which enclose the angles $30°$, $90°$ and $150°$ with $a_1$, and the same rotations with reflections on the lines, which enclose the angles $0°$, $60°$ and $120°$ with $a_1$. We denote the two possibilities with $\mathcal{C}_{3v}^{(l)}$ and $\mathcal{C}_{3v}^{(s)}$, where $l$ stands for long and $s$ stands for short, and they are shown in Fig. 7.18 (with dashed reflection lines). Convince yourself that the symmetries are generated by the elements

$$\mathcal{C}_{3v}^{(l)} : \{c_3, \sigma_1 c_6\} \quad \text{or} \quad \mathcal{C}_{3v}^{(s)} : \{c_3, \sigma_1 c_3\}$$

The vectors $a_1$ and $a_2$ span a rhombus with a short and a long diagonal, see Fig. 7.19. The group $\mathcal{C}_{3v}^{(s)}$ contains the reflection on the short diagonal and $\mathcal{C}_{3v}^{(l)}$ the one on the long diagonal. Show that the groups $\mathcal{C}_{3v}^{(s)}$ and $\mathcal{C}_{3v}^{(l)}$ in $\mathrm{GL}(2, \mathbb{Z})$ are not conjugate. Thus, the corresponding wallpaper groups are not isomorphic.

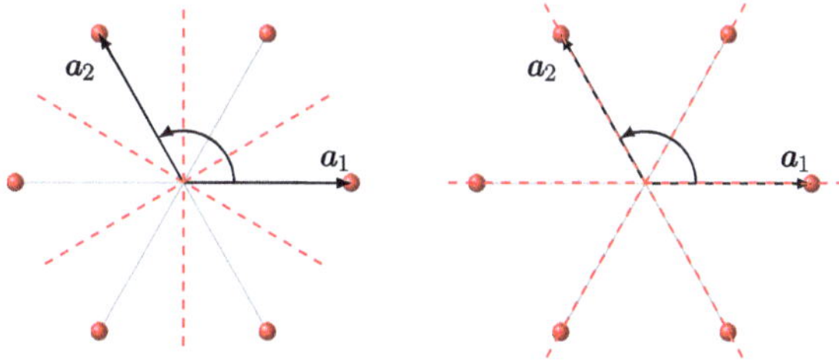

**Fig. 7.18** Action of the group $\mathcal{C}_{3v}$ on the basis vectors

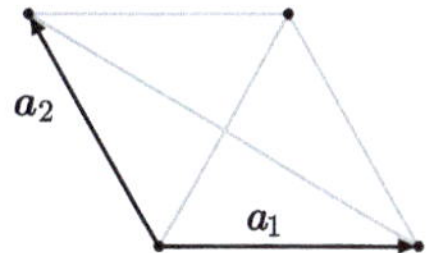

**Fig. 7.19** $\mathcal{C}_{3v}^{(s)}$ and $\mathcal{C}_{3v}^{(l)}$ contain reflections on different diagonals of the rhombus

# Lie Groups

8



In contrast to discrete groups, the elements of continuous groups can be transformed into one another through continuous changes. The simplest continuous group is the group U(1) of the unimodular complex numbers. A group element is represented by a complex number $e^{i\alpha}$. A less simple example is the group SO(3) of proper rotations in Euclidean space.

The elements of continuous groups can be characterized by parameters, $g(\alpha)$ with $\alpha = (\alpha_1, \ldots, \alpha_n)$ and

$$g(\alpha)g(\beta) = g(\gamma), \qquad \gamma = m(\alpha, \beta), \qquad \alpha, \beta, \gamma \in \mathbb{R}^n . \tag{8.1}$$

The number $n$ of necessary real parameters is equal to the dimension of the group. Depending on the type of group, there are various restrictions on the parametrization. For example, the angle $\alpha \in [0, 2\pi]$ parametrizes the elements of U(1). Although $\alpha$ seems to jump at the edge of the interval, the group element changes continuously when exceeding $2\pi$, since the angle functions are $2\pi$-periodic. In other words: In the vicinity of each group element, coordinates can be introduced, on which the group elements continuously depend. If a continuous group is described in this way by coverings in parameter space and these coverings fulfill some relatively natural requirements, then it is a *Lie group*. Many textbooks on symmetries deal with continuous Lie groups. Important in applications are matrix groups, which for example in [37] receive special attention.

A Lie group is characterized by the fact that the functions $m(\alpha, \beta)$ in (8.1), which express the group multiplication in the parameter space, are differentiable functions (for a fixed degree of differentiability) in their arguments. As we have

© The Author(s), under exclusive license to Springer-Verlag GmbH, DE, part of Springer Nature 2026
A. Wipf, *Symmetries in Physics*, https://doi.org/10.1007/978-3-662-72675-4_8

seen with U(1), appropriate coverings of the parameter space should be allowed. A $n$-parameter Lie group is compact if the parameter range is compact (bounded and closed). The groups U(1) and SO(3) are compact Lie groups. On the other hand, the *translations* in $\mathbb{R}^3$,

$$T(a) : x \mapsto x' = x + a \quad \text{with} \quad T(a)T(b) = T(a+b),$$

form a non-compact Lie group. In general, one defines:

**Definition 8.1 (Lie Group)**  A Lie group $G$ is a group that is also a differentiable manifold, so that the multiplication map

$$G \times G \mapsto G, \qquad (g_1, g_2) \mapsto g_1 g_2$$

and the inversion map

$$G \mapsto G, \qquad g \mapsto g^{-1}$$

are both continuous and differentiable mappings.

We first recall the definition of a (differentiable) manifold.

## 8.1   Differentiable Manifolds

A $n$-dimensional manifold can be locally identified with an open set of $\mathbb{R}^n$ (see textbooks on differential geometry [38, 39] and global aspect of Lie groups [40]). A bit more precise:

**Definition 8.2 (Manifold)**  An $n$-dimensional manifold $M$ is a topological space with the following properties

1. it is Hausdorff,
2. it has a countable basis,
3. it is locally Euclidean.

The first property means that two different points have disjoint neighborhoods and can therefore be separated. The second condition states that there is a set $\mathcal{B}$ of countably many open sets in $M$, so that every non-empty open set is a union of those from $\mathcal{B}$. Characteristic is the last property, which we discuss a bit more in detail. It requires that for *every point* $p \in M$ there is a neighborhood $U$ and a homeomorphism (bijective and both-sided continuous),

$$\varphi : U \mapsto \varphi(U) \subset \mathbb{R}^n \text{ open}.$$

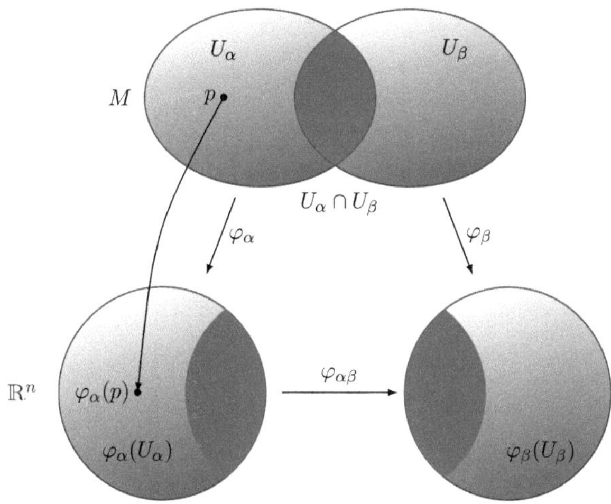

**Fig. 8.1** The group manifold is locally described by maps



Every point therefore has a neighborhood that is homeomorphic to an open set in $\mathbb{R}^n$. The pair $(U, \varphi)$ is often called chart of $M$, $U$ is called the coordinate domain and $\varphi$ the coordinate map. A set of charts $\{(U_\alpha, \varphi_\alpha)\,|\,\alpha \in A\}$ with domains $U_\alpha$ and maps $\varphi_\alpha$ is called an atlas of $M$, if

$$\bigcup_{\alpha \in A} U_\alpha = M\,. \tag{8.2}$$

> **Modeling a Manifold**
> A manifold can be locally described by charts from an atlas.

If the domains $U_\alpha$, $U_\beta$ of two coordinate maps $\varphi_\alpha, \varphi_\beta$ have a non-trivial intersection, $U_{\alpha\beta} := U_\alpha \cap U_\beta \neq \emptyset$, then both maps $\varphi_\alpha, \varphi_\beta$ are defined on this intersection, as sketched in Fig. 8.1.

We obtain a *transition map* (coordinate transformation) $\varphi_{\alpha\beta}$, which is a homeomorphism between open sets of $\mathbb{R}^n$,

$$\varphi_{\alpha\beta} = \varphi_\beta \circ \varphi_\alpha^{-1} : \ \varphi_\alpha(U_{\alpha\beta}) \longmapsto \varphi_\beta(U_{\alpha\beta}), \quad U_{\alpha\beta} = U_\alpha \cap U_\beta\,. \tag{8.3}$$

A manifold is called *differentiable* (of class $C^k$) if $M$ has an atlas such that all occurring transition maps are differentiable[1] (of class $C^k$). According to this definition, all coordinate transformations are diffeomorphisms. We recall the

---

[1] If one demands analytic transition functions then one speaks of analytic manifolds.

**Definition 8.3 (Diffeomorphism)** Let $V, V' \subset \mathbb{R}^n$ be open and $\varphi : V \mapsto V'$ a differentiable bijective mapping. If $\varphi$ has a differentiable inverse, then it is called a diffeomorphism.

> **Linear Lie Groups**
> In this lecture, we often deal with linear Lie groups. These are subgroups of the linear group $GL(n, \mathbb{C})$ or $GL(n, \mathbb{R})$.

When dealing with linear groups, it is sufficient to consider differentiable manifolds in $\mathbb{R}^n$. Then the following theorem (which we do not prove) is useful:

**Theorem 8.1** *Let $U \subset \mathbb{R}^n$ be open and $f : U \mapsto \mathbb{R}^p$ with $p < n$ a differentiable mapping with the property that the Jacobian matrix $(\partial f_i / \partial x_j)$ has maximum rank $p$ whenever $f(x) = 0$. Then $f^{-1}(0)$ is a $(n - p)$-dimensional manifold in $\mathbb{R}^n$.*

In other words: a level set of a differentiable mapping is a manifold, if the Jacobian matrix on the level set has the maximum possible rank $p$.

### Example: Atlas for Sphere $S^n$

From Theorem 8.1 it follows directly, that the sphere $S^n$ is a manifold in $\mathbb{R}^{n+1}$, because $S^n = f^{-1}(0)$ for the function $f : x \to \|x\| - 1$. This also follows from the original definition of a manifold. For this, we cover $S^n$ with two coordinate domains

$$H_+ = \{x \in \mathbb{R}^{n+1} | \ \|x\| = 1, x_{n+1} > -1/2\}$$

$$H_- = \{x \in \mathbb{R}^{n+1} | \ \|x\| = 1, x_{n+1} < 1/2\} \,.$$

$H_+$ can be mapped homomorphically into an open set in $\mathbb{R}^n$ with the stereographic projection from the south pole and $H_-$ with the projection from the north pole, see Fig. 8.2. The transition map (coordinate transformation) is arbitrarily often continuously differentiable. Thus, $S^n$ is a $C^\infty$-manifold. ◄

From differentiable manifolds, one can obtain further differentiable manifolds.

- For example, let $M \sim \{U_\alpha, \varphi_\alpha\}$ and $M' \sim \{U'_\beta, \varphi'_\beta\}$ be differentiable manifolds. Then their product $M \times M' \sim \{U_\alpha \times U'_\beta, \varphi_\alpha \times \varphi'_\beta\}$ is also a differentiable manifold.
- Every open subset $N \subset M$ of a differentiable manifold is a differentiable manifold. Choose as coordinate domains of $N$ the open sets $N \cap U_\alpha$, where the $U_\alpha$ cover the manifold $M$.

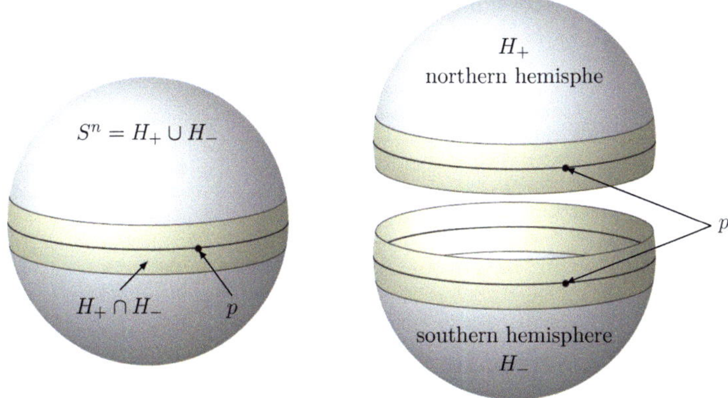


**Fig. 8.2** The sphere can be covered with two coordinate domains. It is a $C^\infty$-manifold

A mapping $f : M \to M'$ between manifolds can be described in local coordinates. Let $f(p) = p' \in M'$ and $(U, \varphi)$ and $(U', \varphi')$ be charts around $p$ and $p'$. Let further $x = \varphi(p)$ and $x' = \varphi'(p')$ be the coordinates of the points $p$ and $p'$. Then

$$(\varphi' \circ f \circ \varphi^{-1})(x) = (\varphi' \circ f)\big(\underbrace{\varphi^{-1}(x)}_{p}\big) = \varphi'(p') = x' .$$

**Differentiable Mappings**
With respect to local coordinates, $f : M \to M'$ is equal to $(\varphi' \circ f \circ \varphi^{-1})$. The mapping is called differentiable, if it is differentiable with respect to local coordinates.

This property is independent of the local coordinates, since the transition maps of a differentiable manifold are diffeomorphisms.

## 8.1.1  Lie Groups

A Lie group $G$ is simultaneously a group and a differentiable manifold, such that the multiplication and inversion maps are differentiable mappings. Each group element in a coordinate domain $U$ is uniquely assigned $n$ real coordinates, $g \to \alpha \in \varphi(U) \subset \mathbb{R}^n$. With respect to these local coordinates, the group multiplication has the representation

$$(\alpha, \beta) \mapsto m(\alpha, \beta) ,$$

where $m$ denotes the function in (8.1). The inversion is

$$g^{-1}(\alpha) = g\big(\mathrm{inv}(\alpha)\big)$$

and has the coordinate representation

$$\alpha \mapsto \mathrm{inv}(\alpha)\,.$$

If the two mappings $m$ and inv are continuously differentiable, then $G$ is a Lie group.

### 8.1.2   (Path-)Connected Lie Groups

To discuss global properties, $G$ does not need to be a Lie group. It is sufficient if $G$ is a topological group. This is a continuous group, endowed with a topology such that the group multiplication and the inversion are continuous mappings. Every Lie group is a topological group.

A continuous mapping $w : [0, 1] \to G$ is called a path in $G$. Two group elements are said to be path-connected, $g_1 \sim g_2$, if they can be connected with a path $w$, $w(0) = g_1$ and $w(1) = g_2$. All $g \in G$ are path-connected to themselves via the constant path. If $g_1$ is path-connected to $g_2$ via the path $w$, then $g_2$ is also path-connected to $g_1$ via the path $\tilde{w}(t) = w(1 - t)$. If $g_1$ is path-connected to $g_2$ through a path $w_1$ and $g_2$ is path-connected to $g_3$ through a path $w_2$, then the composite path

$$(w_1 \circ w_2)(t) = \begin{cases} w_1(2t) & 0 \le t \le 1/2, \\ w_2(2t - 1) & 1/2 \le t \le 1, \end{cases} \tag{8.4}$$

connects the elements $g_1$ and $g_3$. Thus, $\sim$ defines an equivalence relation.

**Definition 8.4 (Connected Components)** The path-connected components of $G$ are the equivalence classes with respect to $\sim$. If $G$ consists of a single component, then $G$ is said to be path-connected.

A path-connected space is always also connected. For manifolds, the converse also holds. The following applies:

**Lemma 8.1** *Let $G_0$ be the connected component of $G$, which contains the identity element $e$. Then $G_0$ is a normal subgroup of $G$ and $G/G_0 = \{connected components of G\}$.*

For the proof, we consider a continuous path $w$, which connects $e$ with $g_0 \in G_0$. Then the continuous path

$$\tilde{w}(t) = g w(t) g^{-1}$$

connects the element $e$ with $g g_0 g^{-1}$. Therefore, with $g_0$ the conjugate $g g_0 g^{-1}$ is also in $G_0$. This proves the first statement. The proof of the second statement I leave to the readers.

### 8.1.3   Lie Subgroups

**Definition 8.5 (Lie Subgroup, Normal Lie Subgroup)** A subgroup $H \leq G$ of a Lie group is called Lie subgroup, if it is simultaneously a submanifold of $G$. If it is additionally a normal subgroup $N \trianglelefteq G$, then it is called normal Lie subgroup.

The following useful theorem, which goes back to JOHN VON NEUMANN (for linear Lie groups) and ELIE CARTAN, holds:

**Theorem 8.2 (Cartan's Subgroup Theorem)** *A subgroup (normal subgroup) $H$ of a Lie group $G$ is exactly a Lie subgroup (normal Lie subgroup), when $H$ is closed.*

Without proof, we note:

1. The direct product $G_1 \times G_2$ of two Lie groups is again a Lie group.
2. If $N \trianglelefteq G$ is a normal Lie subgroup of a Lie group $G$, then $G/N$ is a Lie group.

At this point, it is interesting to know that a connected Lie group can have only a few discrete normal subgroups:

**Lemma 8.2** *For a connected Lie group, a discrete normal subgroup lies in the center of the group.*

Therefore, for example, SO(3) has no discrete normal subgroup. The lemma follows from

**Theorem 8.3** *Let $G$ be a connected Lie group and $U$ a open neighborhood of $e$. Then $G$ is generated by $U$, i.e., every element $g \in G$ is a product $g = g_1 g_2 \cdots g_n$ with $g_i \in U$.*

For the proof, we assume without loss of generality $U = U^{-1}$. Here, $U^{-1}$ is the set of inverse elements of $U$. If $U$ does not have this property, then we choose $U \cap U^{-1}$ as a new open neighborhood of $e$. We first show that the subgroup $H$ of $G$ generated by $U$ is open: For any arbitrary $a \in H$, $aU = \ell_a U \subset H$ is an open neighborhood of $a$, since the left translation $\ell_a$ is a diffeomorphism. Therefore, $H$ is an open set. We now form the *open* cosets $H, bH, b'H, \ldots$. Because

$$H \cup \left( \bigcup_{b \neq e} bH \right) = G$$

$H$ is the complement of an open set and is therefore closed. This means that the subgroup $H \subseteq G$ is open, closed and not empty, and since it is connected, we conclude that $H = G$. We have used that a topological space is connected if and only if the only closed open sets are the empty set and the space itself.

> **Small Neighborhoods of $e$ Determine Global Properties of $G$**
> The neighborhood $U$ of $e$ can be chosen so small that it lies in a coordinate domain of the atlas of $G$. Certain properties that hold in $U$ then also hold for the whole group.

For example, if the multiplication in $U$ is commutative, then the group is abelian.

**Task**

Justify why the Theorem 8.3 implies the Lemma 8.2.

## 8.2    The Lie Groups U(2) and SU(2)

We illustrate the introduced concepts using the (special) unitary group in 2 dimensions. This is the most important group of non-relativistic quantum mechanics and appears in the description of spin, for example of electrons. In Chap. 17 we will show, that the quantum mechanical states with fixed angular momentum form an irreducible representation of the quantum mechanical rotation group SU(2). Furthermore, the group appears as the gauge group of the weak interaction, see Sect. 20.5.

We consider the two-dimensional complex vector space $\mathbb{C}^2$. After choosing a basis $(e_1, e_2)$, every vector $r = x_1 e_1 + x_2 e_2$ is uniquely determined by a complex 2-tuple

$$x = \begin{pmatrix} x_1 \\ x_2 \end{pmatrix} \in \mathbb{C}^2$$

and every linear mapping by a $2 \times 2$ matrix $A = (a_{ij})$. A linear mapping then acts on the tuple $x$ associated with the vector according to

$$x \mapsto Ax, \quad \text{with} \quad (Ax)_i = \sum_j a_{ij} x_j .$$

The unitary matrices (we denote them with $U$, to distinguish them from arbitrary matrices $A$) are characterized by the fact that they preserve the Hermitian scalar product

$$(\boldsymbol{x}, \boldsymbol{y}) = \bar{x}_1 y_1 + \bar{x}_2 y_2, \quad \boldsymbol{x}, \boldsymbol{y} \in \mathbb{C}^2, \tag{8.5}$$

which means $(U\boldsymbol{x}, U\boldsymbol{y}) = (\boldsymbol{x}, \boldsymbol{y})$. A matrix $U$ is thus unitary if and only if

$$U^\dagger U = \mathbb{1} = \begin{pmatrix} \bar{a}_{11} & \bar{a}_{21} \\ \bar{a}_{12} & \bar{a}_{22} \end{pmatrix} \begin{pmatrix} a_{11} & a_{12} \\ a_{21} & a_{22} \end{pmatrix} = \begin{pmatrix} 1 & 0 \\ 0 & 1 \end{pmatrix},$$

i.e., if the column vectors have length one and are orthogonal to each other.

**The Continuous Group U(2)**
The set of all unitary $2 \times 2$ matrices

$$\mathrm{U}(2, \mathbb{C}) \equiv \mathrm{U}(2) = \left\{ U \in \mathrm{Mat}(2, \mathbb{C}) \,|\, U^\dagger U = \mathbb{1} \right\} \tag{8.6}$$

forms a group, the so-called *unitary group* U(2).

The group operation is matrix multiplication or the composition of the linear transformations associated with the matrices. Since U(2) was defined via an invariance property, the product of two unitary matrices and the inverse of a unitary matrix are also unitary. The identity matrix is the identity element of the group.

Since the columns are orthogonal, we can parameterize $U$ as follows,

$$U = \begin{pmatrix} a & \lambda b \\ -\bar{b} & \lambda \bar{a} \end{pmatrix}.$$

Since they have length one, additionally $|a|^2 + |b|^2 = 1$ and $|\lambda| = 1$ must hold. This also ensures that $UU^\dagger = \mathbb{1}$, or $U^\dagger = U^{-1}$, and the two rows are also orthonormal. There are thus four real parameters and therefore U(2) is a four-dimensional Lie group.

## 8.2.1   The Lie Group SU(2)

As an important normal subgroup, U(2) contains the *special unitary group*

$$\mathrm{SU}(2) = \left\{ U = \begin{pmatrix} a & b \\ -\bar{b} & \bar{a} \end{pmatrix} \,\middle|\, a\bar{a} + b\bar{b} = 1 \right\} \lhd \mathrm{U}(2). \tag{8.7}$$

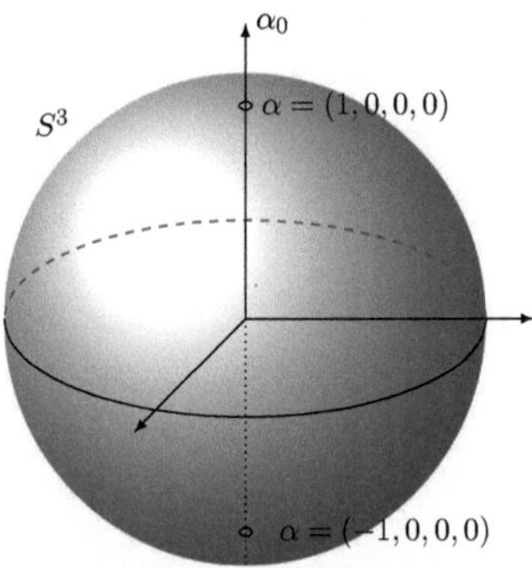

**Fig. 8.3** The group manifold of SU(2) is the sphere $S^3$



This is a closed subgroup and according to the Cartan subgroup theorem a normal Lie subgroup. The assignment

$$\left\{ \alpha = (\alpha_1, \alpha_2, \alpha_3, \alpha_4) = (\Re a, \Im a, \Re b, \Im b) \,\Big|\, \sum \alpha_i^2 \equiv \alpha^2 = 1 \right\} \mapsto \mathrm{SU}(2) \tag{8.8}$$

is bijective and shows that the group manifold is the sphere $S^3$. Figure 8.3 illustrates this in one dimension less.

Every matrix in U(2) has the representation

$$U = \mathrm{e}^{\mathrm{i}\alpha} \cdot U', \qquad U' \in \mathrm{SU}(2) \,,$$

but because $\mathbb{1}, -\mathbb{1}$ both lie in the normal subgroup SU(2), $\mathrm{e}^{\mathrm{i}\alpha} \cdot U'$ and $-\,\mathrm{e}^{\mathrm{i}\alpha} \cdot U'$ define the same SU(2)-cosets in U(2). It follows

$$\mathrm{U}(2)/\mathrm{SU}(2) \cong \left\{ \mathrm{e}^{\mathrm{i}\alpha} \,\big|\, \mathrm{e}^{\mathrm{i}\alpha} \sim -\mathrm{e}^{\mathrm{i}\alpha} \right\} = \mathrm{U}(1)/\mathbb{Z}_2 \,. \tag{8.9}$$

The abelian group U(1) also appears as a subgroup of the diagonal SU(2)-matrices,

$$\left\{ \begin{pmatrix} \mathrm{e}^{\mathrm{i}\alpha} & 0 \\ 0 & \mathrm{e}^{-\mathrm{i}\alpha} \end{pmatrix} \,\bigg|\, \alpha \in [0, 2\pi) \right\} < \mathrm{SU}(2) \,.$$

We may consider the two diagonal elements separately and obtain

$$\mathrm{U}(1) = \left\{ \mathrm{e}^{\mathrm{i}\alpha} \,\big|\, 0 \le \alpha < 2\pi \right\} \,. \tag{8.10}$$

The group manifold is the unit circle $S^1$, see Fig. 8.4. U(1) is also a limit case of the cyclic groups, $\mathrm{U}(1) \sim \lim_{n \to \infty} \mathcal{C}_n$.

**Fig. 8.4** The group U(1) can
be identified with the unit
circle $S^1$

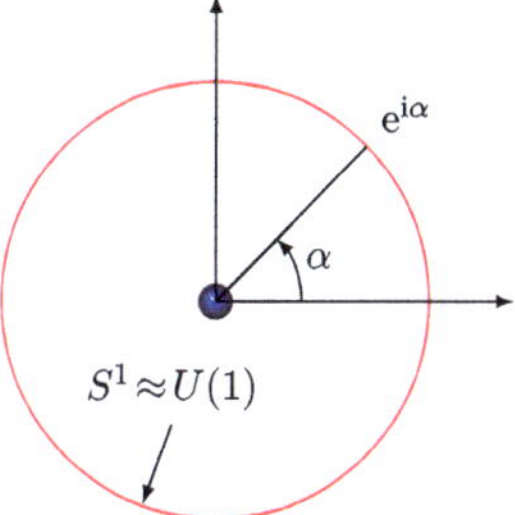



Now investigate the question when two elements in SU(2) are conjugate to each other. From linear algebra we know that every unitary matrix can be diagonalized with a special unitary matrix. Therefore, there is a $V \in$ SU(2) with

$$U = VDV^{-1}, \qquad D = \begin{pmatrix} e^{i\lambda} & 0 \\ 0 & e^{-i\lambda} \end{pmatrix}, \tag{8.11}$$

where $\exp(i\lambda)$ and $\exp(-i\lambda)$ are the eigenvalues of $U$. This means that two elements in SU(2) are conjugate to each other if they have the same eigenvalues. Since there is another conjugation which exchanges the two eigenvalues,

$$\begin{pmatrix} 0 & -1 \\ 1 & 0 \end{pmatrix} \begin{pmatrix} e^{i\lambda} & 0 \\ 0 & e^{-i\lambda} \end{pmatrix} \begin{pmatrix} 0 & 1 \\ -1 & 0 \end{pmatrix} = \begin{pmatrix} e^{-i\lambda} & 0 \\ 0 & e^{i\lambda} \end{pmatrix} \tag{8.12}$$

the order of the eigenvalues does not matter. Therefore, the following holds

**Theorem 8.4 (Conjugate SU(2)-matrices)** *Two matrices $U$ and $U'$ in SU(2) are exactly conjugate to each other if they have the same trace,* $\mathrm{Tr}(U) = \mathrm{Tr}(U')$.

We have shown that $\mathrm{Tr}(U) = \mathrm{Tr}(U')$ implies the similarity of $U$ and $U'$. Conversely, if the traces are different, then the two matrices cannot be conjugate to each other, since $\mathrm{Tr}(VUV^{-1}) = \mathrm{Tr}\,U$.

The *center* of SU(2) contains all matrices $\{z\}$ that commute with all matrices in SU(2). In particular, the center elements must commute with the diagonal matrices $D$ in (8.11). This is only possible if they are diagonal, $z = \mathrm{diag}(a, \bar{a})$. But they also have to commute with the non-diagonal matrix in (8.12), which implies $a = \bar{a}$. Since $a$ also has to have the magnitude 1, it finally follows

**Theorem 8.5 (Center)** *The center of SU(2) is* $Z = \{\mathbb{1}, -\mathbb{1}\}$. *It is isomorphic to* $\mathbb{Z}_2$.

The question of the factor group $\mathrm{SU}(2)/Z$ arises here. We will prove the following theorem:

**Theorem 8.6** *The factor group SU(2)/Z, where Z={$\mathbb{1}, -\mathbb{1}$} is the center of SU(2), is isomorphic to the group SO(3) of proper rotations.*

We will construct this isomorphism explicitly, as it is relevant in quantum mechanics.

### 8.2.2   Pauli Matrices

To establish the direct connection between SU(2) and SO(3), we introduce the Hermitian and traceless *Pauli matrices,*

$$\sigma_1 = \begin{pmatrix} 0 & 1 \\ 1 & 0 \end{pmatrix}, \quad \sigma_2 = \begin{pmatrix} 0 & -i \\ i & 0 \end{pmatrix}, \quad \sigma_3 = \begin{pmatrix} 1 & 0 \\ 0 & -1 \end{pmatrix}. \tag{8.13}$$

Their products are again Pauli matrices up to the identity,

$$\sigma_i \sigma_j = \mathbb{1}\, \delta_{ij} + i \sum_{k=1}^{3} \epsilon_{ijk} \sigma_k . \tag{8.14}$$

Every Hermitian and traceless matrix $A$ is a real linear combination of these matrices,

$$\boldsymbol{a} \cdot \boldsymbol{\sigma} = \sum_{i=1}^{3} a_i \sigma_i = \begin{pmatrix} a_3 & a_1 - i a_2 \\ a_1 + i a_2 & -a_3 \end{pmatrix}, \quad \boldsymbol{a} \in \mathbb{R}^3 , \tag{8.15}$$

and has the determinant

$$\det(\boldsymbol{a} \cdot \boldsymbol{\sigma}) = -\boldsymbol{a}^2 . \tag{8.16}$$

For every unitary matrix $U$, because $U^\dagger = U^{-1}$, $U(\boldsymbol{a} \cdot \boldsymbol{\sigma})U^{-1}$ is also Hermitian and traceless. Therefore, there must be a vector $\boldsymbol{b}$ (dependent on $U$) such that

$$U(\boldsymbol{a} \cdot \boldsymbol{\sigma})U^{-1} = \boldsymbol{b} \cdot \boldsymbol{\sigma} ,$$

where $\boldsymbol{b}$ depends linearly on $\boldsymbol{a}$. Taking the determinant of this equation

$$\det\left(U(\boldsymbol{a} \cdot \boldsymbol{\sigma})U^{-1}\right) = \det(\boldsymbol{a} \cdot \boldsymbol{\sigma}) = -\boldsymbol{a}^2 = \det(\boldsymbol{b} \cdot \boldsymbol{\sigma}) = -\boldsymbol{b}^2 ,$$

we conclude:

**The Group Homomorphism SU(2) $\mapsto$ SO(3)**
The linear mapping $\boldsymbol{a} \mapsto \boldsymbol{b}$ on $\mathbb{R}^3$ is length-preserving. Hence there exists a
$U$-dependent rotation $R$ with

$$U(\boldsymbol{a} \cdot \boldsymbol{\sigma})U^{-1} = \big(R(U)\,\boldsymbol{a}\big) \cdot \boldsymbol{\sigma}, \qquad R^T(U)R(U) = \mathbb{1}_3. \qquad (8.17)$$

In addition, for arbitrary $\boldsymbol{a} \in \mathbb{R}^3$

$$(R(U_1 U_2)\,\boldsymbol{a}) \cdot \boldsymbol{\sigma} = (U_1 U_2)\,(\boldsymbol{a} \cdot \boldsymbol{\sigma})\,(U_1 U_2)^{-1} = U_1 \left(U_2(\boldsymbol{a} \cdot \boldsymbol{\sigma})U_2^{-1}\right) U_1^{-1}$$

$$= U_1 \left(\big(R(U_2)\,\boldsymbol{a}\big) \cdot \boldsymbol{\sigma}\right) U_1^{-1} = (R(U_1)R(U_2)\,\boldsymbol{a}) \cdot \boldsymbol{\sigma}, \qquad (8.18)$$

so that the mapping $U \mapsto R(U)$ is a *group homomorphism* SU(2) $\rightarrow$ O(3),

$$R(\mathbb{1}) = \mathbb{1} \quad \text{and} \quad R(U_1 U_2) = R(U_1)R(U_2). \qquad (8.19)$$

We want to convince ourselves that the $R(U)$ are *proper rotations*. To do this, we
need to prove that matrices in the image of the homomorphism have determinant 1.
Since $R(\mathbb{1}_2) = \mathbb{1}_3$, the three-dimensional unit matrix is obviously in the image of
the homomorphism. This matrix has determinant 1. Now, however, $R(U)$ depends
continuously on the elements of the connected Lie group SU(2). Therefore, the
determinant must be 1 for all matrices in the image of the homomorphism, as it
cannot jump.

**Task**

Show that this homomorphism SU(2) $\mapsto$ SO(3) is surjective.

The question remains about the kernel of the homomorphism or the question
about those SU(2) matrices for which $U(\boldsymbol{a} \cdot \boldsymbol{\sigma})U^{-1} = \boldsymbol{a} \cdot \boldsymbol{\sigma}$ for all $\boldsymbol{a} \in \mathbb{R}^3$.
Obviously, $\pm\mathbb{1}$ are in the kernel. A calculation shows that no further elements are
in the kernel, so that kernel $= \{\mathbb{1}, -\mathbb{1}\}$. According to the first isomorphism theorem,
then

$$\text{SU(2)}/\{\mathbb{1}, -\mathbb{1}\} \cong \text{SO(3)}. \qquad (8.20)$$

> **SU(2) is the Universal Covering of SO(3)**
> The simply connected quantum mechanical rotation group SU(2) is the double universal covering of the group of rotations in three-dimensional space.

At the end of the chapter, we will discuss coverings of non-simply connected Lie groups.

> **Rotations About the $z$-axis**
> The diagonal matrices $U(\varphi) = \mathrm{diag}(e^{-i\varphi}, e^{i\varphi})$ describe a loop in SU(2), since $U(0) = U(2\pi) = \mathbb{1}$. According to (8.17) these matrices are mapped to the rotations $R(e_3, 2\varphi)$ which were introduced in (5.16). When $U(\varphi)$ goes through the loop once, then $R(e_3, 2\varphi)$ describes a rotation by $4\pi$ and goes through its loop twice.

## 8.3    The Matrix Groups GL($n$, $\mathbb{K}$) and Important Subgroups

We choose an orthonormal basis $\{e_i\}$ in the $n$-dimensional $\mathbb{K}$-vector space $\mathcal{V}$ with scalar product. Then a vector $r$ corresponds to the tuple $x$ with components $x_i = (e_i, r)$ and a linear mapping $A : \mathcal{V} \mapsto \mathcal{V}$ corresponds to the matrix with elements $a_{ij} = (e_i, A e_j)$. The linear mapping is then given by (see also Sect. 5.2)

$$x \mapsto x' = Ax, \qquad A = (a_{ij}) \in \mathrm{Mat}(n, \mathbb{K}), \quad x \in \mathbb{K}^n, \tag{8.21}$$

where we use the same symbol for the linear mapping and its matrix. If no further conditions are imposed on the matrices besides invertibility, we obtain the non-compact *general linear group* (general linear group)

$$\mathrm{GL}(n, \mathbb{R}) \quad \text{for} \quad \mathcal{V} = \mathbb{R}^n \quad \text{and} \quad \mathrm{GL}(n, \mathbb{C}) \quad \text{for} \quad \mathcal{V} = \mathbb{C}^n. \tag{8.22}$$

These groups form subsets of $\mathbb{R}^{n^2}$ or $\mathbb{C}^{n^2}$, if we consider a matrix $A$ as a point $(a_{11}, a_{12}, \ldots, a_{nn})$ in $\mathbb{R}^{n^2}$ or $\mathbb{C}^{n^2}$. This identification makes $\mathrm{GL}(n, \mathbb{K})$ a metric space with (squared) distance

$$d(A, B)^2 = \sum_{i,j=1}^{n} |a_{ij} - b_{ij}|^2 = \mathrm{Tr}(A - B)^{\dagger}(A - B) = \|A - B\|_{\mathrm{Frob}}^2. \tag{8.23}$$

On the right is the square of the Frobenius norm of the difference matrix. The groups $GL(n, \mathbb{R})$ and $GL(n, \mathbb{C})$ form open subsets of $\mathbb{R}^{n^2}$ and $\mathbb{C}^{n^2}$. Their (real) dimension is equal to the number of freely selectable matrix elements,

$$\dim\big(GL(n, \mathbb{R})\big) = n^2 \quad \text{and} \quad \dim\big(GL(n, \mathbb{C})\big) = 2n^2. \tag{8.24}$$

Matrix groups are now Lie subgroups of $GL(n, \mathbb{K})$. The following theorem holds:

**Theorem 8.7** *Let $G$ be a subgroup of $GL(n, \mathbb{K})$. Then the multiplication $G \times G \to G$, $(A, B) \to AB$ and the inversion $G \to G$, $A \to A^{-1}$ are continuous mappings.*

***Proof*** The matrix elements of $AB$ are polynomials of the matrix elements of $A$ and $B$ and thus continuous. The matrix elements of $A^{-1}$ are rational functions in the matrix elements of $A$ according to Cramer's rule. The denominator is the polynomial $\det A$, which never vanishes in $GL(n, \mathbb{K})$. Therefore, $A \to A^{-1}$ is also continuous.

$\square$

### 8.3.1  Subgroups of GL($n$, $\mathbb{K}$)

For any subgroup of $GL(n, \mathbb{K})$, the following must always hold (cf. group axioms):

- $\det A$ must not be zero so that $A^{-1}$ exists.
- The $n$-dimensional identity matrix $\mathbb{1}_n$ must be in the subgroup.

By additional requirements on the matrices, which are compatible, for example, with the assumptions in Theorem 8.1 in Sect. 8.1, we obtain further Lie groups as subgroups of $GL(n, \mathbb{K})$.

**Special Linear Groups** are characterized by $\det A = 1$,

$$SL(n, \mathbb{K}) = \{A \in GL(n, \mathbb{K})| \det A = 1\}$$

$$= \mathrm{Kern}\big(\det : GL(n, \mathbb{K}) \to \mathbb{K}^*\big) \lhd GL(n, \mathbb{K}). \tag{8.25}$$

The non-compact group $SL(n, \mathbb{K})$ is a normal subgroup of $GL(n, \mathbb{K})$, as it is equal to the kernel of the homomorphism $A \to \det A$ from $GL(n, \mathbb{K})$ to the multiplicative group $\mathbb{K}^*$. The dimensions of the special linear groups are

$$\dim\big(SL(n, \mathbb{R})\big) = n^2 - 1 \quad \text{and} \quad \dim\big(SL(n, \mathbb{C})\big) = 2n^2 - 2. \tag{8.26}$$

The corresponding linear transformations preserve the volume and the orientation.

The Lie group $SL(2, \mathbb{C})$ plays a prominent role in relativistic quantum mechanics and elementary particle physics, especially in the treatment of fermions. The homomorphism $\mathrm{SL}(2, \mathbb{C}) \mapsto \mathrm{SO}^{\uparrow}_{+}(1, 3)$ generalizes the homomorphism $\mathrm{SU}(2) \mapsto \mathrm{SO}(3)$ in (8.17) and is constructed and discussed in Problem 8.5.

**Orthogonal Groups** contain linear transformations in $\mathbb{R}^n$ that leave the Euclidean scalar product $(\boldsymbol{x}, \boldsymbol{y}) = \sum x_i y_i$ of vectors unchanged,

$$(R\boldsymbol{x}, R\boldsymbol{y}) = (R^T R\boldsymbol{x}, \boldsymbol{y}) = (\boldsymbol{x}, \boldsymbol{y}), \qquad \text{for all} \quad \boldsymbol{x}, \boldsymbol{y} \in \mathbb{R}^n . \tag{8.27}$$

Thus, we have

$$\mathrm{O}(n) = \left\{ R \in \mathrm{GL}(n, \mathbb{R}) | R^T R = R R^T = \mathbb{1}_n \right\} . \tag{8.28}$$

The dimensions of the compact orthogonal groups are

$$\dim\left(\mathrm{O}(n)\right) = n(n-1)/2 . \tag{8.29}$$

For $n = 3$, this is the already frequently discussed 3$-$dimensional group of proper and improper rotations in space. The determinant of an orthogonal matrix is $\pm$ 1. Also of importance are the *pseudo-orthogonal* groups, for example the Lorentz group $\mathrm{O}(1, 3)$ of special relativity. The pseudo-orthogonal group $\mathrm{O}(p, q)$ leaves the following bilinear form on $\mathbb{R}^n$ invariant:

$$(\boldsymbol{x}, \boldsymbol{y}) = \boldsymbol{x}^T \eta \boldsymbol{y}, \quad \eta = \begin{pmatrix} \mathbb{1}_p & 0 \\ 0 & -\mathbb{1}_q \end{pmatrix}, \quad p + q = n . \tag{8.30}$$

The metric tensor $\eta$ with signature $(p, q)$ generalizes the known metric in Minkowski space. It holds

$$\mathrm{O}(p, q) = \left\{ R \in \mathrm{GL}(n, \mathbb{R}) | R^T \eta R = \eta \right\} , \tag{8.31}$$

and $\dim \mathrm{O}(p, q) = \dim \mathrm{O}(n)$ with $n = p + q$.

**Special Orthogonal Groups** are normal subgroups of the orthogonal groups,

$$\mathrm{SO}(n) = \{R \in \mathrm{O}(n) | \det R = 1\} = \mathrm{Kern}\,(\det : \ \mathrm{O}(n) \to \mathbb{Z}_2) \lhd \mathrm{O}(n) . \tag{8.32}$$

They have the same dimensions as the corresponding orthogonal groups and only contain orientation-preserving linear mappings.

**Unitary Groups** generalize the group $\mathrm{U}(2)$ just discussed and leave the Hermitian scalar product of $\mathbb{C}^n$ invariant,

$$\mathrm{U}(n) = \left\{ U \in \mathrm{GL}(n, \mathbb{C}) | U^\dagger U = U U^\dagger = \mathbb{1}_n \right\} . \tag{8.33}$$

The dimension of this compact Lie group is

$$\dim\left(\mathrm{U}(n)\right) = n^2 .\tag{8.34}$$

**Special Unitary Groups** are normal subgroups of the unitary groups. Their elements have the determinant 1,

$$\mathrm{SU}(n) = \{U \in \mathrm{U}(n)\,|\,\det U = 1\} = \mathrm{Kern}\,(\det:\ \mathrm{U}(n) \to \mathrm{U}(1))\ .\tag{8.35}$$

The dimension of the Lie group is

$$\dim\,(\mathrm{SU}(n)) = n^2 - 1 .\tag{8.36}$$

For example, the eight-dimensional group SU(3) is the symmetry group of quantum chromodynamics, see Sect. 20.4. The 24-dimensional group SU(5) was long considered as a candidate for the symmetry of a unified gauge theory (GUT) of elementary particle physics.

**Symplectic Groups**   are known from analytical mechanics. Symplectic matrices belong to linear transformations, which leave the skew-symmetric bilinear form $\langle \boldsymbol{x}, \boldsymbol{y}\rangle := (\boldsymbol{x}, J\boldsymbol{y})$ with

$$J = \begin{pmatrix} 0 & \mathbb{1}_n \\ -\mathbb{1}_n & 0 \end{pmatrix}\tag{8.37}$$

invariant,

$$\mathrm{Sp}(2n, \mathbb{K}) = \{M \in \mathrm{GL}(2n, \mathbb{K})\,|\,M^T JM = J\}.\tag{8.38}$$

The Jacobian matrix of a canonical transformation is an element of $\mathrm{Sp}(2n, \mathbb{R})$. It is evident that every real symplectic matrix has the determinant value $\pm 1$. In fact, using the Pfaffian of $M^T JM$, it can even be proven that $\det(M) = 1$. For example, $\mathrm{Sp}(2, \mathbb{R})$ consists of all real $2 \times 2$-matrices, which fulfill

$$M^T JM = (\det M)J \overset{!}{=} J, \qquad J = \begin{pmatrix} 0 & 1 \\ -1 & 0 \end{pmatrix},$$

i.e. matrices with the determinant 1. Therefore, $\mathrm{Sp}(2, \mathbb{R})$ is isomorphic to $\mathrm{SL}(2, \mathbb{R})$.

We summarize the most important properties of the (simple) matrix groups in a table:

| group | Cartan | condition | dimension |
|---|---|---|---|
| $SL(n+1, \mathbb{R})$ | $A_n$ | $\det A = 1$ | $n(n+2)$ |
| $SU(n+1)$ | $A_n$ | $U^\dagger U = \mathbb{1}_{n+1}$ | $n(n+2)$ |
| $SO(2n+1)$ | $B_n$ | $R^T R = \mathbb{1}_{2n+1}$ | $n(2n+1)$ |
| $Sp(2n, \mathbb{R})$ | $C_n$ | $M^T J_{2n} M = J_{2n}$ | $n(2n+1)$ |
| $SO(2n)$ | $D_n$ | $R^T R = \mathbb{1}_{2n}$ | $n(2n-1)$ |

In the second column, we have given the name of the corresponding group according to the classification of Cartan. These names are also used by the computer program LiE. For example, the commands

```
LiE> dim(A4)
LiE> dim(B4)
LiE> dim(C4)
LiE> dim(D4)
```

lead to the dimensions $\dim(A_4) = 24$, $\dim(B_4) = 36$, $\dim(C_4) = 36$ and $\dim(D_4) = 28$. We will return to the Cartan classification later.

We remind that a simple group $G$ only has the normal subgroups $\{e\}$ and $G$. Therefore, it can neither be the direct nor the semi-direct product[2] of two nontrivial groups. This explains, for example, that the Euclidean group and the Poincaré group are missing in the above list.

The natural question arises about the factor groups

$$GL(n, \mathbb{K})/SL(n, \mathbb{K}), \quad U(n)/SU(n) \quad \text{or} \quad O(n)/SO(n) \,. \tag{8.39}$$

The last one is easy to calculate, $O(n)/SO(n) \cong \mathbb{Z}_2$.

**Task**

Try to determine the other two factor groups.

## 8.4    Global Properties of Lie Groups

The simplest property of a topological space $M$ is the number of its connected components. The Lie groups $SO(n)$, $SU(n)$ and $U(n)$ are connected. On the other hand, the orthogonal group $O(n)$ consists of two components,

$$O(n) = \{R \in O(n)|\det R = 1\} \cup \{R \in O(n)|\det R = -1\}$$

$$= SO(n) \cup \sigma SO(n), \qquad \sigma \in O(n), \quad \det \sigma = -1 \,. \tag{8.40}$$

---

[2] The semi-direct product of groups was introduced in Sect. 3.5.

Global properties are invariant under continuous deformations of the space, which are described by homeomorphisms.

**Definition 8.6** Two topological spaces $M$ and $N$ are called homeomorphic if a homeomorphism $\phi : M \to N$ exists.

Intuitively, two homeomorphic spaces emerge from each other by stretching, compressing, bending, distorting or twisting. Cutting is only allowed if the parts are later precisely reassembled at the cut surface.

**Definition 8.7** A homotopy between two continuous mappings $f, g : M \to N$ is a continuous mapping $H : M \times [0, 1] \to N$ with $H(x, 0) = f(x)$ and $H(x, 1) = g(x)$.

**Homotopy Classes**
If $f$ is homotopic to $g$ one writes $f \sim g$. Homotopy is an equivalence relation, the corresponding equivalence classes are called *homotopy classes*.

Let $f : M \to N$ and $g : N \to M$ be continuous mappings between topological spaces $M$ and $N$. Then the compositions $g \circ f$ and $f \circ g$ are each continuous mappings from $M$ or $N$ onto itself. If there exist $f$ and $g$ with $g \circ f \sim id_M$ and $f \circ g \sim id_N$, then $M$ and $N$ are called *homotopy equivalent or homotopic*. The mappings $f$ and $g$ are then called homotopy equivalences. Homotopy equivalent spaces share most topological properties.

### 8.4.1  Homotopy Groups

We start with the *first homotopy group* of a topological space $M$ and then discuss global properties of (topological) groups. Two paths $w_0$ and $w_1$ (paths were defined in Sect. 8.1.2) that continuously connect two points $p$ and $q$, are called homotopic if one path can be continuously deformed into the other path. More precisely, if there exists a continuous mapping $H : [0, 1] \times [0, 1] \to M$ with the properties

$$H(t, 0) = w_0(t), \quad H(t, 1) = w_1(t), \quad t \in [0, 1]$$

$$H(0, s) = p, \qquad H(1, s) = q, \qquad s \in [0, 1]. \tag{8.41}$$

The parameter $t$ corresponds to the original path parameter, and the parameter $s$ describes the continuous deformation of the path $w_0$ into the path $w_1$, see Fig. 8.5.

A closed path from $p$ to $p$ is called loop with base point $p$. Two *loops* with base point $p$ are homotopic if there is a homotopy between them. Homotopic loops are

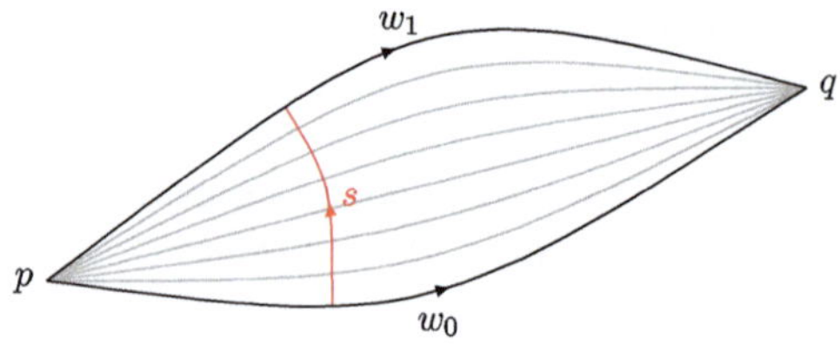

**Fig. 8.5** The paths $w_0$ and $w_1$ can be deformed into each other—they are homotopic

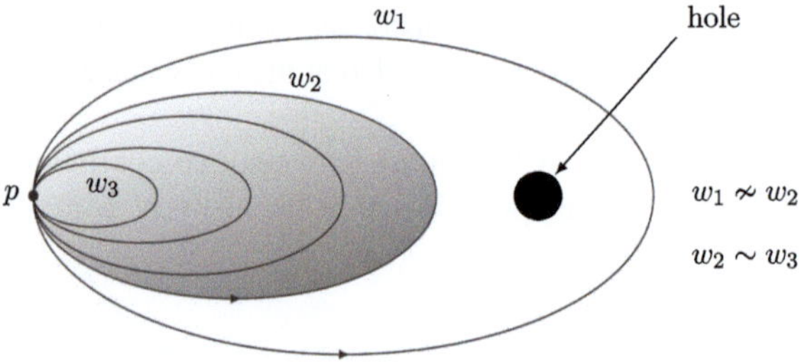

**Fig. 8.6** The loop $w_1$ cannot be continuously deformed into the loop $w_2$. $w_1$ and $w_2$ belong to different homotopy classes. $w_2$ and $w_3$ are homotop and belong to the identity class



considered equivalent and the equivalence classes are called homotopy classes. The set of homotopy classes with base point $p$ is denoted by the symbol $\pi_1(M, p)$.

Two loops can be combined into a third loop by first running through one and then the other, i.e., connecting the end of the first with the beginning of the second. Since homotopies between different representatives can also be used to construct a homotopy between the linked loops, the resulting homotopy class is independent of the choice of representatives. The *neutral element* is the class of loops that can be contracted to the base point. The *inverse class* is obtained by running through the loops of the class in reverse order.

### Fundamental Group

With this method of combining two loops, the set of homotopy classes with base point $p$ becomes a group, the so-called *fundamental group* $\pi_1(M, p)$.

Examples of homotopic and non-homotopic loops in the punctured plane are shown in Fig. 8.6.

Since the loops start and end at $p$, $\pi_1(M, p)$ only measures properties of the connected component in which $p$ lies. For a path-connected $M$, the *choice of base point is irrelevant:* if one chooses another base point $q$, loops from $p$ to $q$ can be shifted by going on a fixed path from $q$ to $p$, then traversing the original loop and then going back the fixed path to $q$. When combining two loops in $q$, the intermediate paths exactly cancel out. The fundamental groups with respect to $p$ and $q$ are therefore isomorphic and one writes $\pi_1(M)$.

**Definition 8.8** A path-connected topological space $M$ with $\pi_1(M) = 0$ is called simply connected.

On a sphere, every loop can be contracted to a point. Therefore, the fundamental group of the sphere is trivial, $\pi_1(S^n) = 0$ for $n > 1$. The two-dimensional plane with a hole $\mathbb{R}^2 \backslash \{0\}$ has the fundamental group $\mathbb{Z}$. The homotopy class of a loop is determined by how many times the loop goes around the hole.

> **Fundamental Groups of $U(1)$ and $SU(2)$**
> $SU(2)$ is simply connected, in contrast to $U(1)$ with $\pi_1(U(1)) = \mathbb{Z}$.

Similar to the fundamental group, the higher homotopy groups $\pi_n(M, p)$ are defined as the set of homotopy classes of continuous pointed mappings $f : (S^n, a) \mapsto (M, p)$, which map a fixed point $a \in S^n$ to $p$. Here, two pointed mappings are similar if they can be deformed into each other with a homotopy (which keeps the base point fixed). Equivalently, we can define the groups $\pi_n(M, p)$ as classes of continuous mappings $g : (I^n, \partial I^n) \to (M, p)$ from the $n$-dimensional unit cube $I^n$ to $M$, with the boundary of the cube going to $p \in M$.

The set of homotopy classes can be endowed with a group structure. The structure of $\pi_n(M, p)$ resembles that of the fundamental group. The group operation is the "gluing" of mappings along one side, i.e., we define the combination of two mappings $f, g : (I^n, \partial I^n) \to (M, p)$ according to

$$(f * g)(t_1, \ldots, t_n) = \begin{cases} f(t_1, \ldots, t_{n-1}, 2t_n) & \text{for } t_n \leq 1/2 \,, \\ g(t_1, \ldots, t_{n-1}, 2t_n - 1) & \text{for } t_n \geq 1/2 \,. \end{cases} \tag{8.42}$$

For a path-connected $M$, $\pi_n(M, p)$, just like $\pi_1(M, p)$, does not depend on the base point $p$ and one writes $\pi_n(M)$. If $M$ and $N$ are path-connected and homeomorphic (or diffeomorphic), then $\pi_n(M) = \pi_n(N)$. The zeroth "homotopy group" $\pi_0$ is generally not a group, but only a set, namely the set of path-connected components of $M$.

In general, it is not easy to calculate the homotopy groups $\pi_n$ of a space. For example, not all groups $\pi_n(S^2)$ are known. But it holds (here unproven)

**Theorem 8.8** *The homotopy group of a product space $M \times N$ is related to the homotopy groups of the factors as follows:*

$$\pi_n(M \times N) = \pi_n(M) \times \pi_n(N), \quad n \geq 1 \,. \tag{8.43}$$

We cite some interesting results about the homotopy groups of Lie groups. We only consider connected groups, $\pi_0(G) = 0$.

**Theorem 8.9 (Weyl)** *For every compact and semi-simple Lie group $G$, $\pi_1(G)$ is finite.*

For example,

$$\pi_1(SU(n)) = 0 \quad \text{and} \quad \pi_1(SO(n)) = \mathbb{Z}_2 \,. \tag{8.44}$$

Non-simple groups, however, can have infinite fundamental groups. For example, $\pi_1(U(1)) = \mathbb{Z}$. Every mapping $S^2 \to G$ is homotopic to the constant mapping, because

**Theorem 8.10 (Cartan)** *For every Lie group $G$, $\pi_2(G) = 0$.*

For the existence of the instanton solutions of the Euclidean Yang-Mills theories, the third homotopy group of the gauge group is relevant. Here, it applies

**Theorem 8.11 (Bott)** *For every compact and simple Lie group $G$, $\pi_3(G) = \mathbb{Z}$.*

The fourth homotopy group of Lie groups characterizes the so-called Witten anomaly. For every compact, simply connected and simple group is $\pi_4(G) = 0$ or $\mathbb{Z}_2$. I am unaware of any physical relevance of the higher homotopy groups.

### 8.4.2  Universal Coverings

The quantum mechanical rotation group $SU(2)$ is the universal covering group of the rotation group $SO(3)$, and the quantum mechanical Lorentz group $SL(2, \mathbb{C})$ is the universal covering of the restricted Lorentz group $SO^\uparrow(1, 3)$. In order to understand these statements, we first define the covering of a topological space.

**Definition 8.9** A covering of a topological space $M$ is a continuous surjective mapping $\pi : C \to M$ from a topological space $C$ to $M$, such that every point $p \in M$ has an open neighborhood $U$ whose preimages $\pi^{-1}(U)$ are the union of disjoint open sets (sheets) in $C$. Each sheet is mapped by $\pi$ homeomorphically onto $U$.

The situation is sketched in Fig. 8.7. For each $p \in M$, the fiber over $p$ is a discrete set in $C$. For each connected component of $M$, the cardinality of the fibers is equal. If each fiber has two elements, we speak of the "double covering".

A covering $\pi : C_0 \to M$ is called *universal*, if $C_0$ is simply connected, i.e. $\pi_1(C_0) = 0$. The term universal comes from the fact that a universal covering covers all connected coverings of $M$.

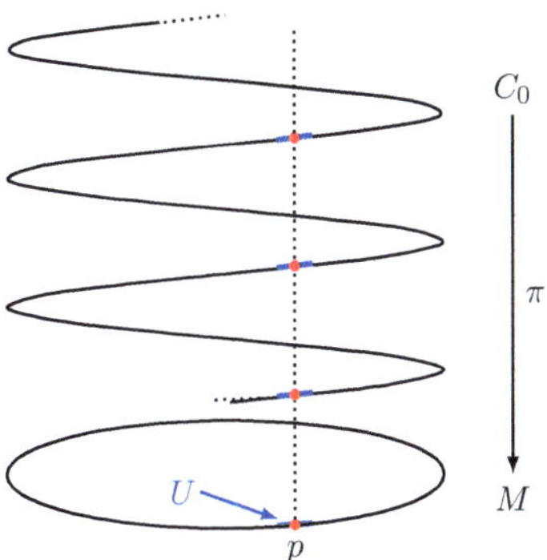

**Fig. 8.7** The universal covering $\pi$ of $M$ projects the simply connected space $C_0$ onto $M$. Shown is the universal covering of U(1) by $\mathbb{R}$

**Uniqueness of the Universal Covering**
If $M$ has a universal covering, then it is unique.

If $\pi : C_0 \to M$ and $\pi' : C_0' \to M$ are two universal coverings of $M$, then there exists a *homeomorphism* $f : C_0 \to C_0'$, such that $\pi' \circ f = \pi$.

**Universal Coverings of U(1) and SO(3)**
The universal covering space of U(1) is $\mathbb{R}$ with $\pi : \alpha \to \exp(i\alpha)$. The fiber over $\exp(i\alpha)$ contains all elements $\alpha + 2\pi\,\mathbb{Z}$ that are mapped onto $\exp(i\alpha)$. The simply connected group SU(2) is the double and universal covering space of the rotation group SO(3).

If $M$ carries additional structures, then these are in many cases inherited by the universal covering. Thus, the universal covering space of a manifold is itself a manifold or the universal covering space of a Lie group is again a Lie group. It is called the universal covering group.

**Theorem 8.12** *If $G$ is a connected Lie group, then there exists a unique (up to isomorphism) universal covering group $\tilde{G}$ with the following properties,*

- $G \cong \tilde{G}/Z$, *where $Z$ is a discrete subgroup of the center of $\tilde{G}$.*
- *If $\pi_1(G) = \{0\}$, then $G$ is isomorphic to $\tilde{G}$.*

The groups SU($n$) are simply connected for $n = 2, 3, \ldots$ and therefore their own covering groups. The covering groups of the SO($n$) are the Spin groups Spin($n$). Recall that Spin(3) $\cong$ SU(2).

---



Justify that $\text{Spin}(4) \cong \text{SU}(2) \times \text{SU}(2)$, $\text{Spin}(5) \cong \text{Sp}(2)$ and $\text{Spin}(6) \cong \text{SU}(4)$.

Herein, $\text{Sp}(n)$ is the compact real form of $\text{Sp}(2n, \mathbb{C})$, defined as invertible quaternion-linear mapping, which leaves the scalar product

$$\langle x, y \rangle = \bar{x}_1 y_1 + \cdots + \bar{x}_n y_n \tag{8.45}$$

defined on the $n$-dimensional quaternionic vector space $\mathbb{H}$, invariant.

---

## 8.5  Exercises for Chap. 8

**Problem 8.1 (The Groups U(1) and SO(2))**  Parametrize the Lie group $\text{SO}(2)$ and show that $\text{SO}(2) \cong \text{U}(1)$.

**Problem 8.2 (Dimensions of Matrix Groups)**  Show that the group $\text{SO}(n)$ has the dimension $n(n-1)/2$ has and $\text{SU}(n)$ has the dimension $n^2 - 1$. For which $n$ do the two groups have the same dimension? Explain this result.

**Problem 8.3 (The Groups U(2) and SU(2))**  Show that every matrix in $\text{SU}(2)$ has the form (8.7) and is uniquely determined by the two complex numbers $a, b$. Why does this prove that $\text{SU}(2)$ is a three-dimensional sphere in $\mathbb{R}^4$? The last remark also shows that $\text{SU}(2)$ is connected and simply connected. Convince yourself that $\text{SU}(2) = [\text{U}(2), \text{U}(2)]$.

**Problem 8.4 (U(n) is not equal to SU(n) × U(1))**  Prove the isomorphism

$$\text{U}(n) = \text{SU}(n) \times \text{U}(1)/\mathbb{Z}_n .$$

Hint: Show that the mapping $\varphi(U, e^{i\lambda}) = e^{i\lambda} U$ defines a surjective homomorphism from $\text{SU}(n) \times \text{U}(1) \mapsto \text{U}(n)$ and determine its kernel.

**Problem 8.5 (The Lie Group SL(2, $\mathbb{C}$))**  The Pauli matrices are given by

$$\mathbb{1} \equiv \sigma_0 = \begin{pmatrix} 1 & 0 \\ 0 & 1 \end{pmatrix}, \quad \sigma_1 = \begin{pmatrix} 0 & 1 \\ 1 & 0 \end{pmatrix}, \quad \sigma_2 = \begin{pmatrix} 0 & -i \\ i & 0 \end{pmatrix}, \quad \sigma_3 = \begin{pmatrix} 1 & 0 \\ 0 & -1 \end{pmatrix} .$$

A linear combination of these matrices with coefficients in $\mathbb{R}^4$ has the form

$$(x, \sigma) = x^\mu \sigma_\mu = \begin{pmatrix} x^0 + x^3 & x^1 - ix^2 \\ x^1 + ix^2 & x^0 - x^3 \end{pmatrix} .$$

- Show that $x^\mu = \frac{1}{2}\mathrm{tr}[\bar{\sigma}^\mu(x,\sigma)]$ holds. Here is

$$\bar{\sigma}_0 = \sigma_0, \quad \bar{\sigma}_i = -\sigma_i, \quad i = 1,2,3.$$

  This means that the mapping $x^\mu \to (x,\sigma)$ is a bijective mapping from $\mathbb{R}^4$ to the linear space of two-dimensional Hermitian matrices.
- Calculate $\det(x,\sigma)$.
- Let $A \in \mathrm{SL}(2,\mathbb{C})$ be a two-dimensional complex matrix with determinant 1. Why is the matrix $A(x,\sigma)A^\dagger$ again a linear combination of the form $(y,\sigma)$?
- Show that the mapping $x \to y$, defined by $(y,\sigma) = A(x,\sigma)A^\dagger$, is linear and can thus be written in the form $y^\mu = \Lambda^\mu{}_\nu x^\nu$. Is this a Lorentz transformation?
- Show that the non-linear mapping $A \to \Lambda(A)$, given by $(\Lambda x,\sigma) = A(x,\sigma)A^\dagger$, is a group homomorphism $\mathrm{SL}(2,\mathbb{C}) \to \mathrm{SO}(1,3)^\uparrow_+$. Is this mapping also an isomorphism?

**Problem 8.6 (Center of Matrix Groups)** What are the centers of the matrix groups $\mathrm{SU}(n)$, $\mathrm{O}(n)$ and $\mathrm{SO}(n)$?

**Problem 8.7 (Fundamental Groups)** Determine the fundamental groups of

- the N-dimensional torus $\mathrm{T}^N$,
- the real projective space $\mathrm{RP}^n = \left(\mathbb{R}^{n+1}\backslash\{0\}\right)/\sim$, where $x \sim \lambda x$ for all $\lambda \in \mathbb{R}\backslash\{0\}$,
- the complex projective space $\mathrm{CP}^n = \left(\mathbb{C}^{n+1}\backslash\{0\}\right)/\sim$, where $x \sim \lambda x$ for all $\lambda \in \mathbb{C}\backslash\{0\}$.

**Problem 8.8 (Global Properties of SU(2) and SO(3))** Explain why $\mathrm{SU}(2)$ is simply connected. This means that every loop in the group can be contracted to a point. Also show that $\mathrm{SO}(3)$ is not simply connected.

**Problem 8.9 (Anti-de Sitter Space)** Consider $\mathbb{R}^5$ with metric tensor

$$(\eta_{\mu\nu}) = \eta = \begin{pmatrix} 1 & 0 & 0 & 0 & 0 \\ 0 & 1 & 0 & 0 & 0 \\ 0 & 0 & -1 & 0 & 0 \\ 0 & 0 & 0 & -1 & 0 \\ 0 & 0 & 0 & 0 & -1 \end{pmatrix}.$$

The elements of the Anti-de Sitter group $\mathrm{SO}(3,2)$ are the linear mappings $\mathbb{R}^5 \to \mathbb{R}^5$, given by $\xi' = \Lambda\xi$, which leave the bilinear form $\xi^T\eta\xi$ invariant and satisfy $\det\Lambda = 1$. The transformation matrices fulfill the relation

$$\Lambda^T\eta\Lambda = \eta.$$

1. The Anti-de Sitter space with 'radius' $R$ is defined by

$$\mathrm{AdS}_4 = \{\xi \in \mathbb{R}^5 | \xi^T \eta \xi = R^2\}.$$

   Visualize this space by the analogous consideration in $\mathbb{R}^3$ with metric $\eta = \mathrm{diag}(1, 1, -1)$.

2. The metric on AdS is the metric induced by the embedding of $\mathrm{AdS}_4 \subset \mathbb{R}^5$. Determine the signature of the metric on AdS. It is sufficient to examine the simpler case in $\mathbb{R}^3$ with metric tensor $\eta = \mathrm{diag}(1, 1, -1)$.

3. $SO(3, 2)$ operates on $\mathrm{AdS}_4$ via

$$SO(3, 2) \times \mathrm{AdS}_4 \to \mathrm{AdS}_4, \quad (\Lambda, \xi) \to \Lambda \xi.$$

   This is an isometry on $(\mathbb{R}^5, G)$, and therefore on the AdS space. Show that all points of the AdS space can be reached by applying a suitable $SO(3, 2)$-"rotation" to the fixed point$(R, 0, 0, 0, 0)$. What does this mean for the action of $SO(3, 2)$ on $\mathrm{AdS}_4$?

4. Construct the closed and timelike geodesics through $(R, 0, 0, 0, 0)$ and spacelike geodesics through $(R, 0, 0, 0, 0)$.

   Hint: no calculation is needed here!

**Problem 8.10 (Factor Groups)** What manifolds do the factor groups describe

$$SO(n)/SO(n - 1) \quad \text{and} \quad SU(n)/SU(n - 1)?$$

Hint: Consider $SO(n - 1)$ as a subgroup of $SO(n)$, for example as the stabilizer of $e_1$:

$$SO(n - 1) = \{R \in SO(n) | R e_1 = e_1\}.$$

Therefore, the coset is just the set $\{R e_1 | R \in SO(n)\}$. A similar approach also leads to success for $SU(n - 1)$-considered as a subgroup of $SU(n)$.

**Problem 8.11 (Symplectic Group)** Show that the non-compact Lie groups $Sp(2, \mathbb{R})$ and $SL(2, \mathbb{R})$ are isomorphic.

**Problem 8.12 (Relationship Between O(4) and SU(2) × SU(2))** The set of matrices

$$x = \begin{pmatrix} a & b \\ -b^* & a^* \end{pmatrix}$$

form the linear space $\mathbb{C}^2 \cong \mathbb{R}^4$ with scalar product

$$\langle x, y \rangle = \mathrm{tr}(x^\dagger y)\,.$$

The matrices with $|a|^2 + |b|^2 = 1$ lie in SU(2) and define the unit sphere in $\mathbb{R}^4$.

1. Let $g_1, g_2$ be two matrices in SU(2) and $x$ a matrix as above. Show that the mapping $R(g_1, g_2)$, defined by

$$R(g_1, g_2)x = g_1 x g_2^{-1}\,,$$

   is linear and preserves the scalar product,

$$\langle R(g_1, g_2)x, R(g_1, g_2)y \rangle = \langle x, y \rangle\,.$$

   Thus, $R(g_1, g_2)$ is a linear mapping $\mathbb{R}^4 \to \mathbb{R}^4$ that preserves the length of vectors.
2. Now show that this mapping defines a group homomorphism $SU(2) \times SU(2) \to SO(4)$.
3. Is this homomorphism faithful?

The group $SU(2) \times SU(2)$ is the symmetry group of the quark-meson model, which describes the degrees of freedom of QCD at low energies.

# Invariant Integration

9



The Lebesgue measure on $\mathbb{R}^n$ is translation invariant. This means that the value of an integral does not change when the domain of definition is translated. More generally, the volume of a body does not change when it is moved in space. Similarly, for most finite, discrete or continuous groups, invariant measures (integrals) can be introduced.

The study of invariant integration over groups goes back to an influential work by A. Hurwitz [41]. In it, explicit expressions for invariant measures of the unitary and orthogonal groups can already be found. Further important contributions come from I. Schur and H. Weyl [42]. The existence and uniqueness of an invariant measure for locally compact topological groups was proven around the same time by J. von Neumann [43] and A. Haar [44]. Invariant measures for groups are also called Haar measures.

To get started, let's consider the average value of a function $f : G \mapsto \mathbb{C}$ on a *finite group*

$$\mathcal{M}(f) = \frac{1}{|G|} \sum_{g \in G} f(g) . \tag{9.1}$$

This averaging has the following properties:

1. linear: $\mathcal{M}(\alpha f_1 + \beta f_2) = \alpha \mathcal{M}(f_1) + \beta \mathcal{M}(f_2)$ for all $\alpha, \beta \in \mathbb{C}$.
2. positive: for real $f$ with $f \geq 0$ it follows that $\mathcal{M}(f) \geq 0$ and $= 0$ only for $f \equiv 0$.

3. normalized: if $f(g) = 1$ for all $g \in G$, then $\mathcal{M}(f) = 1$.
4. left- and right-invariant, $\mathcal{M}(f) = \mathcal{M}(f \circ \ell_a) = \mathcal{M}(f \circ r_a)$.

Linearity, positivity, and normalization of the averaging are obvious. The invariance with respect to the left translation $\ell_a(g) = ag$ with an arbitrary group element $a$ follows from the fact that the sets $\{ag | g \in G\}$ and $\{g | g \in G\}$ are equal:

$$\mathcal{M}(f \circ \ell_a) = \frac{1}{|G|} \sum_{g \in G} f(ag) = \frac{1}{|G|} \sum_{g \in G} f(g) = \mathcal{M}(f).$$

Similarly, one can show the invariance of $\mathcal{M}$ with respect to right translations $r_a(g) = ga$.

Of course, the question arises whether there is also an invariant averaging on a continuous Lie group. If so, then the sum (9.1) would become an integral over the group. For compact groups, the answer to the question is positive:

**Theorem 9.1 (Invariant Measure, Haar Measure)**  *For every compact Lie group, there is a (up to a multiplicative positive constant) unique positive Haar integral*

$$\mathcal{M} : C_0(G) \to \mathbb{C}, \qquad f \to \mathcal{M}(f) = \int_G d\mu(g) f(g), \tag{9.2}$$

*which is left- and right-invariant. For continuous functions, the integral is finite.*

The measure of a set $O \subset G$ is thus equal to the measure of the sets $\ell_a(O)$ and $r_a(O)$ shifted by left and right translation. For compact groups, the measure can be normalized to one, $\mathcal{M}(1) = 1$, and then one speaks of the normalized Haar measure. The normalization determines it *uniquely*. The corresponding averaging then fulfills the above properties: It is linear, positive, normalized, left- and right-invariant.

In 1933, Alfred Haar showed that there is (up to normalizations) also a unique left-invariant measure $d\mu_\ell$ and a unique right-invariant measure $d\mu_r$ for more general locally compact topological groups. The two measures can be different. Groups with $d\mu_\ell(g) = \text{const}\, d\mu_r(g)$ are called unimodular.

---

**Haar Measure on the Multiplicative Group** $\mathbb{R}^*$

The elements of the non-compact group $\mathbb{R}^*$ are the real numbers in $\mathbb{R}^* = \mathbb{R} \setminus \{0\}$ with multiplication as operation. The average of a complex function is

$$\mathcal{M}(f) = \int_{\mathbb{R}^*} \frac{dx}{|x|} f(x). \tag{9.3}$$

(continued)

The averaging is linear, positive, but not normalized. For $y > 0$, the left invariance follows from

$$\mathcal{M}(f \circ \ell_y) = \int_{\mathbb{R}^*} \frac{\mathrm{d}x}{|x|} f(yx) = \int_{\mathbb{R}^*} \frac{\mathrm{d}(yx)}{y|x|} f(yx) = \int_{\mathbb{R}^*} \frac{\mathrm{d}z}{|z|} f(z) = \mathcal{M}(f).$$

(9.4)

The group is abelian and therefore the right invariance follows from the left invariance.

The set $(\mathbb{R}^*, \cdot)$ is a non-compact group with a non-normalizable invariant measure.

**Task**

Show the invariance of the averaging (9.3) also for negative $y$.

## 9.1  Haar Measures on U(1) and SU(2)

In this section, we construct the Haar measure on the compact Lie groups U(1) and SU(2). For SU(2), we will make use of the geometric interpretation of translations on the group manifold $S^3$. Since this section deals with elements of unitary groups, we denote these with $U$.

### 9.1.1  Haar Measure on U(1)

We parameterize the elements of the abelian Lie group U(1) as in (8.10) with an angle $\alpha$. A function $f : \mathrm{U}(1) \to \mathbb{C}$ is a $2\pi$-periodic function of the angle, and

$$\mathcal{M}(f) = \frac{1}{2\pi} \int_0^{2\pi} \mathrm{d}\alpha \, f\left(\mathrm{e}^{\mathrm{i}\alpha}\right).$$

(9.5)

The invariance with respect to left translations is obvious,

$$\mathcal{M}(f \circ \ell_{\mathrm{e}^{\mathrm{i}\beta}}) = \frac{1}{2\pi} \int \mathrm{d}\alpha f\left(\mathrm{e}^{\mathrm{i}\beta} \, \mathrm{e}^{\mathrm{i}\alpha}\right) = \frac{1}{2\pi} \int \mathrm{d}\alpha f\left(\mathrm{e}^{\mathrm{i}(\beta+\alpha)}\right) = \mathcal{M}(f).$$

**Question**

Why is the measure also right-invariant?

### 9.1.2  Haar Measure on SU(2)

What does the invariant Haar measure on SU(2) look like? For this, we clarify the geometric meaning of the left translation on the group manifold $S^3$. In Sect. 8.2, we saw that the group elements can be parameterized as follows,

$$
\alpha \mapsto U(\alpha) = \begin{pmatrix} \alpha_1 + i\alpha_2 & \alpha_3 + i\alpha_4 \\ -\alpha_3 + i\alpha_4 & \alpha_1 - i\alpha_2 \end{pmatrix} \quad \text{with} \quad \alpha = \begin{pmatrix} \alpha_1 \\ \alpha_2 \\ \alpha_3 \\ \alpha_4 \end{pmatrix} \in S^3 . \tag{9.6}
$$

The left translation $U \to \ell_{\tilde U}(U)$ with the group element $\tilde U = U(\beta)$ is then given in the parameter space by

$$
U(\beta)U(\alpha) = U\big(O(\beta)\alpha\big) \quad \text{with} \quad O(\beta)\alpha = \begin{pmatrix} \beta_1 & -\beta_2 & -\beta_3 & -\beta_4 \\ \beta_2 & \beta_1 & -\beta_4 & \beta_3 \\ \beta_3 & \beta_4 & \beta_1 & -\beta_2 \\ \beta_4 & -\beta_3 & \beta_2 & \beta_1 \end{pmatrix} \begin{pmatrix} \alpha_1 \\ \alpha_2 \\ \alpha_3 \\ \alpha_4 \end{pmatrix} . \tag{9.7}
$$

Since $\beta \in S^3$, $O(\beta)$ is an orthogonal matrix, $O^T O = 1$, as can be verified by explicit calculation. Therefore, $O(\beta)\alpha$ is a rotation of $\alpha$. Now, the volume form induced from $\mathbb{R}^4$ to $S^3$ is rotation invariant and thus invariant under left translations (and right translations). If we normalize the volume form to 1, then we obtain the unique Haar measure on SU(2) $\cong S^3$. If we parameterize the points on the unit sphere $S^3$ according to

$$
\begin{pmatrix} \alpha_1 \\ \alpha_2 \\ \alpha_3 \\ \alpha_4 \end{pmatrix} = \begin{pmatrix} \cos\vartheta \\ \sin\vartheta \cos\psi \\ \sin\vartheta \sin\psi \cos\varphi \\ \sin\vartheta \sin\psi \sin\varphi \end{pmatrix} , \tag{9.8}
$$

then the group elements are parameterized as follows,

$$
U(\vartheta, \psi, \varphi) = \begin{pmatrix} \cos\vartheta + i\sin\vartheta \cos\psi & \sin\vartheta \sin\psi\, e^{i\varphi} \\ -\sin\vartheta \sin\psi\, e^{-i\varphi} & \cos\vartheta - i\sin\vartheta \cos\psi \end{pmatrix} , \tag{9.9}
$$

with the following value ranges for the three angles:

$$
0 < \vartheta < \pi, \quad 0 < \psi < \pi \quad \text{and} \quad 0 < \varphi < 2\pi . \tag{9.10}
$$

Using the known volume element on the unit sphere $S^3$ we conclude:

**Haar Measure on SU(2)**
With respect to the angular $(\vartheta, \psi, \varphi)$, the invariant volume element (Haar measure) has the form

$$
\mathrm{d}\mu = \frac{1}{2\pi^2}\sin^2\vartheta \cdot \sin\psi \; \mathrm{d}\vartheta\,\mathrm{d}\psi\,\mathrm{d}\varphi .
\tag{9.11}
$$

### 9.1.3   Reduced Haar Measure on SU(2)

If we consider the average of class functions $f : G \mapsto \mathbb{C}$, this leads directly to the reduced Haar measure. Class functions are special functions, which are constant on each conjugacy class of the group. This means

$$
f(VUV^{-1}) = f(U)
\tag{9.12}
$$

for all group elements $V$. For matrix groups, for example, $\mathrm{Tr}(U)$ is a class function. The trace of the matrix in (9.9) is $\mathrm{Tr}(U) = 2\cos\vartheta$ and its determinant is 1. Therefore, it has the two eigenvalues $\mathrm{e}^{\pm i\vartheta}$ and there is a unitary $V$ with

$$
VUV^{-1} = \begin{pmatrix} \mathrm{e}^{i\vartheta} & 0 \\ 0 & \mathrm{e}^{-i\vartheta} \end{pmatrix} \equiv D(\vartheta) \qquad \vartheta \in (0, \pi) .
$$

The conjugacy class is thus characterized solely by the angle $\vartheta$ and therefore class functions will only depend on this coordinate $\vartheta$. They are $2\pi$-periodic functions of $\vartheta$ and due to

$$
\begin{pmatrix} 0 & 1 \\ -1 & 0 \end{pmatrix} D(\vartheta) \begin{pmatrix} 0 & -1 \\ 1 & 0 \end{pmatrix} = D(-\vartheta)
$$

they must be even functions in $\vartheta$. We conclude that for every class function

$$
f(U) = f(\vartheta) = f(-\vartheta) = f(\vartheta + 2\pi)
$$

must hold. The mean value of a class function is

$$
\mathcal{M}(f) = \frac{1}{2\pi^2} \int \sin^2\vartheta \cdot \sin\psi \cdot f(\vartheta)\,\mathrm{d}\vartheta\,\mathrm{d}\psi\,\mathrm{d}\varphi
$$

$$
= \frac{2}{\pi} \int_0^\pi \sin^2\vartheta\, f(\vartheta)\,\mathrm{d}\vartheta = \int_0^\pi \mathrm{d}\mu_{\mathrm{red}}(\vartheta) f(\vartheta) .
\tag{9.13}
$$

**Reduced Haar Measure on SU(2)**

With respect to the angular coordinates $(\vartheta, \psi, \varphi)$, the reduced Haar measure on SU(2) is

$$\mathrm{d}\mu_{\mathrm{red}} = \frac{2}{\pi}\sin^2\vartheta\,\mathrm{d}\vartheta, \quad \vartheta \in [0,\pi]. \tag{9.14}$$

## 9.2  Haar Measures on Arbitrary Lie Groups

Let $G$ be an $n$-dimensional and not necessarily compact Lie group. We choose a neighborhood $U$ of the neutral element $e$ and parametrize the elements in $U$ and the invariant measure on $U$ with local coordinates $\alpha = (\alpha_1, \ldots, \alpha_n)$. More explicitly, we are looking for the density $\rho_\ell$ of the left-invariant measure for these coordinates, $\mathrm{d}\mu_\ell(g) = \rho_\ell(\alpha)\mathrm{d}\alpha$. We choose the coordinates so that the unit element is at the coordinate origin $\alpha = 0$, as shown in Fig. 9.1. Select an arbitrary (but fixed) group element with the coordinates $\beta$ and perform a left shift of $g(\alpha)$ with this group element, resulting in $g(\beta)g(\alpha) = g(\gamma)$ with $\gamma = m(\beta, \alpha)$. We demand

$$\rho_\ell(\alpha)\mathrm{d}^n\alpha = \rho_\ell(\gamma)\mathrm{d}^n\gamma . \tag{9.15}$$

Because $\gamma = m(\beta, \alpha)$, the volume element transforms according to

$$\mathrm{d}\gamma_1 \cdots \mathrm{d}\gamma_n = \det\begin{pmatrix} \frac{\partial m_1(\beta,\alpha)}{\partial\alpha_1} & \cdots & \frac{\partial m_1(\beta,\alpha)}{\partial\alpha_n} \\ \vdots & & \vdots \\ \frac{\partial m_n(\beta,\alpha)}{\partial\alpha_1} & \cdots & \frac{\partial m_n(\beta,\alpha)}{\partial\alpha_n} \end{pmatrix}_{\alpha=0} \mathrm{d}\alpha_1 \cdots \mathrm{d}\alpha_n = \det J_\ell(\gamma)\,\mathrm{d}^n\alpha .$$

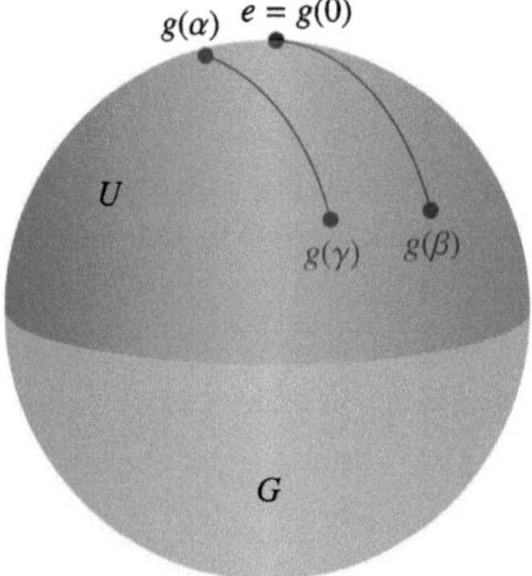


**Fig. 9.1** The density at $e$ is transported to $g(\beta)$ using left translation

For $\alpha = 0$ we have $\beta = \gamma$, so that the determinant of the Jacobian matrix $J_\ell$ is a function of $\gamma$ only. Now it is obvious what the left-invariant measure we are looking for looks like:

$$d\mu_\ell(\gamma) = \rho_\ell(\gamma)\, d^n\gamma = \frac{\rho_\ell(0)}{\det J_\ell(\gamma)}\, d\gamma_1 \cdots d\gamma_n\,. \qquad (9.16)$$

The invariance under left translation results from

$$\rho_\ell(\gamma)\, d^n\gamma = \frac{\rho_\ell(0)}{\det J_\ell(\gamma)}\, d^n\gamma = \rho_\ell(0)d^n\alpha\,. \qquad (9.17)$$

For compact groups, the constant $\rho_\ell(0)$ can be chosen so that $d\mu_\ell$ is normalized to one. To obtain the right-invariant measure we consider the right translation $g(\gamma) = g(\alpha)g(\beta)$ with an arbitrary but fixed $g(\beta)$ and $\gamma = \tilde{m}(\beta, \alpha)$. The measure contains the determinant of the Jacobian matrix for the right translations $J_r$ at the point $\alpha = 0$ in the denominator. We summarize:

---

**Left- and Right-invariant Measures**
The following measures for $G$ are left- or right-invariant:

$$d\mu_\ell(g) = \frac{\rho_\ell(0)}{\det J_\ell(\gamma)}\, d^n\gamma \quad \text{or} \quad d\mu_r(g) = \frac{\rho_r(0)}{\det J_r(\gamma)}\, d^n\gamma\,, \qquad (9.18)$$

where $J_\ell$ and $J_r$ are the Jacobian matrices of the left and right translations (in local coordinates).

---

### Example: Affine Group on $\mathbb{R}$

This subgroup of the conformal transformations contains the dilations and shifts, $x \to \alpha_1 x + \alpha_2$ with $\alpha = (\alpha_1, \alpha_2) \in \mathbb{R}^* \times \mathbb{R}$. In local coordinates, left and right translations of this connected two-dimensional Lie group take the form

$$m_1(\beta, \alpha) = \beta_1\alpha_1, \quad m_2(\beta, \alpha) = \beta_1\alpha_2 + \beta_2 \implies J_\ell(\beta) = \beta_1^2\,, \qquad (9.19)$$

$$\tilde{m}_1(\beta, \alpha) = \beta_1\alpha_1, \quad \tilde{m}_2(\beta, \alpha) = \beta_2\alpha_1 + \alpha_2 \implies J_r(\beta) = \beta_1\,. \qquad (9.20)$$

The invariant measures are different and cannot be normalized,

$$d\mu_\ell(\alpha) = \frac{1}{\alpha_1^2}\, d\alpha_1 \wedge d\alpha_2 \quad \text{and} \quad d\mu_r(\alpha) = \frac{1}{\alpha_1}\, d\alpha_1 \wedge d\alpha_2\,. \qquad (9.21)$$

◀

**Task**

Why is the group neither abelian, nor compact nor semi-simple? Show that the shifts form a non-trivial normal subgroup.

The example shows that for non-compact groups the left- and right-invariant measures can be different. If the group $G$ is abelian or semi-simple, then the two measures (possibly after a rescaling) are identical, $\mathrm{d}\mu_\ell = \mathrm{d}\mu_r$ and it is a unimodular group. For non-compact groups, the invariant measures are not normalizable, i.e.

$$\int_G \mathrm{d}\mu_\ell(g) = \int_G \mathrm{d}\mu_r(g) = \infty.$$

### 9.2.1 Invariant 1-Forms and Invariant Integration

As an introduction, let's take another look at the group of dilations and translations in $\mathbb{R}^*$. These transformations can be represented on $\mathbb{R}^2$ as follows (cf. the analogous construction for the Galilei group in Sect. 5.4)

$$\begin{pmatrix} x \\ 1 \end{pmatrix} \mapsto \begin{pmatrix} \alpha_1 & \alpha_2 \\ 0 & 1 \end{pmatrix} \begin{pmatrix} x \\ 1 \end{pmatrix} = \begin{pmatrix} \alpha_1 x + \alpha_2 \\ 1 \end{pmatrix}. \tag{9.22}$$

The group is thus a subgroup of $\mathrm{GL}(2, \mathbb{R})$,

$$G = \left\{ A = \begin{pmatrix} \alpha_1 & \alpha_2 \\ 0 & 1 \end{pmatrix} \,\middle|\, \alpha_1 \in \mathbb{R}^*, \alpha_2 \in \mathbb{R} \right\}. \tag{9.23}$$

Now we define the matrix-valued 1-forms

$$\omega_\ell(A) = A^{-1}\mathrm{d}A \quad \text{and} \quad \omega_r(A) = \mathrm{d}A A^{-1}. \tag{9.24}$$

These take their values in the associated Lie algebra (more on this in Chap. 13). They are obviously left- or right-invariant, i.e. for each constant $B \in G$ it holds

$$\omega_\ell(A) = \omega_\ell(BA) \quad \text{and} \quad \omega_r(A) = \omega_r(AB). \tag{9.25}$$

For the matrices in (9.23) one finds the explicit expressions

$$\omega_\ell = \frac{1}{\alpha_1} \begin{pmatrix} \mathrm{d}\alpha_1 & \mathrm{d}\alpha_2 \\ 0 & 0 \end{pmatrix} \quad \text{and} \quad \omega_r = \frac{1}{\alpha_1} \begin{pmatrix} \mathrm{d}\alpha_1 & \alpha_1\mathrm{d}\alpha_2 - \alpha_2\mathrm{d}\alpha_1 \\ 0 & 0 \end{pmatrix}. \tag{9.26}$$

These define a left- or a right-invariant matrix-valued 2-form

$$\omega_\ell \wedge \omega_\ell = \frac{d\alpha_1 \wedge d\alpha_2}{\alpha_1^2}\begin{pmatrix} 0 & 1 \\ 0 & 0 \end{pmatrix} \quad , \quad \omega_r \wedge \omega_r = \frac{d\alpha_1 \wedge d\alpha_2}{\alpha_1}\begin{pmatrix} 0 & 1 \\ 0 & 0 \end{pmatrix} . \tag{9.27}$$

The coefficients of the constant matrix are left- or right-invariant volume forms. Not surprisingly, they are the invariant measures given in (9.21). In the following, we will mostly use the left-invariant differential form $\omega_\ell$ and briefly denote it by $\omega$.

## 9.3  Haar Measures for Compact Matrix Groups

Let $G$ be an $n$-dimensional compact matrix group. Using the just introduced left-invariant matrix-valued differential form

$$\omega = g^{-1}dg, \quad dg = \sum \frac{\partial g}{\partial \alpha_i}d\alpha_i , \tag{9.28}$$

one can easily construct left- and right-invariant scalar $m$-forms,

$$\omega_m = \mathrm{Tr}\big(\underbrace{\omega \wedge \omega \wedge \cdots \wedge \omega}_{m \text{ times}}\big) = \mathrm{Tr}(\omega^{\wedge m}), \quad m = 1, \dots, n . \tag{9.29}$$

These are obviously left-invariant under $g \to ag$, since $\omega$ is already left-invariant.

---

**Question**

Why are these $m$-forms also right-invariant?

---

If for $n = \dim(G)$ the form $\omega_n$ is positive or can be chosen positive by multiplication with a constant, then it is proportional to the invariant Haar measure. In this way, one can explicitly and efficiently calculate invariant Haar measures.

**Lemma 9.1 (Haar Measure on Compact Matrix Groups)** *Compact matrix groups are unimodular and the left- and right-invariant Haar measure is given by*

$$d\mu(g) = \mathrm{const} \cdot \omega_n, \quad \omega_n = \mathrm{Tr}\big(\omega^{\wedge n}\big), \quad \omega = g^{-1}dg . \tag{9.30}$$

For the non-compact affine group of dilations and translations, the traces of $\omega_\ell \wedge \omega_\ell$ and $\omega_r \wedge \omega_r$ vanish, so that we (not unexpectedly) could not find a simultaneously left- and right-invariant volume form.

## 9.4    Haar Measures for Unitary Groups

We now examine matrix groups with unitary matrices $U$ in more detail. To calculate the invariant $n$-form $\omega_n$, we diagonalize $U$:

$$U = VDV^{-1}, \quad V \text{ unitary}. \tag{9.31}$$

A class function depends only on the eigenvalues of $U$. These are the matrix elements of the diagonal matrix $D$. We choose these as coordinates of the group. The missing coordinates are provided by the diagonalizing matrix $V$. We have

$$\omega = U^{-1}\mathrm{d}U = V\alpha V^{-1} \quad \text{with} \quad \alpha = D^{-1}V^{-1}\mathrm{d}VD - V^{-1}\mathrm{d}V + D^{-1}\mathrm{d}D. \tag{9.32}$$

For each unitary matrix, the differential form $\omega = U^{-1}\mathrm{d}U$ is an anti-Hermitian matrix (i.e., it is Lie algebra-valued) as one quickly proves:

$$0 = \mathrm{d}(U^{-1}U) = \mathrm{d}U^{-1}U + U^{-1}\mathrm{d}U = \omega^{\dagger} + \omega.$$

We now decompose the matrix-valued form $V^{-1}\mathrm{d}V$ into its diagonal and non-diagonal part. The diagonal part commutes with the diagonal matrix $D$ and cancels in the last sum for $\alpha$ in (9.32). Accordingly, we find

$$\alpha = D^{-1}\beta D - \beta + D^{-1}\mathrm{d}D, \quad \text{where} \quad \beta = (V^{-1}\mathrm{d}V)_{\perp} \tag{9.33}$$

denotes the non-diagonal part of $V^{-1}\mathrm{d}V$. The traces in (9.29) are invariant under conjugation with $V$ and therefore

$$\omega_n = \mathrm{Tr}\left(\omega^{\wedge n}\right) = \mathrm{Tr}\left(\alpha^{\wedge n}\right) = \mathrm{Tr}\left(D^{-1}\beta D - \beta + D^{-1}\mathrm{d}D\right)^{\wedge n}. \tag{9.34}$$

In the explicit calculation of the reduced Haar measure, this formula is very useful and also serves as a starting point for investigations in Chap. 16. We want to illustrate its usefulness with examples.

### 9.4.1    Invariant Differential Forms and Haar Measure on SU(2)

We had already calculated the full and the reduced Haar measure on this group in Sect. 9.1. We will now see that the invariant differential forms lead more quickly to the same result. For SU(2), the anti-Hermitian differential form $U^{-1}\mathrm{d}U$ is traceless. With $D = \mathrm{diag}(\mathrm{e}^{\mathrm{i}\vartheta}, \mathrm{e}^{-\mathrm{i}\vartheta})$ and $\beta = -\beta^{\dagger}$ it follows

$$D^{-1}\mathrm{d}D = \mathrm{i}\begin{pmatrix} \mathrm{d}\vartheta & 0 \\ 0 & -\mathrm{d}\vartheta \end{pmatrix}, \quad \beta = (V^{-1}\mathrm{d}V)_{\perp} = \begin{pmatrix} 0 & \beta_{12} \\ -\beta_{12}^{\dagger} & 0 \end{pmatrix}. \tag{9.35}$$

The Lie algebra-valued 1-form $\alpha$ in (9.33) has in the parametrization (9.9) the independent components

$$\alpha_{11} = \mathrm{i}\,\mathrm{d}\vartheta, \quad \alpha_{12} = -2\mathrm{i}\,\mathrm{e}^{-\mathrm{i}\vartheta} \sin\vartheta\, \beta_{12}\,. \tag{9.36}$$

The trace of the third power is proportional to a positive invariant volume form,

$$\omega_3 = \mathrm{Tr}(\alpha \wedge \alpha \wedge \alpha) = 24\mathrm{i}\sin^2\vartheta\,\mathrm{d}\vartheta\,\beta_{12}^{\dagger}\beta_{12}\,, \tag{9.37}$$

and is therefore proportional to the invariant Haar measure. For comparison with the earlier result (9.11) we choose a $V$, for which $VDV^{-1}$ has the form (9.9),

$$V = \begin{pmatrix} \mathrm{e}^{\mathrm{i}\varphi}\cos\frac{\psi}{2} & \mathrm{i}\sin\frac{\psi}{2} \\ \mathrm{i}\sin\frac{\psi}{2} & \mathrm{e}^{-\mathrm{i}\varphi}\cos\frac{\psi}{2} \end{pmatrix} \implies \beta_{12} = \mathrm{e}^{-\mathrm{i}\varphi}\left( \frac{\mathrm{i}}{2}\,\mathrm{d}\psi - 2\sin\psi\,\mathrm{d}\varphi \right)\,. \tag{9.38}$$

Thus, $\beta_{12}^{\dagger}\beta_{12} = -2\mathrm{i}\sin\psi\,\mathrm{d}\psi\,\mathrm{d}\varphi$ and the invariant measure has the form

$$\mathrm{d}\mu = \mathrm{const}\cdot\omega_3, \quad \omega_3 = 48\sin^2\vartheta\cdot\sin\psi\,\mathrm{d}\vartheta\,\mathrm{d}\psi\,\mathrm{d}\varphi\,. \tag{9.39}$$

With $\mathrm{const} = 1/96\pi^2$ it agrees with the normalized Haar measure in (9.11). The *reduced Haar measure* can now be read off immediately: The angle $\vartheta$ parametrizes the conjugacy classes, and the 2-form $\beta_{12}^{\dagger}\beta_{12}$ is independent of $\vartheta$. We can therefore integrate over equivalent group elements and find the earlier result (9.14).

### 9.4.2   Invariant Forms and Haar Measure on U(2)

For this four-dimensional non-semisimple group, the 4-form $\alpha \wedge \alpha \wedge \alpha \wedge \alpha$ is identically zero and its trace thus does not define an invariant volume form. With a mild modification of the above construction, we can nevertheless extract an invariant Haar measure from $\alpha$. For U(2), the matrix-valued 1-forms matrices $U^{-1}\mathrm{d}U$ are no longer traceless. We choose $D = \mathrm{diag}(\mathrm{e}^{\mathrm{i}\vartheta_1},\, \mathrm{e}^{\mathrm{i}\vartheta_2})$ and obtain

$$D^{-1}\mathrm{d}D = \begin{pmatrix} \mathrm{i}\mathrm{d}\vartheta_1 & 0 \\ 0 & \mathrm{i}\mathrm{d}\vartheta_2 \end{pmatrix}\,. \tag{9.40}$$

The matrix $\beta$ retains the form in (9.35), so that $\alpha$ has the following components,

$$\alpha_{11} = \mathrm{i}\mathrm{d}\vartheta_1, \quad \alpha_{22} = \mathrm{i}\mathrm{d}\vartheta_2, \quad \alpha_{12} = -2\mathrm{i}\,\mathrm{e}^{-\mathrm{i}\vartheta_{12}}\sin(\vartheta_{12})\,\beta_{12}\,, \tag{9.41}$$

where we introduced the half difference angle $\vartheta_{12} = (\vartheta_1 - \vartheta_2)/2$. Because $\alpha \wedge \alpha \wedge \alpha \wedge \alpha$ identically vanishes, one chooses the invariant forms

$$\omega_1 = \mathrm{Tr}\,\omega = \mathrm{Tr}\,\alpha = \mathrm{i}\mathrm{d}(\vartheta_1 + \vartheta_2)\,,$$

$$\omega_3 = \mathrm{Tr}(\alpha \wedge \alpha \wedge \alpha) = 3\mathrm{i}\mathrm{d}(\vartheta_1 - \vartheta_2)\,\alpha_{12}^{\dagger}\alpha_{12}\,,$$

with which we can construct an invariant volume form,

$$d\mu = \omega_1 \wedge \omega_3 = 6d\vartheta_1 d\vartheta_2 \alpha_{12}^\dagger \alpha_{12} = 24 \sin^2(\vartheta_{12})\, d\vartheta_1 d\vartheta_2 \beta_{12}^\dagger \beta_{12}\,. \tag{9.42}$$

From this, one immediately obtains the reduced Haar measure after integration over $\beta_{12}^\dagger \beta_{12}$.

**Reduced Haar Measure on U(2)**
The group U(2) has the reduced normalized Haar measure

$$d\mu_{\text{red}} = \frac{1}{2\pi^2} \sin^2 \frac{\vartheta_1 - \vartheta_2}{2}\, d\vartheta_1 d\vartheta_2, \quad \vartheta_1, \vartheta_2 \in [0, 2\pi)\,. \tag{9.43}$$

Finally, we turn to the unitary groups U($n$) and SU($n$)).

### 9.4.3   Invariant Forms and Haar Measure for U($n$) and SU($n$)

The procedure is similar to that for the groups U(2) and SU(2). First, we parametrize the diagonal matrices of U($n$) according to

$$D = \text{diag}\left(e^{i\vartheta_1}, \ldots, e^{i\vartheta_n}\right) \implies D^{-1}dD = i\,\text{diag}(d\vartheta_1, \ldots, d\vartheta_n)\,. \tag{9.44}$$

The matrix elements of the non-diagonal anti-Hermitian matrix $\beta = (V^{-1}dV)_\perp$ satisfy $\beta_{ji} = -\beta_{ij}^\dagger$ for $i < j$. For the matrix elements of $\alpha$ we find accordingly

$$\alpha_{ii} = id\vartheta_i, \quad \alpha_{ij} = -2i\,e^{-i\vartheta_{ij}} \sin \vartheta_{ij}\, \beta_{ij}, \quad \vartheta_{ij} = \frac{1}{2}(\vartheta_i - \vartheta_j), \quad i < j\,. \tag{9.45}$$

The invariant volume form is a polynomial of the matrix elements of $\alpha$, where the product is the wedge product. In each monomial, a matrix element may only appear once as a factor, since $\alpha_{ij} \wedge \alpha_{ij} = 0$. For U($n$), the invariant volume form is an $n^2$-form. We can construct only one invariant volume form from the $n$ 1-forms $\{d\vartheta_i\}$ and the $n^2 - n$ 1-forms $\{\alpha_{i<j}, \alpha_{i<j}^\dagger\}$ up to a constant, and this is proportional to the *Haar measure*,

$$d\mu \propto \prod_i d\vartheta_i \prod_{i<j} \alpha_{ij}^\dagger \alpha_{ij} = \prod_i d\vartheta_i \prod_{i<j} \left(4\sin^2 \vartheta_{ij}\, \beta_{ij}^\dagger \beta_{ij}\right)\,. \tag{9.46}$$

The forms $\beta_{ij}$ do not depend on the variables $\vartheta_i$ and we conclude

**Reduced Haar Measure on U($n$)**

The reduced and normalized Haar measure on U($n$) has the form

$$d\mu_{\mathrm{red}} = \frac{2^{n(n-2)}}{n!\,\pi^n} \prod_{i<j} \sin^2\left(\frac{\vartheta_i - \vartheta_j}{2}\right) d\vartheta_1 \cdots d\vartheta_n, \quad \vartheta_i \in [0, 2\pi].$$

$$(9.47)$$

The corresponding invariant measure for SU($n$) is obtained by replacing $\vartheta_n$ with $-(\vartheta_1 + \cdots + \vartheta_{n-1})$ and the differential $d\vartheta_n$ with the factor $2\pi$. In this way, for SU(3) we find the reduced and normalized Haar measure

$$d\mu_{\mathrm{red}} = \frac{8}{3\pi^2} \sin^2 \frac{\vartheta_1 - \vartheta_2}{2} \sin^2 \frac{2\vartheta_1 + \vartheta_2}{2} \sin^2 \frac{2\vartheta_2 + \vartheta_1}{2} d\vartheta_1 d\vartheta_2. \qquad (9.48)$$

The density of the measure is shown in Fig. 9.2. Finally, we note that for U($n$) any permutation of the angles $\vartheta_i$ in

$$D = \mathrm{diag}\left(e^{i\vartheta_1}, \ldots, e^{i\vartheta_n}\right)$$

yields a matrix conjugate to $D$. Thus, a class function $f(\vartheta_1, \ldots, \vartheta_n)$ is completely symmetric in its arguments and the reduced Haar measure is invariant under a permutation of the angles. For SU($n$), this statement holds accordingly if one replaces $\vartheta_n$ with $-(\vartheta_1 + \cdots + \vartheta_{n-1})$. In Chap. 16 we will use results about the Lie

**Fig. 9.2** Density of the reduced Haar measure on the group SU(3)

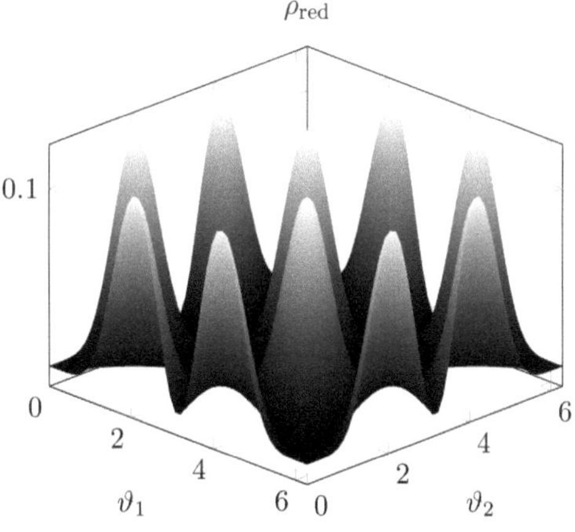

algebra of a Lie group to find a generalization of (9.47) for any connected compact Lie group.

## 9.5  Invariant Integration on SU(1, 1) and SL(2, $\mathbb{R}$)

When the $n$-form $\omega_n = \mathrm{Tr}(\omega^{\wedge n})$ does not vanish, then it defines a left- and right-invariant Haar measure on the group. This also applies to non-compact groups. As an example, we consider the simple and isomorphic groups $SU(1, 1)$ and $SL(2, \mathbb{R})$. These have the Anti-de Sitter space $AdS_3$ as a group manifold. For the elements in $SU(1, 1)$ we choose the parametrization

$$U = \begin{pmatrix} \cosh \rho \ \mathrm{e}^{\mathrm{i}t} \ \sinh \rho \ \mathrm{e}^{-\mathrm{i}\phi} \\ \sinh \rho \ \mathrm{e}^{\mathrm{i}\phi} \ \cosh \rho \ \mathrm{e}^{-\mathrm{i}t} \end{pmatrix} \quad \text{with} \quad \omega = U^{-1}\mathrm{d}U = \omega_\rho \mathrm{d}\rho + \omega_t \mathrm{d}t + \omega_\phi \mathrm{d}\phi,$$

$$(9.49)$$

and find the following coefficients of the left-invariant 1-form

$$\omega_\rho = \begin{pmatrix} 0 & \mathrm{e}^{-\mathrm{i}(t+\phi)} \\ \mathrm{e}^{\mathrm{i}(t+\phi)} & 0 \end{pmatrix}, \quad \omega_t = \mathrm{i}\cosh \rho \begin{pmatrix} \cosh \rho & \sinh \rho \ \mathrm{e}^{-\mathrm{i}(t+\phi)} \\ -\sinh \rho \ \mathrm{e}^{\mathrm{i}(t+\phi)} & -\cosh \rho \end{pmatrix},$$

$$\omega_\phi = -\mathrm{i}\sinh \rho \begin{pmatrix} \sinh \rho & \cosh \rho \ \mathrm{e}^{-\mathrm{i}(t+\phi)} \\ -\cosh \rho \ \mathrm{e}^{\mathrm{i}(t+\phi)} & -\sinh \rho \end{pmatrix}.$$

$$(9.50)$$

Therefore, the invariant line element is

$$\mathrm{d}s^2 = -\frac{1}{2}\mathrm{Tr}\omega^2 = \cosh^2 \rho \ \mathrm{d}t^2 - \left(\mathrm{d}\rho^2 + \sinh^2 \rho \ \mathrm{d}\phi^2\right).$$

$$(9.51)$$

The left- and right-invariant 3-form associated with the left-invariant 1-form is

$$\omega_3 = \mathrm{Tr}\left(\omega^{\wedge 3}\right) = 3\,\mathrm{Tr}\left([\omega_t, \omega_\rho]\,\omega_\phi\right)\mathrm{d}t \wedge \mathrm{d}\rho \wedge \mathrm{d}\phi$$

$$= -12 \cosh \rho \sinh \rho \ \mathrm{d}t \wedge \mathrm{d}\rho \wedge \mathrm{d}\phi.$$

$$(9.52)$$

Either with the help of the metric or the 3-form we conclude:

**Invariant Measure for SU(1, 1)**

The Haar measure on the group $SU(1, 1)$ with parametrization (9.49) is

$$\mathrm{d}\mu(U) = \mathrm{const} \cdot \cosh \rho \sinh \rho \ \mathrm{d}t \wedge \mathrm{d}\rho \wedge \mathrm{d}\phi.$$

$$(9.53)$$

If one chooses the parametrization for the isomorphic group $SL(2, \mathbb{R})$ as

$$A = \cosh \rho \begin{pmatrix} \cos t & \sin t \\ -\sin t & \cos t \end{pmatrix} + \sinh \rho \begin{pmatrix} \cos \phi & -\sin \phi \\ -\sin \phi & -\cos \phi \end{pmatrix} , \tag{9.54}$$

then one obtains the (natural) invariant line element

$$\mathrm{d}s^2 = \frac{1}{2} \mathrm{Tr} \left( A^{-1} \mathrm{d}A \right)^2 = \mathrm{d}\rho^2 - \cosh^2 \rho \, \mathrm{d}t^2 + \sinh^2 \rho \, \mathrm{d}\phi^2 . \tag{9.55}$$

The invariant 3-form agrees with the 3-form for $SU(1, 1)$ in (9.52). Th equivalence $SU(1, 1) \cong SL(2, \mathbb{R})$ and the underlying group manifold $AdS_3$ are discussed in Problem 9.3. This group plays a prominent role in string theory and the AdS-CFT correspondence (see [45] for a competent introduction into the subject). The propagation of bosonic strings (in the limit of weak string coupling) on this group manifold can be exactly solved [46].

## 9.6 Modular Function

Let $G$ be a group with *left-invariant* measure $\mu$, i.e. $\mu(g\mathcal{O}) = \mu(\mathcal{O})$ for all measurable sets $\mathcal{O}$ in $G$ and each $g \in G$. In general, $\mu$ is however not right-invariant and $\mu(\mathcal{O}g)$ can be different from $\mu(\mathcal{O})$. But for each fixed group element $h, \mu_h(\mathcal{O}) = \mu(\mathcal{O}h)$ is also left-invariant, and since a left-invariant measure is unique up to a multiplicative constant, there exists a positive number $m_G(h)$ with

$$\mu(\mathcal{O}h) = m_G(h)\mu(\mathcal{O}), \quad m_G(h) \in \mathbb{R}^+ , \tag{9.56}$$

for all $\mathcal{O}$. Because the left-invariant measure is unique up to a normalization, the number $m_G(h)$ does not depend on the measure.

The mapping $h \to m_G(h)$ defines a homomorphism from $G$ into the multiplicative group $\mathbb{R}^+$,

$$m_G(h_1 h_2)\mu(\mathcal{O}) = \mu(\mathcal{O}h_1 h_2) = m_G(h_2)\mu(\mathcal{O}h_1) = m_G(h_2)m_G(h_1)\mu(\mathcal{O}) , \tag{9.57}$$

or $m_G(h_1 h_2) = m_G(h_1)m_G(h_2)$. The homomorphism $m_G : G \mapsto \mathbb{R}^+$ is called *modular function* of $G$.

**Definition 9.1 (Unimodular Groups)** If $m_G(h) = 1$ for all $h \in G$, then $G$ is called unimodular.

For a unimodular group, the left-invariant measure is also right-invariant. Finite, discrete, abelian and compact groups are all unimodular. The $n$-form $\mathrm{Tr}(w^{\wedge n})$ defined with the left-invariant 1-form $\omega$ in (9.30) is left- and right-invariant (see

Problem 9.4). If it does not vanish, it defines a left-right-invariant measure. This is the case for the non-compact groups $SL(n, \mathbb{R})$. Also $GL(2, \mathbb{R})$ is unimodular with Haar measure

$$\mathrm{d}\mu(g) = \mathrm{Tr}(g^{-1}\mathrm{d}g)\,\mathrm{Tr}\big((g^{-1}\mathrm{d}g)^{\wedge 3}\big) = \frac{6}{(\det g)^2} \prod_{i,j} \mathrm{d}g_{ij}, \quad g = (g_{ij}), \qquad (9.58)$$

or more generally $GL(n, \mathbb{R})$ with Haar measure

$$\mathrm{d}\mu(g) = \frac{\mathrm{const}}{|\det g|^n} \prod_{i,j} \mathrm{d}g_{ij}. \qquad (9.59)$$

Now we want to relate the left-invariant measure $\mathrm{d}\mu(g)$ with $\mathrm{d}\mu_{\mathrm{inv}}(g) = \mathrm{d}\mu(g^{-1})$. First we notice that $\mathrm{d}\mu_{\mathrm{inv}}$ is a right-invariant measure,

$$\mathrm{d}\mu_{\mathrm{inv}}(gh) = \mathrm{d}\mu(h^{-1}g^{-1}) = \mathrm{d}\mu(g^{-1}) = \mathrm{d}\mu_{\mathrm{inv}}(g).$$

Actually, the quotient of $\mathrm{d}\mu$ and $\mathrm{d}\mu_{\mathrm{inv}}$ is equal to the above introduced modular function:

**Lemma 9.2** *For every left-invariant measure* $\mathrm{d}\mu$ *we have* $\mathrm{d}\mu_{\mathrm{inv}}(g) \equiv \mathrm{d}\mu(g^{-1}) = m_G^{-1}(g)\mathrm{d}\mu(g)$.

First, we note that $\mathrm{d}\mu'(g) = m_G^{-1}(g)\mathrm{d}\mu(g)$ is right invariant:

$$\mathrm{d}\mu'(gh) = m_G^{-1}(gh)\mathrm{d}\mu(gh) = m_G^{-1}(g)m_G^{-1}(h)m_G(h)\mathrm{d}\mu(g) = \mathrm{d}\mu'(g).$$

Therefore, there exists a positive constant $C$ with $\mathrm{d}\mu' = C\mathrm{d}\mu_{\mathrm{inv}}$, or with

$$\mathrm{d}\mu(g) = m_G(g)\mathrm{d}\mu'(g) = Cm_G(g)\mathrm{d}\mu_{\mathrm{inv}}(g). \qquad (9.60)$$

We still have to show that $C = 1$. This follows directly from

$$\mathrm{d}\mu(g) = Cm_G(g)\mathrm{d}\mu(g^{-1}) = C^2 m_G(g)m_G(g^{-1})\mathrm{d}\mu(g) = C^2\mathrm{d}\mu(g).$$

This proves the lemma. The following corollary will be needed in Sect. 11.2.4:

**Corollary 9.1** *For a unimodular group,* $\mathrm{d}\mu(g^{-1}) = \mathrm{d}\mu(g)$, *or equivalently*

$$\int_G \mathrm{d}\mu(g)f(g) = \int_G \mathrm{d}\mu(g)f(g^{-1}). \qquad (9.61)$$

The invariance of the Haar measure on unimodular groups under inversion will be useful in the discussion of projectors in Sect. 11.2.4.

**Modular Function of the Affine Group on $\mathbb{R}$**

The affine group discussed in Sect. 9.2 is not unimodular, since $\mathrm{d}\mu_\ell \neq$ const $\mathrm{d}\mu_r$. So let's determine the modular function $m_G$ of the group. Under a right translation (9.20), the left-invariant measure in (9.21) transforms according to

$$\mathrm{d}\mu_\ell(\alpha) = \frac{1}{\alpha_1^2}\,\mathrm{d}\alpha_1 \wedge \mathrm{d}\alpha_2 \longmapsto \frac{1}{(\alpha_1\beta_1)^2}\,\mathrm{d}(\beta_1\alpha_1) \wedge \mathrm{d}(\beta_2\alpha_1+\alpha_2) = \frac{1}{\beta_1}\mathrm{d}\mu_\ell(\alpha)\,,$$

and therefore the modular function is $m(\beta) = 1/\beta_1$. The inverse transformation $\alpha^{-1}$ has the parameters $(1/\alpha_1, -\alpha_2/\alpha_1)$, so that

$$\mathrm{d}\mu_\ell(\alpha^{-1}) = -\alpha_1^2\,\mathrm{d}\!\left(\frac{1}{\alpha_1}\right) \wedge \mathrm{d}\!\left(\frac{\alpha_2}{\alpha_1}\right) = \frac{1}{\alpha_1}\mathrm{d}\alpha_1 \wedge \mathrm{d}\alpha_2 = \alpha_1\mathrm{d}\mu_\ell(\alpha)\,,$$

which exactly corresponds to the transformation behavior in Lemma 9.2.

## 9.7  Exercises for Chap. 9

**Problem 9.1 (Conjugacy Classes of SU(3))** Try to characterize the conjugacy classes of the Lie group SU(3).
Hint: Consider the eigenvalues of the matrices.

**Problem 9.2 (Invariant Integration on SU(2))** Calculate the Haar measure on SU(2) for the parametrization

$$U = \begin{pmatrix} \cos\vartheta\ \mathrm{e}^{\mathrm{i}\zeta} & -\sin\vartheta\ \mathrm{e}^{\mathrm{i}\eta} \\ \sin\vartheta\ \mathrm{e}^{-\mathrm{i}\eta} & \cos\vartheta\ \mathrm{e}^{-\mathrm{i}\zeta} \end{pmatrix},$$

and normalize it so that $\mathrm{Vol}(\mathrm{SU}(2)) = 1$.

**Problem 9.3 (The Non-compact Lie Group SU(1, 1))** This group is isomorphic to SL(2, $\mathbb{R}$) and appears in hyperbolic geometry, string theory, in the quantization of the radiation field and in the parametrization of coherent states.

1. Try to find a similar parametrization for SU(1, 1) as for SU(2),

$$U = \begin{pmatrix} a & b \\ * & * \end{pmatrix}, \quad \text{with} \quad |a|^2 - \ldots\ .$$

2. Use the parametrization $a = \cosh(r)\, e^{i\phi}$ and $b = \sinh(r)\, e^{i\psi}$ and extract the metric coefficients $g_{ij}$ from $ds^2 = \frac{1}{2}\mathrm{Tr}\,(U^{-1}dU)^2$.
3. What are the left-invariant differential form $\omega = U^{-1}dU$ and the (invariant) volume element $d\mu = \sqrt{-g}\, dr\, d\phi\, d\psi$?
4. We expand $\omega = \omega_r dr + \omega_\phi d\phi + \omega_\psi d\psi$ with matrices $\omega_r$, $\omega_\phi$, and $\omega_\psi$. Prove

$$\frac{1}{3}\mathrm{Tr}\,(\omega \wedge \omega \wedge \omega) = \mathrm{Tr}\,\left(\omega_r[\omega_\phi, \omega_\psi]\right) dr \wedge d\phi \wedge d\psi \,,$$

and calculate the invariant volume form $\omega_3$.

5. Show that the group manifold of $SU(1,1)$ is equal to the AdS$_3$ space. Its embedding in $\mathbb{R}^4$ is defined by

$$\mathrm{AdS}_3 = \{X \in \mathbb{R}^4 \mid X_1^2 + X_2^2 - X_3^2 - X_4^2 = 1\}\,.$$

6. What do you learn from this about the fundamental group of $SU(1,1)$?
7. Convince yourself that $SU(1,1) \cong SL(2,\mathbb{R})$.
   Hint: There exists a matrix $C$ such that $CUC^{-1}$ is real for all $U \in SU(1,1)$.

**Problem 9.4 (Invariant Measure and Inversion)** Show that the measure (9.30) has the property

$$d\mu(g^{-1}) = \pm d\mu(g)\,.$$

Hint: use that $d(g^{-1}g) = d(gg^{-1}) = 0$.

**Problem 9.5 (The Heisenberg Group)** The smallest Heisenberg group contains the real matrices of the form

$$\begin{pmatrix} 1 & x & z \\ 0 & 1 & y \\ 0 & 0 & 1 \end{pmatrix}\,.$$

Show that this group is unimodular.

# Representations of Groups

10



When the elements of a group and their group operations are homomorphically mapped onto a concrete algebraic structure, we speak of a realization of the group. If this is done through linear mappings on a vector space $\mathcal{V}$, it is called a *representation* of the group. However, one should distinguish between a group and its representations. The group properties are universal and independent of the representations. A group will generally have many representations. For example, rotations in Euclidean space $E_3$ can be represented by three-dimensional matrices; however, these are not the group elements themselves, but only one of infinitely many representations.

## 10.1 Representations

In physics, representations of groups usually appear, rather than the abstract groups themselves. For example, the three-dimensional representation of the rotation group in classical mechanics or all representations of the rotation group, the Lorentz group, or the crystallographic groups in quantum mechanics. Individual representations can have additional properties (e.g., unitarity) that do not exist for the group. However, properties of the group can "get lost" in a (non-faithful) representation.

Representations are realizations of the group elements as invertible linear mappings in $\mathrm{GL}(\mathcal{V})$. After choosing a basis in the $n$-dimensional vector space $\mathcal{V}$, $\mathrm{GL}(\mathcal{V})$ can be identified with the general linear group $\mathrm{GL}(n,\mathbb{K})$.

**Definition 10.1 (Representation)** A representation $D$ of a group $G$ on a linear space $\mathcal{V}$ is a group homomorphism

$$D : G \mapsto GL(\mathcal{V}), \qquad g \mapsto D(g) \,.$$

**A Representation Is a Homomorphism**
A representation respects the group structure,

$$D(e) = e, \quad D(g_1 g_2) = D(g_1)D(g_2) \Longrightarrow D(g^{-1}) = D^{-1}(g) \,. \qquad (10.1)$$

The dimension of the vector space $\mathcal{V}$ is the *dimension of the representation D*. A representation is a specific left group action that acts on a linear space. Accordingly, an injective representation is called *faithful* and is a group isomorphism $D : G \to$ image$(G) \leq GL(\mathcal{V})$ with kernel$(D) = e \in G$. After choosing a basis in $\mathcal{V}$, the $D(g)$ become matrices.

There is always the *trivial representation*, which maps all group elements to the identity matrix. In this case, all information about the group is lost. The other extreme case are the *faithful representations*, where no information is lost.

## 10.1.1  Representations of the Dihedral Group $\mathcal{D}_3$

This group was introduced in Sect. 2.3 as symmetry transformations of the equilateral triangle, these are linear mappings $\mathbb{R}^2 \mapsto \mathbb{R}^2$. It is generated by the rotation $c_3$ with angle $\varphi_3 = 2\pi/3$ and the reflection $\sigma$ at a symmetry axis, which are sketched in Fig. 10.1.

In fact, we are considering a two-dimensional representation $D_2$ of the abstract group defined by the group table in Sect. 2.3. It is sufficient to know the matrices $D_2(g)$ for generating elements, as those of the other group elements can be calculated with (10.1).

**Fig. 10.1** The symmetries of
the equilateral triangle.

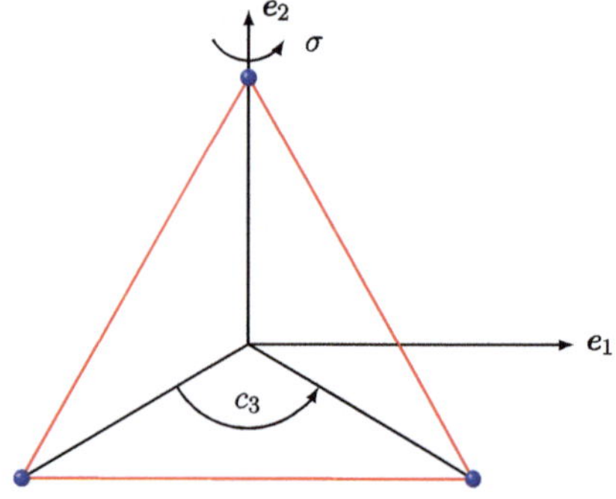

We choose a Cartesian basis $e_1$, $e_2$ in $\mathbb{R}^2$ and set (see, for example (5.7))

$$e_i \mapsto \sum_j e_j D_{ji}(g)\,.$$

The columns of the matrix $(D_{ji})$ contain the images of the basis vectors. For example, to the rotation by $120°$ corresponds the matrix

$$D_2(c_3) = \begin{pmatrix} \cos\varphi_3 & -\sin\varphi_3 \\ \sin\varphi_3 & \cos\varphi_3 \end{pmatrix}, \quad \varphi_n = \frac{2\pi}{n}\,, \tag{10.2}$$

and to the reflection at the line perpendicular to $e_1$ through the center corresponds the matrix

$$D_2(\sigma) = \begin{pmatrix} -1 & 0 \\ 0 & 1 \end{pmatrix}\,. \tag{10.3}$$

The two-dimensional representation $D_2$ (the index indicates the dimension of the representation) is faithful. Furthermore, there is always the *one-dimensional trivial representation* $D_1^1$

$$D_1^1(c_3) = D_1^1(\sigma) = 1 \implies D \equiv 1\,, \tag{10.4}$$

and the also one-dimensional *sign representation* $D_1^2 = \det D_2(g)$ with

$$D_1^2(c_3) = 1 \quad \text{and} \quad D_1^2(\sigma) = -1\,. \tag{10.5}$$

The sign representation assigns 1 to rotations and $-1$ to reflections.

**One-dimensional Determinant Representation**
For every representation $g \mapsto D(g)$, the mapping $g \mapsto \det D(g)$ is also a representation:

$$\det D(g_1 g_2) = \det D(g_1) D(g_2) = \det D(g_1)\det D(g_2)\,.$$

## 10.2   Regular Representation

The regular representation $g \mapsto \mathcal{R}(g)$ of a finite group contains all representations of the group and is therefore of particular interest for its representation theory. The representation $\mathcal{R}$ is constructed as follows: The $n = |G|$ columns of the group table are rearranged so that the neutral element $e$ only appears on the diagonal:

| $G$ | $e$ | $g_2^{-1}$ | $g_3^{-1}$ | $\cdots$ | $g_n^{-1}$ |
|---|---|---|---|---|---|
| $e$ | $e$ | $g_2^{-1}$ | $g_3^{-1}$ | $\cdots$ | $g_n^{-1}$ |
| $g_2$ | $g_2$ | $e$ | $g_2 \circ g_3^{-1}$ | $\cdots$ | $g_2 \circ g_n^{-1}$ |
| $g_3$ | $g_3$ | $g_3 \circ g_2^{-1}$ | $e$ | $\cdots$ | $g_3 \circ g_n^{-1}$ |
| $\vdots$ | $\vdots$ | $\vdots$ | $\vdots$ | $\cdots$ | $\vdots$ |
| $g_n$ | $g_n$ | $g_n \circ g_2^{-1}$ | $g_n \circ g_3^{-1}$ | $\cdots$ | $e$ |

Now, the element $g_i$ is assigned the $n \times n$-matrix $\mathcal{R}(g_i)$ that is obtained when in the above group table everywhere $g_i$ is replaced by 1 and the other elements by 0. In particular, $\mathcal{R}(e) = \mathbb{1}_n$.

## Example: Regular Representation of the Dihedral Group $\mathcal{D}_3$

The (reordered) group table of $\mathcal{D}_3$ was introduced in Sect. 2.3 and has the following form:

| $\mathcal{D}_3$ | $e$ | $c_3^2$ | $c_3$ | $\sigma_v^{(1)}$ | $\sigma_v^{(2)}$ | $\sigma_v^{(3)}$ |
|---|---|---|---|---|---|---|
| $e$ | $e$ | $c_3^2$ | $c_3$ | $\sigma_v^{(1)}$ | $\sigma_v^{(2)}$ | $\sigma_v^{(3)}$ |
| $c_3$ | $c_3$ | $e$ | $c_3^2$ | $\sigma_v^{(3)}$ | $\sigma_v^{(1)}$ | $\sigma_v^{(2)}$ |
| $c_3^2$ | $c_3^2$ | $c_3$ | $e$ | $\sigma_v^{(2)}$ | $\sigma_v^{(3)}$ | $\sigma_v^{(1)}$ |
| $\sigma_v^{(1)}$ | $\sigma_v^{(1)}$ | $\sigma_v^{(3)}$ | $\sigma_v^{(2)}$ | $e$ | $c_3$ | $c_3^2$ |
| $\sigma_v^{(2)}$ | $\sigma_v^{(2)}$ | $\sigma_v^{(1)}$ | $\sigma_v^{(3)}$ | $c_3^2$ | $e$ | $c_3$ |
| $\sigma_v^{(3)}$ | $\sigma_v^{(3)}$ | $\sigma_v^{(2)}$ | $\sigma_v^{(1)}$ | $c_3$ | $c_3^2$ | $e$ |

Accordingly, the rotation $c_3$ and reflection $\sigma_v^{(1)}$ are represented as follows:

$$
\mathcal{R}(c_3) = \begin{pmatrix} 0 & 0 & 1 & 0 & 0 & 0 \\ 1 & 0 & 0 & 0 & 0 & 0 \\ 0 & 1 & 0 & 0 & 0 & 0 \\ 0 & 0 & 0 & 0 & 1 & 0 \\ 0 & 0 & 0 & 0 & 0 & 1 \\ 0 & 0 & 0 & 1 & 0 & 0 \end{pmatrix} , \quad
\mathcal{R}(\sigma_v^{(1)}) = \begin{pmatrix} 0 & 0 & 0 & 1 & 0 & 0 \\ 0 & 0 & 0 & 0 & 1 & 0 \\ 0 & 0 & 0 & 0 & 0 & 1 \\ 1 & 0 & 0 & 0 & 0 & 0 \\ 0 & 1 & 0 & 0 & 0 & 0 \\ 0 & 0 & 1 & 0 & 0 & 0 \end{pmatrix} .
$$

It is easy to verify that these matrices represent the group $\mathcal{D}_3$,

$$
\mathcal{R}^3(c_3) = \mathcal{R}^2(\sigma_v^{(1)}) = \mathbb{1}_6 \quad \text{and} \quad \mathcal{R}(\sigma_v^{(1)})\mathcal{R}(c_3)\mathcal{R}(\sigma_v^{(1)})\mathcal{R}(c_3) = \mathbb{1}_6 .
$$

◀

After this example, we return to the general situation.

Since different columns of the group table contain different permutations of the group elements, the regular representation is faithful. Each column contains exactly one 1 and all other elements are 0 and thus has the norm 1. In different columns, the 1 is at different places, and the column vectors are therefore orthogonal to each other. This means that all matrices $\mathcal{R}(g)$ are orthogonal.

**Lemma 10.1 (Regular Representation)** *The regular representation $g \to \mathcal{R}(g)$ is a $|G|$-dimensional, orthogonal and faithful representation of the group.*

**Proof** To prove the representation property we note that $R_{kj}(g_i)$ is exactly 1 when $g_k \circ g_j^{-1} = g_i$ or when $g_k = g_i \circ g_j$ applies. We conclude

$$g_i \circ g_j = \sum_{p=1}^{n} \mathcal{R}_{pj}(g_i) g_p \,, \tag{10.6}$$

and with the *associativity* of the group operation follows

$$g_i \circ (g_j \circ g_k) = g_i \circ \sum_{p} \mathcal{R}_{pk}(g_j) g_p = \sum_{p} \mathcal{R}_{pk}(g_j) g_i \circ g_p = \sum_{pq} \mathcal{R}_{pk}(g_j) \mathcal{R}_{qp}(g_i) g_q$$

$$= (g_i \circ g_j) \circ g_k = \sum_{q} \mathcal{R}_{qk}(g_i \circ g_j) g_q \Longrightarrow \mathcal{R}(g_i) \mathcal{R}(g_j) = \mathcal{R}(g_i \circ g_j) \,.$$

$$\tag{10.7}$$

Therefore, the mapping $g \mapsto \mathcal{R}(g)$ is a faithful representation of the group $G$ into the group $\mathrm{SO}(|G|, \mathbb{Z}_2)$. $\qquad\square$

## 10.2.1 Equivalent Representations

To describe the symmetries of the triangle, we can use a rotated basis $f_1, f_2$, see Fig. 10.2. The bases $\{e_i\}$ and $\{f_i\}$ are connected by an invertible linear transformation

**Fig. 10.2** Two Cartesian bases.

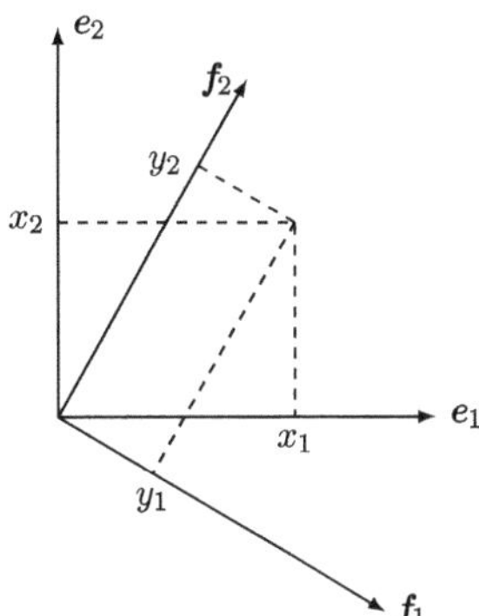

$$e_i = f_j S^j_i \,, \tag{10.8}$$

where the summation convention is used. It follows that the coordinates for $r = x^i e_i = y^i f_i$ are related as follows:

$$y = Sx \,. \tag{10.9}$$

In the basis $e_i$, the group element $g$ is assigned the matrix $D(g)$,

$$x \mapsto D(g)x \,. \tag{10.10}$$

Then it follows

$$y = Sx \mapsto SD(g)x = SD(g)S^{-1}y \equiv \tilde{D}(g)y \,. \tag{10.11}$$

As expected, the representation matrices $\tilde{D}(g)$ are obtained from the $D(g)$ by a conjugation,

$$\tilde{D}(g) = SD(g)S^{-1} \,. \tag{10.12}$$

The conjugated matrices also represent the group,

$$\tilde{D}(g_1)\tilde{D}(g_2) = SD(g_1)D(g_2)S^{-1} = SD(g_1g_2)S^{-1} = \tilde{D}(g_1g_2) \,,$$

and conjugated representations define an equivalence class. Equivalent representations describe the same linear mappings and should be identified. Conversely, if a representation is given, we always have infinitely many equivalent representations of the form $SD(g)S^{-1}$. In the classification of representations, it can therefore only be about the classification of inequivalent representations, i.e., representations that are not conjugated to each other as in (10.12).

**Natural Three-Dimensional Representation of $\mathcal{D}_3$**

Here we introduce a three-dimensional representation of the dihedral group $\mathcal{D}_3$, which is known to be isomorphic to the symmetric group $\mathcal{S}_3$, see Table 4.1. We assume that the permutations in $\mathcal{S}_3$ act on the three basis vectors $\{e_1, e_2, e_3\}$ of a Cartesian basis of $\mathbb{R}^3$. Under $c_3$, $e_1$ goes into $e_2$, $e_2$ into $e_3$ and $e_3$ into $e_1$. This permutation is a rotation around the axis $e_1 + e_2 + e_3$ with angle $\varphi_3$ and this is shown on the left in Fig. 10.3. $D_3(\sigma)$, on the other hand, swaps $e_2$ and $e_3$ and is thus a reflection on the gray shaded plane in Fig. 10.3, spanned by $e_1$ and $e_2 + e_3$. One obtains ($\mathcal{D}_n$ denotes a dihedral group and $D_n$ representations of groups)

$$D_3(c_3) = \begin{pmatrix} 0 & 0 & 1 \\ 1 & 0 & 0 \\ 0 & 1 & 0 \end{pmatrix} \quad \text{and} \quad D_3(\sigma) = \begin{pmatrix} 1 & 0 & 0 \\ 0 & 0 & 1 \\ 0 & 1 & 0 \end{pmatrix} \,. \tag{10.13}$$

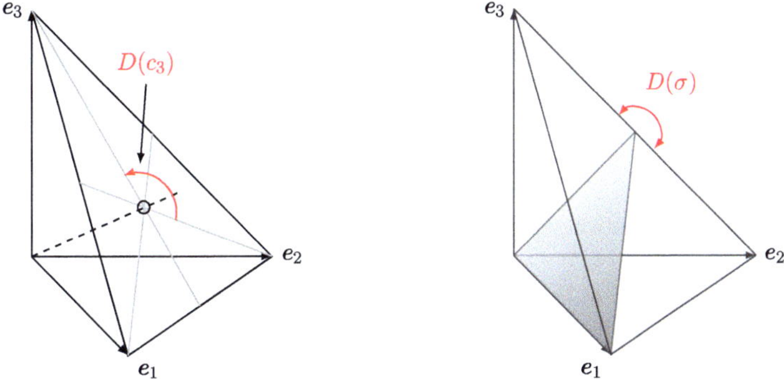

**Fig. 10.3** On the three-dimensional representation of the dihedral group $\mathcal{D}_3$.

The remaining representation matrices can be obtained from those of the generating elements. All matrices $D_3(g)$ are orthogonal and unimodular.

**Representations of $\mathcal{D}_3$**
The dihedral group $\mathcal{D}_3$ has, among others, the following representations:

- the one-dimensional unit representation $D_1^1$,
- the one-dimensional sign representation $D_1^2$,
- the two-dimensional representation $D_2$,
- the three-dimensional natural representation $D_3$,
- the regular representation $\mathcal{R}$.

## 10.3   Reducible and Irreducible Representations

Using the three-dimensional representation of $\mathcal{D}_3$, we explain the concept of a reducible representation. Under the rotation $D_3(c_3)$ and reflection $D_3(\sigma)$, and thus under all $D_3(g)$, the one-dimensional subspace $\mathcal{V}_1$ spanned by $e_1 + e_2 + e_3$ is invariant. We normalize the vector and complete it to an orthonormal basis

$$f_1 = \frac{1}{\sqrt{3}}\,(e_1 + e_2 + e_3)\,, \quad f_2 = \frac{1}{\sqrt{2}}\,(e_2 - e_3)\,, \quad f_3 = \frac{1}{\sqrt{6}}\,(-2e_1 + e_2 + e_3)\,.$$

$$(10.14)$$

The transformation to the new basis is done with a rotation matrix $S$ and in the new basis, the representation matrices $\tilde{D} = SDS^{-1}$ are block-diagonal

$$\tilde{D}_3(c_3) = \begin{pmatrix} 1 & 0 & 0 \\ 0 & \cos\varphi_3 & -\sin\varphi_3 \\ 0 & \sin\varphi_3 & \cos\varphi_3 \end{pmatrix} \quad \text{and} \quad \tilde{D}_3(\sigma) = \begin{pmatrix} 1 & 0 & 0 \\ 0 & -1 & 0 \\ 0 & 0 & 1 \end{pmatrix}. \tag{10.15}$$

**Task**

Determine the rotation matrix $S$ and confirm (10.15).

The linear space $V$ of the three-dimensional representation $D_3$ thus decomposes into two orthogonal subspaces of dimensions 1 and 2,

$$V = V_1 \oplus V_2. \tag{10.16}$$

Each of the two subspaces is mapped onto itself by all $D_3(g)$, i.e. $D_3 : V_1 \to V_1$ and $D_3 : V_2 \to V_2$. These are invariant subspaces of the representation. Restricted to $V_1$, $D_3$ is equal to $D_1^1$ and restricted to $V_2$, $D_3$ is equal to $D_2$. One writes

$$D_3 = D_1^1 \oplus D_2 \quad \text{with} \quad \dim D_3 = \dim D_1^1 + \dim D_2. \tag{10.17}$$

This simple example suggests the following definition:

**Definition 10.2 (Reducible and Irreducible Representations)** A representation is called irreducible if $V$ has no proper subspace that is invariant under all $D(g)$. If a representation is not irreducible, it is called *reducible*.

**Question**

Why is the two-dimensional representation $D_2$ of $\mathcal{D}_3$ irreducible?

## 10.3.1  Reduction of Representations

Now we generalize to representations $D(g) : V \to V$ of arbitrary groups. Let $V_1$ be an $m$-dimensional proper invariant subspace of $V$, i.e. for all $g$ we have $D(g) : V_1 \mapsto V_1$.

**Reducible Representation**
If the basis is chosen so that the first $m$ basis vectors span $V_1$, then the representation matrices of the reducible representation have the form

$$g \mapsto D(g) = \begin{pmatrix} D_1(g) & H(g) \\ 0 & D_2(g) \end{pmatrix}, \tag{10.18}$$

where $D_1$ is an $m$-dimensional and $D_2$ is a $(n-m)$-dimensional matrix.

Specifically, if $H(g) = 0$ for all group elements, then the matrices are block-diagonal,

$$g \mapsto D(g) = \begin{pmatrix} D_1(g) & 0 \\ 0 & D_2(g) \end{pmatrix},$$

and the representation decomposes into two representations $D_1$ and $D_2$. One writes $D = D_1 \oplus D_2$. This procedure can possibly be continued, so that the representing matrices further decompose in a suitable coordinate system:

$$g \mapsto D(g) = \begin{pmatrix} D_1(g) & 0 & 0 \ldots & 0 \\ 0 & D_2(g) & 0 \ldots & 0 \\ \vdots & & \ddots & \vdots \\ 0 & 0 & 0 \ldots & D_m(g) \end{pmatrix}. \tag{10.19}$$

The decomposition stops when no proper invariant subspace exists in any of the invariant subspaces $V_1, \ldots, V_m$.

**Definition 10.3 (Fully Reducible Representation)** A representation that is irreducible or decomposes into purely irreducible representations is called fully reducible.

There are groups with non-fully reducible representations, for which $H(g)$ does not vanish. The above representations of the dihedral group are however unitary $D^\dagger(g) = D^{-1}(g)$, and for such representations the following holds

**Theorem 10.1 (Unitary Representations)** *Every unitary representation is fully reducible.*

**Proof** If $D$ is irreducible, there is nothing to prove. So let $V_1 \subset V$ be a proper invariant subspace and $W = V_1^\perp$ the unitary-orthogonal complement of $V_1$,

$$W = \{w \in V | (w, V_1) = 0\}. \tag{10.20}$$

$V_1$ and its complement span $V$ and every $v \in V$ can be written as a unique sum of a vector from $V_1$ and a vector from $W$. Let $v_1$ and $w$ be arbitrary vectors from $V_1$ and $W$. Then for a *unitary* representation

$$(D(g)w, v_1) = \left(w, D^\dagger(g)v_1\right) = \left(w, D(g^{-1})v_1\right) = 0$$

for all group elements, since $D(g^{-1})v_1$ lies in $V_1$ by assumption. Thus, with $w$ also $D(g)w$ lies in the subspace $W$. Therefore, the orthogonal complement of any invariant subspace is also invariant. This procedure can now be continued, if a true invariant subspace of $\mathcal{W}$ exists. Finally, a decomposition of the representation into purely irreducible components is obtained.                                                      □

## 10.4    Representations of Groups with Invariant Integration

In most applications of group theory in physics, the groups have an invariant averaging. Then the following observation is relevant for representation theory:

**Scalar Product on $L_2(G)$**
The invariant averaging in (9.1) or (9.2) defines an *invariant scalar product* for functions on the group,

$$(f_1, f_2) = \mathcal{M}\left(\bar{f_1} \cdot f_2\right), \quad f_1, f_2 : G \mapsto \mathbb{C}. \tag{10.21}$$

For finite and compact groups, the averaging is invariant with respect to left and right translation on the group.

**Task**

Convince yourself that $(f_1, f_2)$ is an invariant scalar product.

Now we have the useful

**Theorem 10.2** *Every representation of a group with invariant averaging on a vector space with scalar product $(.,.)$ is equivalent to a unitary representation. In other words: We can always find a basis in $\mathcal{V}$, with respect to which the matrices $D(g)$ are unitary.*

**Proof** The idea underlying the elegant proof (sum over of the group elements) goes back to A. Hurwitz. We form the following quadratic form on $\mathcal{V}$:

$$h(x) = \mathcal{M}\left\{\underbrace{\left(D(g)x, D(g)x\right)}_{\geq 0}\right\} = \left(x, \mathcal{M}\{D^{\dagger}(g)D(g)\}x\right) = (x, Qx), \quad x \in \mathcal{V}.$$

As a sum of positive and Hermitian forms, the form $Q$ is positive and Hermitian,

$$h(x \neq 0) > 0 \quad \text{and} \quad Q = Q^{\dagger},$$

and is invariant, due to the translation invariance of the averaging:

$$h\left(D(\tilde{g})x\right) = \mathcal{M}\{(D(g)D(\tilde{g})x, D(g)D(\tilde{g})x)\} = \mathcal{M}\{(D(g\tilde{g})x, D(g\tilde{g})x)\} = h(x).$$

Now we use that for every positive Hermitian form $h(x)$ a basis exists, with respect to which $h$ has the normal form,

$$h(y) = \sum \bar{y}_i y_i \, .$$

After choosing this basis, all representing matrices are unitary, $D^\dagger(g) = D^{-1}(g)$.

$\square$

### 10.4.1  Complex-Conjugated, Real and Pseudo-real Representations

If $D$ is any representation of a group on a complex vector space, then by complex conjugation of the matrices

$$g \mapsto D^*(g) \tag{10.22}$$

we obtain another representation. It is the complex conjugated representation. The proof is simple:

$$D^*(e) = \mathbb{1}^* = \mathbb{1}, \quad D^*(g_1 g_2) = (D(g_1)D(g_2))^* = D^*(g_1)D^*(g_2) \, . \tag{10.23}$$

---

**Question**

Why do $g \mapsto D^\dagger(g)$ and $g \mapsto D^{-1}(g)$ not define representations?

**Definition 10.4 (Real and Pseudo-real Representations)**   If a representation is equivalent to a representation with real matrices, then it is called real. If the representations $D$ and $D^*$ are equivalent, then $D$ is called pseudo-real.

Obviously, every real representation is also pseudo-real. For a pseudo-real representation, there exists a matrix $S$ with

$$D^*(g) = SD(g)S^{-1}, \quad \text{for all} \quad g \in G \, . \tag{10.24}$$

It can be shown that for a unitary representation, the matrix $S$ can be chosen to be symmetric or antisymmetric. This is to be proven in Problem 10.1.

---

**Example: The Defining Representation of SU(2) Is Pseudo-real**

SU(2) is defined by the matrices (8.7) and these have the property

$$\sigma_2 U \sigma_2 = U^* \, , \tag{10.25}$$

which means that the defining representation is pseudo-real. The matrix $S = \sigma_2$ is unitary and antisymmetric. It can be proven that the representation is not real.

◄

## 10.5   Tensor Product of Representations

Let $D$ and $\tilde{D}$ be two representations on vector spaces $\mathcal{V}$ and $\tilde{\mathcal{V}}$ of dimensions $n$ and $\tilde{n}$. We choose basis vectors in the two spaces and identify vectors with tuples. These transform as follows,

$$x \mapsto D(g)x \quad \text{and} \quad y \mapsto \tilde{D}(g)y. \tag{10.26}$$

We now form the second rank tensor $t = x \otimes y$ with components $t_{ip} = x_i y_p$. This transforms as

$$t = x \otimes y \mapsto [D(g)x] \otimes \left[\tilde{D}(g)y\right] \equiv (D \otimes \tilde{D})(g)\,(x \otimes y) = (D \otimes \tilde{D})(g)\,t.$$

For the components of the tensor this means

$$t_{ip} = x_i y_p \mapsto \sum_{j=1}^{n}\sum_{q=1}^{\tilde{n}} \left(D(g)_{ij}\tilde{D}(g)_{pq}\right) x_j y_q = \sum_{j,q} D(g)_{ij}\tilde{D}(g)_{pq}\, t_{jq}. \tag{10.27}$$

This is a linear transformation of the $n \cdot \tilde{n}$ components $t_{ip}$. The tensors form the $n \cdot \tilde{n}$-dimensional vector space $\mathcal{V} \otimes \tilde{\mathcal{V}}$ and the matrices $D(g)_{ij}\tilde{D}(g)_{pq}$ are the components of the tensor representation

$$D \otimes \tilde{D} : \mathcal{V} \otimes \tilde{\mathcal{V}} \mapsto \mathcal{V} \otimes \tilde{\mathcal{V}}. \tag{10.28}$$

---

**Task**

Convince yourself that the tensor product $D \otimes \tilde{D}$ of two representations $D$ and $\tilde{D}$ is also a representation of the group.

---

**Tensor Product in Adapted Basis**
For explicit calculations, it is useful to specify a convention for the numbering of the components of $t$. In the numbering

$$\begin{pmatrix} x_1' y' \\ x_2' y' \\ \vdots \\ x_n' y' \end{pmatrix} = \begin{pmatrix} D_{11}\tilde{D} & D_{12}\tilde{D} & \dots & D_{1n}\tilde{D} \\ D_{21}\tilde{D} & D_{22}\tilde{D} & \dots & D_{2n}\tilde{D} \\ \vdots & \vdots & & \vdots \\ D_{n1}\tilde{D} & D_{n2}\tilde{D} & \dots & D_{nn}\tilde{D} \end{pmatrix} \begin{pmatrix} x_1 y \\ x_2 y \\ \vdots \\ x_n y \end{pmatrix} \tag{10.29}$$

(continued)

the representation matrices $D(g) \otimes \tilde{D}(g)$ are simply constructed from those of the representations $D$ and $\tilde{D}$: Multiply each matrix element of $D(g)$ with the matrix $\tilde{D}(g)$.

## 10.6  Symmetric Groups

The symmetric groups $\mathcal{S}_n$ introduced in Sect. 4.2 have two one-dimensional representations. The trivial representation $S : \pi \mapsto 1$ and the sign representation $A : \pi \mapsto \mathrm{sgn}(\pi)$, which assigns a 1 or a $-1$ to even and odd permutations, respectively. The representations $S$ and $A$ correspond to the one-dimensional representations $D_1^1$ and $D_1^2$ of the dihedral group $\mathcal{D}_3 \cong \mathcal{S}_3$ in Sect. 10.1. Obviously, $A \otimes A = S$.

**Corollary 10.1** *Let $\pi \mapsto D(\pi) \in L(\mathcal{V})$ be a representation of the symmetric group. Then the symmetrizer and anti-symmetrizer*

$$P_S = \sum_\pi D(\pi) \quad and \quad P_A = \sum_\pi \mathrm{sgn}(\pi) D(\pi) \tag{10.30}$$

*are idempotent up to a factor, i.e. $P_S^2 = n!\, P_S$, $P_A^2 = n!\, P_A$, and satisfy $P_S P_A = P_A P_S = 0$.*

For example, since $\pi \mapsto \mathrm{sign}(\pi)$ and $\pi \mapsto D(\pi)$ are representations, the following applies

$$P_A^2 = \sum_\pi \sum_{\pi'} \mathrm{sgn}(\pi)\mathrm{sgn}(\pi') D(\pi) D(\pi')$$

$$= \sum_\pi \sum_{(\pi\pi')} \mathrm{sgn}(\pi\pi') D(\pi\pi') = n!\, P_A \, .$$

The linear operators $P_S$ and $P_A$ are thus projectors that project onto subspaces of the representation space $\mathcal{V}$.

---

**Question**

Which property of the odd and even permutations implies $P_S P_A = 0$?

Similarly, one proves

**Lemma 10.2** *The symmetrizer and anti-symmetrizer satisfy $D(\pi) P_S = P_S D(\pi) = P_S$ and $D(\pi) P_A = P_A D(\pi) = \mathrm{sgn}(\pi) P_A$.*

For example,

$$D(\pi)P_A = \sum_{\pi'} D(\pi)\mathrm{sgn}(\pi')D(\pi') = \sum_{\pi'} \mathrm{sgn}(\pi)\mathrm{sgn}(\pi\pi')D(\pi\pi') = \mathrm{sgn}(\pi)P_A .$$

$$(10.31)$$

These properties mean that $P_S$ and $P_A$ project onto the invariant subspaces of the trivial and sign representations, $D(\pi)P_S v = P_S v$ and $D(\pi)P_A v = \mathrm{sgn}(\pi)P_A v$. In Sect. 11.2.4 we will generalize this construction and introduce projectors onto arbitrary irreducible representations. However, this requires the theory of characters.

## 10.7   Application: Benzene Ring in the Hückel Approximation

We consider the symmetries of the benzene molecule $C_6H_6$ on the left side in Fig. 10.4. This is an equilateral hexagon with double bonds—the benzene ring. The 6 corners represent the CH groups. The symmetry transformations of the benzene ring are the 12 elements of the dihedral group

$$\mathcal{D}_6 = \left\{ c_6, \sigma_d \mid c_6^6 = \sigma_d^2 = \sigma_d c_6 \sigma_d c_6 = e \right\} .$$

$$(10.32)$$

The approximation by E. Hückel provides a simple and effective description of molecular orbitals as a linear combination of atomic orbitals. It is a further simplification of the one-electron approximation, in which an electron moves in the average field of the other electrons and the atomic nuclei. In benzene, we identify the vertices $1, \ldots, 6$ of the benzene (where the atomic orbitals are localized) with orthonormal basis vectors $e_1, \ldots, e_6$ in a six-dimensional effective Hilbert space. A molecular orbital corresponds to a state vector $\psi = \sum \psi_i e_i$, whose coefficient functions $\psi_i$ describe the atomic orbitals. After choosing the basis, we identify the

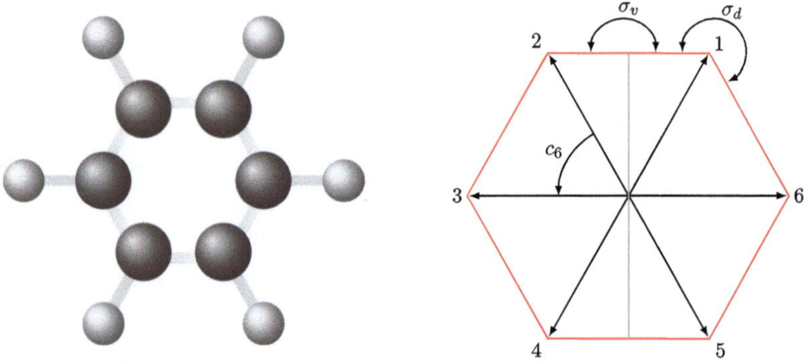


**Fig. 10.4** *Benzene ring and generator of the dihedral group* $\mathcal{D}_6$.

state vector with the 6-component wave function $\psi = (\psi_1, \ldots, \psi_6)^T \in \mathbb{C}^6$ and the Hückel Hamilton operator is the sum of identical effective interaction energies between neighboring C-atoms,

$$H = -\varepsilon \begin{pmatrix} 0 & 1 & 0 & 0 & 0 & 1 \\ 1 & 0 & 1 & 0 & 0 & 0 \\ 0 & 1 & 0 & 1 & 0 & 0 \\ 0 & 0 & 1 & 0 & 1 & 0 \\ 0 & 0 & 0 & 1 & 0 & 1 \\ 1 & 0 & 0 & 0 & 1 & 0 \end{pmatrix} . \tag{10.33}$$

Since neighborhood relations do not change under symmetry transformations, the Hamilton operator $H : \mathbb{C}^6 \mapsto \mathbb{C}^6$ commutes with all these transformations.



Convince yourself that, except for a multiple of the identity, $H$ is proportional to the discretized second derivative on the circular ring.

As sketched in Fig. 10.4, $c_6$ rotates the C-atom at vertex $i$ into the one at vertex $i + 1$, respectively the basis vector $e_i$ into the basis vector $e_{i+1}$ (where $e_7 \equiv e_1$), and the reflection $\sigma_d$ exchanges the vectors $e_2$ and $e_6$ or $e_3$ and $e_5$ and leaves the vectors $e_1$ and $e_4$ unchanged. Hence

$$D(c_6) = \begin{pmatrix} 0 & 0 & 0 & 0 & 0 & 1 \\ 1 & 0 & 0 & 0 & 0 & 0 \\ 0 & 1 & 0 & 0 & 0 & 0 \\ 0 & 0 & 1 & 0 & 0 & 0 \\ 0 & 0 & 0 & 1 & 0 & 0 \\ 0 & 0 & 0 & 0 & 1 & 0 \end{pmatrix}, \quad D(\sigma_d) = \begin{pmatrix} 1 & 0 & 0 & 0 & 0 & 0 \\ 0 & 0 & 0 & 0 & 0 & 1 \\ 0 & 0 & 0 & 0 & 1 & 0 \\ 0 & 0 & 0 & 1 & 0 & 0 \\ 0 & 0 & 1 & 0 & 0 & 0 \\ 0 & 1 & 0 & 0 & 0 & 0 \end{pmatrix} . \tag{10.34}$$

The remaining representation matrices follow using the representation property.

## 10.7.1  Reduction of the 6-Dimensional Representation

The one-dimensional subspace $\mathcal{V}_1$, spanned by

$$f_1 = \frac{1}{\sqrt{6}} \sum_{k=1}^{6} \underbrace{\cos(2\pi k)}_{=1} \, e_k , \tag{10.35}$$

is invariant under all rotations and reflections of the dihedral group. Restricted to $\mathcal{V}_1$, $D$ is the trivial one-dimensional representation $D_1^1$. The vector orthogonal to it

$$f_2 = \frac{1}{\sqrt{6}} \sum_{k=1}^{6} \cos(k\varphi_2)\, e_k, \quad \varphi_n = \frac{2\pi}{n}, \tag{10.36}$$

transforms under the rotation $c_6$ into

$$f_2 \mapsto \frac{1}{\sqrt{6}} \sum_{k=1}^{6} \cos(k\varphi_2)\, e_{k+1} = \frac{1}{\sqrt{6}} \sum_{k=1}^{6} \cos(k\varphi_2 - \varphi_2)\, e_k = -f_2,$$

and under the reflection into $f_2$. Thus, $f_2$ spans another one-dimensional invariant subspace $\mathcal{V}_2$. This carries the representation $D_1^2$, which maps $c_6$ to $-1$ and $\sigma_d$ to 1.

Let now

$$f_3 = \frac{1}{\sqrt{3}} \sum_{k=1}^{6} \cos(k\varphi_3)\, e_k \quad \text{and} \quad f_4 = \frac{1}{\sqrt{3}} \sum_{k=1}^{6} \sin(k\varphi_3)\, e_k \tag{10.37}$$

be two more orthonormal basis vectors. Then

$$c_6: \quad f_3 \mapsto \cos(\varphi_3)\, f_3 + \sin(\varphi_3)\, f_4 \quad , \quad f_4 \mapsto -\sin(\varphi_3)\, f_3 + \cos(\varphi_3)\, f_4 ,$$

$$\sigma_d: \quad f_3 \mapsto \cos(\varphi_3)\, f_3 - \sin(\tfrac{2}{3}\pi)\, f_4 \quad , \quad f_4 \mapsto -\sin(\varphi_3)\, f_3 - \cos(\varphi_3)\, f_4 .$$

Thus, $f_3$ and $f_4$ span a two-dimensional invariant subspace $\mathcal{V}_3$ that carries the representation

$$D_2^1(c_6) = \begin{pmatrix} \cos\varphi_3 & -\sin\varphi_3 \\ \sin\varphi_3 & \cos\varphi_3 \end{pmatrix} \quad \text{and} \quad D_2^1(\sigma_d) = \begin{pmatrix} \cos\varphi_3 & -\sin\varphi_3 \\ -\sin\varphi_3 & -\cos\varphi_3 \end{pmatrix}$$

This is equivalent to the representation of $\mathcal{D}_6$ by the symmetry transformations of the equilateral planar hexagon.

Finally, we supplement the four vectors $f_1, \dots, f_4$ with the orthonormal basis vectors

$$f_5 = \frac{1}{\sqrt{3}} \sum_{k=1}^{6} \cos(\tfrac{1}{3}k\pi)\, e_k \quad \text{and} \quad f_6 = \frac{1}{\sqrt{3}} \sum_{k=1}^{6} \sin(\tfrac{1}{3}k\pi)\, e_k . \tag{10.38}$$

These span another two-dimensional invariant subspace $\mathcal{V}_4$:

$$c_6: \quad f_5 \mapsto \quad\cos(\varphi_6)\, f_5 + \sin(\varphi_6)\, f_6 \quad , \quad f_6 \mapsto -\sin(\varphi_6)\, f_5 + \cos(\varphi_6)\, f_6 ,$$

$$\sigma_d: \quad f_5 \mapsto -\cos(\varphi_6)\, f_5 + \sin(\varphi_6)\, f_6 \quad , \quad f_6 \mapsto \quad\sin(\varphi_6)\, f_5 + \cos(\varphi_6)\, f_6 .$$

The subspace $\mathcal{V}_4$ carries the representation

$$D_2^2(c_6) = \begin{pmatrix} \cos\varphi_6 & -\sin\varphi_6 \\ \sin\varphi_6 & \cos\varphi_6 \end{pmatrix} \quad \text{and} \quad D_2^2(\sigma_d) = \begin{pmatrix} -\cos\varphi_6 & \sin\varphi_6 \\ \sin\varphi_6 & \cos\varphi_6 \end{pmatrix}.$$

The two two-dimensional representations are *inequivalent*, as they have different traces,

$$\text{Tr}\, D_2^1(c_3) = 2\cos\varphi_3 = -1 \quad \text{and} \quad \text{Tr}\, D_2^2(c_3) = 2\cos\varphi_6 = 1.$$

The reduction of the six-dimensional representation $D$ of the group $\mathcal{D}_6$ yields

$$D = D_1^1 \oplus D_1^2 \oplus D_2^1 \oplus D_2^2. \tag{10.39}$$

In the adapted orthonormal basis $\{f_1, \ldots, f_6\}$, the Hamilton operator $H$ in (10.33) is diagonal,

$$H = \epsilon \begin{pmatrix} -2 & 0 & 0 & 0 & 0 & 0 \\ 0 & 2 & 0 & 0 & 0 & 0 \\ 0 & 0 & 1 & 0 & 0 & 0 \\ 0 & 0 & 0 & 1 & 0 & 0 \\ 0 & 0 & 0 & 0 & -1 & 0 \\ 0 & 0 & 0 & 0 & 0 & -1 \end{pmatrix}. \tag{10.40}$$

The eigenvector with minimal energy is the symmetric combination $f_1 \propto \sum e_k$. In an adapted basis, the Hückel Hamiltonian $H$ commuting with all representation matrices is automatically diagonal. In Chap. 11, we will see that this holds more generally for quantum systems with symmetries.

## 10.8  Representations and Special Functions

We now consider the action of an arbitrary group $G$ on a set $M$ (group actions were discussed in detail in Sect. 5.1). We assume that this is a left action. This induces a representation $\Gamma$ of $G$ on the linear space of functions $\mathcal{H} = \{\psi \in \text{maps}(M, \mathbb{C})\}$. In applications, $\mathcal{H}$ will have additional structures, for example, it will be a function space with scalar product, which becomes a Hilbert space after completion. Relevant examples in quantum mechanics will be discussed in the following subsections.

**Group Actions on $M$ Define Representations on $\mathcal{H} = \mathbf{maps}(M, \mathbb{C})$**
Let us assume $G$ acts from the left with $\Phi$ on the set $M$. Then the mapping

$$g \longmapsto \Gamma_g \quad \text{with} \quad \Gamma_g : \mathcal{H} \mapsto \mathcal{H}, \quad \Gamma_g \psi = \psi \circ \Phi_g^{-1} \tag{10.41}$$

defines a representation of $G$ on the linear space $\mathcal{H}$.

The mappings $\Gamma_g : \mathcal{H} \mapsto \mathcal{H}$ are linear and $\Gamma_e$ is the identity operator, because $\Phi_e$ is the identity. To prove the homomorphism property, we need the properties of left actions, as discussed in Sect. 5.1. We conclude

$$\Gamma_{gg'} \psi = \psi \circ \Phi_{gg'}^{-1} = \psi \circ \Phi_{g'}^{-1} \circ \Phi_g^{-1} = \Gamma_g (\psi \circ \Phi_{g'}^{-1}) = \Gamma_g \Gamma_{g'} \psi \,, \tag{10.42}$$

from which the property $\Gamma_{gg'} = \Gamma_g \Gamma_{g'}$ follows.

In general, the representation $g \mapsto \Gamma_g$ will have invariant subspaces in $\mathcal{H}$. On an invariant subspace, spanned by functions $\psi_1, \ldots, \psi_n$, the mappings $\Gamma_g$ have the form

$$\Gamma_g \psi_i = \sum_{j=1}^{n} \psi_j D_{ji}(g), \quad g \in G \,, \tag{10.43}$$

where the matrices $D$ also represent the group $G$,

$$\sum_j \psi_j D_{ji}(gg') = \Gamma_{gg'} \psi_i = \Gamma_g \Gamma_{g'} \psi_i = \Gamma_g \sum_k \psi_k D_{ki}(g')$$

$$= \sum_{jk} \psi_j D_{jk}(g) D_{ki}(g') = \sum_j \psi_j \left( D(g) D(g') \right)_{ji} \,.$$

Thus, $D(e) = \mathbb{1}$ and $D(gg') = D(g)D(g')$, and therefore $D$ is a representation of $G$.

For a transitive group action, we can choose a fixed $p_0$ and connect every point $p$ on $M$ with this point, $p_0 = \Phi_g(p)$, and obtain

$$\psi_i(p) = (\Gamma_g \psi_i)(p_0) = \sum_j \psi_j(p_0) D_{ji}(g) \,. \tag{10.44}$$

If $h$ is from the isotropy group $N_{p_0}$ of $p_0$, then $\Phi_h(p_0) = p_0$ and

$$\psi_i(p_0) = (\Gamma_h \psi_i)(p_0) = \sum_j \psi_j(p_0) D_{ji}(h), \quad \Phi_h(p_0) = p_0 \,. \tag{10.45}$$

Therefore, the $n$-tuple with the components $\psi_i(p_0)$ is a left eigenvector of the matrices $\{D(h)|h \in N_{p_0}\}$ with eigenvalue 1. If we now act with the linear mapping $\Gamma_g$ on the last relation and use (10.43) as well as $\Phi_g^{-1}(p_0) = p$, then we get

$$\psi_i(p) = \sum_j \psi_j(p) D_{ji}(ghg^{-1}). \tag{10.46}$$

In applications, $G$ is the symmetry group of a physical system, and in this situation the $\Gamma_g$ commute with the Hamilton operator $H$ of the quantized system. The diagonalization of $H$ is then much easier if one uses the special functions $\psi_i$ adapted to the symmetry.

### 10.8.1 Spherical Harmonics and SO(3) Representations

We start with an eigenfunction $\psi(x)$ of the Laplace operator (also called Laplacian):

$$-\Delta\psi = \lambda\psi, \qquad \Delta = \sum_{i=1}^{3}\left(\frac{\partial}{\partial x_i}\right)^2. \tag{10.47}$$

This elliptic operator and its eigenvalue problem occur in numerous problems of mechanics, electrostatics, and quantum mechanics.

**Question**

Why is an eigenvalue $\lambda$ real and non-negative?

For every solution $\psi(x)$ of the eigenvalue equation, the transformed function

$$\psi'(x) = (\Gamma_R\psi)(x) = \psi(R^{-1}x) \tag{10.48}$$

with $R \in SO(3)$ is also a solution to the same eigenvalue. To prove this, we make use of

**Lemma 10.3 (Rotation Invariance of the Laplacian)** *The Laplace operator commutes with all $\Gamma_R$, i.e.*

$$\Delta\Gamma_R = \Gamma_R\Delta, \quad R \in SO(3). \tag{10.49}$$

To prove the lemma, we set $Rx = y$, or written in components

$$y_i = \sum_j R_{ij}x_j \quad \text{with} \quad \frac{\partial}{\partial x_i} = \sum_p R_{pi}\frac{\partial}{\partial y_p}, \tag{10.50}$$

and use the chain rule as well as orthogonality of $R$ in

$$(\Gamma_R \Delta_x \Gamma_R^{-1} \psi)(\boldsymbol{x}) = \Gamma_R \Delta_x \psi(\boldsymbol{y}) = \Gamma_R \sum_{i,p,q} R_{pi} R_{qi} \partial_{y_p} \partial_{y_q} \psi(\boldsymbol{y})$$

$$= \Gamma_R \Delta_y \psi(\boldsymbol{y}) = (\Delta_x \psi)(\boldsymbol{x}).$$

This leads to the above statement in the form $\Gamma_R \Delta \Gamma_R^{-1} = \Delta$. It also follows that every $\Gamma_R$ maps an eigenfunction of the Laplace operator into an eigenfunction with the same eigenvalue. To see this, we act with $\Gamma_R$ on the eigenvalue Eq. (10.47) and find

$$\Gamma_R \Delta \psi = \Delta(\Gamma_R \psi) = -\lambda(\Gamma_R \psi). \tag{10.51}$$

The rotations $\boldsymbol{x} \mapsto R\boldsymbol{x}$ define an action of SO(3) on $\mathbb{R}^3$, and according to our general discussion, $R \mapsto \Gamma_R$ is then a representation of the group.

---

**Task**

Verify directly that $(\Gamma_{R'R}\psi)(\boldsymbol{x}) = (\Gamma_{R'}\Gamma_R \psi)(\boldsymbol{x})$ holds. Note that on the right side $R'^{-1}$ acts on $\boldsymbol{x}$ and not on the argument $R^{-1}\boldsymbol{x}$ of the function $\Gamma_R \psi$.

Actually, the $\Gamma_R$ are unitary operators in the Hilbert space $\mathcal{H} = L_2(\mathbb{R}^3)$,

$$(\Gamma_R \phi, \Gamma_R \psi) = \int d^3x \, (\Gamma_R \phi)^*(\boldsymbol{x})(\Gamma_R \psi)(\boldsymbol{x}) \overset{\boldsymbol{x}=R\boldsymbol{y}}{=} \int d^3x \, \phi^*(\boldsymbol{y})\psi(\boldsymbol{y}) = (\phi, \psi).$$

$$\tag{10.52}$$

In the last step, we made use of the rotational invariance of the measure $d^3x$.

The operator $\Delta$ thus commutes with the unitary operators $\Gamma_R$, which are associated with rotations—it is said to be rotationally invariant—similar to the Hückel operator (10.33), which commutes with the representation matrices of its symmetry group $\mathcal{D}_6$. In Problem 10.7 we show, that also a Hamilton operator $H = -\Delta + V(r)$ with $r = |\boldsymbol{x}|$ is rotationally invariant.

The length of $\boldsymbol{x}$ does not change under rotations and the orbits of the SO(3)-action on $\mathbb{R}^3$ are spherical surfaces. We may therefore assume that $\psi$ is a complex function on a spherical surface, on which SO(3) then acts transitively. We choose the radius 1 and consider functions on the unit sphere $S^2 \subset \mathbb{R}^3$. The scalar product is

$$(\phi, \psi) = \int_0^{2\pi} d\varphi \int_0^{\pi} d\theta \, \sin\theta \, \phi^*(\hat{\boldsymbol{x}})\psi(\hat{\boldsymbol{x}}), \qquad \hat{\boldsymbol{x}} = \frac{\boldsymbol{x}}{r} = \begin{pmatrix} \sin\theta\cos\varphi \\ \sin\theta\sin\varphi \\ \cos\varphi \end{pmatrix}.$$

$$\tag{10.53}$$

Then the following holds

**Corollary 10.2 (Unitary Representation of SO(3) on $L_2(S^2)$)** *The mapping* $R \mapsto \Gamma_R$ *in (10.48) defines a unitary representation of the rotation group on the Hilbert space $L_2(S^2)$ of square-integrable functions $S^2 \mapsto \mathbb{C}$.*

It is an infinitely-dimensional representation on $L_2(S^2)$. This representation is fully reducible. For example, the normalized constant function

$$Y_{00} = \frac{1}{\sqrt{4\pi}} \tag{10.54}$$

defines a one-dimensional invariant subspace of $L_2(S^2)$, which carries the trivial representation.

### Homogeneous Polynomials in $x_1$, $x_2$, $x_3$ as Invariant Subspaces

We consider the representation $\Gamma_R$ on the linear space $\mathcal{H}$ of polynomials in the Cartesian coordinates $x_1, x_2, x_3$. If we require $|\boldsymbol{x}| = 1$, then polynomials define normalizable functions in the Hilbert space $L_2(S^2)$. Rotations $\boldsymbol{x} \mapsto R\boldsymbol{x}$ are linear mappings and do not change the degree of a homogeneous polynomial in the variables $x_i$. For example, the homogeneous polynomials $x_1, x_2, x_3$ of degree 1 or the three linear combinations

$$Y_{10} = \sqrt{\frac{3}{4\pi}} \cos\theta \propto x_3, \quad Y_{1\pm1} = \mp\sqrt{\frac{3}{8\pi}} e^{\pm i\varphi} \sin\theta \propto \mp(x_1 \pm x_2) \tag{10.55}$$

span a three-dimensional invariant subspace of $\mathcal{H}$. These are the spherical harmonics known from electrodynamics and quantum mechanics for orbital angular momentum $\ell = 1$.

A polynomial is a linear combination of monomials and the number of monomials $x_1^{p_1} x_2^{p_2} x_3^{p_3}$ of degree $\ell = p_1 + p_2 + p_3$ is equal to

$$\binom{\ell + 2}{2} = \frac{(\ell + 2)(\ell + 1)}{2} . \tag{10.56}$$

The constant monomial 1 of degree 0 spans a one-dimensional invariant subspace of $\mathcal{H}$, which carries the trivial SO(3)-representation $D^0$. The monomials $x_1, x_2, x_3$ of degree 1 span a three-dimensional invariant subspace, which carries the representation $D^1$. The monomials $x_i x_j$ of degree 2 span a six-dimensional invariant subspace. However, this is reducible, since the polynomial $\boldsymbol{x}^2 = 1$ spans a one-dimensional invariant subspace which carries the trivial representation $D^0$. The remaining 5 polynomials

$$x_i x_j - \frac{1}{3}\boldsymbol{x}^2 = x_i x_j - \frac{1}{3} \tag{10.57}$$

transform according to a five-dimensional irreducible representation $D^2$ of SO(3). More generally, if we contract two indices in a monomial of degree $\ell$, e.g.

$$x_{i_1} x_{i_2} x_{i_3} \cdots x_{i_\ell} \mapsto \sum_i x_i x_i \, x_{i_3} \cdots x_{i_\ell} = x_{i_3} \cdots x_{i_\ell} \, , \tag{10.58}$$

then the contracted monomials lie in the invariant subspace of the homogeneous polynomials of degree $\ell - 2$. For example, in the 15 monomials of degree 4, you can contract one pair of indices and obtain the 5 independent monomials of degree 2 transforming according to $D^2$, or you can contract two pairs of indices and obtain the constant monomial which transform according to $D^0$. There are thus $15 - 5 - 1 = 9$ monomials of degree 4 that carry an irreducible representation. More generally, one finds the following iterative relationship for the dimensions of the irreducible representations

$$\dim D^\ell + \dim D^{\ell-2} + \dim D^{\ell-4} + \cdots = \frac{(\ell+1)(\ell+2)}{2} \, , \tag{10.59}$$

with the known dimensions of the representations $D^0$, $D^1$ and $D^2$. It follows that the homogeneous polynomials of degree $\ell$ carry an irreducible representation $D^\ell$ of dimension

$$\dim D^\ell = 2\ell + 1 \tag{10.60}$$

The spherical functions $Y_\ell^m$ are complex linear combinations of these $2\ell + 1$ monomials.


## 10.8.2 Representations of SU(2) on $L_2(\mathbb{C}^2, \rho)$

The representations of the unitary group SU(2) are obtained relatively easily on the linear space of functions of two complex variables, $\psi(\zeta)$ with $\zeta = (z_1, z_2)^t \in \mathbb{C}^2$, equipped with the scalar product

$$(\phi, \psi) = \int d\mu(\zeta)\, \phi^*(\zeta)\psi(\zeta) \quad \text{with} \quad d\mu = -\frac{1}{(2\pi)^2} \prod_{i=1}^{2} d\bar{z}_i dz_i \; e^{-|z_1|^2 - |z_2|^2} \, .$$

$$\tag{10.61}$$

With respect to this weighted scalar product, polynomials have a finite norm. We calculate the scalar product of two monomials $p_{ab} = z_1^a z_2^b$ and $p_{cd}$ in polar coordinates,

$$z_i = x_i + \mathrm{i} y_i = \rho_i \, e^{\mathrm{i}\varphi_i} \quad \text{with} \quad d\bar{z}_i dz_i = 2\mathrm{i}\rho_i d\rho_i d\varphi_i \, . \tag{10.62}$$

One finds for the scalar product

$$(p_{ab}, p_{cd}) = \frac{1}{\pi^2} \int d\rho_1 d\rho_2 \, \rho_1^{1+a+c} \rho_2^{1+b+d} \, e^{-\rho_1^2 - \rho_2^2} \int d\varphi_1 d\varphi_2 \, e^{i\varphi_1(c-a) + i\varphi_2(d-b)}$$

$$= \delta_{ac}\delta_{bd} \int dx dy \, x^a y^b \, e^{-x-y} = a! b! \, \delta_{ac}\delta_{bd} \,.$$

This means that the homogeneous polynomials

$$v_m^j(\zeta) = \frac{z_1^{j+m} z_2^{j-m}}{[(j+m)!(j-m)!]^{1/2}} \quad \text{with} \quad j \in \left\{0, \tfrac{1}{2}, 1, \tfrac{3}{2}, \dots\right\} \tag{10.63}$$

and $m = j, j-1, \dots, -j$ are orthonormal with respect to the scalar product (10.61). For a fixed $j$, the $2j+1$ polynomials $v_m^j$ form a basis in the space of homogeneous polynomials of degree $2j$. Although $j$ and $m$ can be half-integer, the exponents are in $\mathbb{N}_0$.

The group acts with $\Phi_U(\zeta) = U^*\zeta$ from the left on the set $\mathbb{C}^2$, where we followed the usual convention of choosing $U^*$ instead of $U$. This is allowed because $U \mapsto U^*$ represents the group SU(2). We therefore consider the representation

$$(\Gamma_U \psi)(\zeta) = \psi\left(U^{*-1}\zeta\right) = \psi\left(U^T \zeta\right), \quad U = \begin{pmatrix} a & b \\ -\bar{b} & \bar{a} \end{pmatrix}, \quad |a|^2 + |b|^2 = 1\,,$$

$$\tag{10.64}$$

on the space $\mathcal{H} = \text{maps}(\mathbb{C}^2, \mathbb{C})$.

---

### Task

Convince yourself that $\Gamma_U : \mathcal{H} \mapsto \mathcal{H}$ with scalar product (10.61) is unitary. The proof is similar to the proof of the unitarity of the operators $\Gamma_R$ on $L_2(\mathbb{R}^3)$.

The homogeneous polynomials $v_m^j$ of degree $2j$ span an invariant subspace, i.e.

$$(\Gamma_U v_m^j)(\zeta) = v_m^j(U^T \zeta) = \sum_{m'=-j}^{j} v_{m'}^j D_{m'm}^j(U)\,, \tag{10.65}$$

with $(2j+1)$-dimensional matrices $D^j(U)$, which represent the group SU(2). Since the $\Gamma_U$ are unitary and the $v_m^j$ are orthonormal, the representation matrices are unitary.

The functions $v_m^j$ with negative $m$ can be obtained from those with positive $m$,

$$v_{-m}^j(z_1, z_2) = v_m^j(z_2, z_1)\,, \tag{10.66}$$

and it suffices to specify the basis functions with $m \geq 0$. For the representation with $j = 1/2$ one obtains the following monomials and two-dimensional unitary representation:

$$v_{1/2}^{1/2} = z_1 \Rightarrow D^{1/2}(U) = U \,. \tag{10.67}$$

It is the defining representation of SU(2). For $j = 1$ one finds the following monomials and three-dimensional unitary representation:

$$v_1^1 = \frac{z_1^2}{\sqrt{2}}, \ v_0^1 = z_1 z_2 \Rightarrow D^1(U) = \begin{pmatrix} a^2 & \sqrt{2}ab & b^2 \\ -\sqrt{2}a\bar{b} & a\bar{a} - b\bar{b} & \sqrt{2}\bar{a}b \\ \bar{b}^2 & -\sqrt{2}\bar{a}\bar{b} & \bar{a}^2 \end{pmatrix} \,. \tag{10.68}$$

The unit matrix in SU(2) has coordinates $a = 1$ and $b = 0$, so that $D^1(\mathbb{1}_2) = \mathbb{1}_3$. This three-dimensional representation is equivalent to a real orthogonal representation (8.17) through the proper rotations $R(U) \in \mathrm{SO}(3)$, as proven in Sect. (12.2.1).

For $j = 3/2$ one finds the following monomials and representation matrices

$$v_{3/2}^{3/2} = \frac{z_1^3}{\sqrt{6}}, \ v_{1/2}^{3/2} = \frac{z_1^2 z_2}{\sqrt{2}} \Rightarrow D^{3/2}(U)$$

$$= \begin{pmatrix} a^3 & \sqrt{3}a^2b & \sqrt{3}ab^2 & b^3 \\ -\sqrt{3}a^2\bar{b} & a(\bar{a}a - 2b\bar{b}) & b(2a\bar{a} - b\bar{b}) & \sqrt{3}\bar{a}b^2 \\ \sqrt{3}a\bar{b}^2 & \bar{b}(-2a\bar{a} + b\bar{b}) & \bar{a}(a\bar{a} - 2b\bar{b}) & \sqrt{3}\bar{a}^2 b \\ -\bar{b}^3 & \sqrt{3}\bar{a}\bar{b}^2 & -\sqrt{3}\bar{a}^2\bar{b} & \bar{a}^3 \end{pmatrix} \,.$$

To verify the unitarity of the matrix, one uses the condition $|a|^2 + |b|^2 = 1$.

### Traces of the Representation Matrices

In Chap. 11 we will need the traces of the representation matrices $D^j(U)$. It is surprisingly simple to calculate these. We start with the diagonal SU(2) matrices $U_d = \mathrm{diag}(e^{i\vartheta}, e^{-i\vartheta})$, which map the monomials $v_m^j$ onto themselves,

$$\Gamma_{U_d} v_m^j = e^{2im\vartheta} v_m^j, \quad m = j, j - 1, \ldots, -j \,, \tag{10.69}$$

and accordingly lead to diagonal matrices $D^j$,

$$D^j(U_d) = \mathrm{diag}\left(e^{2ij\vartheta}, e^{2i(j-2)\vartheta}, \ldots, e^{-2ij\vartheta}\right) \,. \tag{10.70}$$

Every unitary $U$ can be diagonalized with a unitary matrix $V$, $U = V U_d V^{-1}$. Since $D^j$ is a representation, then also $D^j(V)$ diagonalizes $D^j(U)$. This means that $D^j(U)$ is conjugate to the diagonal matrix $D^j(U_d)$ in (10.70), where $\mathrm{e}^{\pm i\vartheta}$ are the eigenvalues of $U$. From this follows the trace formula

$$\mathrm{Tr}\, D^j(U) = \mathrm{Tr}\, D^j(U_d) = \sum_{m=-2j}^{2j} \mathrm{e}^{2im\vartheta} = \frac{\sin(2j+1)\vartheta}{\sin\vartheta}\,. \tag{10.71}$$

In the following chapter we will use these traces to show that the representations $D^j$ are irreducible.

The irreducible representations of the quantum mechanical Lorentz group $\mathrm{SL}(2,\mathbb{C})$ are obtained analogously as for $\mathrm{SU}(2)$, and the representations of dimensions $2, 3$ and $4$ are determined in Problem 10.9. In fact, there are two non-equivalent representations for each dimension, and these are not unitary.

## 10.9  Exercises for Chap. 10

**Problem 10.1 (Complex Conjugate Representation)** Let $g \mapsto D(g)$ be a representation. In the main text, it was shown that $g \mapsto D^*(g)$ is also a representation.

- Assume that $D$ is irreducible and equivalent to its complex conjugate representation, i.e. there exists an $S$ with $D^*(g) = S D(g) S^{-1}$. Show that then $SS^* = \lambda\mathbb{1}$ holds.
- Further show that for a unitary representation $S$ is either symmetric or antisymmetric and $SS^\dagger = \lambda\mathbb{1}$.

**Problem 10.2 (The Sum of Representations)** Let $V_1$ and $V_2$ be two vector spaces and $V_1 \times V_2$ their Cartesian product, consisting of pairs $(v_1, v_2)$ with $v_i \in V_i$. Scalar multiplication and addition are defined in $V_1 \times V_2$ as follows:

$$\lambda(v_1, v_2) := (\lambda v_1, \lambda v_2), \quad (v_1, v_2) + (v_1', v_2') := (v_1 + v_1', v_2 + v_2')\,.$$

- $V_1$ and $V_2$ are endowed with scalar products. Prove now that

$$\big\langle (v_1, v_2), (v_1', v_2') \big\rangle := \langle v_1, v_1'\rangle_{V_1} + \langle v_2, v_2'\rangle_{V_2}$$

  is a scalar product on $V_1 \times V_2$. The vector space $V_1 \times V_2$, equipped with this scalar product, is the direct sum and is denoted by $V_1 \oplus V_2$.
- Show that $\{(e_m, 0)\}\cup\{(0, f_n)\}$ defines an orthonormal basis of $V_1\oplus V_2$ if $\{e_m\}$ and $\{f_n\}$ are orthonormal bases in $V_1$ and $V_2$. What does this imply for the dimension of $V_1 \oplus V_2$?

- Let $D_1$ and $D_2$ be two representations of the group $G$ on vector spaces $\mathcal{V}_1$ and $\mathcal{V}_2$. Show that then

$$g \to D_1(g) \oplus D_2(g), \quad \text{with} \quad (D_1(g) \oplus D_2(g))(v_1, v_2) = (D_1(g)v_1, D_2(g)v_2)$$

  is a representation of the direct sum $\mathcal{V}_1 \oplus \mathcal{V}_2$.

**Problem 10.3 (A Useful Theorem)** A representation is called faithful if the homomorphism $D : G \mapsto \mathrm{GL}(\mathcal{V})$ is injective. Prove the following theorem:

1. If the group G has a non-trivial normal subgroup $N \lhd G$, then a representation of the factor group $G/N$ is also a representation of G. However, this is not faithful.
2. Conversely: if $D$ is a non-faithful representation of $G$, then G has at least one normal subgroup, so that $D$ is a faithful representation of the corresponding factor group.

**Problem 10.4 (Rotations of Wave Functions)** We consider the wave function of a particle $\psi(x) \in L_2(\mathbb{R}^3)$ in position space. Under rotations, the wave function transforms according to (10.48). Prove that $R \mapsto \Gamma_R$ is a unitary representation of SO(3) on the Hilbert space $L_2(\mathbb{R}^3)$. Since the radius $r = |x|$ does not change under rotations, we can set $r = 1$ and consider $R \mapsto \Gamma_R$ as a representation on $L_2(S^2)$. Which subspaces carry the one-dimensional and three-dimensional irreducible representation?

**Problem 10.5 (Tensor Product 1)** Let $R_1, R_2$ be orthogonal matrices. Show that their tensor product $R_1 \otimes R_2$ is also an orthogonal matrix.

**Problem 10.6 (Tensor Product 2)** Let $A, B$ be square matrices. Prove the properties

$$\mathrm{tr}(A \otimes B) = \mathrm{tr}(A)\mathrm{tr}(B) \quad \text{and} \quad \det(A \otimes B) = (\det A)^m (\det B)^n ,$$

where $n$ and $m$ are dimensions of the matrices $A$ and $B$.

**Problem 10.7 (Rotation Invariant Hamilton Operators)** Prove that the Hamilton operator

$$H = -\frac{\hbar^2}{2m}\Delta + V(r), \quad r = |x|$$

is rotation invariant, i.e. that $[\Gamma_R, H] = 0$ holds.

**Problem 10.8 (Spin-2 Representation of SU(2))** In Sect. 10.8.2, the representations with angular momentum $j = \frac{1}{2}, 1, \frac{3}{2}$ were constructed. Try now to calculate the representation matrices of the five-dimensional representation with $j = 2$. You should obtain the following matrix:

$$D^2(U)$$

$$= \begin{pmatrix} a^4 & 2a^3b & \sqrt{6}a^2b^2 & 2ab^3 & b^4 \\ -2a^3\bar{b} & a^2(\bar{a}a - 3\bar{b}b) & \sqrt{6}ab(\bar{a}a - \bar{b}b) & b^2(3\bar{a}a - \bar{b}b) & 2\bar{a}b^3 \\ \sqrt{6}a^2\bar{b}^2 & -\sqrt{6}ab(\bar{a}a - \bar{b}b) & (\bar{a}a - \bar{b}b)^2 & \sqrt{6}\bar{a}b(\bar{a}a - \bar{b}b) & \sqrt{6}\bar{a}^2b^2 \\ -2a\bar{b}^3 & b^2(3\bar{a}a - \bar{b}b) & -\sqrt{6}\bar{a}b(\bar{a}a - \bar{b}b) & \bar{a}^2(\bar{a}a - 3\bar{b}b) & 2\bar{a}^3b \\ \bar{b}^4 & -2\bar{a}\bar{b}^3 & \sqrt{6}\bar{a}^2\bar{b}^2 & -2\bar{a}^3\bar{b} & \bar{a}^4 \end{pmatrix} .$$

**Problem 10.9 (Spin-2 Representation of SL(2, $\mathbb{C}$))**  Representations of the quantum mechanical Lorentz group SL(2, $\mathbb{C}$) are constructed similarly to those of the quantum mechanical rotation group SU(2) on the space of functions $\mathbb{C}^2 \mapsto \mathbb{C}$. One starts with the action $\Phi_A(\zeta) = \tilde{A}\zeta$ of the group SL(2, $\mathbb{C}$) on $\mathbb{C}^2$. It is advantageous to use here the representation $\tilde{A} = \sigma_2 A \sigma_2$, which is equivalent to the defining representation $A$, and accordingly introduce the representation on maps($\mathbb{C}^2$, $\mathbb{C}$) according to

$$(\Gamma_A \psi)(\zeta) = \psi(\tilde{A}^{-1}\zeta), \quad \zeta = \begin{pmatrix} z_1 \\ z_2 \end{pmatrix}, \quad A = \begin{pmatrix} a & b \\ c & d \end{pmatrix}, \quad ad - bc = 1,$$

1. Choose the basis $v_m^j$ of monomials in (10.63) and define the representation matrices for $\Gamma_A$ as in (10.65). Show that for $j = \frac{1}{2}, 1, \frac{3}{2}$ the representation matrices have the following forms:

$$D^{1/2}(A) = \begin{pmatrix} a & b \\ c & d \end{pmatrix}, \quad D^1(A) = \begin{pmatrix} a^2 & \sqrt{2}ab & b^2 \\ \sqrt{2}ac & ad + bc & \sqrt{2}bd \\ c^2 & \sqrt{2}cd & d^2 \end{pmatrix},$$

$$D^{3/2}(A) = \begin{pmatrix} a^3 & \sqrt{3}a^2b & \sqrt{3}ab^2 & b^3 \\ \sqrt{3}a^2c & a(ad + 2bc) & b(2ad + bc) & \sqrt{3}b^2d \\ \sqrt{3}ac^2 & c(2ad + bc) & d(ad + 2bc) & \sqrt{3}bd^2 \\ c^3 & \sqrt{3}c^2d & \sqrt{3}cd^2 & d^3 \end{pmatrix} .$$

2. Do these matrices have the determinant 1?
3. Explicitly verify that $D^1$ is a representation.
4. Are the above representations of SL(2, $\mathbb{C}$) unitary?
5. These are not all irreducible representations. The missing representations are obtained by adding the complex conjugated representations. Why are $D^j(A)^*$ and $D^j(A)$ inequivalent representations of SL(2, $\mathbb{C}$) but $D^j(U)$ and $D^j(U)^*$ equivalent representations of SU(2)?
   Hint: This exercise requires some routine calculations, for which you can use an algebraic computer program. Try to use one.

**Problem 10.10 (Representations of SU(3))**  We extend the construction of unitary representations of SU(2) to SU(3). For this, we consider monomials

$$p_{abc}(\zeta) = z_1^a z_2^b z_3^c, \quad \zeta = (z_1, z_2, z_3)^t \in \mathbb{C}^3 . \tag{10.72}$$

1. Show that with respect to the scalar product

$$(\phi, \psi) = \int d\mu(\zeta) \, \phi^*(\zeta) \psi(\zeta) \quad \text{with} \quad d\mu = \frac{1}{(2\pi i)^3} \prod_{i=1}^{3} d\bar{z}_i \, dz_i \; e^{-|z_1|^2 - |z_2|^2 - |z_3|^2}$$

the above monomials are orthogonal, $(p_{abc}, p_{def}) = a! b! c! \, \delta_{ad} \, \delta_{be} \, \delta_{cf}$ .
2. Now define, similar to SU(2), the following representation on $\mathcal{H} = \mathrm{maps}(\mathbb{C}^3, \mathbb{C})$,

$$(\Gamma_U \psi)(\zeta) = \psi(U^T \zeta), \quad \zeta \in \mathbb{C}^3, \quad U \in \mathrm{SU}(3) .$$

3. Convince yourself that the homogeneous polynomials of degree $n$ form a $(n + 1)(n + 2)$-dimensional representation. So you get $1, 3, 6, 10, \ldots$-dimensional representations.
4. Why are there two inequivalent representations for each of these dimensions?

Remark: In this simple way, not all irreducible representations are obtained. For example, the eight-dimensional adjoint representation is missing (see Sect. 14.1.4).

# Characters and Schur's Lemma 11

*Lemmas do the work in mathematics: Theorems, like
management, just take the credit.*

—Paul Taylor

The representation theory of groups can be developed either using character theory
or, alternatively for Lie groups, using the Lie algebra associated with the Lie
group. In this chapter, we follow the first approach, which is based on the work
of FERDINAND FROBENIUS and his student ISSAI SCHUR, as well as WILLIAM
BURNSIDE and RICHARD BRAUER. The important contributions of these pioneers
of representation theory are acknowledged in [47]. Applications of the theory of
characters in quantum theory can already be found in the book by Hermann Weyl
[1]. Many works deal with the theory (see e.g. [48]) and its applications [9, 12, 49].

The character of a representation is a class function, i.e., a function $f : G \to \mathbb{C}$,
which is constant on each conjugacy class. These functions were introduced in the
discussion of reduced Haar measures in Sect. 9.1.3. We will see that the characters
of the irreducible representations define a basis in the linear space of class functions.

## 11.1 Character of a Representation

It can be difficult to find the explicit reduction of a given representation. In this
chapter, we will show that a representation $D$ is characterized solely by the traces of
the matrices $D(g)$, and this will significantly simplify the reduction. Since only
the traces matter, it is sufficient to compile a table of traces for the irreducible
representations.



A. Wipf, *Symmetries in Physics*, https://doi.org/10.1007/978-3-662-72675-4_11

**Definition 11.1 (Character)** The character of a representation $g \to D(g)$ is the class function

$$\chi_D(g) = \operatorname{Tr} D(g)\,. \tag{11.1}$$

Each representation thus has its own character $\chi_D$ and this is a class function:

$$\chi_D\!\left(aga^{-1}\right) = \operatorname{Tr} D\!\left(aga^{-1}\right) = \operatorname{Tr} D(a)D(g)D^{-1}(a) = \operatorname{Tr} D(g) = \chi_D(g)\,. \tag{11.2}$$

We made use of $\operatorname{Tr} AB = \operatorname{Tr} BA$.

**Theorem 11.1 (Characters)** *Characters have the following properties:*

1. *Equivalent representations have the same character.*
2. *The dimension of the representation $D$ is $\chi_D(e)$.*
3. *For unitary representations, $\chi_D(g^{-1}) = \chi_D^*(g)$ holds.*
4. *For $D = D_1 \oplus D_2 \oplus \cdots \oplus D_r$ we have $\chi_D = \chi_{D_1} + \cdots + \chi_{D_r}$.*

The first property is equivalent to the well-known fact that the trace does not change under a similarity transformation. The second follows from $\operatorname{Tr} D(e) = \operatorname{Tr} \mathbb{1} = \dim \mathcal{V}$ and the third from

$$\chi_D\!\left(g^{-1}\right) = \operatorname{Tr} D\!\left(g^{-1}\right) = \operatorname{Tr} D^{-1}(g) = \operatorname{Tr} D^{\dagger}(g) = \chi_D^*(g)\,. \tag{11.3}$$

Thus, for groups in which each element is conjugate to its inverse, the characters are real. The last property in the theorem is evident in a basis in which the $D(g)$ are block diagonal:

$$SD(g)S^{-1} = \begin{pmatrix} D_1(g) & 0 & 0 \ldots & 0 \\ 0 & D_2(g) & 0 \ldots & 0 \\ \vdots & & \ddots & \vdots \\ 0 & 0 & 0 \ldots & D_r(g) \end{pmatrix}. \tag{11.4}$$

Just as the character of a direct sum is the sum of the characters, the character of the tensor product of representations is the product of the characters of these representations. This follows directly from the explicit form (10.29) of $D \otimes \tilde{D}$:

$$\chi_{D \otimes \tilde{D}} = \operatorname{Tr}\!\left(D \otimes \tilde{D}\right) = \sum_i D_{ii}\,\operatorname{Tr} \tilde{D} = \operatorname{Tr} D\,\operatorname{Tr} \tilde{D} = \chi_D \cdot \chi_{\tilde{D}}\,. \tag{11.5}$$

Since a character does not depend on the choice of basis, we conclude

**Theorem 11.2 (Character of the Tensor Product)** *The character of the tensor product of two representations is equal to the product of their characters.*

**Example: Characters of $\mathcal{D}_3$**

The representation matrices of the representation $D_2$ for the generating elements $c_3$ and $\sigma$ are given in (10.2) and (10.3). Their tensor product $D_2 \otimes D_2$ is given by

$$c_3 \mapsto \begin{pmatrix} \cos\varphi_3\, D_2(c_3) & -\sin\varphi_3\, D_2(c_3) \\ \sin\varphi_3\, D_2(c_3) & \cos\varphi_3\, D_2(c_3) \end{pmatrix}, \quad \sigma \mapsto \begin{pmatrix} -D_2(\sigma) & 0 \\ 0 & D_2(\sigma) \end{pmatrix}.$$

In accordance with Theorem 11.2 we find

$$\chi_{D_2 \otimes D_2}(c_3) = 4\cos^2\varphi_3 = 1 = \chi_{D_2}(c_3)\chi_{D_2}(c_3),$$

$$\chi_{D_2 \otimes D_2}(\sigma) = 0 = \chi_{D_2}(\sigma)\chi_{D_2}(\sigma),$$

and thus the characters listed in the following table. For class functions, it is sufficient to know the values on the three conjugacy classes.

| $\mathcal{D}_3$ | $\chi_{D_1^1}$ | $\chi_{D_1^2}$ | $\chi_{D_2}$ | $\chi_{D_3}$ | $\chi_{D_2 \otimes D_2}$ |
|---|---|---|---|---|---|
| $K_e$ | 1 | 1 | 2 | 3 | 4 |
| $K_{c_3}$ | 1 | 1 | $-1$ | 0 | 1 |
| $K_\sigma$ | 1 | $-1$ | 0 | 1 | 0 |

In Sect. 10.3 we justified the decomposition $D_3 = D_2 \oplus D_1^1$, which explains the relationship $\chi_{D_3} = \chi_{D_2} + \chi_{D_1^1}$. If the representation $D_2 \otimes D_2$ were reducible, then it would have to be a sum of the lower-dimensional representations $D_1^1$, $D_1^2$ and $D_2$, where the dimensions of the occurring representations add up to 4. From the character table, one can see that only

$$D_2 \otimes D_2 = D_2 \oplus D_1^1 \oplus D_1^2 \tag{11.6}$$

is possible. This correct reduction is achieved much faster if one knows further properties of the characters. ◄

## 11.2  Schur's Lemma

Schur's Lemma is a basic but very useful statement about irreducible representations of groups.

**Lemma 11.1 (Schur)** *Let $D_1$, $D_2$ be two irreducible representations of a group $G$ in vector spaces $\mathcal{V}_1$ and $\mathcal{V}_2$ of dimensions $n_1$ and $n_2$. Furthermore, let*

$$H : \mathcal{V}_1 \mapsto \mathcal{V}_2$$

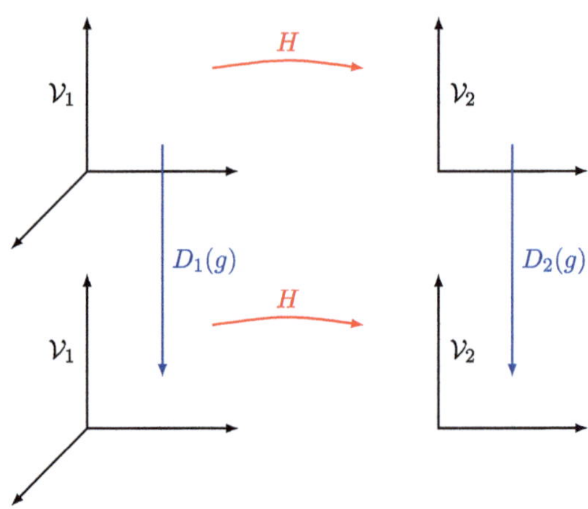

**Fig. 11.1** $HD_1(g)$ and $D_2(g)H$ are supposed to be the same mappings

*be a linear mapping such that*

$$HD_1(g) = D_2(g)H, \qquad \forall g \in G. \tag{11.7}$$

*Then either:*

1. *$H = 0$ or (in the exclusive sense)*
2. *$n_1 = n_2$, $H$ is non-singular and $D_2(g) = HD_1(g)H^{-1}$ for all $g \in G$.*

In the last alternative, the representations $D_1$ and $D_2$ are equivalent to each other and $H$ mediates the conjugation. The situation in the lemma is sketched in Fig. 11.1. In applications in quantum theory, $H$ is often the Hamilton operator and the letter $H$ is supposed to remind us of this.

**Proof** Let $H(\mathcal{V}_1) \subseteq \mathcal{V}_2$ be the image of $\mathcal{V}_1$ under $H$. By assumption

$$D_2 H(\mathcal{V}_1) = HD_1(\mathcal{V}_1) \subseteq H(\mathcal{V}_1).$$

So $H(\mathcal{V}_1) \subset \mathcal{V}_2$ is an invariant subspace of the representation $D_2$. By assumption, $D_2$ is irreducible and there are only the two alternatives

1. $H(\mathcal{V}_1) = 0$. This leads to the first alternative in the lemma.
2. $H(\mathcal{V}_1) = \mathcal{V}_2$. Then $H$ is surjective. We now consider the kernel of the mapping $H$. By assumption

$$HD_1\big(\mathrm{Kern}(H)\big) = D_2 H\big(\mathrm{Kern}(H)\big) = \emptyset,$$

which means that $\mathrm{Kern}(H)$ is an invariant subspace of $D_1$. Therefore, $\mathrm{Kern}(H) = \mathcal{V}_1$ or $\mathrm{Kern}(H) = 0$. We have already dealt with the first case and it remains

$\text{Kern}(H) = 0$. In this case, $H$ is a bijective linear mapping and with the assumption in the lemma

$$H D_1(g) H^{-1} = D_2(g), \qquad \forall g \in G .$$

Note that Schur's Lemma applies to real and complex representation spaces.    □

## 11.2.1 Systems with Invariant Hamilton Operator

We will now draw some remarkable conclusions from the lemma.

**Corollary 11.1** *Let D be an irreducible representation on a vector space $\mathcal{V}$ and H a linear operator that commutes with all representing matrices: $H D(g) = D(g) H$, $\forall g \in G$. Then H is a multiple of the identity.*

**Proof** Let $\lambda$ be an eigenvalue of $H$ (a root of the characteristic polynomial). With $H$, $H - \lambda \mathbb{1}$ also commutes with $D(g)$. Since $\det(H - \lambda \mathbb{1}) = 0$, only the 1st alternative in Schur's Lemma is possible: Thus $H - \lambda \mathbb{1} = 0$ or $H = \lambda \mathbb{1}$.

**Lemma 11.2** *Every irreducible representation of an abelian group is one-dimensional.*

Let $\tilde{g}$ be an arbitrarily selected group element and $\tilde{D} = D(\tilde{g})$. For an abelian group, we have $D(g)\tilde{D} = \tilde{D}D(g)$ for all $g \in G$, so that according to the corollary $\tilde{D} = \lambda \mathbb{1}$. Thus, all representing matrices are diagonal, and only one-dimensional representations can be irreducible.    □

**Corollary 11.2** *A fully reducible representation D on $\mathcal{V}$ decomposes into irreducible, pairwise inequivalent representations,*

$$D = D_1 \oplus D_2 \oplus \cdots \oplus D_r ,$$

*and the linear mapping $H : \mathcal{V} \to \mathcal{V}$ commutes with all $D(g)$:*

$$[H, D(g)] = 0, \quad \forall g \in G . \tag{11.8}$$

*Then the invariant subspaces belonging to the irreducible representations are eigenspaces of $H$.*

**Proof** for $r = 2$ in the corollary: in an adapted basis the $D(g)$ are block diagonal,

$$D(g) = \begin{pmatrix} D_1(g) & 0 \\ 0 & D_2(g) \end{pmatrix} .$$

In analogy to this, we also write $H$ in block matrix form

$$H = \begin{pmatrix} H_1 & Q_1 \\ Q_2 & H_2 \end{pmatrix}.$$

By assumption, it holds

$$\begin{pmatrix} D_1 H_1 & D_1 Q_1 \\ D_2 Q_2 & D_2 H_2 \end{pmatrix} = \begin{pmatrix} H_1 D_1 & Q_1 D_2 \\ Q_2 D_1 & H_2 D_2 \end{pmatrix},$$

so in particular

$$[D_1(g), H_1] = 0 \quad \text{and} \quad [D_2(g), H_2] = 0.$$

According to the previous corollary, $H_1$ and $H_2$ must be proportional to the identity on the representation spaces $\mathcal{V}_1$ and $\mathcal{V}_2$ of the irreducible representations $D_1$ and $D_2$:

$$H_n = \lambda_n \mathbb{1} \quad \text{on} \quad \mathcal{V}_n, \quad n = 1, 2.$$

On the other hand

$$D_1(g)Q_1 = Q_1 D_2(g) \quad \text{and} \quad D_2(g)Q_2 = Q_2 D_1(g), \quad \forall g \in G.$$

Since the two representations are supposed to be inequivalent, only the 1st alternative of Schur's lemma can occur. Therefore, $Q_1 = Q_2 = 0$ and the two representation spaces $\mathcal{V}_1$ and $\mathcal{V}_2$ are eigenspaces of $H$ with eigenvalues $\lambda_1$ and $\lambda_2$.

$$\square$$

## 11.2.2 Orthogonality Relations

The orthogonality relations can be proven for groups with an invariant averaging, i.e., for finite and discrete groups and for unimodular continuous groups. In the following, we assume that such an invariant averaging exists.

Let $D_1$ and $D_2$ be *irreducible representations* of dimensions $n_1$ and $n_2$ and $U$ a linear mapping $\mathcal{V}_1 \mapsto \mathcal{V}_2$. We now consider the following matrix,

$$H = \mathcal{M}\left(D_2(g)U D_1^{-1}(g)\right) : \mathcal{V}_1 \mapsto \mathcal{V}_2, \tag{11.9}$$

where the averaging of matrices means averaging the individual matrix elements. Further, let

$$\tilde{D}_1 = D_1(\tilde{g}) \quad \text{and} \quad \tilde{D}_2 = D_2(\tilde{g}),$$

for a fixed group element $\tilde{g}$. Due to the invariance of the averaging, it follows

$$H\tilde{D}_1 = \mathcal{M}\left(D_2(g)UD_1^{-1}(g)\tilde{D}_1\right) = \tilde{D}_2\mathcal{M}\left(\tilde{D}_2^{-1}D_2(g)UD_1^{-1}(g)\tilde{D}_1\right)$$

$$= \tilde{D}_2\mathcal{M}\left(D_2(\tilde{g}^{-1}g)UD_1^{-1}(\tilde{g}^{-1}g)\right) = \tilde{D}_2\mathcal{M}\left(D_2(g)UD_1^{-1}(g)\right).$$

Thus we have

$$HD_1(\tilde{g}) = D_2(\tilde{g})H, \qquad \forall \tilde{g} \in G.$$

This holds for all elements $\tilde{g}$, so we can apply Schur's lemma:

*1st Alternative:* $D_1$ and $D_2$ are not equivalent.

In this case, $H = 0$ and we have

$$D_1, D_2 \text{ inequivalent} \implies \mathcal{M}\left(D_2(g)UD_1^{-1}(g)\right) = 0.$$

These are $n_1 \cdot n_2$ relations. If we now choose for $U$ the rectangular matrix, which contains a 1 only at the position $(p, q)$ and otherwise only zeros, then it follows

**Inequivalent Irreducible Representations**

For two irreducible and inequivalent representations $D_1$ and $D_2$ we find

$$\mathcal{M}\left(D_2(g)_{ip}D_1(g^{-1})_{qj}\right) = 0. \tag{11.10}$$

Finally we choose $p = i$ and $q = j$ and sum over $i$ and $j$. Then, with $\operatorname{Tr} D_1^{-1} = \chi_{D_1}^*$, the orthogonality relation leads to

$$0 = \mathcal{M}\left(\chi_{D_1}^* \chi_{D_2}\right) = \int d\mu_{\text{red}}(g)\, \chi_{D_1}^*(g)\chi_{D_2}(g) = (\chi_{D_1}, \chi_{D_2}), \tag{11.11}$$

where, for a finite or discrete group, integration over the reduced Haar measure becomes a sum over the elements of the group. If $D_1, \ldots, D_r$ is a list of all possible irreducible representations and $\chi_n \equiv \chi_{D_n}$ their characters, then we have

**Orthogonality of Characters**

The characters $\chi_m$ and $\chi_n$ of two inequivalent irreducible representations are orthogonal, $(\chi_m, \chi_n) = 0$.

*2nd alternative:* The two $n$-dimensional representations $D_1$ and $D_2$ are equivalent.

In suitable chosen bases $D_1 = D_2 = D$, so that $D(\tilde{g})H = HD(\tilde{g})$ for all $\tilde{g} \in G$ and according to the above corollary, $H = \lambda \mathbb{1}$. The constant $\lambda$ follows from

$$\operatorname{Tr} H = \lambda \dim(D) = \operatorname{Tr} \mathcal{M}\left(DUD^{-1}\right) = \mathcal{M}\left(\operatorname{Tr}\left(DUD^{-1}\right)\right) = \operatorname{Tr} U \,,$$

where we made use of $\mathcal{M}(1) = 1$. Thus, we have

$$H \equiv \mathcal{M}\left(DUD^{-1}\right) = \frac{\operatorname{Tr} U}{\dim D} \, \mathbb{1} \,. \tag{11.12}$$

The elements of the square matrix $U$ should now all vanish, except for $U_{jp} = 1$. For this matrix, $\operatorname{Tr} U = \delta_{jp}$ and we conclude:

**Averaged Product of Matrix Elements**
The average of the product of the functions $g \to D_{ij}(g)$ and $g \to D_{pq}^{-1}(g)$ is

$$\mathcal{M}\left(D_{ij}D_{pq}^{-1}\right) = \frac{1}{\dim D} \, \delta_{jp}\delta_{iq} \,. \tag{11.13}$$

We now set $i = j$, $p = q$ and sum over $i$ and $p$.

**Normalization**
The character of each irreducible representation is normalized, $\mathcal{M}\left(\chi_D^*\chi_D\right) = 1$.

We summarize the results for groups with an invariant average:

**Theorem 11.3 (Orthonormality)** *The characters of the irreducible representations are orthonormal in the linear space of complex-valued class functions with scalar product*

$$(\chi_m, \chi_n) = \int d\mu_{\text{red}}(g) \, \chi_m^*(g)\chi_n(g) = \delta_{mn} \,. \tag{11.14}$$

### 11.2.3  Reduction of Arbitrary Group-Representations

An arbitrary representation of a group (with invariant averaging) is a direct sum of irreducible representations,

$$D = \bigoplus_{n=1}^{r} c_n D_n, \quad \text{so that} \quad \chi_D(g) = c_1\chi_1(g) + \cdots + c_r\chi_r(g). \tag{11.15}$$

Equivalent representations are identified here and $2D_2$ means, for example, that the irreducible representation $D_2$ appears twice in the decomposition of $D$. Taking the scalar product of the last sum with the character $\chi_n$ of an irreducible representation, yields the important reduction formula:

**Lemma 11.3 (Reduction Formula)** *The irreducible representation $D_n$ appears exactly $c_n = (\chi_n, \chi_D)$ times in an arbitrary representation $D$.*

Now we can draw the following interesting conclusion:

**Theorem 11.4** *Two arbitrary representations are equivalent if and only if they have the same character.*

Two equivalent representations obviously have the same character. The converse is less obvious because the characteristic polynomial of a matrix contains significantly more information than its trace. But because of the representation property, we also know the traces of its powers.

**Lemma 11.4** *From the traces of all matrices of a representation, the characteristic polynomials of its matrices can be calculated.*

***Proof*** Let $\lambda_1, \ldots, \lambda_d$ be the eigenvalues of $D = D(g)$. Each eigenvalue is listed as often as its algebraic multiplicity. The characteristic polynomial of $D$ is

$$\det(\lambda\mathbb{1} - D) = (\lambda - \lambda_1)(\lambda - \lambda_2)\cdots(\lambda - \lambda_d) = \lambda^d - s_1\lambda^{d-1} + \cdots + (-)^d s_d, \tag{11.16}$$

where the coefficients are the *elementary symmetric polynomials* of the eigenvalues:

$$s_1 = \sum \lambda_j, \quad s_2 = \sum_{i\neq j} \lambda_i\lambda_j, \quad s_3 = \sum_{i\neq j\neq k} \lambda_i\lambda_j\lambda_k, \ldots, \quad s_d = \lambda_1\lambda_2\cdots\lambda_d. \tag{11.17}$$

On the other hand, $D^2 = D(g^2), \ldots, D^d = D(g^d)$ are also matrices in the given representation and their traces are given by the *power sums*:

$$\sigma_p = \operatorname{Tr} D^p = \sum_i \lambda_i^p \,. \tag{11.18}$$

$\square$

> **Elementary Symmetric Polynomials and Power Sums**
> With Newton's identities, each elementary symmetric polynomial can be uniquely represented as a polynomial of power sums. For example, we have
>
> $$s_1 = \sigma_1, \quad s_2 = \sigma_1^2 - \sigma_2, \quad s_3 = \sigma_1^3 - 3\sigma_1\sigma_2 + 2\sigma_3, \ldots$$

We conclude: If two representations $D_1$ and $D_2$ have the same character, i.e. the same $\sigma_p$, then the representing matrices $D_1(g)$ and $D_2(g)$ have identical characteristic polynomials for all $g \in G$ and are therefore similar. Thus, $D_1$ and $D_2$ are equivalent.


## 11.2.4 Method of Projection Operators

A representation $D$ (of a group with invariant averaging) decomposes into irreducible representations and their multiplicities can be determined with the reduction formula (Lemma 11.3). Now we want to construct the projection operators onto the invariant subspaces.

**Theorem 11.5 (Projectors)** *The representation $D$ decomposes into irreducible unitary representations $D = c_1 D_1 \oplus \cdots \oplus c_r D_r$ and accordingly the representation space $\mathcal{V}$ into a sum of pairwise orthogonal subspaces*

$$\mathcal{V} = \bigoplus_{n=1}^{r} \mathcal{V}_n^{\oplus c_n}, \qquad \mathcal{V}_n^{\oplus c_n} = \underbrace{\mathcal{V}_n \oplus \cdots \oplus \mathcal{V}_n}_{c_n - \text{times}} \,. \tag{11.19}$$

*Let $\chi_n$ be the character of $D_n$. Then the linear operator $P_n : \mathcal{V} \mapsto \mathcal{V}$, given by*

$$P_n = \dim(D_n)(\chi_n, D) \tag{11.20}$$

*is a projector onto the linear subspace $\mathcal{V}_n^{\oplus c_n} \subseteq \mathcal{V}$.*

Here, $(\chi_n, D)$ is the matrix with matrix elements $(\chi_n, D)_{ij} = (\chi_n, D_{ij}) = \int d\mu(g)\, \chi_n^*(g) D_{ij}(g)$.

***Proof***  If we take the trace of $D^{-1}$ in the orthogonality relations (11.10) and (11.13), we find

$$(\chi_m, D_n) = \frac{\delta_{mn}}{\dim(D_n)}\, \mathbb{1}_{V_n}\,,\qquad (11.21)$$

from which it immediately follows that $\sum P_n = \mathbb{1}_V$. Furthermore, we get

$$\int d\mu(h)\chi_m^*(h)D_n(hg^{-1}) = \int d\mu(h)\chi_m^*(h)D_n(h)D_n^{-1}(g) = \frac{\delta_{mn}}{\dim(D_n)}\, D_n^{-1}(g)\,,$$

$$(11.22)$$

or after taking the trace

$$\int d\mu(h)\chi_m^*(h)\chi_n^*(gh^{-1}) = \frac{\delta_{mn}}{\dim(D_n)}\, \chi_n^*(g)\,.\qquad (11.23)$$

With this, we can now prove that the $P_n$ in (11.20) are projectors. Because $D(g)D(h) = D(gh)$ and the invariance of the integration (or summation for finite and discrete groups), we have

$$P_n P_m = \dim(D_n)\dim(D_m)\int d\mu(g)d\mu(h)\chi_n^*(g)\chi_m^*(h)D(gh)$$

$$= \dim(D_n)\dim(D_m)\int d\mu(g)\left\{\int d\mu(h)\chi_n^*(gh^{-1})\chi_m^*(h)\right\}D(g)$$

$$\stackrel{(11.23)}{=} \dim(D_n)\,\delta_{mn}\int d\mu(g)\chi_n^*(g)D(g) = \delta_{mn}P_n\,.$$

For unimodular groups the $P_n$ are *orthogonal projectors*. This follows from the property (9.61) of the measure:

$$P_n^\dagger = \dim(D_n)\int d\mu(g)\chi_n(g)D^\dagger(g) = \dim(D_n)\int d\mu(g)\chi_n(g)D(g^{-1}) = P_n\,.$$

$$(11.24)$$

The projector onto the subspace that carries the trivial representation—this is the space of singlets—is particularly simple,

$$P_S = \int d\mu(g)D(g)\,.\qquad (11.25)$$

For representations of the symmetric groups, this is the projector (10.30).  $\qquad\square$

**Example: Regular Representation**

The regular representation introduced in Sect. 10.2 contains the trivial representation once (this will be proven in the next section), and the projector onto the singlet state is

$$P_S = \frac{1}{|G|} \sum_{g \in G} \mathcal{R}(g) = \frac{1}{|G|} \begin{pmatrix} 1 & \cdots & 1 \\ \vdots & \cdots & \vdots \\ 1 & \cdots & 1 \end{pmatrix}. \tag{11.26}$$

◀

## 11.3   All Representations of a Finite Group

First, we determine how many inequivalent irreducible representations a finite group has. Since the characters are constant on conjugacy classes, there can be at most as many independent characters as the group has classes. We will see, that the number of inequivalent irreducible representations is actually equal to the number of conjugacy classes.

How do we find *all* irreducible representations of a finite group? We recall that the representation matrices $\mathcal{R}(g)$ of the regular representation introduced in Sect. 10.2 have only zeros on the diagonal for $g \neq e$. Therefore, the character of the regular representation is

$$\chi_{\text{reg}}(g) = \text{Tr}\,\mathcal{R}(g) = \begin{cases} |G| & \text{if } g = e, \\ 0 & \text{otherwise.} \end{cases} \tag{11.27}$$

Let $\chi_n$ be the character of any *irreducible* representation $D_n$. Due to (11.27), the reduction formula for the multiplicity $c_n$, with which $D_n$ appears in $\mathcal{R}$, yields

$$c_n = \left( \chi_n, \chi_{\text{reg}} \right) = \frac{1}{|G|} \sum_{g \in G} \chi_n^*(g) \cdot \chi_{\text{reg}}(g) = \frac{1}{|G|} \chi_n^*(e) \cdot \chi_{\text{reg}}(e).$$

With $\chi_n(e) = \dim(D_n)$ and $\chi_{\text{reg}}(e) = |G|$, we obtain the

**Theorem 11.6 (Burnside)**   *Every irreducible representation $D_n$ appears in the $|G|$-dimensional regular representation $\mathcal{R}$ exactly $\dim(D_n)$ times:*

$$\mathcal{R} = \bigoplus_{n=1}^{r} \dim(D_n) D_n\,, \quad \text{so that} \quad |G| = \sum_{n=1}^{r} (\dim D_n)^2. \tag{11.28}$$

**Every Irreducible Representation Is Contained in the Regular Representation**

Every one-dimensional irreducible representation appears in the regular representation once, every two-dimensional irreducible representation twice, etc. There is no irreducible representation $D$ with $\dim(D) \geq |G|^{1/2}$.

**Example: Irreducible Representations of $\mathcal{D}_6$**

For $\mathcal{D}_6$, the relationship (11.28) reads as follows:

$$|\mathcal{D}_6| = 12 = \sum_{n=1}^{r}(\dim D_n)^2 = 1^2+1^2+2^2+\sum_{n=4}^{r}(\dim D_n)^2 = 6+\sum_{n=4}^{r}(\dim D_n)^2 \,,$$

where we used that $\mathcal{D}_6$ has at least two one-dimensional and one two-dimensional irreducible representation. Thus, $\mathcal{D}_6$ can only have 6 one-dimensional or one two-dimensional and 2 one-dimensional irreducible representations in addition to these known representations. We will soon show that the latter is the case. ◀

## 11.3.1 The Character Matrix of Finite Groups

We examine a finite group $G$ with $r$ irreducible and inequivalent representations $D_1, \ldots, D_r$ and $k$ conjugacy classes $K_1, \ldots, K_k$. A character is a function on the conjugacy classes. The values of the characters of all irreducible representations $D_n$ on the conjugacy classes $K_i$ of the group are the entries of the *character table*:

| class | $K_1$ | $K_2$ | $\ldots$ | $\ldots$ | $K_k$ |
|-------|-------|-------|----------|----------|-------|
| $\chi_1$ | 1 | 1 | $\ldots$ | $\ldots$ | 1 |
| $\chi_2$ | $\chi_2(K_1)$ | $\chi_2(K_2)$ | $\ldots$ | $\ldots$ | $\chi_2(K_k)$ |
| $\vdots$ | $\vdots$ | $\vdots$ | $\ldots$ | $\ldots$ | $\vdots$ |
| $\chi_r$ | $\chi_r(K_1)$ | $\chi_r(K_2)$ | $\ldots$ | $\ldots$ | $\chi_r(K_k)$ |

In Appendix 11.5, the character tables of important point groups are given. From the above character table, we now construct the so-called *character matrix*:

| class | $K_1$ | $K_2$ | $\cdots$ | $\cdots$ | $K_k$ |
|-------|-------|-------|----------|----------|-------|
| $\chi_1$ | $h_1 \cdot 1$ | $h_2 \cdot 1$ | $\cdots$ | $\cdots$ | $h_k \cdot 1$ |
| $\chi_2$ | $h_1 \cdot \chi_2(K_1)$ | $h_2 \cdot \chi_2(K_2)$ | $\cdots$ | $\cdots$ | $h_k \cdot \chi_2(K_k)$ |
| $\vdots$ | $\vdots$ | $\vdots$ | $\cdots$ | $\cdots$ | $\vdots$ |
| $\chi_r$ | $h_1 \cdot \chi_r(K_1)$ | $h_2 \cdot \chi_r(K_2)$ | $\cdots$ | $\cdots$ | $h_k \cdot \chi_r(K_k)$ |

The occurring positive factors

$$h_i = \left( \frac{|K_i|}{|G|} \right)^{1/2} \tag{11.29}$$

contain the number of elements $|K_i|$ of the conjugacy class $K_i$ and were chosen so that the *rows of the character matrix* are unitarily orthogonal to each other:

$$\sum_{i=1}^{k} h_i^2 \, \chi_m^*(K_i) \cdot \chi_n(K_i) = \frac{1}{|G|} \sum_{i=1}^{k} |K_i| \, \chi_m^*(K_i) \cdot \chi_n(K_i) = (\chi_m, \chi_n) = \delta_{mn} \, .$$

From this it follows that the number of rows must be less than or equal to the number of columns, i.e., $r \le k$.

To see that $r = k$, we multiply (11.14) with $\chi_m(\tilde{g})$ and sum over $m$. For a finite group, the average is the arithmetic mean, so that

$$\chi_n(\tilde{g}) = \sum_{m=1}^{r} \chi_m(\tilde{g})(\chi_m, \chi_n) = \frac{1}{|G|} \sum_{g,m} \chi_m(\tilde{g}) \chi_m^*(g) \chi_n(g) \, . \tag{11.30}$$

The characters are class functions and the sum over the group elements becomes a sum over the $k$ conjugacy classes $K_j$ of the group. With $\tilde{g} \in K_i$ it follows

$$\sum_{j=1}^{k} \left( \frac{1}{|G|} \sum_{m=1}^{r} |K_j| \, \chi_m(K_i) \chi_m^*(K_j) \right) \chi_n(K_j) = \chi_n(K_i) \, . \tag{11.31}$$

From this follows directly the orthogonality relation

$$\sum_{m=1}^{r} h_i \chi_m(K_i) \, h_j \chi_m^*(K_j) = \delta_{ij}, \quad i, j = 1, \ldots, k \, , \tag{11.32}$$

and we conclude:

> **Unitarity of the Character Matrix**
> The columns of the character matrix are orthogonal to each other, and
> therefore the matrix is square and unitary.

From this follows the number of irreducible representations:

**Theorem 11.7** *There are exactly as many irreducible representations of a finite group as there are conjugacy classes. Each class function is a linear combination of the orthonormal characters.*

---

**Example: Irreducible Representations of $\mathcal{S}_4$ (Tetrahedral Group)**

The group $\mathcal{S}_4$ has $P(4) = 5$ conjugacy classes and thus 5 irreducible representations. Because of

$$\sum_{n=1}^{5}(\dim D_n)^2 = 24$$

there are, in addition to the trivial and sign representation, one two-dimensional and two three-dimensional irreducible representations: $1^2 + 1^2 + 2^2 + 3^3 + 3^2 = 24$.

The group $\mathcal{D}_6$ of order 12 has 6 conjugacy classes. Therefore, it has four one-dimensional and two two-dimensional irreducible representations. ◄

---

## 11.4   Representations of Symmetric Groups

With the help of character theory, we continue the analysis of symmetric groups started in Sects. 4.2 and 10.6. For a finite group, the group algebra $\mathcal{A}(G)$ is defined as a $\mathbb{C}$-vector space, spanned by the elements of the group, with multiplication inherited from group multiplication. The group algebra carries the regular representation of the group, which contains all irreducible representations. According to (10.6), the regular representation of $\mathcal{S}_n$ acts on the basis vectors $|\pi\rangle$, where $\pi \in \mathcal{S}_n$ is a permutation, as follows:

$$|\pi\,\pi'\rangle = \sum_{\pi''} \mathcal{R}_{\pi''\pi'}(\pi)|\pi''\rangle \equiv \mathcal{R}(\pi)|\pi'\rangle . \tag{11.33}$$

Recall that in the product $\pi\pi'$ of permutations, $\pi'$ acts first and then $\pi$, e.g. $(12)(13) = (132)$.

**Question**

Why does $\pi \mapsto \mathcal{R}(\pi)$ define a representation of $\mathcal{S}_n$?

As an example, let us consider the group $\mathcal{S}_3 \cong \mathcal{D}_3$ once again.

**Example: Regular Representation of $\mathcal{S}_3$**

If the 6 elements of $\mathcal{S}_3$ are arranged in the order $e$, (123), (132), (12), (13), (23), then the six-dimensional matrices $\mathcal{R}(\pi)$ of the regular representation have the following form: the even permutations are represented by block-diagonal matrices,

$$\mathcal{R}(\pi) = \sigma_0 \otimes A_\pi, \quad A_e = \begin{pmatrix} 1 & 0 & 0 \\ 0 & 1 & 0 \\ 0 & 0 & 1 \end{pmatrix}, \quad A_{(123)} = \begin{pmatrix} 0 & 0 & 1 \\ 1 & 0 & 0 \\ 0 & 1 & 0 \end{pmatrix}, \quad A_{(132)} = \begin{pmatrix} 0 & 1 & 0 \\ 0 & 0 & 1 \\ 1 & 0 & 0 \end{pmatrix},$$

and the odd permutations by block-off-diagonal matrices,

$$\mathcal{R}(\pi) = \sigma_1 \otimes A_\pi, \quad A_{(12)} = \begin{pmatrix} 1 & 0 & 0 \\ 0 & 0 & 1 \\ 0 & 1 & 0 \end{pmatrix}, \quad A_{(13)} = \begin{pmatrix} 0 & 1 & 0 \\ 1 & 0 & 0 \\ 0 & 0 & 1 \end{pmatrix}, \quad A_{(23)} = \begin{pmatrix} 0 & 0 & 1 \\ 0 & 1 & 0 \\ 1 & 0 & 0 \end{pmatrix}.$$

In the regular representation $\pi \mapsto \mathcal{R}(\pi)$, the basis vectors in $\mathbb{C}^6$ are characterized by the permutations. For the explicit form of the matrices, we chose the basis

$$e_1 = |e\rangle, \quad e_2 = |123\rangle, \quad e_3 = |132\rangle, \quad e_4 = |12\rangle, \quad e_5 = |13\rangle, \quad e_6 = |23\rangle, \tag{11.34}$$

where $e$ is the identity element. The regular representation decomposes into the trivial and sign representation $S$ and $A$, and 2 additional equivalent two-dimensional irreducible representations. According to Theorem 11.20, the projection operators

$$P_S = \frac{1}{6}(\sigma_0 + \sigma_1) \otimes B \quad \text{and} \quad P_A = \frac{1}{6}(\sigma_0 - \sigma_1) \otimes B, \quad B = \begin{pmatrix} 1 & 1 & 1 \\ 1 & 1 & 1 \\ 1 & 1 & 1 \end{pmatrix}, \tag{11.35}$$

project onto the subspaces that carry the trivial and sign representations. ◀

General projection operators onto subspaces that carry irreducible representations of $\mathcal{S}_n$, are constructed using Young symmetrizers. To prepare the ground we discuss some general properties of representations of the groups $\mathcal{S}_n$.

**Definition 11.2 (Associated Representation)**  If $A$ is the sign representation of $\mathcal{S}_n$ and $D$ any representation, then the representation $D_A = D \otimes A$ is associated with $D$.

The sign representation is one-dimensional, which implies $\dim D_A = \dim D$. For the trivial representation $S$, $A = S \otimes A$ holds, so that $S$ and $A$ are associated.

**Corollary 11.3**  *Because* $\chi_{D_A} = \chi_D \chi_A$ *we conclude:*

*1. $D_A$ is irreducible if and only if $D$ is irreducible.*
*2. If $D$ is self-associated, i.e. $\chi_D(\pi) = 0$ for odd $\pi$, then $D_A$ and $D$ are equivalent.*

**Proof**  With the orthogonality of the characters, it is immediately proven that $D_A$ is irreducible if and only if $D$ is:

$$(\chi_D, \chi_D) = 1 \iff (\chi_{D_A}, \chi_{D_A}) = (\chi_D, \chi_A^2 \chi_D) \overset{\chi_A = \pm 1}{=} (\chi_D, \chi_D) = 1 \, . \qquad (11.36)$$

If $D$ is self-associated, then $\chi_{D_A}(\pi) = \mathrm{sgn}(\pi)\chi_D(\pi) = \chi_D(\pi)$. This is the second claim in the corollary.  $\qquad\qquad\square$

**Lemma 11.5**  *Let $D_1$ and $D_2$ be two irreducible representations of $\mathcal{S}_n$. If these are*

*1. equivalent (inequivalent), then $D_1 \otimes D_2$ contains the representation $S$ once (not),*
*2. associated (not associated), then $D_1 \otimes D_2$ contains the representation $A$ once (not).*

**Proof**  We note that the characters of all representations of $\mathcal{S}_n$ are real, since $\pi$ and $\pi^{-1}$ are conjugate to each other, and therefore $D(\pi^{-1}) = D^*(\pi) = D(\pi)$ holds. We determine the multiplicity $c_S$ of $S$ in the tensor product $D_1 \otimes D_2$. Writing briefly $\chi_i = \chi_{D_i}$, then

$$c_S = (\chi_S, \chi_{D_1 \otimes D_2}) = (\chi_S, \chi_1 \chi_2) \overset{\chi_S = 1}{=} (\chi_1, \chi_2) \, , \qquad (11.37)$$

and the first statement follows from the orthonormality of the irreducible characters.  $\qquad\square$

---

**Task**

Prove the second statement in the lemma.

### 11.4.1 Young Diagrams and Young Tableaux

The symmetric group $\mathcal{S}_n$ has as many irreducible representations as it has conjugacy classes or Young diagrams, see Sect. 4.2.2. So each irreducible representation corresponds to a Young diagram, and indeed one can assign to each Young diagram a projector onto an irreducible representation. This will be outlined further below.

The $n$ cells or boxes of a *Young diagram* are arranged from top to bottom and left-aligned so that their number in each new row does not increase. For example, the partition $\lambda = [\lambda_1, \lambda_2, \lambda_3, \lambda_4] = [4, 2, 2, 1]$ of the number 9 has the diagram $T_\lambda$ with 9 empty boxes as shown on the left.

The Young diagram filled with the numbers $1, 2, \ldots, n$ is called a *Young tableau*. An example is shown on the right. The symmetric group $S_n$ acts on the set of tableaux of the form $\lambda$. This action is used to rearrange the entries in the columns and rows of a Young tableau using operators.

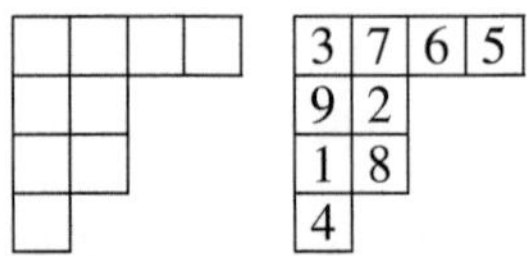

**Definition 11.3 (Normal and Standard Tableaux)**  A Young tableau, in which the entries in each column increase from top to bottom and in each row from left to right, is called a standard tableau. If the numbers increase along the first row and then the second row etc., then it is called a normal tableau.

Every normal tableau is a standard tableau, but not vice versa.

$$
\begin{array}{|c|c|c|c|}
\hline
1 & 2 & 4 & 5 \\
\hline
\end{array}
\;\;\;
\begin{array}{|c|c|c|}
\hline
1 & 3 & 5 \\
\hline
\end{array}
$$

standard tableaux

$$
\begin{array}{|c|c|c|c|}
\hline
1 & 2 & 3 & 4 \\
\hline
\end{array}
\;\;\;
\begin{array}{|c|c|c|}
\hline
1 & 2 & 3 \\
\hline
\end{array}
$$

normal tableaux

For a Young diagram or normal Young tableau, we write briefly $T_\lambda$. Any tableau can be obtained by permuting the elements of the corresponding normal tableau

$$T_\lambda^\pi = \pi T_\lambda . \tag{11.38}$$

For example,

$$(124)\;\begin{array}{|c|c|c|}\hline 1 & 2 & 3 \\ \hline 4 & 5 & \\ \hline\end{array} \;=\; \begin{array}{|c|c|c|}\hline 2 & 4 & 3 \\ \hline 1 & 5 & \\ \hline\end{array} \; .$$

The following relationship follows

$$\pi' T_\lambda^\pi = T_\lambda^{\pi'\pi} , \tag{11.39}$$

**Fig. 11.2** Illustrating the
Hook length formula

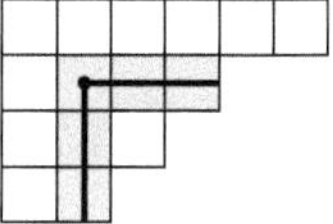

as illustrated by the following example:

$$(23)(124) \; \boxed{\begin{array}{ccc} 1 & 2 & 3 \\ 4 & 5 \end{array}} \;=\; (2,3)\; \boxed{\begin{array}{ccc} 2 & 4 & 3 \\ 1 & 5 \end{array}} \;=\; \boxed{\begin{array}{ccc} 3 & 4 & 2 \\ 1 & 5 \end{array}} \;=\; (1324)\; \boxed{\begin{array}{ccc} 1 & 2 & 3 \\ 4 & 5 \end{array}} \; .$$

**Lemma 11.6 (Hook Length Formula)** *The number of standard tableaux of a Young diagram $T_\lambda$ with n boxes in r rows is given by*

$$d_\lambda = \frac{n!}{|T_\lambda|}, \quad where \quad |T_\lambda| = \prod_{\text{box } k} h(k) \tag{11.40}$$

*is the product of the hook lengths $h(k)$ of all boxes in the diagram. $h(k)$ is equal to the number of boxes to the right and below the cell, including the cell itself. The following applies*

$$\sum_\lambda \frac{1}{|T_\lambda|^2} = \frac{1}{n!} \implies \sum_\lambda d_\lambda^2 = n! = |\mathcal{S}_n| . \tag{11.41}$$

In Fig. 11.2 the hook for the box in row 2 and column 2 is drawn. The hook length is $h(k) = 5$. For the remarkably simple hook length formula (11.40) by Frame et al. [50] there exist a number of alternative proofs, based on very different methods. An older formula for $d_\lambda$ is due to FROBENIUS, who related the number of standard tableaux to the Frobenius determinant.

---

**Example: Standard Tableaux for the Partition $\lambda = [3, 2]$**

There exist $d_\lambda = 5$ standard tableaux for this partition,

$$\boxed{\begin{array}{ccc} 1 & 2 & 3 \\ 4 & 5 \end{array}} \quad \boxed{\begin{array}{ccc} 1 & 2 & 4 \\ 3 & 5 \end{array}} \quad \boxed{\begin{array}{ccc} 1 & 2 & 5 \\ 3 & 4 \end{array}} \quad \boxed{\begin{array}{ccc} 1 & 3 & 4 \\ 2 & 5 \end{array}} \quad \boxed{\begin{array}{ccc} 1 & 3 & 5 \\ 2 & 4 \end{array}} \; ,$$

and the hook lengths of the boxes are entered in the following tableau

$$\boxed{\begin{array}{ccc} 4 & 3 & 1 \\ 2 & 1 \end{array}} \; . \tag{11.42}$$

With the hook length formula (11.40) one naturally also gets $d_\lambda = 5$. ◄

Another formula for the number of standard tableaux for a diagram $T_\lambda$ is:

**Lemma 11.7** *The number of standard tableaux of $T_\lambda$ with n boxes in r rows is*

$$d_\lambda = \frac{\prod_{i<j}(\ell_i - \ell_j)}{\prod_i \ell_i!} \, n! \, , \tag{11.43}$$

*where $\ell_i = \lambda_i + r - i$ and $1 \leq i, j \leq r$.*

For the diagram [3, 2], $(\ell_1, \ell_2) = (4, 2)$ and accordingly $d_\lambda = 5! \, 2/(2! \, 4!) = 5$.


### 11.4.2 Young Symmetrizers

Let $S_{\lambda,i}^\pi$ be the symmetrizer of the $i$'th row of a general Young tableau $T_\lambda^\pi$.
Their product $S_\lambda^\pi = S_{\lambda,1}^\pi S_{\lambda,2}^\pi \cdots$ is the row symmetrizer of the tableau. Note
that the $S_{\lambda,i}^\pi$ act on different sets and thus commute. Similarly, let $A_{\lambda,i}^\pi$ be the
antisymmetrizer of the $i$'th column of $T_\lambda^\pi$. Then their product $A_\lambda^\pi = A_{\lambda,1}^\pi A_{\lambda,2}^\pi \cdots$
is the column antisymmetrizer of the tableau. Multiplying the antisymmetrizer $A_\lambda^\pi$
and the symmetrizer $S_\lambda^\pi$, one obtains the so-called Young symmetrizer

$$Y_\lambda^\pi = \frac{1}{|T_\lambda|} \cdot A_\lambda^\pi \, S_\lambda^\pi \quad \text{with} \quad Y_\lambda^\pi \, Y_\lambda^\pi = Y_\lambda^\pi \, , \tag{11.44}$$

also called Young operator. If one now forms the Young symmetrizer for the
standard tableaux of a fixed partition $\lambda$, one obtains basis vectors, which span an
irreducible subrepresentation. With these vectors, the decomposition of the $\mathcal{R}(\pi)$
can be determined. The symmetrizers, antisymmetrizers and Young symmetrizers
are linear mappings on the vector space that carries the regular representation and
thus are elements of the group algebra $\mathcal{A}(\mathcal{S}_n)$. As usual we write briefly $\pi_1 + \pi_2$,
even though we mean $D(\pi_1) + D(\pi_2)$.

The symmetrizer for a row is the sum of all horizontal permutations that act
within the row. The column antisymmetrizer for a column, on the other hand, is the
alternating sum of the vertical permutations that act within the column. The set of
all horizontal permutations $H$ and all vertical permutations $V$ have only the element
$e$ in common, i.e. $H \cap V = \{e\}$. This leads to

$$Y_\lambda^\pi \propto \sum_{\pi_v \in V} \sum_{\pi_h \in H} \text{sgn}(\pi_v) \pi_v \pi_h \, . \tag{11.45}$$

We now show that all summands on the right side are different. If $\pi_v \pi_h = \pi_{v'} \pi_{h'}$,
then it would follow that

$$\pi_{v'}^{-1} \pi_v = \pi_{h'} \pi_h^{-1} \in H \cap V = \{e\} \quad \text{or} \quad \pi_{v'} = \pi_v, \, \pi_{h'} = \pi_h \, , \tag{11.46}$$

and since a permutation appears at most once in $A_\lambda^\pi$ and in $S_\lambda^\pi$, the statement follows.
It further follows that

$$Y_\lambda^\pi \propto e + \text{terms proportional to } \pi \in \mathcal{S}_n \setminus \{e\} \,. \tag{11.47}$$

For a Young diagram or the corresponding normal Young tableau $T_\lambda$, we write the
symmetrizers, antisymmetrizers, and Young symmetrizers as $S_\lambda$, $A_\lambda$ and $Y_\lambda$.

**Corollary 11.4**  *If $Y_\lambda^\pi$ is the Young symmetrizer for the tableau $T_\lambda^\pi$, then*

$$Y_\lambda^\pi = \pi Y_\lambda \pi^{-1} \,. \tag{11.48}$$

**Proof**  It is sufficient to prove the statement for transpositions $(ij) = (ij)^{-1}$. $Y_\lambda$ is a
sum of permutations, and the conjugated permutations $\pi_k' = (ij)\,\pi_k\,(ij)$ in

$$(ij)Y_\lambda(ij) \propto (ij)\big(e + c_1\pi_1 + c_2\pi_2 + \ldots\big)(ij) = e + c_1\,\pi_1' + c_2\,\pi_2' + \ldots,$$

are equal to the permutations $\pi_k$ with $i$ and $j$ swapped, see Sect. 4.2.2. Since every
permutation is the product of transpositions, the corollary follows.                     $\square$

### Irreducible Representations of $\mathcal{S}_3$

We illustrate the general results using the example $\mathcal{S}_3$. There exist three Young
diagrams with corresponding normal tableaux,

$$
\begin{array}{ccccccc}
\square\square\square & \;\; \text{⊞-shape} \;\; & \;\; \text{vertical} \;\; & , & \boxed{1\,2\,3} & \boxed{\begin{smallmatrix}1\,2\\3\end{smallmatrix}} & \boxed{\begin{smallmatrix}1\\2\\3\end{smallmatrix}} \,.
\end{array} \tag{11.49}
$$

The left diagram has only one row and

$$Y_{[3]} = \frac{1}{3\cdot 2}AS = \frac{1}{6}S_1 = \frac{1}{6}\big(e + (123) + (132) + (12) + (13) + (23)\big)\,. \tag{11.50}$$

Applied to the vector $|e\rangle$ we get

$$|\alpha\rangle \equiv Y_{[3]}\,|e\rangle = \frac{1}{6}\big(|e\rangle + |123\rangle + |132\rangle + |12\rangle + |13\rangle + |23\rangle\big)\,. \tag{11.51}$$

In fact, $Y_{[3]}$ maps *every* vector to a multiple of $|\alpha\rangle$—in particular $|\alpha\rangle$ itself. The
one-dimensional invariant subspace defined by $|\alpha\rangle$ carries the trivial representation.

Now we consider the third diagram with $\lambda = [1, 1, 1]$ in (11.49). Obviously, $S$ is
the identity, $A$ is the antisymmetrizer $A_1$ and thus

$$Y_{[1,1,1]} = \frac{1}{3\cdot 2}AS = \frac{1}{6}A_1 = \frac{1}{6}\big(e + (123) + (132) - (12) - (13) - (23)\big)\,. \tag{11.52}$$

This Young symmetrizer maps every vector to a multiple of

$$|\beta\rangle \equiv Y_{[1,1,1]}|e\rangle = \frac{1}{6}\big(|e\rangle + |123\rangle + |132\rangle - |12\rangle - |13\rangle - |23\rangle\big) \qquad (11.53)$$

The invariant subspace spanned by $|\beta\rangle$ carries the sign representation.

In addition to the one-dimensional representations, $\mathcal{S}_3 \cong \mathcal{D}_3$ has a two-dimensional representation. The projection onto the corresponding invariant subspace is performed by the Young symmetrizer for the middle standard tableau in (11.49). The product of the hook lengths is 3, so that

$$Y_{[2,1]} = \frac{1}{3}A_1 S_1 = \frac{1}{3}\big(e-(13)\big)\big(\mathbb{1}+(12)\big) = \frac{1}{3}\big(e+(12)-(13)-(123)\big). \qquad (11.54)$$

This operator projects onto the two-dimensional subspace, spanned by

$$|\gamma\rangle = |e\rangle - |123\rangle + |12\rangle - |13\rangle \quad \text{and} \quad |\delta\rangle = |e\rangle - |132\rangle + |12\rangle - |23\rangle. \qquad (11.55)$$

---

**Task**

Verify that the Young symmetrizers $Y_{[3]}$, $Y_{[1,1,1]}$ and $Y_{[2,1]}$ are idempotent.

In this invariant subspace, the representation matrices have the form

$$e \mapsto \begin{pmatrix} 1 & 0 \\ 0 & 1 \end{pmatrix}, \quad (123) \mapsto \begin{pmatrix} -1 & -1 \\ 1 & 0 \end{pmatrix}, \quad (132) \mapsto \begin{pmatrix} 0 & 1 \\ -1 & -1 \end{pmatrix},$$

$$(12) \mapsto \begin{pmatrix} 0 & 1 \\ 1 & 0 \end{pmatrix}, \quad (13) \mapsto \begin{pmatrix} -1 & -1 \\ 0 & 1 \end{pmatrix}, \quad (23) \mapsto \begin{pmatrix} 1 & 0 \\ -1 & -1 \end{pmatrix}. \qquad (11.56)$$

and define the two-dimensional irreducible representation.

---

**Task**

Convince yourself that in an adapted orthonormal basis the transformed matrices are symmetry transformations of the equilateral triangle.

Different tableaux generally have different Young symmetrizers. For example, the standard tableau

$$\begin{array}{|c|c|} \hline 1 & 3 \\ \hline 2 \\ \cline{1-1} \end{array}$$

is associated with a symmetrizer that projects onto a two-dimensional subspace orthogonal to the $Y_\lambda$, which carries a representation equivalent to (11.56), see Problem 11.8.

The representation (11.56) is self-associated while $A$ and $S$ are associated with each other, as indicated in the following figure,

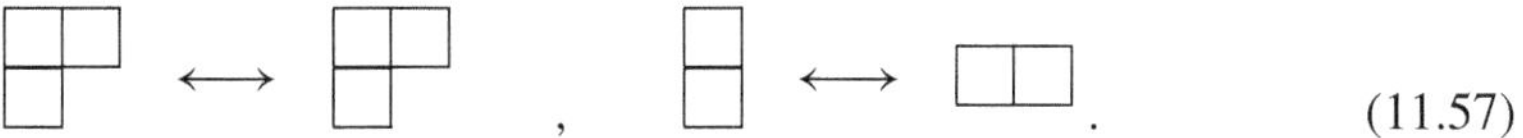

$$\tag{11.57}$$

According to Burnside's theorem, we have found all irreducible representations of $\mathcal{S}_3$ with the two one-dimensional and the two-dimensional representation. According to the theorem, the regular representation contains the two-dimensional representation twice. The Young symmetrizers for the two standard tableaux to $\lambda = [2, 1]$ project onto these two representations. This already suggests that the number of standard tableaux for a partition $\lambda$ relates to the dimension of the irreducible representation and at the same time the multiplicity of this representation in the regular representation.

---

**Example: Representations of $\mathcal{S}_4$**

The Young diagrams of $\mathcal{S}_4$ with entered hook lengths are

$$\tag{11.58}$$

and the corresponding number of standard tableaux is 1, 1, 3, 3 and 2. The square sum of these numbers is $|\mathcal{S}_4| = 24$, in agreement with (11.41). This supports our conjecture about the connection between the number of standard tableaux and the dimensions and multiplicities of the irreducible representations. ◄

We will see that the assignment between standard tableaux and Young symmetrizers is unique. For each partition $\lambda = [\lambda_1, \ldots, \lambda_n]$ of $n$ we can assign the vector space $V_\lambda$ of dimension $d_\lambda$, which carries an irreducible representation. If again $\mathcal{A}(\mathcal{S}_n)$ denotes the group algebra of $\mathcal{S}_n$, then we call the irreducible representation $D^\lambda$ on the subspace $V_\lambda(\pi) = \mathcal{A}(\mathcal{S}_n)Y_\lambda|\pi\rangle$ the representation associated with $T_\lambda$.

**Lemma 11.8 (Young Symmetrizers)** *Let $\lambda$ and $\lambda'$ be different partitions of $n$. Then:*

1. *The irreducible representations for $T_\lambda$ and $T_\lambda^\pi$ are equivalent.*
2. *The irreducible representations for $T_\lambda$ and $T_{\lambda'}$ are inequivalent.*
3. *The (appropriately normalized) Young operators $Y_\lambda^\pi$ are idempotent and project onto invariant subspaces that carry an irreducible representation.*

***Proof*** We had explicitly checked the properties for the small group $\mathcal{S}_3$. Here we provide the proof for the most important statements in the lemma. A more detailed and complete proof can be found for example in [51]. The first statement in the lemma follows immediately from $Y_\lambda^\pi = \pi Y_\lambda \pi^{-1}$. We prove a mild generalization of the second statement, namely that for two Young symmetrizers with $\lambda \neq \lambda'$ and for an arbitrary permutation $\pi$ the product $Y_\lambda Y_{\lambda'}^\pi = A_\lambda S_\lambda A_{\lambda'}^\pi S_{\lambda'}^\pi$ vanishes. This is obviously the case, if $S_\lambda A_{\lambda'}^\pi$ is zero, which due to $\pi A_{\lambda'} \pi^{-1} = A_{\lambda'}^\pi$ in turn is equivalent to $S_\lambda \pi A_{\lambda'} = 0$. $\qquad\square$

**Corollary 11.5** *A sufficient condition for* $Y_\lambda Y_{\lambda'}^\pi = 0$ *is*

$$\left( S_{\lambda,1} S_{\lambda,2} \cdots \right) \left( A_{\lambda',1}^\pi A_{\lambda',2}^\pi \cdots \right) = 0 . \tag{11.59}$$

Note that the $S_{\lambda,i}$ with different $i$ commute, just like the $A_{\lambda',i}^\pi$.

---

**Task**

Show that also the condition

$$\left( A_{\lambda,1} A_{\lambda,2} \cdots \right) \left( S_{\lambda',1}^\pi S_{\lambda',2}^\pi \cdots \right) = 0 \quad \text{for all} \quad \pi \tag{11.60}$$

is sufficient for $Y_\lambda Y_{\lambda'}^\pi$ to vanish. Hint: $S_\lambda A_{\lambda'}^\pi$ is a sum of permutations.

---

We proceed inductively and use that a product $S_{\lambda,k} A_{\lambda',k'}^\pi$ in (11.59) vanishes when row $k$ of $T_\lambda$ and column $k'$ of $T_{\lambda'}^\pi$ have more than one element in common. In particular, the numbers $1, 2, \ldots, \lambda_1$ in the first row of $T_\lambda$ must appear in different columns of $T_{\lambda'}^\pi$, so that none of the products $S_{\lambda,1} A_{\lambda',k}$ is zero. This is only possible if the number of columns of $T_{\lambda'}^\pi$ is greater than or equal to $\lambda_1$, which means $\lambda_1' \geq \lambda_1$. For $\lambda_1' > \lambda_1$ the statement is proven. If, on the other hand, $\lambda_1' = \lambda_1$, then we continue with the second row of $T_\lambda$ with the elements $\lambda_1 + 1, \lambda_1 + 2, \ldots, \lambda_1 + \lambda_2$. These elements must be distributed to $\lambda_2$ unoccupied boxes in different columns of $T_{\lambda'}^\pi$, which is only possible for $\lambda_2' \geq \lambda_2$. If $\lambda_2' > \lambda_2$, then the statement is proven. But if $\lambda_2' = \lambda_2$, then one continues with the iteration until one reaches the last row. At the end of the iteration, one concludes that $Y_\lambda Y_{\lambda'}^\pi$ can only be nonzero for $\lambda = \lambda'$.

Now we sketch the argument for $Y_\lambda Y_\lambda = k Y_\lambda$. For this, we write $A_\lambda S_\lambda$ as a linear combination of permutations, starting with $e$, and find

$$Y_\lambda Y_\lambda \propto A_\lambda (e + c_1 \pi_1 + c_2 \pi_2 + \ldots) S_\lambda = Y_\lambda + \sum_{\pi_k \neq e} c_k A_\lambda \pi_k S_\lambda . \tag{11.61}$$

According to Lemma 10.2, $\pi S_\lambda = S_\lambda$ for every row-preserving permutation, and $A_\lambda \pi = \mathrm{sgn}(\pi) A_\lambda$ for every column-preserving permutation. Thus, for a permutation $\pi$ that either preserves the rows or the columns, $A_\lambda \pi S_\lambda = \pm Y_\lambda$. For other permutations, $A_\lambda \pi S_\lambda$ vanishes. This means that $Y_\lambda Y_\lambda = c Y_\lambda$. If the constant $c$ were zero, then $Y_\lambda$ would be nilpotent and would have a vanishing trace. But due to (11.47), $\mathrm{Tr}(Y_\lambda) \neq 0$. With the normalization factor $1/|T_\lambda|$ as in (11.44), the Young symmetrizers are even idempotent [52]. With the same arguments, one concludes $Y_\lambda \pi Y_\lambda \propto Y_\lambda$. The dimension of the representation $D_\lambda$ for the partition $\lambda$ is

$$\dim D_\lambda = \mathrm{Tr}\, Y_\lambda = \frac{1}{|T_\lambda|} \cdot \mathrm{Tr}(\mathbb{1}) = \frac{n!}{|T_\lambda|} = d_\lambda \,, \qquad (11.62)$$

This results in the following statement:

> **Irreducible Representations of $\mathcal{S}_n$**
> The $d_\lambda$ standard tableaux for a fixed partition $\lambda$ project onto invariant subspaces, all of which (up to equivalence) carry the same irreducible representation of dimension $d_\lambda$. The representations for the $P(n)$ different Young diagrams are inequivalent. There is a bijection between the $P(n)$ normal Young tableaux with $n$ boxes and the irreducible representations of $\mathcal{S}_n$.

Finally, we want to answer the question, what are the Young diagrams of (in the sense of the Definition 11.2) associated representations. The representations $D$ and $D_A$ are inequivalent for non-self-associated representations, but have the same dimension. This means that the associated Young diagrams $T_\lambda$ and $T_{\lambda'}$ must have the same product of hook lengths. This is the case when the two diagrams are conjugate (associated, transposed), which means that the $i$'th column of $T_{\lambda'}$ is as long as the $i$'th row of $T_\lambda$.

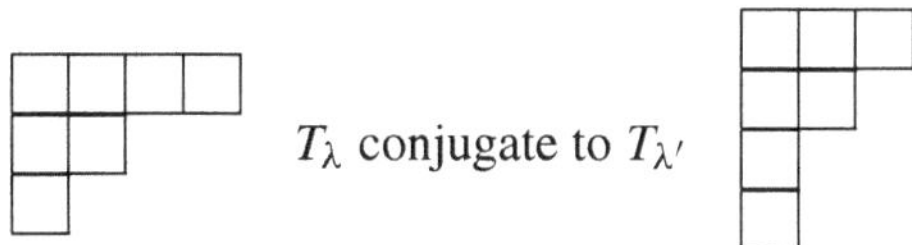


If the two diagrams are equal, then they are called self-conjugate or self-associated and belong to the self-associated representations with $\chi(\pi) = 0$ for odd permutations.

## 11.5    Appendix A: Character Tables of Point Groups

The number of elements and conjugation classes of the proper and improper point groups has already been listed in Table 6.1. We remind here of the symmetry operations or symmetry elements of the point groups:

**Identity**   $E$
**Rotation**   $c_n$ around $n-$fold rotation axes

- $c_n^m$: rotation by $2\pi m/n$
- highest order rotation axis is placed in the $z$-direction
- $c_n'$, $c_n''$: rotation axes not in $z$-direction

**Reflection**   $\sigma$ on a mirror plane

- $\sigma_v$ : reflection plane contains highest order axis
- $\sigma_h$ : reflection plane perpendicular to the highest order axis
- $\sigma_d$ : reflection planes bisect the angle between two rotation axes $c_2'$

**Inversion**   $i$

- point reflection at the inversion center

**Rotoreflection**   $s_n$ on a rotation-reflection axis

- $s_n$ rotation by $2\pi/n$ and subsequent reflection on the plane perpendicular to the rotation axis
- $s_1 = \sigma_h$, $s_2 = i$.

**Notation**   in the group tables we use $\omega_n = \mathrm{e}^{2\pi i/n}$

**Characters of $\mathcal{C}_n$**   The generating element of this abelian group of order $n$ is $c_n$. It has $n$ conjugacy classes and $n$ one-dimensional irreducible representations. The character table has the following form:

| $\mathcal{C}_n$ | $e$ | $c_n$ | $c_n^2$ | $c_n^3$ | $\ldots$ | $c_n^{n-1}$ |
|---|---|---|---|---|---|---|
| $D_1^0$ | 1 | 1 | 1 | 1 | $\ldots$ | 1 |
| $D_1^1$ | 1 | $\omega_n$ | $\omega_n^2$ | $\omega_n^3$ | $\ldots$ | $\omega_n^{n-1}$ |
| $D_1^2$ | 1 | $\omega_n^2$ | $\omega_n^4$ | $\omega_n^6$ | $\ldots$ | $\omega_n^{2n-2}$ |
| $\vdots$ | $\vdots$ | $\vdots$ | $\vdots$ | $\vdots$ | $\ldots$ | $\vdots$ |
| $D_1^{n-1}$ | 1 | $\omega_n^{n-1}$ | $\omega_n^{2n-2}$ | $\omega_n^{3n-3}$ | $\ldots$ | $\omega_n^{(n-1)(n-1)}$ |

**Character of $\mathcal{C}_{nh}$**   This group of order $2n$ is abelian and isomorphic to $\mathcal{C}_n \times \{\mathbb{1}, -\mathbb{1}\}$ for even $n$ and isomorphic to $\mathcal{C}_n \times \{\mathbb{1}, \sigma_h\}$ for odd $n$. It has the generating elements $\{c_n, -\mathbb{1}\}$ or $\{c_n, \sigma_h\}$. The number of irreducible representations or conjugacy classes is $2n$. For even $n$, $\mathcal{C}_{nh}$ has an inversion center and for odd $n$, $\mathcal{C}_{nh}$ is isomorphic to $S_n$. For odd $n$, the character table has the form

| $\mathcal{C}_{nh}$ | $e$ | $c_n$ | $c_n^2$ | $\ldots$ | $\sigma_h$ | $\sigma_h c_n$ | $\sigma_h c_n^2$ | $\ldots$ |
|---|---|---|---|---|---|---|---|---|
| $D_1^0$ | 1 | 1 | 1 | $\ldots$ | 1 | 1 | 1 | $\ldots$ |
| $D_1^1$ | 1 | $\omega_n$ | $\omega_n^2$ | $\ldots$ | 1 | $\omega_n$ | $\omega_n^2$ | $\ldots$ |
| $\vdots$ | $\vdots$ | $\vdots$ | $\vdots$ | $\ldots$ | $\vdots$ | $\vdots$ | $\vdots$ | $\ldots$ |
| $D_1^{n-1}$ | 1 | $\omega_n^{n-1}$ | $\omega_n^{2n-2}$ | $\ldots$ | 1 | $\omega_n^{n-1}$ | $\omega_n^{2n-2}$ | $\ldots$ |
| $D_1'^{0}$ | 1 | 1 | 1 | $\ldots$ | $-1$ | $-1$ | $-1$ | $\ldots$ |
| $D_1'^{1}$ | 1 | $\omega_n$ | $\omega_n^2$ | $\ldots$ | $-1$ | $-\omega_n$ | $-\omega_n^2$ | $\ldots$ |
| $\vdots$ | $\vdots$ | $\vdots$ | $\vdots$ | $\ldots$ | $\vdots$ | $\vdots$ | $\vdots$ | $\ldots$ |
| $D_1'^{n-1}$ | 1 | $\omega_n^{n-1}$ | $\omega_n^{2n-2}$ | $\ldots$ | $-1$ | $-\omega_n^{n-1}$ | $-\omega_n^{2n-2}$ | $\ldots$ |

**Character of $\mathcal{C}_{nv} \cong \mathcal{D}_n$**   This non-abelian point group of order $2n$ is isomorphic to $\mathcal{C}_n \times \{\mathbb{1}, \sigma_v\}$ and has the generating elements $c_n$ and $\sigma_v$. The number of conjugacy classes is $\frac{n}{2} + 3$ for even $n$ and $\frac{n+3}{2}$ for odd $n$. If $n$ is even, then $\mathcal{C}_{nv}$ has an inversion center. In the following character table of $\mathcal{D}_6$, $3c_2'$ stands for the conjugacy class of $c_2'$ with 3 elements.

| $\mathcal{D}_6$ | $e$ | $2c_6$ | $2c_3$ | $c_2$ | $3c_2'$ | $3c_2''$ |
|---|---|---|---|---|---|---|
| $D_1^1$ | 1 | 1 | 1 | 1 | 1 | 1 |
| $D_1^2$ | 1 | 1 | 1 | 1 | $-1$ | $-1$ |
| $D_1^3$ | 1 | $-1$ | 1 | $-1$ | 1 | $-1$ |
| $D_1^4$ | 1 | $-1$ | 1 | $-1$ | $-1$ | 1 |
| $D_2^1$ | 2 | 1 | $-1$ | $-2$ | 0 | 0 |
| $D_2^2$ | 2 | $-1$ | $-1$ | 2 | 0 | 0 |

**Characters of $\mathcal{T}$**   It has 12 elements, 4 conjugacy classes, three one-dimensional representations and one three-dimensional irreducible representation.

| $\mathcal{T}$ | $e$ | $4c_3$ | $4c_3^2$ | $3c_2'$ |
|---|---|---|---|---|
| $D_1^1$ | 1 | 1 | 1 | 1 |
| $D_1^2$ | 1 | $\omega_3$ | $\omega_3^2$ | 1 |
| $D_1^3$ | 1 | $\omega_3^2$ | $\omega_3$ | 1 |
| $D_3$ | 3 | 0 | 0 | $-1$ |

**Characters of $\mathcal{O}$**   The group has 24 elements and 5 classes. There are two one-dimensional, one two-dimensional and two three-dimensional irreducible representations.

| $\mathcal{O}$ | $e$ | $6c_4$ | $3c_2$ | $8c_3$ | $6c_2'$ |
|---|---|---|---|---|---|
| $D_1^1$ | 1 | 1 | 1 | 1 | 1 |
| $D_1^2$ | 1 | $-1$ | 1 | 1 | $-1$ |
| $D_2$ | 2 | 0 | 2 | $-1$ | 0 |
| $D_3^1$ | 3 | 1 | $-1$ | 0 | $-1$ |
| $D_3^2$ | 3 | $-1$ | $-1$ | 0 | 1 |

**Characters of $\mathcal{Y}$** The group has 60 elements and 5 classes. It has one one-dimensional, two three-dimensional, one four-dimensional and one five-dimensional representation.

| $\mathcal{Y}$ | $e$ | $12c_5$ | $12c_5^2$ | $20c_3$ | $15c_2$ |
|---|---|---|---|---|---|
| $D_1^1$ | 1 | 1 | 1 | 1 | 1 |
| $D_3^1$ | 3 | $-2\cos 2\varphi_5$ | $2\cos\varphi_5$ | 0 | $-1$ |
| $D_3^2$ | 3 | $-2\cos\varphi_5$ | $2\cos 2\varphi_5$ | 0 | $-1$ |
| $D_4$ | 4 | $-1$ | $-1$ | 1 | 0 |
| $D_5$ | 5 | 0 | 0 | $-1$ | 1 |

For the corresponding tables of further point groups, I refer to the literature, e.g. the book by Wagner [16].

## 11.6   Exercises for Chap. 11

**Problem 11.1 (The Trace Is Invariant Under Cyclic Permutation)** Show that the trace of a product of $n$-dimensional matrices $A_1, \ldots, A_k$ does not change under cyclic permutation of the matrices, for example $\mathrm{Tr}\,(A_1 A_2 \cdots A_{k-1} A_k) = \mathrm{Tr}\,(A_k A_1 A_2 \cdots A_{k-1})$. What does this mean for the trace of the product of two or three matrices? What follows for the trace of the matrix $UAU^{-1}$ conjugated to matrix $A$?

**Problem 11.2 (Commuting Matrices)** Show that two matrices can be simultaneously diagonalized if and only if they commute.
Hint: Two matrices are simultaneously diagonalizable, if they can be diagonalized using the same matrix. That simultaneous diagonalizability implies commutativity is easy to show. To prove the converse, bring one of the matrices into diagonal form using a conjugation.

**Problem 11.3 (Conjugacy Classes of the Cubic Group)** In lattice field theory, the space-time continuum is often replaced by a hyper-cubic lattice. The symmetries of the lattice are then from the cubic subgroup of the rotation group and the states on the lattice are classified by the irreducible representations of that subgroup. Show that the cubic group contains the following five conjugacy classes:

1. the trivial class with the identity element $e$,
2. the class $C_2$ with the $\pi$-rotations about the axes connecting opposite faces,
3. the class $C_3$ with the $2\pi/3$-rotations about the body diagonals in the cube,
4. the class $C_2'$ with the $\pi$-rotations about the axes connecting opposite edges,
5. the class $C_4$ of the $\pm\pi/2$-rotations about axes connecting opposite faces.

Thus, the group has five irreducible representations. What are their dimensions?

**Problem 11.4 (Regular Representations of $S_3$)** This group is known to have 6 elements and is isomorphic to $\mathcal{D}_3$.

- Determine the six-dimensional regular representation $\mathcal{R}$ of $S_3$.
- Show that the regular representation contains the three irreducible representations with the expected multiplicities.

**Problem 11.5 (Irreducible Representations of the Dodecahedron Group)** The proper symmetry transformations of the dodecahedron form one of the Platonic groups. It has 60 elements and 5 conjugacy classes. The group has one one-dimensional, no two-dimensional and *two* three-dimensional irreducible representations (two of these representations you should know). What are the dimensions of the remaining irreducible representations?

**Problem 11.6 (Buckyball)** The fullerene $C_{60}$ (often called buckyball), as well as some quasicrystals, possess the icosahedron group $\mathcal{Y}_h$ as symmetry group. Determine the character table and convince yourself that Burnside's theorem applies.

The buckyball shown in the illustration was first produced at Rice University in 1985. Harold Kroto, Robert Curl, and Richard Smalley were awarded the 1996 Nobel Prize in Chemistry for this work.

Buckyball (Taken from
Wikipedia)

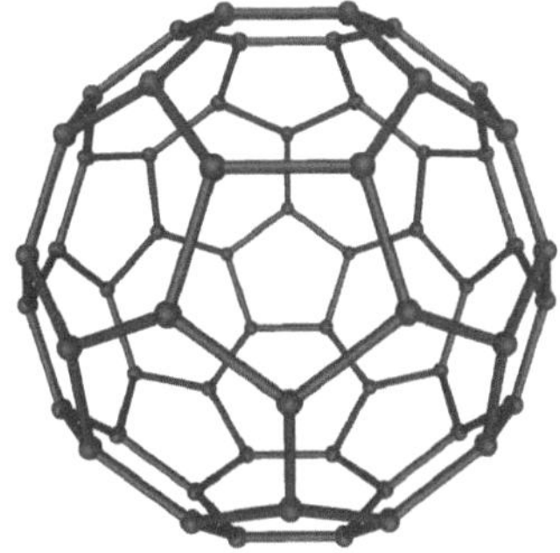

**Problem 11.7 (Young Operators 1)** Apply the Young symmetrizer $Y_{[2,1]}$ in (11.54) to the 6 vectors $|123\rangle$, $|213\rangle$, $\ldots$ and show that each image vector is

a linear combination of $|\gamma\rangle$ and $|\delta\rangle$ in (11.55). How does the Young symmetrizer act on the basis elements $|\gamma\rangle$ and $|\delta\rangle$?

**Problem 11.8 (Young Operators 2)** Determine the Young symmetrizer for the second standard tableau in

$$\begin{array}{|c|c|}\hline 1 & 2 \\\hline 3 \\\cline{1-1}\end{array} \qquad \begin{array}{|c|c|}\hline 1 & 3 \\\hline 2 \\\cline{1-1}\end{array}$$

and show that the resulting two-dimensional representation is equivalent to the representation of the first standard tableau.

# Irreducible Representations of Lie Groups  12

> *It seems to be one of the fundamental features of nature that fundamental physical laws are described in the form of a mathematical theory of great beauty and power, ...*
>
> —Paul Adrien Maurice Dirac

After the representations of finite groups, we now consider representations of continuous groups with averaging. The focus is on the characters and their completeness. The latter property is the content of the theorem by Peter and Weyl, according to which the characters of the irreducible representations form an orthonormal basis in the space of square-integrable class functions. We start with the simplest compact Lie group, which is the abelian group U(1). The theory of group representations using characters is presented in the books [48,53].

## 12.1   Characters of U(1) and Theorem of Peter and Weyl

For the abelian group U(1), each element forms a conjugacy class and therefore all functions are also class functions. According to Lemma 11.2 in Sect. 11.2, all irreducible representations are one-dimensional and according to Theorem 10.2 in a suitable basis unitary. Thus, each irreducible representation has the form

$$D\!\left(e^{i\vartheta}\right) = e^{ih(\vartheta)} \quad \text{and} \quad e^{ih(\vartheta+2\pi)} = e^{ih(\vartheta)}, \quad h(\vartheta)\ \text{real}. \tag{12.1}$$

The representation property implies

$$e^{ih(\vartheta_1+\vartheta_2)} = e^{ih(\vartheta_1)+ih(\vartheta_2)}.$$



A. Wipf, *Symmetries in Physics*, https://doi.org/10.1007/978-3-662-72675-4_12

Thus, $h(\vartheta)$ is a linear function with $h(\vartheta + 2\pi) = h(\vartheta) + 2\pi n$, and the irreducible representations have the form

$$D_n : \mathrm{e}^{\mathrm{i}\vartheta} \mapsto D_n\big(\mathrm{e}^{\mathrm{i}\vartheta}\big) = \mathrm{e}^{\mathrm{i}n\vartheta}, \qquad n \in \mathbb{Z}. \tag{12.2}$$

So there exist countably many irreducible representations with characters

$$\chi_n(\vartheta) = \mathrm{Tr}\, D_n\big(\mathrm{e}^{\mathrm{i}\vartheta}\big) = \mathrm{e}^{\mathrm{i}n\vartheta}. \tag{12.3}$$

In accordance with Theorem 11.3, these are orthonormal with respect to averaging,

$$(\chi_m, \chi_n) = \mathcal{M}\left(\chi_m^* \chi_n\right) = \frac{1}{2\pi} \int \mathrm{d}\vartheta \; \mathrm{e}^{-\mathrm{i}m\vartheta}\, \mathrm{e}^{\mathrm{i}n\vartheta} = \delta_{mn}. \tag{12.4}$$

Every class function $f(\vartheta)$ is $2\pi$-periodic and according to the theory of Fourier series a linear combination of the exponential functions (12.3). So, as for finite groups:

---

**The Characters of U(1) Are Complete**

Every class function is a linear combination of the characters,

$$f(\vartheta) = \sum_n c_n \chi_n(\vartheta), \qquad c_n = \frac{1}{2\pi} \int_0^{2\pi} \mathrm{d}\vartheta \; \mathrm{e}^{-\mathrm{i}n\vartheta}\, f(\vartheta) = (\chi_n, f).$$

$$\tag{12.5}$$

---

The theory of Fourier series is thus a special case of the theory of characters. The completeness of characters for finite groups and for the Lie group U(1) holds more generally for continuous groups with averaging, whose irreducible representations are generally no longer one-dimensional. The following theorem from harmonic analysis, which is proven for example in [5, 6, 54], applies:

**Theorem 12.1 (Theorem of Peter and Weyl)** *For every compact (topological) group G, there exist infinitely many characters. These form a complete orthonormal basis in the space of class functions.*

There are several versions of theorem of Peter and Weyl, and we have cited the one for class functions whose scalar product is calculated using the reduced Haar measure. A more general version provides a decomposition of $L_2(G)$, the space of square-integrable functions on $G$ with respect to the Haar measure and it is the generalization of Burnside's theorem in Sect. 11.3:

**Theorem 12.2 (Theorem of Peter and Weyl)**  *Let $G$ be a compact group. Then the regular representation $\mathcal{R}$ on $L_2(G)$, given by*

$$\left(\mathcal{R}(a)\psi\right)(g) = \psi\left(a^{-1}g\right), \quad a, g \in G \quad \text{and} \quad \psi \in L_2(G), \tag{12.6}$$

*is isomorphic to the direct sum of irreducible representations. In fact, one has the decomposition*

$$\mathcal{R} = \bigoplus_n \dim(D_n) D_n, \tag{12.7}$$

*where $n$ enumerates the irreducible unitary representations of $G$ (up to isomorphism).*

The irreducible unitary representations of a compact group are finite-dimensional, $\dim(D_n) < \infty$, and $D_n$ occurs $\dim(D_n)$ times in the decomposition of the regular representation. Hence the decomposition is almost identical as for finite groups. The difference is that for continuous groups $\mathcal{R}$ is an infinite-dimensional representation and the sum over the irreducible representations has infinitely many terms. The regular representation is a special group action (see Sect. 10.8), in which the group acts on itself. The mapping $a \mapsto \mathcal{R}(a)$ is a representation, and the proof was given in Sect. 10.8.

## 12.2  Irreducible Representations of SU(2)

In Sect. 10.8.1, we obtained the irreducible representations of SU(2) from the left action of the group on $\mathbb{C}^2$. Now we take a different approach and use the theorem of Peter and Weyl for class functions to construct these representations. We start with the trivial one-dimensional representation $D_1$ and the two-dimensional defining representation $D_2$ of SU(2) in (8.7) and (9.9). These have characters

$$\chi_1(U) = 1 \quad \text{and} \quad \chi_2(U) = a + \bar{a} = 2\cos\vartheta = \mathrm{e}^{\mathrm{i}\vartheta} + \mathrm{e}^{-\mathrm{i}\vartheta}. \tag{12.8}$$

With the help of the integrals

$$\int \mathrm{d}\mu_{\mathrm{red}} \cos^2\vartheta = \frac{1}{4} \quad \text{and} \quad \int \mathrm{d}\mu_{\mathrm{red}} \cos\vartheta = 0 \tag{12.9}$$

one finds for the scalar products calculated with the reduced Haar measure (9.14)

$$(\chi_1, \chi_1) = (\chi_2, \chi_2) = 1 \quad \text{and} \quad (\chi_1, \chi_2) = 0,$$

in accordance with the general theory. We will see later that these are the only one- and two-dimensional representations of the group SU(2).

**Task**

Prove the integral formula

$$\int d\mu_{\text{red}}(\vartheta)\cos^{2p}\vartheta = \frac{1}{4^p}\frac{(2p)!}{p!\,(p+1)!}\,. \tag{12.10}$$

## 12.2.1 The Three-Dimensional Representation SO(3)

We start with the four-dimensional tensor product representation

$$D = D_2 \otimes D_2 \quad \text{with} \quad \chi_D(\vartheta) = \chi_2(\vartheta)\cdot\chi_2(\vartheta) = 4\cos^2\vartheta\,. \tag{12.11}$$

The scalar products of $\chi_D$ with the characters in (12.8) are

$$(\chi_1, \chi_D) = 1 \quad \text{and} \quad (\chi_2, \chi_D) = 0\,,$$

so that $D$ contains the one-dimensional representation exactly once. Thus, the tensor product decomposes into the trivial representation $D_1$ and a three-dimensional representation,

$$D_2 \otimes D_2 = D_1 \oplus D_3 \quad \text{with} \quad \chi_3(\vartheta) = \chi_{D_3} = e^{2i\vartheta} + 1 + e^{-2i\vartheta}\,. \tag{12.12}$$

Because $(\chi_3, \chi_3) = 1$, $D_3$ is an *irreducible* representation of dimension $\chi_3(e) = 3$. To see that $D_3$ is equivalent to the group SO(3) of rotations in $\mathbb{R}^3$, we take a closer look at the tensor representation $D_2 \otimes D_2$. Under $D_2$, $x \in \mathbb{C}^2$ and $y \in \mathbb{C}^2$ transform into $Ux$ and $Uy$. In the parametrization (8.7), we then have

$$\begin{pmatrix} x_1 y_1 \\ x_1 y_2 \\ x_2 y_1 \\ x_2 y_2 \end{pmatrix} \mapsto \begin{pmatrix} a^2 & ab & ba & b^2 \\ -a\bar{b} & a\bar{a} & -b\bar{b} & b\bar{a} \\ -\bar{b}a & -\bar{b}b & \bar{a}a & \bar{a}b \\ \bar{b}^2 & -\bar{b}\bar{a} & -\bar{a}\bar{b} & \bar{a}^2 \end{pmatrix}\begin{pmatrix} x_1 y_1 \\ x_1 y_2 \\ x_2 y_1 \\ x_2 y_2 \end{pmatrix}\,. \tag{12.13}$$

If we use the anti-symmetric and three symmetric combinations

$$Y_{00} = \frac{1}{\sqrt{2}}(x_1 y_2 - x_2 y_1) \quad \text{and} \quad Y = \begin{pmatrix} Y_{11} \\ Y_{10} \\ Y_{1-1} \end{pmatrix} = \begin{pmatrix} x_1 y_1 \\ \frac{1}{\sqrt{2}}(x_1 y_2 + x_2 y_1) \\ x_2 y_2 \end{pmatrix}\,,$$

then we find for these the transformation behavior

$$Y_{00} \mapsto Y_{00}, \quad Y \mapsto D^1(U)\,Y \tag{12.14}$$

with $D^1(U)$ from (10.68). The one-dimensional invariant subspace is thus spanned by the anti-symmetric combination $Y_{00}$ and the three-dimensional invariant subspace by the symmetric combinations $Y_{1m}$, i.e. by the spherical harmonics with angular momentum 1.

Finally, if we set

$$(X_1, X_2, X_3) = \left( \frac{1}{\sqrt{2}}(Y_{11} - Y_{1-1}), \frac{1}{i\sqrt{2}}(Y_{11} + Y_{1-1}), Y_{10} \right), \qquad (12.15)$$

then in the parametrization (9.9) the three-dimensional representation takes the form

$$X \mapsto (e, X)e + e \wedge X \sin 2\vartheta - e \wedge (e \wedge X)\cos 2\vartheta = R(2\vartheta, e)X, \qquad (12.16)$$

with unit vector

$$e = \begin{pmatrix} \sin \psi \cos(\pi + \varphi) \\ \sin \psi \sin(\pi + \varphi) \\ \cos \psi \end{pmatrix}. \qquad (12.17)$$

The mapping (12.16) describes a proper rotation of $X$ about the axis $e$ with angle $2\vartheta$. This means that $U \to R(U)$ represents the group SU(2) by rotations in $\mathbb{R}^3$. Because $R(U) = R(-U)$, this representation is not faithful.

## 12.2.2 Higher-Dimensional Representations

Similar to $D_2 \otimes D_2$, we can reduce the six-dimensional tensor product representation $D_3 \otimes D_2$ with character

$$\chi_3 \cdot \chi_2 = e^{3i\vartheta} + 2\,e^{i\vartheta} + 2\,e^{-i\vartheta} + e^{-3i\vartheta} = \chi_2 + \chi_4$$

Obviously, it contains the representation $D_2$ exactly once. Since $(\chi_4, \chi_4) = 1$, the four-dimensional representation $D_4$ is irreducible. Similarly, from $D_4 \otimes D_2$ we obtain a new irreducible representation $D_5$ of dimension 5. If we iterate this process, we obtain infinitely many irreducible representations $D_n$ with $\dim(D_n) = n$.

**Characters of the Irreducible SU(2) Representations**

For each $n \in \mathbb{N}$, there is a an irreducible representation $D_n$ of SU(2). It has the character

$$\chi_n \equiv \chi_{D_n} = \sum_{k=1-n,3-n,\ldots}^{n-1} e^{ik\vartheta} = \frac{\sin n\vartheta}{\sin \vartheta}. \qquad (12.18)$$

Because $\chi_n(e) = n$, $D_n$ has the dimension $\dim(D_n) = n$.

We have already encountered this form of characters earlier through the action of the group on $\mathbb{C}^2$, see (10.71). The $\chi_n$ form an orthonormal system of class functions,

$$(\chi_m, \chi_n) = \frac{2}{\pi} \int_0^\pi \sin m\vartheta \, \sin n\vartheta \, d\vartheta = \delta_{mn} \,,$$

which proves that the $D_n$ are irreducible and inequivalent representations of SU(2).

The question arises as to which of the $D_n$ are faithful. A representation is known to be faithful if its kernel contains only the identity element.

---

**Task**

Show that of all $U \in \mathrm{SU}(n)$ only $\mathbb{1}_n$ has the trace $n$.

---

For the unitary representations $D_n$ this means

$$U \in \mathrm{Ker}(D_n) \iff n = \chi_n(U) = \frac{\sin n\vartheta}{\sin \vartheta}, \quad \text{with} \quad \vartheta \in [0, \pi] \,. \tag{12.19}$$

For $\vartheta \to 0$, $\chi_n$ tends towards $n$ (use l'Hospital's rule), so that as expected $D_n(\mathbb{1}_2) = \mathbb{1}_n$. For odd $n$, however, the character also tends towards $n$ for $\vartheta \to \pi$. For this value of $\vartheta$, $U$ in (9.9) is $-\mathbb{1}_2$, so that

$$D_n(\mathbb{1}_2) = D_n(-\mathbb{1}_2) = \mathbb{1}_n \quad \text{for odd} \quad n \,. \tag{12.20}$$

The irreducible representations with odd $n$ are therefore not faithful, while those with even $n$ are faithful.

According to the Peter-Weyl theorem, every class function $f(\vartheta)$ of SU(2) can be written as a linear combination of the characters of the irreducible representations,

$$f(\vartheta) = \sum_{n=0}^\infty c_n \frac{\sin n\vartheta}{\sin \vartheta}$$

$$c_n = \int d\mu_{\mathrm{red}}(\vartheta) \chi_n(\vartheta) f(\vartheta) = \frac{2}{\pi} \int \sin \vartheta \, \sin n\vartheta \, f(\vartheta) d\vartheta. \tag{12.21}$$

It can also be seen in other ways that the characters $\chi_n$ form a complete system of functions on

$$L_2\left([0, \pi], d\mu_{\mathrm{red}}(\vartheta)\right)$$

This proves that we constructed *all* irreducible SU(2) representations.

> **Clebsch-Gordan Formula**
> It can be shown (see Problem 12.2) that for $n \geq m$ the following reduction formula applies:
>
> $$D_n \otimes D_m = D_{n+m-1} \oplus D_{n+m-3} \oplus D_{n+m-5} \oplus \cdots \oplus D_{n-m+1} \,. \qquad (12.22)$$

In quantum mechanics the SU(2) representations are denoted by the half-integer valued angular momentum $j$. The representation for angular momentum $j$ has the dimension $n = 2j + 1$ and the reduction formula (12.22) reads

$$D_j \otimes D_{j'} = D_{j+j'} \oplus D_{j+j'-1} \oplus D_{j+j'-2} \oplus \cdots \oplus D_{|j-j'|} \,. \qquad (12.23)$$

In the LiE program, an irreducible representation is characterized by its highest weight vector $v$ (more on this later). Accordingly, the dimension of the representation is queried with $\mathtt{dim(v)}$. The reduction of the tensor product of two irreducible representations with weight vectors $v$ and $w$ is done with the call $\mathtt{tensor(v, w)}$.

```
LiE> setdefault A1
LiE> dim([1]) -> 2
LiE> dim([2]) -> 3
LiE> dim([3]) -> 4
LiE> tensor([1],[1]) -> 1X[0] +1X[2]
LiE> p_tensor(3,[1]) -> 2X[1] +1X[3]
```

The second last line is equivalent to $D_2 \otimes D_2 = D_1 \oplus D_3$ and the last one to $D_2 \otimes D_2 \otimes D_2 = 2D_2 + D_4$. Here, the representations are indexed with their dimension.

## 12.3  Representations of SU(3)

We have already encountered some representations of the group in Problem 10.10. A unitary matrix can be diagonalized and has unimodular numbers as eigenvalues. For the matrices from SU($n$), their product must be equal to one. Hence, every SU(3) matrix is conjugate to a diagonal matrix

$$U = \begin{pmatrix} e^{i\vartheta_1} & 0 & 0 \\ 0 & e^{i\vartheta_2} & 0 \\ 0 & 0 & e^{-i(\vartheta_1+\vartheta_2)} \end{pmatrix}, \qquad (12.24)$$

with real angles $\vartheta_1$ and $\vartheta_2$. The three-dimensional defining representation is denoted by 3. Its character is

$$\chi_3(\vartheta_1, \vartheta_2) = e^{i\vartheta_1} + e^{i\vartheta_2} + e^{-i(\vartheta_1+\vartheta_2)} \,. \qquad (12.25)$$

The defining representation is irreducible, since

$$\chi_3^* \chi_3 = 3 + 2\cos(\vartheta_1 - \vartheta_2) + 2\cos(2\vartheta_1 + \vartheta_2) + 2\cos(2\vartheta_2 + \vartheta_1) \qquad (12.26)$$

integrates to one,

$$(\chi_3, \chi_3) = \int d\mu_{\mathrm{red}}(\vartheta_1, \vartheta_2)\, |\chi_3(\vartheta_1, \vartheta_2)|^2 = 1\,. \qquad (12.27)$$

The reduced Haar measure of SU(3) appearing here was determined in (9.48).

The complex conjugate representation $U \mapsto U^*$ is denoted by $\bar{3}$. It is not equivalent to the representation 3, since $\chi_{\bar{3}} = \chi_3^* \neq \chi_3$. It is also irreducible due to $|\chi_{\bar{3}}|^2 = |\chi_3|^2$. The nine-dimensional tensor product $3 \times \bar{3}$ with character $\chi_{3\otimes\bar{3}} = \chi_3\chi_3^*$ contains the trivial representation once due to

$$(\chi_{3\otimes\bar{3}}, \chi_1) = \int d\mu_{\mathrm{red}}\, |\chi_3|^2 = 1 \qquad (12.28)$$

If we set $3 \otimes \bar{3} = 1 \oplus 8$, then the eight-dimensional representation is irreducible, since $\chi_8 = \chi_{3\otimes\bar{3}} - \chi_1$ has the norm 1

$$(\chi_8, \chi_8) = (\chi_3^2, \chi_3^2) - 2(\chi_3, \chi_3) + (\chi_1, \chi_1) = 1\,. \qquad (12.29)$$

---

**Task**

Calculate the scalar products in (12.29).

---

**Generating Function for Scalar Products**

For the scalar products $(\chi_3^n, \chi_3^m)$ there exists a generating function:

$$\int d\mu_{\mathrm{red}}\, e^{u\chi_3 + v\chi_3^*} = \sum_{m,n=0}^{\infty} (\chi_3^n, \chi_3^m)\, \frac{u^m v^n}{m!\, n!} \qquad (12.30)$$

$$= \sum_{p,q=0}^{\infty} \frac{2}{(p+q+1)!\,(p+q+2)!\,q!} \binom{3(p+q+1)}{p} (uv)^p (u^3 + v^3)^q$$

$$= 1 + uv + \frac{u^3 + v^3}{6} + \frac{u^2 v^2}{2} + \frac{11 u^3 v^3}{72} + \frac{91 u^4 v^4}{2880} + \frac{uv(u^3 + v^3)}{8} + \cdots$$

$$\qquad (12.31)$$

The proof can be found in the publication [55]. In particular, by comparing the coefficients of $u^2 v^2$ the missing first integral on the right side of (12.29) follows:

$$(\chi_{\bar{3}}^2, \chi_{\bar{3}}^2) = 2 \,. \tag{12.32}$$

The tensor product

$$3 \otimes \bar{3} = 1 \oplus 8 \tag{12.33}$$

thus contains an eight-dimensional irreducible representation. It is the adjoint representation 8. Every Lie group has an adjoint representation of dimension $\dim(G)$, and these representations are further discussed in Sect. 14.1.2.

We now consider the tensor product $3 \otimes 3$ with the character $\chi_3^2$. Comparing the coefficients of $u^3$ in (12.30) and (12.31), we find

$$(\chi_{\bar{3}}, \chi_{3\otimes 3}) = \int d\mu_{\mathrm{red}} \, \chi_3^3 = 1 \,, \tag{12.34}$$

and thus the irreducible representation $\bar{3}$ is once contained in $3 \otimes 3$. Therefore

$$3 \otimes 3 = 6 \oplus \bar{3} \implies \chi_6 = \chi_3^2 - \chi_3^* \,, \tag{12.35}$$

with a six-dimensional, and due to $(\chi_6, \chi_6) = 1$ irreducible representation. In the next step follows

$$3 \otimes 3 \otimes 3 = (3 \otimes 3) \otimes 3 = (6 \otimes 3) \oplus (\bar{3} \otimes 3) = (6 \otimes 3) \oplus 8 \oplus 1 \,. \tag{12.36}$$

Now we are still missing $6 \otimes 3$ with the character

$$\chi_{6\otimes 3} = \chi_6 \chi_3 = \chi_3^3 - \chi_3^* \chi_3 \,. \tag{12.37}$$

Without explicit proof, we note that this representation decomposes into two irreducible representations, $6 \otimes 3 = 10 \otimes 8$, so that

$$3 \otimes 3 \otimes 3 = 10 \oplus 8 \oplus 8 \oplus 1 \,. \tag{12.38}$$

The results (12.33) and (12.38) are relevant for the quark model of elementary particle physics, which is introduced in Sect. 20.4. Quarks appear in three colors, which transform under the defining representation 3 of the color group SU(3). Their antiparticles, the antiquarks, transform under the complex conjugated representation $\bar{3}$.

Using the LiE program, tensor products of SU(3) representations can be reduced:

```
LiE> setdefault A2
LiE> dim([0,0]) -> 1
LiE> dim([1,0]) -> 3
LiE> dim([0,1]) -> 3
LiE> dim([2,0]) -> 6
LiE> dim([0,2]) -> 6
LiE> dim([1,1]) -> 8
LiE> dim([3,0]) -> 10
LiE> tensor([1,0],[0,1])  -> 1X[0,0] +1X[1,1]
LiE> p_tensor(3,[1,0]) -> 1X[0,0] +2X[1,1] +1X[3,0]
```

The call in the second-to-last line leads to the reduction (12.33) and the call in the last line to (12.38).

---

## 12.4   Exercises for Chap. 12

**Problem 12.1 (Irreducible Representations of SU(2))**  In the main text, the integration formula

$$\int d\mu_{\mathrm{red}}(\vartheta)(\cos\vartheta)^{2p} = \frac{2}{\pi}\int_0^{\pi} d\vartheta\,(\sin\vartheta)^2(\cos\vartheta)^{2p} = \frac{1}{4^p}\frac{(2p)!}{p!(p+1)!}$$

was used to calculate the scalar products of SU(2) characters. Prove this formula.

**Problem 12.2 (Tensor Product of SU(2) Representations)**  Let $D_n$ be the irreducible SU(2) representations of dimensions $n = 1, 2, 3, \ldots$. Prove for $m \le n$ the reduction formula

$$D_m \otimes D_n = D_{m+n-1} \oplus D_{m+n-3} \oplus \cdots \oplus D_{m-n+1}\,.$$

Hint: Consider the characters $\chi_n$ of the $D_n$ and remember what the reduction of the tensor product means for the characters. It may be helpful to recall (12.18).

**Problem 12.3 (Irreducible Representations of G$\times$ G)**  It is relatively easy to construct all irreducible representations of SO(4) from those of SU(2). In fact, this also yields all irreducible representations of the Lorentz group SO(1, 3). Here we consider general Lie groups.

- Let $D_1$ and $D_2$ be two irreducible representations of G. Show that $D(g_1, g_2) = D_1(g_1) \otimes D_2(g_2)$ is an irreducible representation of $G \times G$.
- What are the irreducible representations of SU(2) $\times$ SU(2) ?
- What are the irreducible representations of SU(2) $\times$ SU(2)$/\mathbb{Z}_2$ ?
  Hint: Two pairs in SU(2) $\times$ SU(2), which are identified, must be mapped into the same $D(g_1, g_2)$.

# Theory of Lie Algebras

**13**

> *As time goes on, it becomes increasingly evident that the rules which the mathematician finds interesting are the same as those which Nature has chosen.*
>
> —Paul Adrien Maurice Dirac

This chapter introduces the theory of Lie algebras. A well-known example of a Lie algebra is the vector space $\mathbb{R}^3$ with the cross product as multiplication. The cross product is bilinear and antisymmetric,

$$a \times (\beta b + \gamma c) = \beta a \times b + \gamma a \times c \quad \text{and} \quad a \times b = -b \times a, \quad (\beta, \gamma \in \mathbb{R}), \tag{13.1}$$

and satisfies the Jacobi identity

$$(a \times b) \times c + (b \times c) \times a + (c \times a) \times b = 0. \tag{13.2}$$

Lie algebras already appear in classical Hamilton mechanics.: the Poisson bracket endows the linear space of real $C^\infty$ functions on a phase space $\Gamma$ with the structure of a Lie algebra. For three $C^\infty$ functions $f, g, h : \Gamma \to \mathbb{R}$, the following hold

$$
\begin{aligned}
\text{bilinearity:} \quad & \{f, \beta g + \gamma h\} = \beta\{f, g\} + \gamma\{f, h\}, \quad (\beta, \gamma \in \mathbb{R}), \\
\text{antisymmetry:} \quad & \{f, g\} = -\{g, f\}, \\
\text{Jacobi identity:} \quad & 0 = \{f, \{g, h\}\} + \{g, \{h, f\}\} + \{h, \{f, g\}\}.
\end{aligned}
\tag{13.3}
$$

Lie algebras often show up as infinitesimal symmetry operations. A well-known example is the components $L_1, L_2, L_3$ of angular momentum, which generate rotations in space. In classical mechanics, these form a Lie algebra with respect


A. Wipf, *Symmetries in Physics*, https://doi.org/10.1007/978-3-662-72675-4_13

to the Poisson bracket and in quantum mechanics with respect to the commutator of operators,

$$[L_a, L_b] = i\hbar \epsilon_{abc} L_c \,. \tag{13.4}$$

The representation theory of angular momentum operators is of importance for the classification of states and the *selection rules* in atomic and molecular physics. More generally, with the help of the theory of Lie algebras and their representations, structural insights into a physical theory can be gained without understanding or even solving the underlying (often complex) dynamics. A detailed discussion of Lie algebras (and their representations) can be found, for example, in [13, 17, 56–59].

## 13.1   Introduction to Lie Algebras

The properties of a Lie algebra are modeled after those of the cross product on $\mathbb{R}^3$. In general, a Lie algebra is defined as follows:

**Definition 13.1 (Lie Algebra)**  A Lie algebra is a $\mathbb{K}$-vector space $\mathfrak{g}$, equipped with a bilinear mapping (Lie bracket, Lie product) $[\,,\,] : \mathfrak{g} \times \mathfrak{g} \longmapsto \mathfrak{g}$, which

- is antisymmetric:  $\qquad\qquad [X, Y] = -[Y, X]\,,$
- satisfies the Jacobi identity:  $\quad [X, [Y, Z]] + [Z, [X, Y]] + [Y, [Z, X]] = 0\,.$

---

**Structure Constants**

Introducing a basis $X_1, \ldots, X_n$ in the $n$-dimensional vector space $\mathfrak{g}$, the commutator of two basis elements is a linear combination of the basis elements,

$$[X_a, X_b] = f_{ab}{}^c X_c \,. \tag{13.5}$$

The $n^3$ expansion coefficients $f_{ab}{}^c \in \mathbb{K}$ are called *structure constants* of $\mathfrak{g}$.

---

The *rank* of a Lie algebra is equal to the maximum number of commuting basis elements. If the Lie product vanishes for every pair $X, Y$ in $\mathfrak{g}$, then all structure constants vanish and $\mathfrak{g}$ is called *abelian*. For abelian Lie algebras, $\mathrm{rank}(\mathfrak{g}) = \dim(\mathfrak{g})$.

---

**Example: The Lie Algebra** $\mathfrak{gl}(n, \mathbb{K})$

This contains all $n \times n$ matrices with elements in $\mathbb{K}$ and is equipped with the Lie product

$$[X, Y] = XY - YX, \tag{13.6}$$

where $XY$ is the product of matrices. It is a Lie algebra because the Jacobi identity is automatically satisfied

$$
\begin{aligned}
& XYZ - XZY - YZX + ZYX \\
& + ZXY - ZYX - XYZ + YXZ \\
& + YZX - YXZ - ZXY + XZY = 0.
\end{aligned}
\tag{13.7}
$$

◄

From the properties of the Lie product, corresponding properties of the structure constants can be deduced. For example, from the antisymmetry of the product follows the antisymmetry of the structure constants in the lower indices,

$$f_{ab}{}^{c} = -f_{ba}{}^{c}, \qquad a, b, c = 1, \ldots, \dim(\mathfrak{g}). \tag{13.8}$$

From the Jacobi identity for the basis elements, one obtains

$$[X_a, [X_b, X_c]] + \text{cyclic} = 0 \implies f_{ap}{}^{q} f_{bc}{}^{p} + f_{bp}{}^{q} f_{ca}{}^{p} + f_{cp}{}^{q} f_{ab}{}^{p} = 0. \tag{13.9}$$

The structure constants determine the Lie products of the basis elements. With bilinearity, the Lie products for all pairs $X, Y \in \mathfrak{g}$ follow. For each set of $n^3$ constants $f_{ab}{}^{c} \in \mathbb{K}$, which satisfy (13.8) and (13.9), there is a Lie algebra with these $f_{ab}{}^{c}$ as structure constants.

---

## 13.2  Lie Subalgebras

In many applications, Lie algebras appear as subalgebras of known Lie algebras, e.g., of $\mathfrak{gl}(n, \mathbb{K})$. A Lie subalgebra is defined as follows:

**Definition 13.2 (Lie Subalgebra)**  A linear subspace $\mathfrak{h}$ of $\mathfrak{g}$, which is closed under the Lie product, is called a Lie subalgebra of $\mathfrak{g}$. We write $\mathfrak{h} \leq \mathfrak{g}$.

A Lie subalgebra is itself a Lie algebra since $[\mathfrak{h}, \mathfrak{h}] \subseteq \mathfrak{h}$ holds and the Jacobi identity of $\mathfrak{g}$ is inherited. The trivial Lie subalgebras $\mathfrak{h} = \emptyset$ and $\mathfrak{h} = \mathfrak{g}$ always exist. If $\mathfrak{h}$ is neither of these two trivial Lie subalgebras, it is called a proper Lie subalgebra.

If $\mathfrak{h}_i$ are two Lie subalgebras, then the set

$$[\mathfrak{h}_1, \mathfrak{h}_2] = \text{span}\big\{[H_1, H_2] \,|\, H_i \in \mathfrak{h}_i\big\} \tag{13.10}$$

is also a Lie subalgebra, where span$\{M\}$ is the linear span of $M \subseteq \mathfrak{g}$. The set of commutators is generally not a linear space, and then the linear span is needed. In particular, one calls

$$\mathfrak{g}' \equiv [\mathfrak{g}, \mathfrak{g}] \leq \mathfrak{g} \tag{13.11}$$

the *derived Lie algebra* of $\mathfrak{g}$. For an abelian Lie algebra, $\mathfrak{g}' = \emptyset$.

---

**Task**

Convince yourself that $[\mathfrak{h}_1, \mathfrak{h}_2]$ is a Lie subalgebra.

---

**Example: The Derived Lie Algebra $\mathfrak{sl}(n, \mathbb{K})$ of $\mathfrak{gl}(n, \mathbb{K})$**

The traceless matrices form the Lie subalgebra $\mathfrak{sl}(n, \mathbb{K}) \leq \mathfrak{gl}(n, \mathbb{K})$. For two arbitrary matrices $X, Y$, $\mathrm{Tr}[X, Y] = 0$ holds, and therefore $\mathfrak{gl}'(n, \mathbb{K}) = \mathfrak{sl}(n, \mathbb{K})$.  ◄

## 13.2.1  Ideals (Invariant Lie Subalgebras)

There is a subclass of Lie subalgebras important for the classification of Lie algebras, which are invariant under the action of $\mathfrak{g}$. These invariant Lie subalgebras are called ideals.

**Definition 13.3 (Ideal)**  A Lie subalgebra $\mathcal{I} \leq \mathfrak{g}$ is called an ideal of $\mathfrak{g}$, if

$$[\mathfrak{g}, \mathcal{I}] \subseteq \mathcal{I}. \tag{13.12}$$

We write $\mathcal{I} \trianglelefteq \mathfrak{g}$ or $\mathcal{I} \triangleleft \mathfrak{g}$ if $\mathcal{I}$ is strictly smaller than $\mathfrak{g}$.

For example, the derived Lie algebra $\mathfrak{g}'$ is always an ideal of $\mathfrak{g}$ because $[\mathfrak{g}', \mathfrak{g}] \subseteq [\mathfrak{g}, \mathfrak{g}] = \mathfrak{g}'$. Specifically, $\mathfrak{sl}(n, \mathbb{K}) \triangleleft \mathfrak{gl}(n, \mathbb{K})$. From two ideals $\mathcal{I}_1, \mathcal{I}_2$, one can construct further ideals in various ways (see Problem 13.4):

**Lemma 13.1 (Ideals from Ideals)**  *The intersection of two ideals $\mathcal{I}_1 \cap \mathcal{I}_2$, their sum $\mathcal{I}_1 + \mathcal{I}_2 = \{X_1 + X_2 \mid X_i \in \mathcal{I}_i\}$, and their product $[\mathcal{I}_1, \mathcal{I}_2] = span\{[X_1, X_2] \mid X_i \in \mathcal{I}_i\}$ are ideals of $\mathfrak{g}$.*

We convince ourselves here of the last statement: the linear span of $[\mathcal{I}_1, \mathcal{I}_2]$ is, as we just argued, a Lie subalgebra. To prove that it is an invariant Lie subalgebra, one uses the Jacobi identity. Let $X_1 \in \mathcal{I}_1$, $X_2 \in \mathcal{I}_2$ and $X \in \mathfrak{g}$ be arbitrary. Then

$$[X, [X_1, X_2]] = [X_2, [X_1, X]] - [X_1, [X_2, X]] \in [\mathcal{I}_1, \mathcal{I}_2],$$

which means that $[\mathcal{I}_1, \mathcal{I}_2]$ is invariant.

---

**Task**

Prove the other two statements.

## 13.2.2 Simple and Semi-Simple Lie Algebras

Analogous to the classification of groups based on their normal subgroups, Lie algebras can be distinguished based on their ideals:

**Definition 13.4 (Simple Lie Algebras)** A Lie algebra $\mathfrak{g}$ is called simple if it is non-abelian and has no (non-trivial) ideal. It is called semi-simple if $\{0\}$ is its only abelian ideal.

In other words, a semi-simple Lie algebra can only have non-abelian ideals. Every simple Lie algebra is semi-simple and every abelian Lie algebra is neither simple nor semi-simple.

Similar to a group with normal subgroups, for a Lie algebra with ideals, one can define the *quotient space* $\mathfrak{g}/\mathcal{I}$ for each ideal $\mathcal{I} \trianglelefteq \mathfrak{g}$. Its elements are the equivalence classes for the relation

$$X \sim Y \iff X - Y \in \mathcal{I} \trianglelefteq \mathfrak{g}. \tag{13.13}$$

A representative $X$ defines the class $X + \mathcal{I}$. The classes define a linear space with the class $\mathcal{I}$ as the zero vector. The commutator is independent of the representative since $\mathcal{I}$ is an ideal: $X' - X \in \mathcal{I} \implies [X' - X, Y] \in \mathcal{I} \sim 0$.

**Lemma 13.2** *For each ideal $\mathcal{I} \trianglelefteq \mathfrak{g}$, the quotient space $\mathfrak{g}/\mathcal{I}$ defines a Lie algebra.*

For example, $\mathfrak{gl}(n, \mathbb{K})/\mathfrak{sl}(n, \mathbb{K})$ is the abelian Lie algebra $\{\lambda\mathbb{1} | \lambda \in \mathbb{K}\}$.

### Center, Centralizers, and Normalizers

The *center* $\mathfrak{z} \leq \mathfrak{g}$ contains those elements that commute with *all* elements of $\mathfrak{g}$,

$$\mathfrak{z} := \{Z \in \mathfrak{g} \,|\, [Z, \mathfrak{g}] = 0\}. \tag{13.14}$$

It is a very special ideal in $\mathfrak{g}$. For abelian Lie algebras, $\mathfrak{z} = \mathfrak{g}$.

---

**Question**

What are the centers of $\mathfrak{gl}(n, \mathbb{K})$ and $\mathfrak{sl}(n, \mathbb{K})$?

Finally, we consider two more classes of subalgebras.

- The *centralizer of a subset* $\mathfrak{m} \subseteq \mathfrak{g}$ contains all elements of $\mathfrak{g}$ whose Lie products with all elements from $\mathfrak{m}$ vanish,

$$\mathfrak{z}_{\mathfrak{g}}(\mathfrak{m}) = \{X \in \mathfrak{g} \,|\, [X, \mathfrak{m}] = 0\}. \tag{13.15}$$

$\mathfrak{z}_\mathfrak{g}(\mathfrak{m})$ is a vector space, and with the Jacobi identity, one proves that it is closed under the Lie product. Thus, $\mathfrak{z}_\mathfrak{g}(\mathfrak{m})$ is a Lie subalgebra. The centralizer of $\mathfrak{g}$ is equal to the center of the Lie algebra, $\mathfrak{z}_\mathfrak{g}(\mathfrak{g}) = \mathfrak{z}$.

- The *normalizer of a Lie subalgebra* $\mathfrak{h} \leq \mathfrak{g}$ is defined by

$$\mathfrak{n}_\mathfrak{g}(\mathfrak{h}) = \{X \in \mathfrak{g} \mid [X, \mathfrak{h}] \subseteq \mathfrak{h}\}. \qquad (13.16)$$

It is a vector space, and for two vectors $X_1, X_2 \in \mathfrak{n}_\mathfrak{g}(\mathfrak{h})$ one has

$$[[X_1, X_2], \mathfrak{h}] = [X_1, [X_2, \mathfrak{h}]] - [X_2, [X_1, \mathfrak{h}]] \subseteq \mathfrak{h},$$

so that their Lie product is also in the normalizer. $\mathfrak{n}_\mathfrak{g}(\mathfrak{h})$ is the largest Lie subalgebra of $\mathfrak{g}$ that contains $\mathfrak{h}$ as an ideal. The normalizer of an ideal is equal to $\mathfrak{g}$.

## 13.3   Lie Algebra Homomorphisms and Representations

The Lie subalgebras of Lie algebras are closely related to the homomorphisms of Lie algebras.

**Definition 13.5** A homomorphism $\phi : \mathfrak{g}_1 \to \mathfrak{g}_2$ of Lie algebras is a structure-preserving linear map,

$$\phi(\alpha X + \beta Y) = \alpha\phi(X) + \beta\phi(Y) \quad \text{and} \quad \phi([X, Y]) = [\phi(X), \phi(Y)]. \qquad (13.17)$$

In the second relation, the Lie product on the left is in $\mathfrak{g}_1$ and on the right in $\mathfrak{g}_2$.

The composition of two Lie algebra homomorphisms is again a Lie algebra homomorphism.

**Lemma 13.3** *Let* $\phi : \mathfrak{g}_1 \to \mathfrak{g}_2$ *be a Lie algebra homomorphism. Then the kernel and the image of* $\phi$ *are Lie subalgebras of* $\mathfrak{g}_1$ *and* $\mathfrak{g}_2$.

The kernel of a linear map is a linear subspace of $\mathfrak{g}_1$. It defines a subalgebra because, due to the homomorphism property, with $X, Y$, also $[X, Y]$ lies in the kernel of $\phi$, $\phi([X, Y]) = [\phi(X), \phi(Y)] = 0$.

**Task**

Show that the image of $\phi$ is a Lie subalgebra of $\mathfrak{g}_2$.

As is well known, the kernel of any group homomorphism defines a normal subgroup. The same applies to the kernel of a Lie algebra homomorphism:

**Lemma 13.4** *The kernel of any Lie algebra homomorphism $\phi$ is an ideal of $\mathfrak{g}_1$, Kernel$(\phi) \trianglelefteq \mathfrak{g}_1$.*

The statement follows directly from

$$\phi([\mathrm{Kernel}(\phi), \mathfrak{g}_1]) = [\phi(\mathrm{Kernel}(\phi)), \phi(\mathfrak{g}_1)]$$

$$= 0 \implies [\mathrm{Kernel}(\phi), \mathfrak{g}_1] \subset \mathrm{Kernel}(\phi). \tag{13.18}$$

### 13.3.1 Representations of a Lie Algebra

Let $V$ be an $n$-dimensional vector space and $L(V)$ the set of linear mappings $V \to V$. These form a Lie algebra with the commutator of two linear mappings as the Lie product. A representation of $\mathfrak{g}$ on $V$ is defined as follows:

**Definition 13.6 (Representation of a Lie Algebra)**  An $n$-dimensional representation $\mathfrak{D}$ of a Lie algebra $\mathfrak{g}$ is a Lie algebra homomorphism

$$\mathfrak{D} : \mathfrak{g} \mapsto L(V), \quad X \mapsto \mathfrak{D}(X) \in L(V), \quad \dim(V) = n.$$

For a representation, $\mathfrak{g}_2$ in Definition 13.5 is equal to $L(V)$. After choosing a basis in $V$, a linear mapping is given by a matrix, and the Lie product of two linear mappings is the commutator of the associated matrices. Thus, a representation $\mathfrak{D}$ is a homomorphism $\mathfrak{g} \mapsto \mathfrak{gl}(n, \mathbb{K})$.

### 13.3.2 Irreducible Representations

Similar to the representation theory of groups, the focus is on representations $\mathfrak{D}$ of Lie algebras that do not have invariant subspaces in $V$. Such representations $\mathfrak{D}$ are called irreducibel. $\mathfrak{D}$ is called fully reducible if it is irreducible or the direct sum of irreducible representations. The classification of all irreducible inequivalent representations of a Lie algebra is an important goal of representation theory and will occupy us in Chap. 16.

In Sect. 11.2, we proved and discussed Schur's lemma for irreducible representations of groups. The corresponding lemma for irreducible representations of Lie algebras is similar, and the proofs are closely related (see Problem 13.6).

**Lemma 13.5 (Schur)** *Let $\mathfrak{D}_1$ and $\mathfrak{D}_2$ be two irreducible representations of a Lie algebra $\mathfrak{g}$ on vector spaces $V_1$ and $V_2$, and let $H : V_1 \mapsto V_2$ be a linear mapping* *(a morphism) with*

$$H\mathfrak{D}_1 = \mathfrak{D}_2 H \,. \tag{13.19}$$

*Then the following statements hold:*

1. *H is either 0 or (in the exclusive sense) H is invertible.*
2. *If $\mathfrak{D}_1 = \mathfrak{D}_2 \equiv \mathfrak{D}$, then H is proportional to the identity $\mathbb{1}_V$.*

If $H$ is invertible in the first statement, then the irreducible representations $\mathfrak{D}_1$ and $\mathfrak{D}_2$ are equivalent. The second statement can also be formulated as follows: If a linear operator $H : V \mapsto V$ commutes with all matrices $\mathfrak{D}(X)$ of an irreducible representation, then $H = \lambda\mathbb{1}$.

### 13.3.3 Adjoint Representation

A Lie algebra $\mathfrak{g}$ always has the *adjoint representation* on the linear space $\mathfrak{g}$. This representation assigns to each element $X \in \mathfrak{g}$ a linear mapping $\mathrm{ad}_X : \mathfrak{g} \mapsto \mathfrak{g}$, i.e., an element of $L(\mathfrak{g})$:

$$\mathrm{ad} : \mathfrak{g} \mapsto L(\mathfrak{g}), \qquad \mathrm{ad}_X Y = [X, Y] \,. \tag{13.20}$$

The Jacobi identity implies $[\mathrm{ad}_X, \mathrm{ad}_Y](Z) = \mathrm{ad}_{[X,Y]}(Z)$, so that:

**Corollary 13.1 (Adjoint Representation)** *The mapping $X \to \mathrm{ad}_X$ defines a representation of $\mathfrak{g}$ on $L(\mathfrak{g})$. Its dimension is equal to $\dim(\mathfrak{g})$.*

---

**Example: Adjoint Representation of the Lie Algebra $\mathfrak{su}(2)$**

The Lie algebra $\mathfrak{su}(2)$ is the linear space of two-dimensional anti-Hermitian traceless matrices. As a basis we choose

$$X_1 = \frac{i}{2}\sigma_1, \quad X_2 = \frac{i}{2}\sigma_2, \quad X_3 = \frac{i}{2}\sigma_3 \,. \tag{13.21}$$

Due to the relations $[X_a, X_b] = -\epsilon_{abc}X_c$, the following matrices are assigned to the $X_a$:

$$\mathrm{ad}_{X_1} = \begin{pmatrix} 0 & 0 & 0 \\ 0 & 0 & 1 \\ 0 & -1 & 0 \end{pmatrix}, \quad \mathrm{ad}_{X_2} = \begin{pmatrix} 0 & 0 & -1 \\ 0 & 0 & 0 \\ 1 & 0 & 0 \end{pmatrix}, \quad \mathrm{ad}_{X_3} = \begin{pmatrix} 0 & 1 & 0 \\ -1 & 0 & 0 \\ 0 & 0 & 0 \end{pmatrix} \,. \tag{13.22}$$

These have the same commutation relations as the $X_a$: $[\mathrm{ad}_{X_a}, \mathrm{ad}_{X_b}] = -\epsilon_{abc}\,\mathrm{ad}_{X_c}$. ◄

The subspaces of $\mathfrak{g}$ invariant under the adjoint representation are the ideals of $\mathfrak{g}$. In particular, $\mathfrak{g}$ is simple if and only if the adjoint representation is irreducible.

The kernel of the adjoint representation is equal to the center of $\mathfrak{g}$,

$$\text{Kernel(ad)} = \{Z \in \mathfrak{g} \mid [Z, \mathfrak{g}] = 0\} = \mathfrak{z} . \tag{13.23}$$

From this follows the important property

**Lemma 13.6** *The image* $\text{ad}_{\mathfrak{g}}$ *is isomorphic to* $\mathfrak{g}/\mathfrak{z}$. *For* $\mathfrak{z} = 0$, $\text{ad}_{\mathfrak{g}} \cong \mathfrak{g}$.

---

**Task**

The Lie algebra $\mathfrak{su}(n)$ generalizes $\mathfrak{su}(2)$ and contains the traceless anti-Hermitian $n \times n$ matrices. Show that $\text{ad}(\mathfrak{su}(n)) \cong \mathfrak{su}(n)$ holds.

---

## 13.4 Invariant Killing Form

With the help of the adjoint representation, we introduce the bilinear and ad-invariant Killing form $K(., .)$, which plays an important role in the structure theory of semi-simple Lie algebras. For example, we will see that for a semisimple Lie algebra $\mathfrak{g}$, the Killing form is not degenerate and $\mathfrak{g}^{\perp} = \{X \in \mathfrak{g} \mid K(X, \mathfrak{g}) = 0\}$ is an ideal containing only the zero vector.

**Definition 13.7 (Killing Form)** The Killing form of a Lie algebra is given by

$$K(X, Y) = \text{Tr}\,(\text{ad}_X \text{ad}_Y) . \tag{13.24}$$

It is also of interest because it is an invariant form under to the adjoint representation:

**Lemma 13.7** *The Killing form is symmetric, bilinear, and ad-invariant,*

$$K(\text{ad}_Z X, Y) + K(X, \text{ad}_Z Y) = 0 . \tag{13.25}$$

***Proof*** The first property follows from the invariance of the trace under cyclic permutation of the arguments and the second from the linearity of $X \to \text{ad}_X$. The ad-invariance is proven as follows:

$$K\,(\text{ad}_Z X, Y) = K([Z, X], Y) = \text{Tr}\,\left(\text{ad}_{[Z,X]}\text{ad}_Y\right) = \text{Tr}\,([\text{ad}_Z, \text{ad}_X]\text{ad}_Y)$$

$$= \text{Tr}\,(\text{ad}_X[\text{ad}_Y, \text{ad}_Z]) = \text{Tr}\,\left(\text{ad}_X \text{ad}_{[Y,Z]}\right) = -K(X, \text{ad}_Z Y) . \tag{13.26}$$

An important consequence of the invariance (13.25) is the

**Lemma 13.8 (Orthogonal Complement of an Ideal)** *The orthogonal complement* $\mathcal{I}^{\perp}$ *of an ideal* $\mathcal{I}$ *of* $\mathfrak{g}$*, given by* $\mathcal{I}^{\perp} = \{Y \in \mathfrak{g} \mid K(Y, \mathcal{I}) = 0\}$*, is also an ideal of* $\mathfrak{g}$*.*

***Proof*** Let $X \in \mathfrak{g}$ be arbitrary and $Y \in \mathcal{I}^{\perp}$. We want to prove that then $[X, Y]$ is in $\mathcal{I}^{\perp}$. Indeed, it holds

$$K([X, Y], \mathcal{I}) = K(\mathrm{ad}_X Y, \mathcal{I}) = -K(Y, \mathrm{ad}_X \mathcal{I}) = 0 , \tag{13.27}$$

because $\mathrm{ad}_X \mathcal{I} \subseteq \mathcal{I}$. Thus, with $\mathcal{I}$, $\mathcal{I}^{\perp}$ is also an ideal of $\mathfrak{g}$. In particular, $\mathfrak{g}^{\perp} \trianglelefteq \mathfrak{g}$ is an ideal.

**Lemma 13.9** *If* $\mathcal{I} \trianglelefteq \mathfrak{g}$ *is an ideal, and* $X, Y \in \mathcal{I}$*, then*

$$K_{\mathfrak{g}}(X, Y) = K_{\mathcal{I}}(X, Y), \qquad X, Y \in \mathcal{I}, \tag{13.28}$$

*where* $K_{\mathfrak{g}}$ *and* $K_{\mathcal{I}}$ *are the Killing forms of the Lie algebras* $\mathfrak{g}$ *and* $\mathcal{I} \trianglelefteq \mathfrak{g}$*.*

***Proof*** We choose an adapted basis, starting with the elements of $\mathcal{I}$ and extending to a basis of $\mathfrak{g}$. For $X, Y \in \mathcal{I}$, the linear mapping $\mathrm{ad}_X \mathrm{ad}_Y$ has the block form

$$\mathrm{ad}_X \mathrm{ad}_Y = \begin{pmatrix} A & B \\ 0 & 0 \end{pmatrix}, \quad X, Y \in \mathcal{I},$$

and therefore $\mathrm{Tr}_{\mathfrak{g}}\,(\mathrm{ad}_X \mathrm{ad}_Y) = \mathrm{Tr}_{\mathcal{I}}\,(\mathrm{ad}_X \mathrm{ad}_Y)$.

Now we want to address the question for which $\mathfrak{g}$ the Killing form is degenerate. By definition, a non-semi-simple Lie algebra has an abelian ideal $\mathfrak{a} \trianglelefteq \mathfrak{g}$. For every $X \in \mathfrak{g}$, then $\mathrm{ad}_X(\mathfrak{a}) \subseteq \mathfrak{a}$, so that for every $A \in \mathfrak{a}$ we have $\mathrm{ad}_A \mathrm{ad}_X(\mathfrak{a}) = 0$. Furthermore, it naturally holds $\mathrm{ad}_A \mathrm{ad}_X(\mathfrak{g}) \subseteq \mathfrak{a}$. In an adapted basis, starting with the elements of $\mathfrak{a}$ and extending to a basis of $\mathfrak{g}$, $\mathrm{ad}_A \mathrm{ad}_X$ is an upper triangular matrix,

$$\mathrm{ad}_A \mathrm{ad}_X = \begin{pmatrix} 0 & B \\ 0 & 0 \end{pmatrix}, \quad A \in \mathfrak{a}, \ X \in \mathfrak{g}.$$

It follows for every abelian ideal $\mathfrak{a} \trianglelefteq \mathfrak{g}$ the property $K(\mathfrak{a}, \mathfrak{g}) = 0$ or the result

**Lemma 13.10 (Degenerate Killing Form)** *The Killing form of a non-semi-simple Lie algebra is degenerate.*

**Example: Lie Algebra of Affine Transformations on a Line**

This is the two-dimensional Lie algebra belonging to the Lie group of affine transformations on $\mathbb{R}$, as introduced in Sect. 9.2 (more on this in Chap. 14). As a basis, we choose

$$X_1 = \begin{pmatrix} 1 & 0 \\ 0 & 0 \end{pmatrix}, \quad X_2 = \begin{pmatrix} 0 & 1 \\ 0 & 0 \end{pmatrix} \quad \text{with} \quad \mathrm{ad}_{X_1} = \begin{pmatrix} 0 & 0 \\ 0 & 1 \end{pmatrix}, \quad \mathrm{ad}_{X_2} = \begin{pmatrix} 0 & 0 \\ -1 & 0 \end{pmatrix}.$$

The Killing form is degenerate, since

$$K(X_1, X_1) = 1, \quad K(X_1, X_2) = 0, \quad K(X_2, X_2) = 0,$$

so that for all $X$ in the Lie algebra $K(X, X_2)$ vanishes. ◄

On the other hand, the Killing form of a semi-simple Lie algebra is non-degenerate, as proven in Theorem 13.6 in the appendix to this chapter. We summarize:

**Theorem 13.1 (Simple Lie Algebras)** *The following statements are equivalent:*

*1. $\mathfrak{g}$ is semi-simple,*
*2. the Killing form of $\mathfrak{g}$ is non-degenerate,*
*3. $\mathfrak{g} = \mathfrak{g}_1 \oplus \cdots \oplus \mathfrak{g}_m$ with simple ideals $\mathfrak{g}_k \trianglelefteq \mathfrak{g}$. This decomposition is unique, and the only ideals of $\mathfrak{g}$ are direct sums of $\mathfrak{g}_k$.*

To prove the third statement, consider an ideal $\mathcal{I} \trianglelefteq \mathfrak{g}$ with the smallest positive dimension. If $\mathcal{I} = \mathfrak{g}$, then $\mathfrak{g}$ is simple, and there is nothing to prove. Otherwise, one notes that the restriction of the Killing form to the ideal $\mathcal{I} \cap \mathcal{I}^\perp$ vanishes, so that $\mathcal{I} \cap \mathcal{I}^\perp$ is solvable (solvable Lie algebras are introduced and discussed in the appendix to this chapter). For a semi-simple $\mathfrak{g}$, $\mathcal{I} \cap \mathcal{I}^\perp$ must then be $\{0\}$. Since $\dim \mathcal{I} + \dim \mathcal{I}^\perp = \dim \mathfrak{g}$, it follows that $\mathcal{I} \oplus \mathcal{I}^\perp = \mathfrak{g}$. Inductively, one proves the third statement in the theorem.

## 13.4.1 Cartan-Killing Metric and Structure Constants

We choose a basis $\{X_a\}$ of $\mathfrak{g}$ and consider the corresponding components of the Killing form, $g_{ab} = -K(X_a, X_b)$. The invariant matrix $(g_{ab})$ is also called the Cartan-Killing metric.

**Invertible Cartan-Killing Metric**
A Lie algebra is semi-simple if and only if $\det(g_{ab}) \neq 0$.

Due to

$$\mathrm{ad}_{X_b} X_c = f_{bc}{}^d X_d \quad \text{and} \quad \mathrm{ad}_{X_a} \mathrm{ad}_{X_b} X_c = f_{ad}{}^e f_{bc}{}^d X_e$$

the linear mapping $\mathrm{ad}_{X_a} \mathrm{ad}_{X_b}$ corresponds to the matrix $\sum_d f_{ad}{}^e f_{bc}{}^d$, and the coefficients of the metric are quadratic in the structure constants,

$$g_{ab} = -\mathrm{Tr}\,(\mathrm{ad}_{X_a} \mathrm{ad}_{X_b}) = -f_{ad}{}^c f_{bc}{}^d . \tag{13.29}$$

With the metric, we can lower indices. The structure constants with lower indices

$$f_{abc} \equiv f_{ab}{}^d g_{dc} = -f_{bac} \tag{13.30}$$

are not only antisymmetric in the indices $a, b$, but completely antisymmetric in $a, b$, and $c$. In the proof, one uses in

$$K(\mathrm{ad}_{X_a} X_b, X_c) = f_{ab}{}^d K(X_d, X_c) = -f_{abc} ,$$

the ad-invariance of the Killing form and its symmetry,

$$K(\mathrm{ad}_{X_a} X_b, X_c) = -K(\mathrm{ad}_{X_a} X_c, X_b) \quad \text{or} \quad f_{abc} = -f_{acb} ,$$

which proves the antisymmetry in the last two indices—and thus all indices.

---

**Example: The Lie Algebra $\mathfrak{sl}(2, \mathbb{R})$**

This Lie algebra of real and traceless $2 \times 2$ matrices has the basis

$$X_1 = \frac{1}{2}\sigma_1, \quad X_2 = \frac{1}{2i}\sigma_2, \quad X_3 = \frac{1}{2}\sigma_3 . \tag{13.31}$$

Their commutators $\mathrm{ad}_{X_a} X_b = [X_a, X_b]$ are

$$\mathrm{ad}_{X_1} X_2 = X_3, \quad \mathrm{ad}_{X_1} X_3 = X_2, \quad \text{etc.,}$$

so that ad maps the basis elements to the following matrices,

$$\mathrm{ad}_{X_1} = \begin{pmatrix} 0 & 0 & 0 \\ 0 & 0 & 1 \\ 0 & 1 & 0 \end{pmatrix}, \quad \mathrm{ad}_{X_2} = \begin{pmatrix} 0 & 0 & 1 \\ 0 & 0 & 0 \\ -1 & 0 & 0 \end{pmatrix}, \quad \mathrm{ad}_{X_3} = \begin{pmatrix} 0 & -1 & 0 \\ -1 & 0 & 0 \\ 0 & 0 & 0 \end{pmatrix} .$$

The corresponding metric $(g_{ab}) = \mathrm{diag}(-2, 2, -2)$ is invertible and thus $\mathfrak{sl}(2, \mathbb{R})$ is simple. For $\mathfrak{su}(2)$ (with the basis in Sect. 13.3.3) the metric $g_{ab} = -2\delta_{ab}$ is even negative definite. ◄

> **Compact Lie Algebras**
> Lie algebras with a negative definite Killing form are called compact Lie
> algebras. $\mathfrak{su}(2)$ is compact and $\mathfrak{sl}(2, \mathbb{R})$ is non-compact.

We will see in Sect. 14.2 that the Lie algebra belonging to a semi-simple *compact*
Lie group is compact, i.e., it has a positive definite ad-invariant metric $(g_{ab})$.

## 13.4.2 Killing Forms for Matrix Lie Algebras

For matrix Lie algebras, the Killing form can be explicitly specified. To this end,
we define for two matrices of a Lie algebra the symmetric, bilinear, and ad-invariant
trace form

$$(X, Y) = \mathrm{Tr}(XY). \tag{13.32}$$

> **Question**
>
> Why does the negative trace form define a scalar product on $\mathfrak{su}(n)$?

Because the trace form is ad-invariant, one infers with Schur's lemma (see [57])

**Corollary 13.2 (Trace Forms and Killing Form)**  *For a simple matrix Lie algebra,
the trace form* (13.32) *is proportional to the Killing form,*

$$K(X, Y) = \lambda \, \mathrm{Tr}(X, Y). \tag{13.33}$$

We prove this statement by considering the ad-invariant bilinear form $B(X, Y) =
K(X, Y) - \lambda \, \mathrm{Tr}(X, Y)$. The characteristic polynomial of the symmetric matrix
$B_{ab} = K(X_a, X_b) - \lambda \, \mathrm{Tr}(X_a, X_b)$ vanishes for a $\lambda$, and the corresponding $B$ is
degenerate. Because $B$ is ad-invariant, $\mathfrak{g}^{\perp} \neq \{0\}$ is an ideal in $\mathfrak{g}$, where $\mathfrak{g}^{\perp}$ means
"perpendicular with respect to the $B$-form." Since $\mathfrak{g}$ is simple, we have $\mathfrak{g}^{\perp} = \mathfrak{g}$.
This means, however, that for a root $\lambda$ of the characteristic polynomial, the $B$-form
vanishes, i.e., $K(X, Y) = \lambda \, \mathrm{Tr}(X, Y)$.

To calculate the factor $\lambda$ for the simple Lie algebra $\mathfrak{sl}(n, \mathbb{R})$, we determine
$K(H, H)$ for the traceless matrix $H = \mathrm{diag}(1, 0, \ldots, 0, -1)$. Let $E_{ab}$ be the matrix
with all elements equal to zero except for the element in row $a$ and column $b$, which
is equal to one. Then the following matrices are eigenvectors of $\mathrm{ad}_H^2$ with non-zero
eigenvalues:

$$\lambda = 4 : \ E_{1n}, \ E_{n1}; \quad \lambda = 1 : \ E_{1a}, \ E_{a1}, \ E_{na}, \ E_{an}, \qquad a = 2, \ldots, n - 1. \tag{13.34}$$

Therefore, $K(H, H) = 2 \cdot 4 + 4 \cdot (n - 2) = 4n$ and it follows with $\text{Tr}(HH) = 2$, that

$$K(X, Y) = 2n\,\text{Tr}(XY), \qquad X, Y \in \mathfrak{sl}(n, \mathbb{R}). \tag{13.35}$$

For $\mathfrak{su}(n)$, one chooses the anti-Hermitian traceless matrix $H = \text{diag}(i, 0, \ldots, 0, -i)$ and finds, up to a sign, the same eigenvalues as for $\mathfrak{sl}(n, \mathbb{R})$. The eigenvectors are the anti-Hermitian linear combinations of the eigenvectors (13.34), for example, $E_{1n} - E_{n1}$ and $iE_{1n} + iE_{n1}$. Because $\text{Tr}(HH) = -2$, one finds the same result as for $\mathfrak{sl}(n, \mathbb{R})$,

$$K(X, Y) = 2n\,\text{Tr}(XY), \qquad X, Y \in \mathfrak{su}(n). \tag{13.36}$$

The simple orthogonal Lie algebra $\mathfrak{so}(n)$ contains the anti-symmetric real matrices, see Sect. 14.2. The anti-symmetric matrix $H = E_{1n} - E_{n1}$ has the following $2n - 4$ eigenvectors for the only non-zero eigenvalue $\lambda = -1$:

$$E_{1a} - E_{a1}, \ E_{na} - E_{an}, \quad a = 2, \ldots, n - 1. \tag{13.37}$$

Accordingly, $K(H, H) = 4 - 2n$ and

$$K(X, Y) = (n - 2)\,\text{Tr}(XY), \qquad X, Y \in \mathfrak{so}(n). \tag{13.38}$$

$\mathfrak{gl}(n, \mathbb{R})$ is not simple, and the Killing form does not need to be proportional to $\text{Tr}(X, Y)$. But the subalgebra $\mathfrak{sl}(n, \mathbb{R})$ is an ideal, and according to (13.28) the Killing form is given by (13.35). Since the identity matrix is orthogonal to $\mathfrak{gl}(n, \mathbb{R})$, the bilinear Killing form is given by

$$K(X, Y) = 2n\,\text{Tr}(X, Y) - 2\,\text{Tr}(X)\text{Tr}(Y), \qquad X, Y \in \mathfrak{gl}(n, \mathbb{R}). \tag{13.39}$$

In Table 13.1, you can find the Killing forms of important matrix Lie algebras. These also include the symplectic Lie algebras $\mathfrak{sp}(n, \mathbb{K})$, which are introduced in Sect. 14.2.

---

**Task**

Try to express the Killing form for $\mathfrak{sp}(n, \mathbb{R})$ in terms of traces.

---

## 13.5 Universal Enveloping Algebra of a Lie Algebra

In the theory of the hydrogen atom, the squared angular momentum operator plays a prominent role because it commutes with all components of the angular momentum and with the Hamiltonian operator, and can be diagonalized simultaneously with the latter. In an abstract Lie algebra, however, the product of two elements is not defined.

**Table 13.1** The Killing form of $\mathfrak{gl}(n, \mathbb{R})$ and those of the simple matrix groups

| $\mathfrak{g}$ | $K(X, Y)$ |
|---|---|
| $\mathfrak{gl}(n, \mathbb{R})$ | $2n\,\mathrm{Tr}(XY) - 2\mathrm{Tr}(X)\mathrm{Tr}(Y)$ |
| $\mathfrak{sl}(n, \mathbb{R})$ | $2n\,\mathrm{Tr}(XY)$ |
| $\mathfrak{su}(n, \mathbb{R})$ | $2n\,\mathrm{Tr}(XY)$ |
| $\mathfrak{so}(n, \mathbb{R})$ | $(n-2)\,\mathrm{Tr}(XY)$ |
| $\mathfrak{sp}(n, \mathbb{R})$ | $(2n+2)\,\mathrm{Tr}(XY)$ |
| $\mathfrak{sp}(n, \mathbb{C})$ | $(2n+2)\,\mathrm{Tr}(XY)$ |

Therefore, its universal enveloping algebra is introduced to define generalizations of the squared angular momentum. The enveloping algebra is an associative algebra in which the Lie product can always be regarded as a commutator, even for Lie algebras that are not subalgebras of $\mathfrak{gl}(n, \mathbb{K})$.

**Definition 13.8 (Universal Enveloping Algebra)** Let $\mathfrak{g}$ be an (abstract) Lie algebra. An associative algebra $\mathfrak{U}_{\mathfrak{g}}$ with unity is called the universal enveloping algebra of $\mathfrak{g}$, if

1. there is a linear mapping $\varphi : \mathfrak{g} \mapsto \mathfrak{U}_{\mathfrak{g}}$ such that

$$\varphi([X, Y]) = \varphi(X)\varphi(Y) - \varphi(Y)\varphi(X), \quad X, Y \in \mathfrak{g}, \tag{13.40}$$

2. for every associative algebra $\mathfrak{A}$ with unity and linear mapping $\alpha : \mathfrak{g} \mapsto \mathfrak{A}$, which also satisfies the relation (13.40), there exists exactly one algebra homomorphism $\phi : \mathfrak{U}_{\mathfrak{g}} \mapsto \mathfrak{A}$ with $\phi(1) = 1$ and $\alpha = \phi \circ \varphi$.

In the second property, $\phi$ is an algebra homomorphism and not just a Lie algebra homomorphism. Because of this property, $\mathfrak{U}_{\mathfrak{g}}$ is called universal. In the definition, it is naturally assumed that $\mathfrak{g}$ is a Lie algebra and $\mathfrak{U}_{\mathfrak{g}}$, $\mathfrak{A}$ are algebras over the same field $\mathbb{K}$. If $\{X_a\}$ is a basis of $\mathfrak{g}$ and $f_{ab}{}^c$ are the structure constants with respect to this basis, then the following relations hold in $\mathfrak{U}_{\mathfrak{g}}$

$$x_a x_b - x_b x_a = f_{ab}{}^c x_c, \qquad x_a = \varphi(X_a). \tag{13.41}$$

**Example: Universal Enveloping Algebra of $\mathfrak{sl}(2, \mathbb{C})$**

As a basis of this three-dimensional complex Lie algebra, we choose

$$X = \begin{pmatrix} 0 & 1 \\ 0 & 0 \end{pmatrix}, \quad H = \begin{pmatrix} 1 & 0 \\ 0 & -1 \end{pmatrix}, \quad Y = \begin{pmatrix} 0 & 0 \\ 1 & 0 \end{pmatrix}. \tag{13.42}$$

These basis vectors have the commutation relations

$$[H, X] = 2X, \quad [H, Y] = -2Y \quad \text{and} \quad [X, Y] = H. \tag{13.43}$$

Let $x, y, h$ be the images of $X, Y, H$ in the enveloping algebra, then this algebra is freely generated by $x, y, h$ modulo the relations

$$hx - xh = 2x, \quad hy - yh = -2y \quad \text{and} \quad xy - yx = h. \tag{13.44}$$

The enveloping algebra has the following vector space basis

$$\{y^a h^b x^c \,|\, a, b, c \in \mathbb{N}_0\}. \tag{13.45}$$

To prove this, one first shows that every monomial (every word) in the generators can be written as a linear combination of the monomials in (13.45). For example, $hxy = hyx + hh = yhx - 2yx + h^2$. Then, it is necessary to prove the linear independence of the basis elements (13.45), which is somewhat more difficult, see Chap. 17 in [57]. Note that, unlike $X^2$, the monomial $x^2$ is not zero. Also, $\dim \mathfrak{U}_{\mathfrak{sl}(2,\mathbb{C})} = \infty$. ◄

The universal enveloping algebra $\mathfrak{U}_{\mathfrak{g}}$ is independent of the chosen basis. In other words, if the construction is carried out for two different bases, the resulting algebras are isomorphic.

**Theorem 13.2 (Poincaré-Birkhoff-Witt Theorem)**  *For a Lie algebra $\mathfrak{g}$ with basis $\{X_1, \ldots, X_n\}$, the universal enveloping algebra $\mathfrak{U}_{\mathfrak{g}}$ has the vector space basis*

$$\{x_1^{a_1} x_2^{a_2} \cdots x_n^{a_n} \,|\, a_1, \ldots, a_n \in \mathbb{N}_0\}, \qquad x_a = \varphi(X_a). \tag{13.46}$$

*In particular, the elements $x_1, \ldots, x_n$ are linearly independent, and the subspace $\varphi(\mathfrak{g})$ of $\mathfrak{U}_{\mathfrak{g}}$ can be identified with $\mathfrak{g}$.*

The proof of the PBW theorem is essentially inductive. It exploits the fact that, with the help of relations (13.41), every word can be sorted, whereby the "error term" is a linear combination of words of shorter length.


### 13.5.1 Quadratic Casimir Invariants

A Casimir invariant is a special element from the center of the universal enveloping algebra. We will show that the universal enveloping algebra of a semi-simple Lie algebra, whose Cartan-Killing metric is non-degenerate, possesses a Casimir invariant $C_2$ that is quadratic in the generators. Since $\varphi(\mathfrak{g}) \subseteq \mathfrak{U}_{\mathfrak{g}}$ can be identified with $\mathfrak{g}$, we will no longer distinguish in notation between $x_i = \varphi(X_i)$ and $X_i$.

In the investigation of invariants, we make use of the fact that in an associative algebra, the commutator $[X, Y] = XY - YX$ satisfies the derivation rule, $[XY, Z] = X[Y, Z] + [X, Z]Y$. The quadratic Casimir invariant is $C_2 = g^{ab} X_a X_b$, where $g^{ab}$

is the inverse Cartan-Killing metric. It commutes with the generating elements of $\mathfrak{U}_\mathfrak{g}$,

$$[C_2, X_c] = g^{ab}[X_a X_b, X_c] = g^{ab} f_{ac}{}^d X_d X_b + g^{ab} f_{bc}{}^d X_a X_d$$

$$= f_{acd} (X^d X^a + X^a X^d) = 0, \tag{13.47}$$

where contravariant indices are lowered with the Killing metric $g_{ab}$ and covariant indices are raised with the inverse metric $g^{ab}$. In the last step, we used the fact that $f_{acd}$ is totally antisymmetric.

If one chooses a representation $\mathfrak{D}$ for $\alpha$ in Definition 13.8, then due to the existence of the algebra homomorphism $\phi : \mathfrak{U}_\mathfrak{g} \mapsto \mathfrak{D}(\mathfrak{g})$, the properties of the universal enveloping also apply to the representation.

**Lemma 13.11 (Quadratic Casimir Invariant)** *Let $\mathfrak{D}$ be a representation of a semi-simple $\mathfrak{g}$. Then the quadratic Casimir operator in this representation commutes with all matrices of the representation,*

$$[C_2(\mathfrak{D}), \mathfrak{D}(X)] = 0, \qquad C_2(\mathfrak{D}) = g^{ab}\, \mathfrak{D}(X_a)\mathfrak{D}(X_b)\,. \tag{13.48}$$

With Schur's lemma, it follows that in every irreducible representation, $C_2$ is a multiple of $\mathbb{1}_\mathcal{V}$,

$$C_2(\mathfrak{D}) = \lambda_\mathfrak{D}\mathbb{1}\,. \tag{13.49}$$

Taking the trace of (13.49) in the adjoint representation, we obtain $\mathrm{Tr}\,C_2(\mathrm{ad}) = \lambda_{\mathrm{ad}}\dim \mathfrak{g}$. On the other hand, using the definition (13.48) of the quadratic Casimir invariants, we find $\mathrm{Tr}\,C_2(\mathrm{ad}) = g^{ab} g_{ab} = \dim(\mathfrak{g})$, so we conclude $C_2(\mathrm{ad}) = 1$. The normalization of the Killing form changes with a scaling of the basis vectors,

$$X_a \to \alpha X_a \implies f_{ab}{}^c \to \alpha f_{ab}{}^c, \;\; g_{ab} \to \alpha^2 g_{ab}, \;\; g^{ab} \to \alpha^{-2} g^{ab}\,. \tag{13.50}$$

Mathematicians often choose a basis normalized with respect to the Killing form for compact simple Lie algebras with $K(X_a, X_b) = -\delta_{ab}$. Physicists prefer the normalizations $\mathrm{Tr}\,X_a X_b = -2\delta_{ab}$ or $\mathrm{Tr}\,X_a X_b = -\frac{1}{2}\delta_{ab}$ in the defining representation and normalize the quadratic Casimir invariant according to

$$C_2'(\mathfrak{g}) = \mathrm{Tr}\,X_a X_a\,. \tag{13.51}$$

With similar arguments as after (13.33), one proves that for simple groups, the ad-invariant bilinear forms $\mathrm{Tr}\,\mathfrak{D}(X)\mathfrak{D}(Y)$ are proportional to each other for different irreducible representations. If one chooses $\mathrm{Tr}(X_a X_b) \propto \delta_{ab}$, then for the quadratic Casimir invariant in the representation $\mathfrak{D}$, one finds the expression

$$C_2'(\mathfrak{D}) = \frac{\dim(\mathfrak{g})}{\dim(\mathfrak{D})} \mathrm{Tr}\,\mathfrak{D}^2(X_1)\, \mathbb{1}_\mathfrak{D}\,. \tag{13.52}$$

---

**Task**

Try to prove this expression for the Casimir invariant.

---

**Example: Casimir Invariant for Representations of $\mathfrak{su}(2)$**

In Chap. 16, we will see that the diagonal matrix $X_3 = \frac{i}{2}\sigma_3 \in \mathfrak{su}(2)$ in the representation for half-integer spin $j$ of dimension $2j+1$ has the form $\mathfrak{D}_j(X_3) = \mathrm{diag}(j, j-1, \ldots, -j)$. In this representation the quadratic Casimir invariant in (13.52) reads

$$C_2' = \frac{3}{2j+1} \sum_{k=-j}^{j} k^2\, \mathbb{1}_{\mathfrak{D}_j} = j(j+1)\, \mathbb{1}_{\mathfrak{D}_j} \,. \tag{13.53}$$

◄

## 13.5.2 Higher-Order Casimir Invariants

In general, a Lie algebra has several independent Casimir invariants. Their number is equal to the rank, denoted by $\mathrm{rang}(\mathfrak{g})$ and introduced in Sect. 13.1. The additional invariants are no longer quadratic, but of higher order: an invariant of order $p$ has the form

$$C_p = \kappa^{a_1 a_2 \cdots a_p}\, X_{a_1} X_{a_2} \cdots X_{a_p}, \quad \text{with} \tag{13.54}$$

$$\kappa^{a_1 a_2 \cdots a_p} = f^{a_1 b_1}{}_{b_2} f^{a_2 b_2}{}_{b_3} \cdots f^{a_p b_p}{}_{b_1}, \quad f^{ab}{}_c = g^{ar} g^{bs} g_{ct} f_{rs}{}^{t} \,. \tag{13.55}$$

An exception is the (non-simple) Lie algebra $\mathfrak{so}(4)$—it has two quadratic Casimir invariants. The unitary Lie algebra $\mathfrak{su}(n)$ has invariants $C_p$ of orders $2, 3, \ldots, n$. Table 13.2 shows the Casimir invariants for the simple matrix Lie algebras.

**Table 13.2** The Casimir invariants $C_p$ of simple matrix Lie algebras. $C_p$ is an invariant of order $p$. Every Lie algebra has the quadratic invariant $C_2 = g_{ab} X^a X^b$

| $\mathfrak{g}$ | Casimir invariants |
|---|---|
| $\mathfrak{sl}(n, \mathbb{R})$, $\mathfrak{su}(n)$ | $C_2, C_3, \ldots, C_n$ |
| $\mathfrak{so}(2n+1, \mathbb{R})$ | $C_2, C_4, \ldots, C_{2n}$ |
| $\mathfrak{usp}(2n, \mathbb{R})$ | $C_2, C_4, \ldots, C_{2n}$ |
| $\mathfrak{so}(2n, \mathbb{R})$ | $C_2, C_4, \ldots, C_{2n-2}; C_n$ |

## 13.6  Appendix A: Nilpotent and Soluble Lie Algebras

Lie algebras are classified according to their ideals and their nesting. For example, simple or semi-simple Lie algebras have no ideals or no abelian ideals other than the trivial ones. To define further classes of Lie algebras, one introduces recursively defined series. In particular, the lower central series and derived series, which lead to nilpotent and solvable Lie algebras. An exhaustive description of these classes of Lie algebras can be found, for example, in [57].

**Definition 13.9 (Nilpotent Lie Algebra)**  The (descending) lower central series is given by

$$\mathfrak{g}_{(0)} = \mathfrak{g} \quad \text{und} \quad \mathfrak{g}_{(n+1)} = [\mathfrak{g}, \mathfrak{g}_{(n)}] \le \mathfrak{g}_{(n)} \,. \tag{A.1}$$

A Lie algebra $\mathfrak{g}$ is called nilpotent of level $k$ if $\mathfrak{g}_{(k)} = 0$ and $\mathfrak{g}_{(k-1)} \ne 0$.

More explicitly, this means that $[\mathfrak{g}, [\mathfrak{g}, [\mathfrak{g}, \ldots [\mathfrak{g}, \mathfrak{g}] \ldots ]]] = 0$, respectively

$$\mathrm{ad}_{X_1}\mathrm{ad}_{X_2} \cdots \mathrm{ad}_{X_k} Y = 0 \quad \forall\, X_1, X_2, \ldots, X_k, Y \in \mathfrak{g} \,. \tag{A.2}$$

According to Lemma 13.1 the commutator of two ideals, for example the derived Lie algebra, is again an ideal. Furthermore, Lie subalgebras of nilpotent Lie algebras are also nilpotent.

---

**Nilpotent Ideals and Subalgebras of Nilpotent Lie Algebras**
The members of the lower central series are ideals $\mathfrak{g}_{(n)} \trianglelefteq \mathfrak{g}$. Lie subalgebras and homomorphic images of a nilpotent Lie algebra are nilpotent. Quotients of a nilpotent Lie algebra and one of its ideals are nilpotent. Abelian Lie algebras are nilpotent.

---

**Task**

Let $\mathfrak{g}$ be nilpotent and $\mathfrak{h} \trianglelefteq \mathfrak{g}$ an ideal. Show that $\mathfrak{g}/\mathfrak{h}$ is also nilpotent.

The set of strict upper triangular matrices (these have zeros on the diagonal) are a prototype of a nilpotent Lie algebra.

---

**Example: Heisenberg Algebra of Position and Momentum Operators**

The Heisenberg Lie algebra $\mathfrak{h}_n$ is the $(2n + 1)$-dimensional real Lie algebra with basis elements $\{q_1, \ldots, q_n, p_1, \ldots, p_n, z\}$, which obey the commutation relations

$$[p_a, q_b] = \delta_{ab}\, z, \quad [p_a, z] = [q_a, z] = [p_a, p_b] = [q_a, q_b] = [z, z] = 0 \,. \tag{A.3}$$

It is the Lie algebra of the position and momentum operators in quantum mechanics and a central extension of the Poisson algebra of coordinates in phase space by the central element $z$. This generates the one-dimensional center $\mathfrak{z} = \{\lambda z | \lambda \in \mathbb{R}\}$ and thus $\mathfrak{h}_n$ is not semi-simple. The base elements of $\mathfrak{h}_1$ are represented by triangular matrices,

$$p = E_{12}, \quad q = E_{23}, \quad z = E_{13}. \tag{A.4}$$

In this representation, any element of $\mathfrak{h}_1$ is a real upper triangular matrix,

$$\mathfrak{h}_1 = \begin{pmatrix} 0 & \alpha & \gamma \\ 0 & 0 & \beta \\ 0 & 0 & 0 \end{pmatrix}, \quad \text{mit} \quad \mathfrak{h}_{1(1)} = [\mathfrak{h}_1, \mathfrak{h}_1] = \begin{pmatrix} 0 & 0 & \alpha \\ 0 & 0 & 0 \\ 0 & 0 & 0 \end{pmatrix} = \mathfrak{z}, \tag{A.5}$$

and $\mathfrak{h}_{1(2)} = [\mathfrak{h}_1, [\mathfrak{h}_1, \mathfrak{h}_1]] = 0$. This means that $\mathfrak{h}_1$ is nilpotent of level 2. ◄

In addition to abstract Lie algebras $\mathfrak{g}$, in what follows we also consider endomorphisms $X \in \mathfrak{gl}(\mathcal{V})$ of a complex vector space $\mathcal{V}$, for example the linear mappings of a representation of $\mathfrak{g}$. An endomorphism $X$ is called nilpotent if $X^p = 0$ for a $p \in \mathbb{N}$.

**Lemma 13.12** *Let $X \in \mathfrak{gl}(\mathcal{V})$ be a nilpotent endomorphism. Then $\mathrm{ad}_X$ is also nilpotent.*

**Proof** The adjoint representation is the difference of left and right multiplication, $\mathrm{ad}_X = L_X - R_X$. The endomorphisms $L_X : Y \mapsto XY$ and $R_X : Y \mapsto YX$ commute with each other and are nilpotent, since with $X^p = 0$ also $(L_X)^p = L_{X^p} = 0$ and $(R_X)^p = 0$. The binomial theorem then provides for $(\mathrm{ad}_X)^{2p}$ a sum in which either the exponent of $L_X$ or the exponent of $R_X$ is at least $p$ in each summand. Therefore, each summand disappears and $(\mathrm{ad}_X)^{2p} = 0$

**Theorem 13.3 (Theorem of Engel)**   *Let $\mathfrak{g}$ be a Lie subalgebra of $\mathfrak{gl}(\mathcal{V})$. If all $X \in \mathfrak{g}$ are nilpotent, then there exists a non-zero element $v \in \mathcal{V}$ with $Xv = 0$ for all $X$.*

**Proof** We prove the theorem using complete induction on $n = \dim \mathfrak{g}$. For $n = 1$ the statement is obviously true. We therefore assume that it is true for all $n < \dim \mathfrak{g}$. For the induction step, we recall the normalizer $\mathfrak{n}_{\mathfrak{g}}(\mathfrak{h})$ of a subalgebra of $\mathfrak{h}$, see (13.16). It is the largest Lie subalgebra of $\mathfrak{g}$ that contains $\mathfrak{h}$ as an ideal, $\mathfrak{h} \trianglelefteq \mathfrak{n}_{\mathfrak{g}}(\mathfrak{h}) \leq \mathfrak{g}$.

We first show that if $\mathfrak{h}$ is smaller than $\mathfrak{g}$, $\mathfrak{h} < \mathfrak{g}$, then $\mathfrak{h} \triangleleft \mathfrak{n}_{\mathfrak{g}}(\mathfrak{h})$ also holds. To prove this, we consider an $H$ in the Lie subalgebra $\mathfrak{h}$. According to the condition in Theorem 13.3, $H$ is nilpotent and according to Lemma 13.12 the same applies to $\mathrm{ad}_H \in L(\mathfrak{g})$. Hence, the subalgebra $\mathfrak{h}$ also acts on the vector space $\mathfrak{g}/\mathfrak{h}$ with dimension less than $n$ by nilpotent endomorphisms. According to the induction assumption, there is a vector $X_\mathfrak{h} = X + \mathfrak{h} \in \mathfrak{g}/\mathfrak{h}$ with $X \notin \mathfrak{h}$ (because $X_\mathfrak{h}$ is not the zero vector) and $\mathrm{ad}_H(X_\mathfrak{h}) = 0 \in \mathfrak{g}/\mathfrak{h}$. Thus $\mathrm{ad}_X(H) = -\mathrm{ad}_H(X) \in \mathfrak{h}$ applies to all $H \in \mathfrak{h}$, which means $X \in \mathfrak{n}_\mathfrak{g}(\mathfrak{h})$. Since $X$ is not in $\mathfrak{h}$, this proves the statement.

Now let $\mathfrak{h} < \mathfrak{g}$ be a maximal Lie subalgebra. According to the previous result, $\mathfrak{h}$ can only be maximal for $\mathfrak{n}_\mathfrak{g}(\mathfrak{h}) = \mathfrak{g}$, and then $\mathfrak{h} \lhd \mathfrak{g}$. A one-dimensional subspace of the Lie algebra $\mathfrak{g}/\mathfrak{h}$ is a Lie subalgebra, and its pre-image in $\mathfrak{g}$ is a subalgebra of $\mathfrak{g}$, which contains $\mathfrak{h}$ as proper Lie subalgebra. For a maximal $\mathfrak{h}$, this pre-image is all of $\mathfrak{g}$. Thus $\mathfrak{g}/\mathfrak{h}$ is one-dimensional.

> **Maximal Lie Subalgebras**
> If all endomorphisms of $\mathfrak{g} \leq \mathfrak{gl}(\mathcal{V})$ are nilpotent, then a maximal Lie subalgebra is an ideal of codimension 1.

Having said this, we now consider the subspace

$$W = \{ v \in \mathcal{V} \,|\, Hv = 0 \ \text{ for all } H \in \mathfrak{h} \} \subseteq \mathcal{V}. \tag{A.6}$$

From $\dim \mathfrak{h} < \dim \mathfrak{g}$ follows according to the induction assumption $W \neq 0$. Because of $\mathfrak{h} \lhd \mathfrak{g}$, the subspace $W$ is stable under $\mathfrak{g}$,

$$H \in \mathfrak{h}, \ X \in \mathfrak{g}, \ v \in W \Longrightarrow HXv = XHv - [X, H]v = 0.$$

In the last step, we used $[X, H] \in \mathfrak{h}$. We now choose $Y \in \mathfrak{g} - \mathfrak{h}$. The endomorphism $Y \in \mathfrak{gl}(\mathcal{V})$ is nilpotent and $Y : W \mapsto W$. According to the induction assumption, there is a vector $0 \neq v \in W$ with $Yv = 0$. Then $Xv = 0$ for all $X \in \mathfrak{g} = \mathfrak{h} + \mathbb{C}Y$.

**Lemma 13.13** *The center $\mathfrak{z}_\mathfrak{g}$ of a (non-trivial) nilpotent Lie algebra $\mathfrak{g}$ is non-trivial. If the quotient $\mathfrak{g}/\mathfrak{z}_\mathfrak{g}$ of a Lie algebra $\mathfrak{g}$ is nilpotent, then $\mathfrak{g}$ is nilpotent as well.*

***Proof*** If $k \geq 1$ is the level of the nilpotent algebra, then $\mathfrak{g}_{(k-1)} \neq \emptyset$ and $[\mathfrak{g}_{(k-1)}, \mathfrak{g}] = 0$. This means that $\mathfrak{g}_{(k-1)} \trianglelefteq \mathfrak{z}_\mathfrak{g}$ which proves the first statement. Now let $\mathfrak{g}$ be a Lie algebra with nilpotent quotient $\mathfrak{g}/\mathfrak{z}_\mathfrak{g}$ of level $k$. Then every commutator of length $k$ vanishes in $\mathfrak{g}/\mathfrak{z}_\mathfrak{g}$,

$$0 = [X_1 + \mathfrak{z}_\mathfrak{g}, [X_2 + \mathfrak{z}_\mathfrak{g}, \dots [X_{k-1} + \mathfrak{z}_\mathfrak{g}, X_k + \mathfrak{z}_\mathfrak{g}] \dots ]]$$

$$= [X_1, [X_2, \dots [X_{k-1}, X_k] \dots ]] + \mathfrak{z}_\mathfrak{g}.$$

Therefore, $[X_1, [X_2, \dots [X_{k-1}, X_k] \dots ]] \in \mathfrak{z}_\mathfrak{g} \trianglelefteq \mathfrak{g}$ and every commutator of length $k + 1$ is zero, i.e. $\mathfrak{g}_{(k+1)} = 0$.

**Theorem 13.4 (Adjoint Representation of Nilpotent $\mathfrak{g}$)** *A Lie algebra $\mathfrak{g}$ is nilpotent if and only if the $\mathrm{ad}_X$ are nilpotent for all $X \in \mathfrak{g}$.*

***Proof*** For all elements $X$ of a nilpotent Lie algebra of level $k$ $(\mathrm{ad}_X)^k \mathfrak{g} = 0$, i.e. $\mathrm{ad}_X$ is nilpotent. Conversely, we show again by induction on $\dim(\mathfrak{g})$ that $\mathfrak{g}$ is nilpotent if all $X$ are ad-nilpotent. The image of the adjoint representation $\mathrm{ad}_\mathfrak{g} \cong \mathfrak{g}/\mathfrak{z}_\mathfrak{g}$ in $L(\mathfrak{g})$ fulfills the conditions in Theorem 13.3, and therefore there exists a $Z \in \mathfrak{g}$ with $\mathrm{ad}_X Z = [X, Z] = 0$ for all $X \in \mathfrak{g}$. Thus $\mathfrak{g}$ has a non-trivial center $\mathfrak{z}_\mathfrak{g} \trianglelefteq \mathfrak{g}$. The quotient $\mathfrak{g}/\mathfrak{z}_\mathfrak{g}$ has a smaller dimension than $\mathfrak{g}$ and consists of ad-nilpotent elements. Thus, according to the induction assumption $\mathfrak{g}/\mathfrak{z}_\mathfrak{g}$ is nilpotent and thus according to Lemma 13.13 also $\mathfrak{g}$ is nilpotent.

**Corollary 13.3** *Let $\mathfrak{g}$ be a Lie subalgebra of $\mathfrak{gl}(\mathcal{V})$. If every $X \in \mathfrak{g}$ is nilpotent, then there exists a basis of $\mathcal{V}$, for which all $X$ are represented by strict upper triangular matrices. In particular the Killing form of a nilpotent Lie algebra vanishes.*

***Proof*** See Problem 13.7.

---

**Question**

There are nilpotent $\mathfrak{g}$ whose elements are not described in any base by strict upper triangular matrices. Do you know an example? Why does this not contradict the corollary?

---

As a further important class, we now introduce the solvable Lie algebras.

**Definition 13.10 (Solvable Lie Algebra)** The derived series is given by

$$\mathfrak{g}^{(0)} = \mathfrak{g} \quad \text{and} \quad \mathfrak{g}^{(n+1)} = [\mathfrak{g}^{(n)}, \mathfrak{g}^{(n)}] \leq \mathfrak{g}^{(n)} . \tag{A.7}$$

A Lie algebra $\mathfrak{g}$ is called solvable of level $k$ if there exists a $k$ such that $\mathfrak{g}^{(k)} = 0$ and $\mathfrak{g}^{(k-1)} \neq 0$.

The terms $\mathfrak{g}^{(n)}$ of the derived series are all ideals of $\mathfrak{g}$, and since $\mathfrak{g}^{(n)} \subseteq \mathfrak{g}_{(n)}$, every nilpotent Lie algebra is also solvable. If $\mathfrak{g}$ is solvable, then every subalgebra and every homomorphic image of $\mathfrak{g}$ is also solvable. The nilpotent Heisenberg algebras $\mathfrak{h}_n$ are therefore solvable.

---

**Example: Triangular Matrices**

Let $\mathfrak{g}$ be the set of upper triangular matrices

$$\mathfrak{g} = \begin{pmatrix} * & * & * & \cdots & * \\ 0 & * & * & \cdots & * \\ 0 & 0 & \ddots & \ddots & \vdots \\ \vdots & \vdots & \ddots & \ddots & * \\ 0 & 0 & \cdots & 0 & * \end{pmatrix} \quad \text{with} \quad \mathfrak{g}_{(1)} = \mathfrak{g}^{(1)} = \begin{pmatrix} 0 & * & * & \cdots & * \\ 0 & 0 & * & \cdots & * \\ 0 & 0 & \ddots & \ddots & \vdots \\ \vdots & \vdots & \ddots & \ddots & * \\ 0 & 0 & \cdots & 0 & 0 \end{pmatrix}$$

$$\mathfrak{g}_{(2)} = \begin{pmatrix} 0 & * & * & \cdots & * \\ 0 & 0 & * & \cdots & * \\ 0 & 0 & \ddots & \ddots & \vdots \\ \vdots & \vdots & \ddots & \ddots & * \\ 0 & 0 & \cdots & 0 & 0 \end{pmatrix} \quad \text{and} \quad \mathfrak{g}^{(2)} = \begin{pmatrix} 0 & 0 & * & \cdots & * \\ 0 & 0 & 0 & \cdots & * \\ 0 & 0 & \ddots & \ddots & \vdots \\ \vdots & \vdots & \ddots & \ddots & 0 \\ 0 & 0 & \cdots & 0 & 0 \end{pmatrix}.$$

It can be seen that $\mathfrak{g}_{(n)} = \mathfrak{g}_{(1)}$ and $\mathfrak{g}^{(n)} = 0$ for a sufficiently large $n$. Thus, $\mathfrak{g}$ is solvable but not nilpotent.  ◄

The second term $\mathfrak{g}^{(1)}$ of the derived series of a solvable Lie algebra of level $k$ is the derived Lie algebra $\mathfrak{g}'$ in (13.11). The last term $\mathfrak{g}^{(k-1)}$ is a non-trivial abelian ideal in $\mathfrak{g}$, and we conclude that a solvable Lie algebra cannot be semi-simple.

---

**Task**

Show: If $\mathcal{I}$ is a solvable ideal of $\mathfrak{g}$ such that $\mathfrak{g}/\mathcal{I}$ is solvable, then $\mathfrak{g}$ is also solvable.

In the following, we shall further explore the differences between solvable, nilpotent, and semi-simple Lie algebras.

**Theorem 13.5 (Lie's Theorem)**   *Let $\mathfrak{g} \leq \mathfrak{gl}(V)$ be a complex solvable Lie algebra. Then there exists a simultaneous eigenvector $0 \neq v \in V$, for which $Xv = \lambda_X v$ for all $X \in \mathfrak{g}$ with $\lambda_X \in \mathbb{C}$.*

***Proof***  The proof is similar to that of Engel's Theorem. For a solvable $\mathfrak{g}$, $[\mathfrak{g}, \mathfrak{g}] \lhd \mathfrak{g}$ (not just $\trianglelefteq$) and $\mathfrak{a} = \mathfrak{g}/[\mathfrak{g}, \mathfrak{g}]$ is an abelian Lie algebra. The preimage $\mathfrak{h}$ in $\mathfrak{g}$ of a subspace of $\mathfrak{a}$ of codimension one is then an ideal of codimension one in $\mathfrak{g}$. One assumes as induction hypothesis that there is a simultaneous eigenvector $v_0 \in V$ for all $X \in \mathfrak{h}$. The respective eigenvalues are $\lambda_X$. One then considers the subspace $W \subset V$ of all vectors with the same set of simultaneous eigenvalues, i.e., $W = \{v \in V : Xv = \lambda_X v \ \forall X \in \mathfrak{h}\}$. Let $Y \in \mathfrak{g}$ be any element not in $\mathfrak{h}$. It suffices to show that there is a vector $v \in W$ that is mapped by $Y$ to a multiple of itself. To do this, it is again sufficient to show that $Y$ maps the space $W$ into itself. This follows from the following lemma:

**Lemma 13.14**  *Let $\mathfrak{h}$ be an ideal of $\mathfrak{g} \leq \mathfrak{gl}(V)$ and $\lambda : \mathfrak{h} \mapsto \mathbb{C}$ a linear function. Then the subspace*

$$W = \{v \in V \mid Hv = \lambda_H v \ \forall H \in \mathfrak{h}\} \tag{A.8}$$

*is stable under $\mathfrak{g}$, i.e., $X(W) \subset W$ for all $X \in \mathfrak{g}$.*

***Proof*** For $H \in \mathfrak{h}$, $X \in \mathfrak{g}$ and $v \in \mathcal{W}$, one has

$$HXv = XHv + [H, X]v = \lambda_H Xv + \lambda_{[H,X]}v$$

because $[H, X] \in \mathfrak{h}$. If $\lambda_{[H,X]}$ were to vanish, then $Xv$ would also be an eigenvector of $H \in \mathfrak{h}$ with eigenvalue $\lambda_H$. To see this, consider the subspace $\mathcal{V}_0 = \mathrm{span}\{v, Xv, X^2v, \dots\}$ of $\mathcal{V}$. This subspace is (by construction) mapped into itself by $X$ and also by all $H \in \mathfrak{h}$. The last statement follows by induction: $H$ obviously maps the vectors $v$ and $Xv$ into vectors in $\mathcal{V}_0$. From $HX^nv = XHX^{n-1}v + [H, X]X^{n-1}v$, the statement for $X^n$ follows if it is true for $X^{n-1}$. In the basis $\{v, Xv, X^2v, \dots\}$ of $\mathcal{V}_0$, $X$ acts through an upper triangular matrix with diagonal elements $\lambda_X$. In particular, $\mathrm{Tr}_{\mathcal{V}_0} X = \lambda_X \dim(\mathcal{V}_0)$. On the other hand, the commutator $[H, X]$ also operates on $\mathcal{V}_0$ and $\mathrm{Tr}_{\mathcal{V}_0}[H, X] = 0$. Therefore, $\lambda_{[H,X]} = 0$. In the literature, the existence of an adapted basis is also referred to as Lie's Theorem:

**Consequence of Lie's Theorem**

Let $\mathfrak{g} \le \mathfrak{gl}(\mathcal{V})$ be solvable. Then there exists a basis in the complex vector space $\mathcal{V}$, with respect to which all elements of $\mathfrak{g}$ are represented by upper triangular matrices.

There is a simple relationship between solvable and nilpotent Lie algebras:

**Lemma 13.15** *A Lie algebra $\mathfrak{g}$ over $\mathbb{C}$ is solvable if and only if $\mathfrak{g}' = [\mathfrak{g}, \mathfrak{g}]$ is nilpotent.*

***Proof*** If $\mathfrak{g}' = [\mathfrak{g}, \mathfrak{g}]$ is nilpotent, then it is solvable and thus $\mathfrak{g}$ is also solvable. Conversely, let $\mathfrak{g}$ be solvable. The image of $\mathfrak{g}$ under the adjoint representation is a solvable subalgebra of $L(\mathfrak{g})$. According to Lie's theorem, there is an adapted basis of $\mathfrak{g}$, with respect to which the mappings $\mathrm{ad}_X$ are upper triangular matrices. Since $[\mathrm{ad}_X, \mathrm{ad}_Y] = \mathrm{ad}_{[X,Y]}$ is described by a strict upper triangular matrix with respect to this basis, it is nilpotent. Therefore, all $Z \in \mathfrak{g}'$ are ad-nilpotent and according to Engel's theorem, $[\mathfrak{g}, \mathfrak{g}]$ is nilpotent.

Without proof, we note a lemma from linear algebra, which is proven in Sect. 4.3 of [57]:

**Lemma 13.16** *Let $A \subset B$ be two subspaces of $L(\mathcal{V})$ and $M = \{Z \in L(\mathcal{V}) \mid [Z, B] \subset A\}$. Then an element $Z \in M$ with $\mathrm{Tr}(ZX) = 0$ for every $X \in M$ is nilpotent.*

The following lemma provides a useful criterion for the solvability of a Lie algebra.

**Lemma 13.17** *A Lie algebra $\mathfrak{g}$ is solvable if and only if $K(\mathfrak{g}, \mathfrak{g}') = K(\mathfrak{g}, [\mathfrak{g}, \mathfrak{g}]) = 0$.*

***Proof*** If $\mathfrak{g}$ is solvable, then in an adapted basis, the $\mathrm{ad}_X$ are upper triangular matrices and the $\mathrm{ad}_{[Y,Z]} = [\mathrm{ad}_X, \mathrm{ad}_Y]$ are strict upper triangular matrices, and thus $\mathrm{Tr}(\mathrm{ad}_X \mathrm{ad}_{[Y,Z]}) = 0$. For the converse, it suffices according to Lemma 13.15 to show that the ideal $\mathfrak{g}' = [\mathfrak{g}, \mathfrak{g}]$ is nilpotent. We assume $\mathfrak{g} \leq \mathfrak{gl}(\mathcal{V})$ and define $\mathfrak{m} = \{Z \in \mathfrak{gl}(\mathcal{V}) \,|\, [Z, \mathfrak{g}] \subseteq \mathfrak{g}'\}$. Obviously, $\mathfrak{g}' \leq \mathfrak{g} \leq \mathfrak{m}$. We now prove $K(\mathfrak{m}, \mathfrak{g}') = 0$. If $X, Y \in \mathfrak{g}$ and $M \in \mathfrak{m}$, then

$$K([X, Y], M) = -K(Y, [M, X]) = 0, \tag{A.9}$$

since $K(\mathfrak{g}, \mathfrak{g}') = 0$ was assumed. According to Lemma 13.16, $[X, Y] \in \mathfrak{g}'$ is nilpotent and according to Lemma 13.15, $\mathfrak{g}$ is nilpotent. Since we assumed $\mathfrak{g} \leq \mathfrak{gl}(\mathcal{V})$, the lemma is first proven for the adjoint representation. But since $\mathrm{ad}_{\mathfrak{g}} = \mathfrak{g}/\mathfrak{z}_{\mathfrak{g}}$, it also holds for the Lie algebra $\mathfrak{g}$.

The following criterion provides a useful characterization of semi-simple Lie algebras.

**Theorem 13.6 (Cartan's Criterion)**   *A Lie algebra $\mathfrak{g}$ over $\mathbb{C}$ is semi-simple if and only if its Killing form is non-degenerate.*

We had proven in Lemma 13.10 that the Killing form of a semi-simple Lie algebra is non-degenerate. Here we prove the converse. Let the Killing form be non-degenerate and $\mathfrak{a}$ be any abelian ideal of $\mathfrak{g}$. For $A \in \mathfrak{a}$ and $X \in \mathfrak{g}$, it holds that $(\mathrm{ad}_A \mathrm{ad}_X)(\mathfrak{g}) \subseteq \mathfrak{a}$. Thus, $(\mathrm{ad}_A \mathrm{ad}_X)^2(\mathfrak{g}) \subseteq [\mathfrak{a}, \mathfrak{a}] = \{0\}$, so $(\mathrm{ad}_A \mathrm{ad}_X)$ is nilpotent. It follows that $K(A, X) = \mathrm{Tr}(\mathrm{ad}_A \mathrm{ad}_X) = 0$. Since $X \in \mathfrak{g}$ was arbitrary, we see that $\mathfrak{a} \subseteq \mathfrak{g}^{\perp} = \{0\}$. Thus, $\mathfrak{g}$ has no non-trivial abelian ideal and is therefore semi-simple.

**Definition 13.11**   A radical $\mathfrak{r}_{\mathfrak{g}} \trianglelefteq \mathfrak{g}$ is a solvable ideal in $\mathfrak{g}$ with maximal dimension.

**Corollary 13.4**   *A radical $\mathfrak{r}_{\mathfrak{g}}$ contains all solvable ideals of $\mathfrak{g}$ and is unique.*

***Proof*** If the ideal $\mathfrak{h} \trianglelefteq \mathfrak{g}$ is solvable, then the ideal $\mathfrak{h} + \mathfrak{r}_{\mathfrak{g}}$ is also solvable. Since $\mathfrak{r}_{\mathfrak{g}}$ has maximal dimension, $\mathfrak{h} + \mathfrak{r}_{\mathfrak{g}} = \mathfrak{r}_{\mathfrak{g}}$ and $\mathfrak{h} \leq \mathfrak{r}_{\mathfrak{g}}$. If there were two different radicals of $\mathfrak{g}$, then the sum of these ideals would have to be equal to both ideals, which would be a contradiction.

**Lemma 13.18 (Radical)**   *A Lie algebra $\mathfrak{g}$ is semi-simple if its radical is trivial, $\mathfrak{r}_{\mathfrak{g}} = \{0\}$. More generally, $\mathfrak{g}/\mathfrak{r}_{\mathfrak{g}}$ is a semi-simple Lie algebra.*

***Proof*** We prove only the first statement here. The ideal $\mathfrak{g}^{\perp}$ introduced in Sect. 13.4 is solvable due to Lemma 13.17, because

$$K_{\mathfrak{g}}(\mathfrak{g}^{\perp}, \mathfrak{g}) = 0 \Rightarrow K_{\mathfrak{g}}(\mathfrak{g}^{\perp}, \mathfrak{g}^{\perp}) = K_{\mathfrak{g}^{\perp}}(\mathfrak{g}^{\perp}, \mathfrak{g}^{\perp}) = 0. \tag{A.10}$$

If $\mathfrak{r}_{\mathfrak{g}} = \{0\}$, then $\mathfrak{g}^{\perp}$ must also be $\{0\}$, which was to be proven.

## 13.7   Exercises for Chap. 13

**Problem 13.1 (Baker-Campbell-Hausdorff Formula)**  Assume that two matrices (operators) $A$, $B$ satisfy the relation $[A, [A, B]] = [B, [A, B]] = 0$. Show that in this case the following identity holds:

$$\exp(A) \exp(B) = \exp\left(A + B + \frac{1}{2}[A, B]\right).$$

Hint: Find a differential equation for $D(t) = e^{tA} e^{tB} e^{-t(A+B)}$.

**Problem 13.2 (Heisenberg Algebra)**  The Heisenberg algebra $\mathfrak{h}_1$ is a three-dimensional real Lie algebra with basis elements $p, q, r$ which obey $[p, q] = r$ and $[p, r] = [q, r] = 0$.

1. Show that $\mathfrak{h}_1$ is represented by the following matrices (we denote the matrices with the same symbols):

$$p = \begin{pmatrix} 0 & 1 & 0 \\ 0 & 0 & 0 \\ 0 & 0 & 0 \end{pmatrix}, \quad q = \begin{pmatrix} 0 & 0 & 0 \\ 0 & 0 & 1 \\ 0 & 0 & 0 \end{pmatrix}, \quad r = \begin{pmatrix} 0 & 0 & 1 \\ 0 & 0 & 0 \\ 0 & 0 & 0 \end{pmatrix}.$$

2. Thus, $\mathfrak{h}_1$ is a Lie subalgebra of $\mathfrak{sl}(3, \mathbb{R})$. Let $A, B \in \mathfrak{h}$. Determine $C(t) \in \mathfrak{h}$ such that

$$e^{tA} e^{tB} = e^{C(t)}.$$

**Problem 13.3 (Finite and Infinitesimal Rotations)**  Hermitian operators $J_1, J_2$ and $J_3$, which satisfy the commutation relations $[J_a, J_b] = i\epsilon_{abc} J_c$, are called angular momentum operators (we set $\hbar = 1$ here). The corresponding unitary operators

$$R(\boldsymbol{\phi}) = e^{i\boldsymbol{\phi}\cdot\boldsymbol{J}}, \quad \boldsymbol{\phi} \in \mathbb{R}^3,$$

describe rotations with the angle $\phi = |\boldsymbol{\phi}|$ around the axis $\hat{\boldsymbol{\phi}}$.

1. In Sect. 5.2, the matrices for rotations around the axes $e_1, e_2$ and $e_3$ were introduced. For $\boldsymbol{\phi} = \phi\, e_1$, for example, $\boldsymbol{\phi} \cdot \boldsymbol{J} = \phi\, J_1$. Find the matrices $J_1, J_2$ and $J_3$ (for example by differentiation with respect to $\phi$). Check the identity $\boldsymbol{J}^2 = j(j+1)$.
2. The angular momentum operators for a spinor $u \in \mathbb{C}^2$ are the Pauli matrices $J_a^{(1/2)} = \frac{1}{2}\sigma_a$. The corresponding finite rotations are denoted by $D^{(1/2)}(\boldsymbol{\phi})$. Show that $(\boldsymbol{J}^{(1/2)})^2 = j(j+1)$ with $j = 1/2$. Find the explicit form of the rotation matrices for rotations around the coordinate axes.

**Problem 13.4 (Ideals)** Let $\mathcal{I}_1$ and $\mathcal{I}_2$ be two ideals of $\mathfrak{g}$. Show that the intersection $\mathcal{I}_1 \cap \mathcal{I}_2$, their sum $\mathcal{I}_1 + \mathcal{I}_2$, and their product $[\mathcal{I}_1, \mathcal{I}_2]$ are also ideals of $\mathfrak{g}$.
Hint: The sum and product of two Lie subalgebras were introduced in Sect. 13.2.

**Problem 13.5 (Normal Subalgebras)** Let $\mathfrak{g}$ be a Lie algebra and $\mathfrak{h} \le \mathfrak{g}$ a Lie subalgebra. The normalizer of $\mathfrak{h}$ in $\mathfrak{g}$ is

$$\mathfrak{n}_\mathfrak{g}(\mathfrak{h}) = \{X \in \mathfrak{g} : [X, \mathfrak{h}] \in \mathfrak{h}\}.$$

- Show that $\mathfrak{n}_\mathfrak{g}(\mathfrak{h})$ is a Lie subalgebra of $\mathfrak{g}$ that contains $\mathfrak{h}$.
- Let $\mathfrak{g} = \mathfrak{gl}(n, \mathbb{C})$ and $\mathfrak{h}$ contain all diagonal matrices in $\mathfrak{g}$. Determine $\mathfrak{n}_\mathfrak{g}(\mathfrak{h})$.

**Problem 13.6 (Schur's Lemma for Representations of Lie Algebras)** Prove Schur's Lemma in Sect. 13.3
Hint: You can almost copy the proof of the lemma for group representations in Sect. 11.2. Consider again the image and the kernel of the morphism $H$ and use the irreducibility of the representations. For the second statement consider the endomorphism $(\lambda \mathbb{1}_\mathcal{V} - H)$ with $H$-eigenvalue $\lambda$.

**Problem 13.7 (Proof of Corollary 13.3)** Prove Corollary 13.3 using complete induction.
Hints: The induction should proceed on to the dimension of $\mathcal{V}$. Use the Theorem 13.3, according to which there exists a vector $v \in \mathcal{V}$ not equal to zero for which $Xv = 0$ for all $X \in \mathfrak{g}$. Consider $\mathcal{V}_1 = \mathrm{span}\{v\}$ and the linear map $\bar{X}$ induced by $X$ on $\mathcal{V}/\mathcal{V}_1$. The map $\mathfrak{g} \mapsto \mathfrak{gl}(\mathcal{V}/\mathcal{V}_1)$ is a Lie algebra homomorphism, whose image is a subalgebra of $\mathfrak{gl}(\mathcal{V}/\mathcal{V}_1)$ consisting of nilpotent maps. Now use the induction hypothesis.

**Problem 13.8 (Jordan Normal Form of Matrices)** Use the Jordan normal form to show that every $n \times n$ matrix $M$ can be expressed as $M = S + N$, with $S$ being diagonalizable (over $\mathbb{C}$) and $N$ being nilpotent with $SN = NS$. Recall that in the normal form $M$ is block diagonal, and each block has the following form:

$$B = \begin{pmatrix} \lambda & & * \\ & \ddots & \\ 0 & & \lambda \end{pmatrix}$$

**Problem 13.9 (Irreducible Representations of Solvable Lie Algebras)** Use Lie's theorem to prove that every irreducible representation of a solvable Lie algebra is one-dimensional.

# Lie Algebras of Lie Groups

14

*Calculus required continuity, and continuity was supposed to require the infinitely little; but nobody could discover what the infinitely little might be.*

—Bertrand Russell

A fruitful strategy, going back to SOPHUS LIE (1842–1899), for investigating a Lie group consists of linearizing it at the identity element. The infinitesimal deviations of the group elements from the identity element form the linear tangent space at the identity element—this is the Lie algebra $\mathfrak{g}$ associated with a Lie group $G$. One of the main results is that the Lie algebra $\mathfrak{g}$ encodes the information about the local structure of the Lie group in a neighborhood of the identity element. Linear spaces are significantly simpler objects than manifolds, and therefore many properties of Lie groups and their representations are more easily accessible through the theory of Lie algebras, which was introduced in Chap. 13. The transition from a Lie algebra to its Lie group generalizes the transition from the linear space $(\mathbb{R}, +)$ to the multiplicative group $(\mathbb{R}, \cdot)$, which is established by the exponential function.

In a geometric approach, one notices that the vector fields on a Lie group $G$ with the Lie bracket form an infinite-dimensional Lie algebra. The subspace of left-invariant vector fields is closed with respect to the Lie bracket and defines the Lie algebra $\mathfrak{g}$ of $G$. This vector space is isomorphic to the tangent space $T_eG$ at the identity element, and thus it holds that $\dim(\mathfrak{g}) = \dim(T_eG) = \dim(G)$. The geometric approach via the left-invariant vector fields is sketched at the end of the chapter.

According to a theorem by the Russian mathematician IGOR ADO, every finite-dimensional complex Lie algebra is isomorphic to a subalgebra of the Lie algebra $\mathfrak{gl}(n, \mathbb{C})$ of complex $n \times n$ matrices for sufficiently large $n$. This means that one can represent every finite-dimensional complex Lie algebra as a Lie algebra of matrices.

We will therefore assume at the beginning of the chapter that the group elements are matrices. In the appendix, we will return to the more general case.

The theory of Lie algebras of Lie groups can be found in many textbooks on continuous symmetries and their representations, e.g., in [5, 8, 14, 17, 57, 58].

## 14.1   Lie Algebra of Infinitesimal Generators

We consider an $n$-dimensional matrix group, i.e., a Lie subgroup of $GL(n, \mathbb{K})$, and identify the tangent space at the identity element $e$ as the Lie algebra of the group. This is sketched in Fig. 14.1. Let $U$ be a coordinate neighborhood of $e$ and $x = (x^1, \ldots, x^n)$ local coordinates of a chart of $U$. We choose the coordinates such that

$$g(x = 0) = e \,, \tag{14.1}$$

and examine the group elements in the neighborhood of $e$. For this purpose, we expand them in a Taylor series around $x = 0$,

$$g(x) = g(0) + \sum_a x^a \left. \frac{\partial g(x)}{\partial x^a} \right|_{x=0} + O(x^2), \quad g(0) = e \,. \tag{14.2}$$

The first derivatives describe the infinitesimal changes of the group elements in the direction of the coordinate axes. They define vectors in the tangent space at $e$ and are called *generators* or *infinitesimal generators* of the Lie group,

$$X_a \equiv \left. \frac{\partial g(x)}{\partial x^a} \right|_{x=0} \,, \quad a = 1, 2, \ldots, n \,. \tag{14.3}$$

For example, the vector $X_1$ is tangent to the coordinate line $g(x^1, 0, \ldots, 0)$ through $e$. Now, if $x(t)$ is any differentiable curve with $x(0) = 0$, then $g(x(t)) \equiv g(t)$ describes a curve in the group with $g(0) = e$ and with tangent vector

$$X = \left. \frac{dg(t)}{dt} \right|_0 \equiv \dot{g}(0) = \sum_a \left. \frac{\partial g(x)}{\partial x^a} \right|_{x=0} \dot{x}^a(0) = \sum_a X_a \, \dot{x}^a(0) \,. \tag{14.4}$$

**Fig. 14.1** Tangent space at the identity

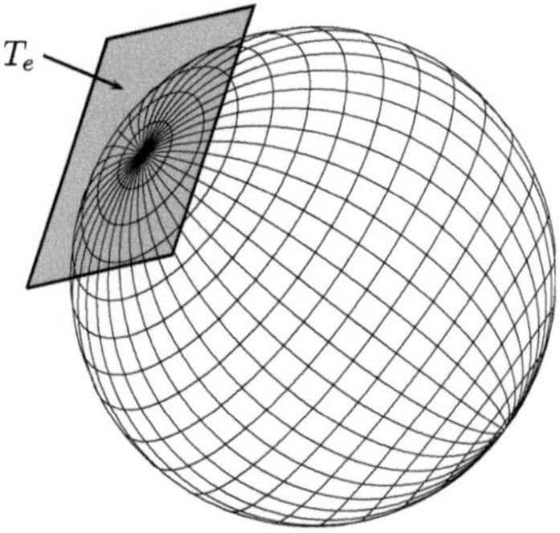

Thus, any real linear combination of the tangent vectors $\{X_a\}$ is also tangent to a curve through $e$ and therefore also an infinitesimal generator of the Lie group. Now, if $g(t)$ is a curve with tangent vector $X$ at $e$, then due to $g(t)g^{-1}(t) = e$, it holds that

$$0 = \frac{\mathrm{d}}{\mathrm{d}t}\left[g(t)g^{-1}(t)\right]\bigg|_0 = X + \frac{\mathrm{d}g^{-1}(t)}{\mathrm{d}t}\bigg|_0 \implies \frac{\mathrm{d}g^{-1}(t)}{\mathrm{d}t}\bigg|_0 = -X. \tag{14.5}$$

The curves $g(t)$ and $g^{-1}(t)$ therefore have anti-parallel tangent vectors at $e$. If a curve $g(t)$ does not pass through the neutral element, one can instead consider the curve $g^{-1}(t)g(t+s)$, which for $s = 0$ passes through $e$. Its derivative with respect to $s$ at the point $s = 0$ lies in the tangent space $T_eG$, and we conclude:

**Corollary 14.1** *For a (differentiable) curve $g(t)$ in $G$, the vector*

$$X = g^{-1}(t)\,\frac{\mathrm{d}g(t)}{\mathrm{d}t} \tag{14.6}$$

*lies in the tangent space $T_eG$ of $G$.*

---

**$T_eG$ and Left-invariant Vector Fields**

Let $g(t)$ be a smooth curve through $e$, as sketched in Fig. 14.2, and $\dot{g}(0) = X$ the corresponding tangent vector in $T_eG$. Then the left-translated curve $L_a g(t)$ describes a curve through $a$ with tangent vector $(L_a)_*X \in T_aG$ (the pushforward map $(L_a)_*$ is discussed in Sect. 14.8). A vector field $g \mapsto X_g$ is left-invariant if $X_a = (L_a)_*X_e$ or $X_{hg} = (L_h)_*X_g$. The linear map $(L_h)_* : T_gG \mapsto T_{hg}G$ defines an isomorphism. Thus, the set of left-invariant vector fields can be identified with $T_eG$. This is further explained at the end of Sect. 14.8.

---

**Fig. 14.2** For a left-invariant vector field $g \mapsto X_g \in T_gG$, the vector at $a \in G$ is equal to the vector at $e$, transported with the pushforward map $(L_a)_*$ to the left translation $L_a$

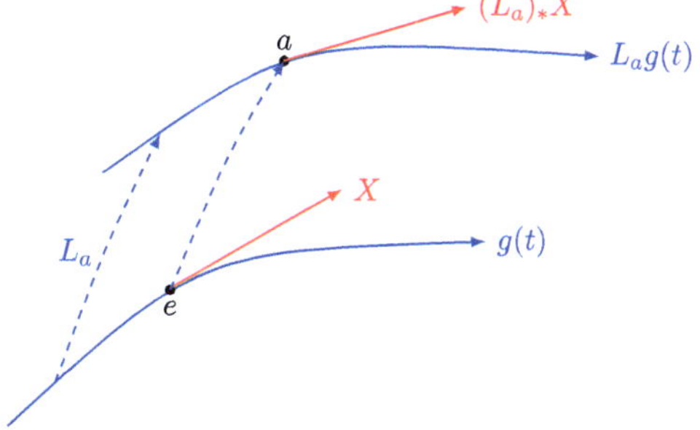

### 14.1.1  The Generators Form a Lie Algebra

We now show that the infinitesimal generators in $T_e G$ form a Lie algebra—it is the Lie algebra associated with $G$. For this, we consider two curves $g(t)$ and $h(t)$, which pass through the identity for $t = 0$ and have tangent vectors $X$ and $Y$ there. Then $g(\alpha t)h(\beta t)$ is also such a curve for arbitrary real constants $\alpha, \beta$. This curve has the tangent vector

$$\frac{\mathrm{d}}{\mathrm{d}t}\big(g(\alpha t)h(\beta t)\big)\big|_{t=0} = \alpha X + \beta Y \,. \tag{14.7}$$

This means that the tangent space $T_e G$ (as expected) is a *linear space*. The number of independent infinitesimal generators is equal to the number of independent real parameters of the Lie group, i.e., its dimension. For matrix groups, the infinitesimal generators are also matrices.

---

**Example: Infinitesimal Generators of Rotations in Space**

The proper rotations $R(e_i, \varphi)$ around the axes defined by the basis vectors $e_i$ were introduced earlier in (5.16) and (5.17). For $\varphi = 0$, these curves pass through the identity matrix. Therefore, the three skew-symmetric matrices

$$\Omega_i = \frac{\mathrm{d}}{\mathrm{d}\varphi}\bigg|_{\varphi=0} R(e_i, \varphi) \implies (\Omega_i)_{jk} = -\epsilon_{ijk} \tag{14.8}$$

are infinitesimal generators of SO(3). It is easy to verify that they form a basis of the Lie algebra $T_e\mathrm{SO}(3) = \mathfrak{so}(3)$ of skew-symmetric real $3 \times 3$ matrices. ◀

Let $h(t)$ again be a curve through $e$ with tangent vector $Y \in T_e G$. The curve conjugated with $g \in G$, $gh(t)g^{-1}$, also passes through $e$ and has the tangent vector

$$\frac{\mathrm{d}}{\mathrm{d}t}\big(gh(t)g^{-1}\big)\big|_{t=0} = gYg^{-1} \equiv \mathrm{Ad}(g)Y \,. \tag{14.9}$$

This holds for all group elements $g$, and therefore, with every vector $Y$, the curve $g(t)Yg^{-1}(t)$ also lies in the linear space $T_e G$. Due to linearity, the derivative

$$\frac{\mathrm{d}}{\mathrm{d}t}\big(g(t)Yg^{-1}(t)\big)\big|_{t=0} = XY - YX \equiv [X, Y], \qquad X = \dot{g}(0)\,, \tag{14.10}$$

also lies in $T_e G$. Here, we used the result (14.5) for the derivative of $g^{-1}(t)$. This proves that for $X, Y \in T_e G$, their commutator $[X, Y]$ also lies in $T_e G$. The Lie product of $X$ and $Y$ is equal to the commutator of the two matrices which satisfies the well-known *Jacobi identity*

$$[X, [Y, Z]] + [Z, [X, Y]] + [Y, [Z, X]] = 0 \,. \tag{14.11}$$

This proves that the infinitesimal generators in the tangent space $T_e G$ form a Lie algebra. This statement also holds for more general Lie groups, whose elements are no longer matrices. However, the corresponding arguments require knowledge from differential geometry, and in particular about left-invariant vector fields on manifolds. More on this can be found in the appendix to this chapter.

### 14.1.2  Adjoint Representation and Center of a Group

The right-hand side of Eq. (14.9) assigns to each $g \in G$ a linear map $\mathrm{Ad}(g) : T_e G \mapsto T_e G$. We want to examine this more closely because it will play an important role in the following.

**Definition 14.1 (Adjoint Representation of a Lie Group)**  The linear mapping

$$\mathrm{Ad} : G \mapsto L(T_e G) \quad \text{with} \quad \mathrm{Ad}(g)X = gXg^{-1} \tag{14.12}$$

defines the adjoint representation of $G$ on the tangent space $T_e G$.

**Lie Algebra and Lie Group**
One should distinguish between the adjoint action (or conjugation) $\mathrm{Ad}_g$ acting on $G$, see (3.2), the just introduced adjoint representation of the group $\mathrm{Ad} : g \mapsto \mathrm{Ad}(g)$ on the linear space $T_e G$ and finally the adjoint representation ad of a Lie algebra on $\mathfrak{g}$, see (13.20). That these representations bear the same name is, of course, no coincidence (more on this below).

The representation properties of $\mathrm{Ad}$ follow from $\mathrm{Ad}(e) = \mathbb{1}$ and from

$$\mathrm{Ad}(gh)X = ghXh^{-1}g^{-1} = g\,(\mathrm{Ad}(h)X)\,g^{-1} = \mathrm{Ad}(g)\mathrm{Ad}(h)X . \tag{14.13}$$

The *adjoint representation* exists for every Lie group and its dimension is equal to that of the Lie group.

In general, the homomorphism $\mathrm{Ad}$ in (14.12) is not faithful. It is only faithful if the kernel$(\mathrm{Ad}) = \{e\}$ or if there is no group element $g \neq e$ with

$$\mathrm{Ad}(g)X \equiv gXg^{-1} = X \quad \text{for all} \quad X \in T_e G . \tag{14.14}$$

We recall that the *center* $\mathcal{Z}$ of a group contains those elements that commute with all other elements. Specifically, for $z \in \mathcal{Z}$, it holds that $zg(t)z^{-1} = g(t)$ for every curve in $G$. The derivative with respect to the curve parameter at $t = 0$ then leads to

$$zXz^{-1} = X . \tag{14.15}$$

In fact, the elements in $\mathcal{Z}$ are the only elements with this property. Therefore, $\mathrm{Ad}(g) = \mathbb{1}$ if and only if $g$ lies in the center of the group.

$$\mathrm{Kernel}(\mathrm{Ad}) = \mathcal{Z} \quad \text{or} \quad \mathrm{Ad}(g_1) = \mathrm{Ad}(g_2) \Longleftrightarrow g_2 = g_1 z, \quad z \in \mathcal{Z}. \tag{14.16}$$

**Lemma 14.1 (Kernel of Ad)** *The kernel of the adjoint representation is equal to the center of the group, Kernel*$(\mathrm{Ad}) = \mathcal{Z} \leq G$. *Thus, according to the isomorphism theorem,* $\mathrm{Ad}(G) = G/\mathcal{Z}$.

**Task**

Show that $\mathrm{Ad}(\mathrm{SU}(2)) = \mathrm{SU}(2)/\mathbb{Z}_2 = \mathrm{SO}(3)$.

The center of SU(2) consists of $\{\mathbb{1}, -\mathbb{1}\}$, and therefore the three-dimensional adjoint representation maps the two elements $\pm U$ to the same rotation.

**Example: The Lie Algebras $\mathfrak{u}(2)$ and $\mathfrak{su}(2)$ of U(2) and SU(2)**

Let $U(t) \in \mathrm{U}(2)$ be a curve with $U(0) = \mathbb{1}$. Differentiating the relation $U(t)U^\dagger(t) = \mathbb{1}_2$ with respect to $t$ at $t = 0$ leads to the following requirement for the generators $X = \dot{U}(0)$:

$$\dot{U}(0)\, U^\dagger(0) + U(0)\dot{U}^\dagger(0) = X + X^\dagger = 0. \tag{14.17}$$

Therefore, the elements of $\mathfrak{u}(2)$ are anti-Hermitian $2 \times 2$ matrices. If the curve lies in the subgroup $\mathrm{SU}(2) < \mathrm{U}(2)$, then additionally $\det U(t) = 1$ holds. Using

$$\frac{\mathrm{d}}{\mathrm{d}t} \log \det U = \mathrm{Tr}\left(U^{-1}\dot{U}\right) \tag{14.18}$$

the additional condition $\mathrm{Tr}\, X = 0$ follows, so that

$$\mathfrak{su}(2) = \{X \in \mathrm{Mat}(2, \mathbb{C})\,|\,X = -X^\dagger,\ \mathrm{Tr}\, X = 0\}. \tag{14.19}$$

Clearly, the commutator of two elements in $\mathfrak{su}(2)$ is again in $\mathfrak{su}(2)$,

$$[X, Y]^\dagger = -[X, Y] \quad \text{and} \quad \mathrm{Tr}[X, Y] = 0.$$

Anti-Hermitian and traceless $2 \times 2$ matrices are linear combinations of the Pauli matrices with imaginary coefficients,

$$\mathfrak{su}(2) = \left\{X = i\boldsymbol{a} \cdot \boldsymbol{\sigma}\,\middle|\,\boldsymbol{a} \in \mathbb{R}^3\right\}. \tag{14.20}$$

From the commutation rules for the Pauli matrices it follows that the Lie algebra $\mathfrak{su}(2)$ has rank one. It has only the trivial ideals $\emptyset$ and $\mathfrak{su}(2)$ and is thus simple.

The three-dimensional adjoint representation acts on the linear space of traceless anti-Hermitian matrices. It is equal to the rotation group in Euclidean space $E_3$.
◀

### 14.1.3  Representations of $G$ Induce Representations of $\mathfrak{g}$

Every representation $D$ of a Lie group induces a representation $\mathfrak{D}$ of its Lie algebra on the same vector space $\mathcal{V}$. To understand this, we again consider a curve $g(t)$ in $G$ through $e$ with image curve $D(g(t))$ in $\mathrm{GL}(\mathcal{V})$. The curve $g(t)$ has the tangent vector $X \in T_e G$. Due to the representation property, the image curve for $t = 0$ is the identity mapping, $D(g(0)) = D(e) = \mathbb{1}$. The vector tangent to the image curve at $t = 0$ is then the image of $X$ under the *induced representation* $\mathfrak{D}$ of the Lie algebra,

$$\frac{\mathrm{d}}{\mathrm{d}t} g(t)\Big|_{t=0} = X \implies \frac{\mathrm{d}}{\mathrm{d}t} D(g(t))\Big|_{t=0} = \mathfrak{D}(X). \tag{14.21}$$

$\mathfrak{D}(X)$ characterizes the infinitesimal change of $D(g)$ with an infinitesimal change of $g$ away from the neutral element, $D(e+\varepsilon X) \approx \mathbb{1}+\varepsilon\mathfrak{D}(X)$. From the representation property of $D$, it follows that $\mathfrak{D}$ is indeed a representation of the Lie algebra $T_e G$. The linearity results from (14.7):

$$\mathfrak{D}(\alpha X + \beta Y) = \frac{\mathrm{d}}{\mathrm{d}t} D\big(g(\alpha t)h(\beta t)\big)\Big|_{t=0}$$

$$= \frac{\mathrm{d}}{\mathrm{d}t} D\big(g(\alpha t)\big)\Big|_{t=0} D(e) + D(e)\frac{\mathrm{d}}{\mathrm{d}t} D\big(h(\beta t)\big)\Big|_{t=0}$$

$$= \alpha\mathfrak{D}(X) + \beta\mathfrak{D}(Y). \tag{14.22}$$

To prove that $\mathfrak{D}$ is a representation of the Lie algebra, we first note that

$$\mathfrak{D}\big(gYg^{-1}\big) = \frac{\mathrm{d}}{\mathrm{d}t} D\big(gh(t)g^{-1}\big)\Big|_{t=0}$$

$$= D(g)\frac{\mathrm{d}}{\mathrm{d}t} D\big(h(t)\big)\Big|_{t=0} D\big(g^{-1}\big) = D(g)\mathfrak{D}(Y)D^{-1}(g). \tag{14.23}$$

This relation holds for all elements of a curve $g(t)$ through $e$ with $\dot{g}(0) = X$. Now we differentiate (14.23) with respect to $t$ at the point $t = 0$. Due to the linearity of $\mathfrak{D}$, we may move the derivative past $\mathfrak{D}$ on the left side and obtain the expression $\mathfrak{D}([X, Y])$. The derivative of the right side, on the other hand, results in the commutator of $\mathfrak{D}(X)$ and $\mathfrak{D}(Y)$. Thus, we have proven that $\mathfrak{D}$ is a Lie algebra homomorphism:

$$\mathfrak{D}([X, Y]) = [\mathfrak{D}(X), \mathfrak{D}(Y)]. \tag{14.24}$$

We summarize these important results for representation theory:

**Induced Representation**

If $D : G \mapsto \mathrm{GL}(\mathcal{V})$ is a representation of a Lie group $G$, then $\mathfrak{D} : T_e G \mapsto L(\mathcal{V})$ defined in (14.21) is a representation of its Lie algebra $T_e G$, i.e.

$$\mathfrak{D} \text{ is linear,} \quad \mathfrak{D}\left([X, Y]\right) = [\mathfrak{D}(X), \mathfrak{D}(Y)] \quad \forall X, Y \in T_e G. \tag{14.25}$$

Thus, every representation of a Lie group induces a representation of its Lie algebra on the same vector space. As an intermediate result, we have derived a useful relation in (14.23), which we want to record as we will need it soon:

$$\mathfrak{D}\left(gYg^{-1}\right) = D(g)\mathfrak{D}(Y)D^{-1}(g), \qquad g \in G, \ Y \in T_e G. \tag{14.26}$$

Let $\{X_a\}$ be a basis of $T_e G$ with structure constants $f_{ab}{}^c$. Then the image vectors $\mathfrak{D}(X_a)$ obviously define a basis of the Lie algebra $\mathfrak{D}(T_e G)$ with the same structure constants,

$$[\mathfrak{D}(X_a), \mathfrak{D}(X_b)] = \mathfrak{D}\left([X_a, X_b]\right) = \mathfrak{D}\left(f_{ab}{}^c X_c\right) = f_{ab}{}^c \mathfrak{D}(X_c). \tag{14.27}$$

In physics, one often constructs representations of a group by studying the representations of its Lie algebra, see the discussions in Chaps. 17, 18, 19, 20, and 21.

**Example: Three-dimensional Representation of $\mathfrak{su}(2)$**

We choose the parameterization (9.9) for the matrices in SU(2) and choose $\vartheta$ as the curve parameter. Since $U(\vartheta = 0) = \mathbb{1}$, the derivative with respect to the curve parameter at $\vartheta = 0$ (for fixed $\psi$ and $\varphi$) is in $\mathfrak{su}(2)$. If the prime denotes the derivative with respect to the curve parameter, then we find the following linearly independent elements of the Lie algebra,

$$U'(0)\big|_{\psi=\varphi=\frac{\pi}{2}} = i\sigma_1, \quad U'(0)\big|_{\psi=\frac{\pi}{2},\varphi=0} = i\sigma_2, \quad U'(0)\big|_{\psi=\varphi=0} = i\sigma_3. \tag{14.28}$$

A three-dimensional representation of these basis vectors is found by differentiating the representation matrices (10.68) with respect to $\vartheta$ at $\vartheta = 0$ with values of $\psi, \varphi$ as in (14.28)

$$\mathfrak{D}(i\sigma_1) = \begin{pmatrix} 0 & i & 0 \\ i & 0 & i \\ 0 & i & 0 \end{pmatrix}, \quad \mathfrak{D}(i\sigma_2) = \begin{pmatrix} 0 & 1 & 0 \\ -1 & 0 & 1 \\ 0 & -1 & 0 \end{pmatrix}, \quad \mathfrak{D}(i\sigma_3) = \begin{pmatrix} i & 0 & 0 \\ 0 & 0 & 0 \\ 0 & 0 & -i \end{pmatrix}. \tag{14.29}$$

The change of basis from (12.15) to (12.16) leads to the equivalent real representation

$$\mathfrak{D}(i\sigma_1) = \begin{pmatrix} 0 & 0 & 0 \\ 0 & 0 & 2 \\ 0 & -2 & 0 \end{pmatrix}, \quad \mathfrak{D}(i\sigma_2) \begin{pmatrix} 0 & 0 & 2 \\ 0 & 0 & 0 \\ -2 & 0 & 0 \end{pmatrix}, \quad \mathfrak{D}(i\sigma_3) \begin{pmatrix} 0 & -2 & 0 \\ 2 & 0 & 0 \\ 0 & 0 & 0 \end{pmatrix}.$$

(14.30)

The latter generate the rotations (12.16) around the coordinate axes in Euclidean space. ◄

**Task**

Show that the generators $i\sigma_a$ and $\mathfrak{D}(i\sigma_a)$ satisfy identical commutation relations.

Representations of groups with invariant averaging can be chosen to be unitary, see Theorem 10.2 in Sect. 10.4. Then the generators $\mathfrak{D}(X)$ are anti-Hermitian (the proof is identical to the argument (14.17)) and we can define a scalar product on $\mathfrak{g}$,

$$(X, Y) = \mathrm{Tr}\left(\mathfrak{D}^\dagger(X)\mathfrak{D}(Y)\right) = -\mathrm{Tr}\left(\mathfrak{D}(X)\mathfrak{D}(Y)\right) \geq 0.$$

(14.31)

In particular, the adjoint representation is unitary (even orthogonal), as can be verified with the help of (14.26). This is referred to as the Ad-invariance of the scalar product,

$$(\mathrm{Ad}(g)X, \mathrm{Ad}(g)Y) = (X, Y), \quad \mathrm{Ad}(g)X = gXg^{-1}.$$

(14.32)

The representation of the Lie algebra on $T_e G$ induced by $\mathrm{Ad} : G \mapsto \mathrm{GL}(T_e G)$ is, according to (14.10), equal to the adjoint representation ad. The associated scalar product (14.31) is the Killing form $K(X, Y)$ introduced in Sect. 13.4. More generally, for a semi-simple Lie group, the scalar product (14.31) is proportional to the non-degenerate Killing form for every irreducible representation. On the Lie algebra of a compact (not necessarily semi-simple) Lie group, $K(X, Y)$ is negative semidefinite.

## 14.1.4  Tensor Product of Representations

In Sect. 10.5, we convinced ourselves that the tensor product of two arbitrary representations $D_1 : G \mapsto \mathrm{GL}(\mathcal{V}_1)$ and $D_2 : G \mapsto \mathrm{GL}(\mathcal{V}_2)$ of a group $G$ defines a representation

$$D = D_1 \otimes D_2 : g \mapsto D_1(g) \otimes D_2(g)$$

(14.33)

on the tensor product $\mathcal{V}_1 \otimes \mathcal{V}_2$ of the two representation spaces. In most cases, this representation is reducible, even if $D_1$ and $D_2$ were irreducible. If $G$ is a Lie group, then the two representations of the group induce representations $\mathfrak{D}_1$ and $\mathfrak{D}_2$ of the associated Lie algebra $T_e G$ on $\mathcal{V}_1$ and $\mathcal{V}_2$. To characterize the representation induced by the tensor product $D_1 \otimes D_2$, we consider a curve $g(s) \subset G$ with tangent vector $X$ at $g(0) = e$. When differentiating $D(g(s))$ at the point $s = 0$, we use the product rule as well as $D_i(e) = \mathbb{1}$ and find

$$(\mathfrak{D}_1 \otimes \mathfrak{D}_2)(X) = \frac{\mathrm{d}}{\mathrm{d}s}\bigg|_{s=0} D_1(g(s)) \otimes D_2(g(s)) = \mathfrak{D}_1(X) \otimes \mathbb{1}_{\mathcal{V}_2} + \mathbb{1}_{\mathcal{V}_1} \otimes \mathfrak{D}_2(X) .$$

$$(14.34)$$

We summarize:

**Corollary 14.2** *If the representations $D_1$ and $D_2$ of $G$ induce the representations $\mathfrak{D}_1$ and $\mathfrak{D}_2$ of the Lie algebra of $G$, then $D = D_1 \otimes D_2$ induces the representation $\mathfrak{D}_1 \otimes \mathfrak{D}_2$, given by*

$$X \mapsto (\mathfrak{D}_1 \otimes \mathfrak{D}_2)(X) = \mathfrak{D}_1(X) \otimes \mathbb{1}_{\mathcal{V}_2} + \mathbb{1}_{\mathcal{V}_1} \otimes \mathfrak{D}_2(X) . \tag{14.35}$$

In applications, the tensor products with the identities are often not written out, and one writes briefly $(\mathfrak{D}_1 \otimes \mathfrak{D}_2)(X) = \mathfrak{D}_1(X) + \mathfrak{D}_2(X)$, where the first summand acts on $\mathcal{V}_1$ and the second on $\mathcal{V}_2$. In this sense, a generator in the tensor product $\mathfrak{D}_1 \otimes \mathfrak{D}_2$ of two representations is equal to the sum of the generators in the individual representations.

---

**Task**

Prove that the following identity for the representation $\mathfrak{D} = \mathfrak{D}_1 \otimes \mathfrak{D}_2$ of $\mathfrak{g}$ holds,

$$[\mathfrak{D}(X), \mathfrak{D}(Y)] = [\mathfrak{D}_1(X) + \mathfrak{D}_2(X), \mathfrak{D}_1(Y) + \mathfrak{D}_2(Y)] = \mathfrak{D}([X, Y]) . \tag{14.36}$$

---

Tensor products play an important role in applications—for example, in many-particle physics—see Sect. 17.1. The vector spaces $\mathcal{V}_i$ are then the Hilbert spaces of subsystems of a total system with Hilbert space $\mathcal{V}_1 \otimes \mathcal{V}_2$.

---

## 14.2   Lie Algebras of Classical Matrix Groups

The Lie algebra of $\mathrm{GL}(n, \mathbb{C})$ is denoted by $\mathfrak{gl}(n, \mathbb{C})$. If $g(t)$ is a curve in $\mathrm{GL}(n, \mathbb{C})$ with $g(0) = \mathbb{1}$, then the corresponding tangent vector $X = \dot{g}(0)$ is an arbitrary complex $n \times n$ matrix. Therefore, the Lie algebra

$$\mathfrak{gl}(n, \mathbb{C}) = \{X = \mathrm{Mat}(n, \mathbb{C})\} \tag{14.37}$$

has the real dimension $2n^2$. The Lie product for two matrices is equal to the commutator. Similarly, for the Lie algebra of the general linear group with *real* coefficients, one finds

$$\mathfrak{gl}(n, \mathbb{R}) = \{X = \mathrm{Mat}(n, \mathbb{R})\}, \quad \dim \mathfrak{gl}(n, \mathbb{R}) = n^2 . \tag{14.38}$$

Matrix groups (classical Lie groups) are closed subgroups of $G = \mathrm{GL}(n, \mathbb{C})$, see Sect. 8.3.1. If $H$ is such a subgroup, then the tangent vectors of curves through $e \in H \leq G$ define a linear subspace $T_e H$ of $T_e G$. Repeating the arguments in Sect. 14.1.1 for elements and curves in $H$, one sees that $T_e H$ is a Lie subalgebra of $T_e G$.

If the subgroup is a normal subgroup $N \trianglelefteq G$, then the Lie subalgebra $T_e N$ is even an ideal of $T_e G$. To see this, consider a curve $n(t) \in N$ through $e$. Then for each group element, the conjugate curve $gn(t)g^{-1}$ also lies in $N$ and

$$g(T_e N)g^{-1} \subseteq T_e N . \tag{14.39}$$

This relation holds for all elements $g(t)$ of a curve through in $G$. Differentiating with respect to $t$ leads, due to the linearity of $T_e N$, to

$$[T_e G, T_e N] \subseteq T_e N , \tag{14.40}$$

which means that $T_e N$ is an ideal in $T_e G$, i.e. $T_e N \trianglelefteq T_e G$.

**Lemma 14.2 (Tangent Spaces of Lie Subgroups and Normal Subgroups)** *If $H$ is a Lie subgroup of $G$, then $T_e H$ is a Lie subalgebra of $T_e G$. If $N \trianglelefteq G$ is a normal Lie subgroup, then $T_e N$ is an ideal of $T_e G$.*

> **Generalization for General Lie Groups**
> We have shown the lemma for the subgroups and normal subgroups of matrix groups. However, it also holds for Lie subgroups and normal Lie subgroups of a general Lie group (this is the content of Theorems 14.3 and 14.4 in Sect. 14.7).

**The Lie Algebra $\mathfrak{sl}(n, \mathbb{K})$ of $\mathbf{SL}(n, \mathbb{K})$**

According to Lemma 14.2, the Lie algebras of the matrix groups are Lie subalgebras of $\mathfrak{gl}(n, \mathbb{K})$. Accordingly, we now consider a curve $g(t)$ in the special linear group $\mathrm{SL}(n, \mathbb{K})$. The condition $\det g(t) = 1$ implies for the infinitesimal generator $X = \dot{g}(0)$ the condition $\mathrm{Tr}\, X = 0$ due to (14.18). Therefore, the Lie algebra contains the traceless matrices,

$$\mathfrak{sl}(n, \mathbb{K}) = \{X \in \mathrm{Mat}(n, \mathbb{K}) | \mathrm{Tr}\, X = 0\} . \tag{14.41}$$

For $\mathbb{K} = \mathbb{R}$, the (real) dimension of this Lie algebra is equal to $\dim(\mathfrak{sl}(n, \mathbb{R})) = n^2 - 1$. Since $\mathrm{SL}(n, \mathbb{K})$ is a normal subgroup of $\mathrm{GL}(n, \mathbb{K})$, $\mathfrak{sl}(n, \mathbb{K})$ is an ideal of $\mathrm{GL}(n, \mathbb{K})$. We had already established this in Sect. 13.2.

**The Lie Algebra $\mathfrak{so}(n)$ of $\mathbf{SO}(n)$**
We consider a differentiable curve $R(t)$ in $\mathrm{O}(n)$. A continuous curve with $R(0) = \mathbb{1}$ will lie for all $t$ in the connected component $\mathrm{SO}(n) < \mathrm{O}(n)$. A generator satisfies

$$0 = \frac{\mathrm{d}}{\mathrm{d}t}\left(R^T(t)R(t)\right)\Big|_{t=0} = X^T + X \,,$$

and the Lie algebra of $\mathrm{SO}(n)$ is equal to the set of antisymmetric real matrices,

$$\mathfrak{so}(n) = \left\{X \in \mathrm{Mat}(n, \mathbb{R}) | X = -X^T\right\}. \tag{14.42}$$

Its dimension is equal to the number of linearly independent skew-symmetric matrices, $\dim(\mathfrak{so}(n)) = n(n - 1)/2$.

**The Lie Algebra $\mathfrak{u}(n)$ of $\mathbf{U}(n)$**
Due to $U^\dagger U = \mathbb{1}$, the infinitesimal generators satisfy $X^\dagger + X = 0$, and

$$\mathfrak{u}(n) = \left\{X \in \mathrm{Mat}(n, \mathbb{C}) | X = -X^\dagger\right\}. \tag{14.43}$$

The dimension of the Lie algebra is equal to the number of linearly independent anti-Hermitian matrices, thus $\dim(\mathfrak{u}(n)) = n^2$.

**The Lie Algebra $\mathfrak{su}(n)$ of $\mathbf{SU}(n)$**
This Lie algebra contains all generators of $\mathrm{U}(n)$ with vanishing trace,

$$\mathfrak{su}(n) = \{X \in \mathfrak{u}(n) | \mathrm{Tr}\, X = 0\} \,. \tag{14.44}$$

It is an ideal in $\mathfrak{u}(n)$ and has the dimension $\dim(\mathfrak{su}(n)) = n^2 - 1$. We studied the special case $n = 2$ in the last section.

**The Lie Algebra $\mathfrak{sp}(2n, \mathbb{K})$ of $\mathbf{Sp}(2n, \mathbb{K})$ with $\mathbb{K} = \mathbb{R}, \mathbb{C}$**
By differentiating the defining relation $M(t)^T J M(t) = J$ with respect to $t$, we obtain the following set of infinitesimal generators,

$$\mathfrak{sp}(2n, \mathbb{K}) = \left\{X \in \mathrm{Mat}(2n, \mathbb{K}) | X^T J + J X = 0\right\}. \tag{14.45}$$

The real Lie algebra $\mathfrak{sp}(2n, \mathbb{R})$ has the dimension $n(2n + 1)$.

The results obtained for the classical Lie groups are summarized in Table 14.1.

**Table 14.1** The matrix groups and their Lie algebras. In the second-last column, the generators are characterized. In the last column, one finds the real dimension of the group

| Group | Lie algebra | Generators | Real dimension |
|---|---|---|---|
| $GL(n, \mathbb{C})$ | $\mathfrak{gl}(n, \mathbb{C})$ | $X$ complex | $2n^2$ |
| $GL(n, \mathbb{R})$ | $\mathfrak{gl}(n, \mathbb{R})$ | $X$ real | $n^2$ |
| $SL(n, \mathbb{C})$ | $\mathfrak{sl}(n, \mathbb{C})$ | $\operatorname{Tr} X = 0$ | $2n^2 - 2$ |
| $SL(n, \mathbb{R})$ | $\mathfrak{sl}(n, \mathbb{R})$ | $X$ real, $\operatorname{Tr} X = 0$ | $n^2 - 1$ |
| $U(n)$ | $\mathfrak{u}(n)$ | $X + X^\dagger = 0$ | $n^2$ |
| $SU(n)$ | $\mathfrak{su}(n)$ | $X + X^\dagger = 0$, $\operatorname{Tr} X = 0$ | $n^2 - 1$ |
| $O(n), SO(n)$ | $\mathfrak{so}(n)$ | $X$ real, $X + X^T = 0$ | $n(n-1)/2$ |
| $Sp(2n, \mathbb{C})$ | $\mathfrak{sp}(2n, \mathbb{C})$ | $JX + X^T J = 0$ | $2n(2n + 1)$ |
| $Sp(2n, \mathbb{R})$ | $\mathfrak{sp}(2n, \mathbb{R})$ | $X$ real, $JX + X^T J = 0$ | $n(2n + 1)$ |

## 14.3   The Exponential Map

The Lie algebra of a Lie group is the tangent space at the identity element, and its elements, the infinitesimal generators, are obtained by differentiating curves that pass through the group identity. The reverse route from the Lie algebra to the Lie group leads via the exponential map. This can be defined for arbitrary Lie groups. Here we discuss this map for matrix groups.

Let $X$ be an $n$-dimensional real or complex matrix. We define the exponential $\exp(X)$ of this matrix via the series expansion

$$e^X = \sum_{n=0}^{\infty} \frac{X^n}{n!} \,. \tag{14.46}$$

This definition is meaningful because

**Lemma 14.3** *For any matrix $X \in Mat(n, \mathbb{K})$, the series (14.46) converges absolutely and is a continuous function of $X$.*

We denote the vectors in $\mathbb{K}^n$ by $\varphi$. A possible norm for matrices is

$$\|X\| = \sup_{\varphi \neq 0} \frac{\|X\varphi\|}{\|\varphi\|} \,. \tag{14.47}$$

For any matrix, this norm is finite, and the inequalities

$$\|XY\| \leq \|X\| \, \|Y\| \quad \text{and} \quad \|X + Y\| \leq \|X\| + \|Y\| \,.$$

hold. Iterating the first inequality, we obtain $\|X^n\| \leq \|X\|^n$, so that

$$\sum_n \frac{1}{n!} \|X^n\| \leq \sum_n \frac{1}{n!} \|X\|^n = e^{\|X\|} < \infty \,. \tag{14.48}$$

This means that the series (14.46) is a Cauchy series and therefore converges absolutely. The matrix exponential function has the following easily proven properties:

**Lemma 14.4** *Let $X, Y \in Mat(n, \mathbb{K})$. Then the following hold:*

1. *If $[X, Y] = 0$, then $\mathrm{e}^{X+Y} = \mathrm{e}^X \mathrm{e}^Y = \mathrm{e}^Y \mathrm{e}^X$.*
2. *The element $\mathrm{e}^X$ is invertible and has $\mathrm{e}^{-X}$ as its inverse.*
3. *For an invertible matrix $g$, the following applies:  $\mathrm{e}^{gXg^{-1}} = g\,\mathrm{e}^X g^{-1}$.*
4. *The inequality $\|\mathrm{e}^X\| \le \mathrm{e}^{\|X\|}$ holds.*
5. *$g(t) = \mathrm{e}^{tX}$ is a smooth curve in $GL(n, \mathbb{K})$ and $\frac{\mathrm{d}}{\mathrm{d}t}\,g(t) = X\,\mathrm{e}^{tX}$.*

The first property is verified by multiplying the power series. The second property follows immediately from it when setting $Y = -X$. The third property results from $gX^n g^{-1} = (gXg^{-1})^n$, and the fourth is (essentially) the inequality (14.48). The last property follows from the term-by-term differentiation of the power series for $\exp(tX)$.

If $X$ is a diagonalizable matrix, then there exists an invertible matrix $g$ such that

$$X = gDg^{-1}, \quad \text{with} \quad D = \mathrm{diag}(\lambda_1, \lambda_2, \ldots, \lambda_n). \tag{14.49}$$

Using the third property in the lemma, we now obtain

$$\mathrm{e}^X = g\,\mathrm{e}^D g^{-1}, \quad \mathrm{e}^D = \mathrm{diag}\left(\mathrm{e}^{\lambda_1}, \mathrm{e}^{\lambda_2}, \ldots, \mathrm{e}^{\lambda_n}\right). \tag{14.50}$$

Even for real matrices $X$, both $g$ and $D$ can be complex. Now let us consider a special class of curves $g(t)$ that pass through the neutral element for $t = 0$:

**Definition 14.2 (One-parameter Subgroups)** A function $g : \mathbb{R} \to G$ is called a one-parameter subgroup of $G$ if

1. $g$ is continuous,
2. $g(0) = e$ and $g(t + s) = g(t)g(s) = g(s)g(t)$  for all  $t, s \in \mathbb{R}$.

From the last property we conclude $g^{-1}(t) = g(-t)$. The following lemma applies:

**Lemma 14.5** *For a one-parameter subgroup of $G$, there exists a unique $X \in T_e G$ with $g(t) = \exp(tX)$.*

**Proof** A one-parameter subgroup satisfies the ordinary differential equation

$$\frac{\mathrm{d}}{\mathrm{d}t}g(t) = \frac{\mathrm{d}}{\mathrm{d}s}g(s + t)\Big|_{s=0} = \frac{\mathrm{d}}{\mathrm{d}s}g(s)\Big|_{s=0}\,g(t) = Xg(t) \quad \text{with} \quad X = \dot{g}(0) \in T_e G \tag{14.51}$$

and initial condition $g(0) = \mathbb{1}$. The unique solution to this initial value problem is $g(t) = \exp(tX)$.

**Lemma 14.6** *If D is a representation of G and $g = \exp(X)$, then*

$$D(g) = D\left(e^X\right) = e^{\mathfrak{D}(X)} \, . \tag{14.52}$$

***Proof*** The argument is similar to the previous one: We set $g = \exp(X)$ and consider the one-parameter subgroup $g(t) = \exp(tX)$ and $D(t) = D(g(t))$. Then

$$\frac{\mathrm{d}}{\mathrm{d}t}D(t) = \frac{\mathrm{d}}{\mathrm{d}s}D(s)|_{s=0}D(t) = \mathfrak{D}(X)D(t), \quad D(0) = \mathbb{1} \, . \tag{14.53}$$

The unique solution to the initial value problem is $D(t) = \exp(t\,\mathfrak{D}(X))$. Setting $t = 1$, the statement of the lemma follows.

---

### Example: Exponential Mapping for SU(2)

To exponentiate the matrices $X = i\boldsymbol{v} \cdot \boldsymbol{\sigma}$ in $\mathfrak{su}(2)$, we use $(\boldsymbol{v}\boldsymbol{\sigma})^2 = v^2\mathbb{1}$, where $v$ denotes the magnitude of $\boldsymbol{v}$, and obtain

$$\exp(X) = \exp(i\boldsymbol{v}\boldsymbol{\sigma}) = \cos v \, \sigma_0 + i \sin v \, \hat{\boldsymbol{v}}\boldsymbol{\sigma}, \quad \hat{\boldsymbol{v}} = \frac{\boldsymbol{v}}{v} \, . \tag{14.54}$$

In this formula, we set

$$\cos v + i\hat{v}_3 \sin v = a \quad \text{and} \quad \left(i\hat{v}_1 - \hat{v}_2\right)\sin v = b \, , \tag{14.55}$$

then $\exp(X)$ is the SU(2)-matrix in (8.7). The normalization condition $|a|^2 + |b|^2 = 1$ is automatically satisfied for the parameterization (14.55). We conclude that every SU(2)-matrix can be written as the exponent of an anti-Hermitian and traceless matrix. Thus, for SU(2), we have demonstrated the connection between the Lie algebra and the Lie group with an explicit calculation. The one-parameter subgroup for $X$ is

$$\exp(tX) = \cos(tv) \, \sigma_0 + i \sin(tv) \, \hat{\boldsymbol{v}}\boldsymbol{\sigma} \, . \tag{14.56}$$

and, as expected, we have

$$\frac{\mathrm{d}}{\mathrm{d}t} \exp(tX)\Big|_{t=0} = i\boldsymbol{v} \cdot \boldsymbol{\sigma} = X \, . \tag{14.57}$$

◄

## 14.4   Weyl Group and Maximal Tori

A maximal compact, connected, and abelian subgroup of a compact Lie group is called a maximal torus. Such a subgroup is isomorphic to an $n$-dimensional torus $T = \mathbb{R}^n / \mathbb{Z}^n$ [60]. The dimension $n$ of a maximal torus $T \leq G$ is called the rank of the Lie group $G$ and is often denoted by $r$. If $T$ is a maximal torus, then the conjugate Lie subgroup $gTg^{-1}$ is also an abelian subgroup of $G$. Thus, there is not just one maximal torus. More precisely, the following holds:

**Theorem 14.1 (Maximal Tori)** *Two maximal tori $T$ and $T'$ in a compact and connected Lie group $G$ are conjugate to each other, $T' = gTg^{-1}$, and every element of $G$ is contained in a maximal torus.*

From the theorem, it follows that two maximal tori have the same dimension.

---

**Example: Maximal Tori for U($n$),  SU($n$) and SO($n$)**

For U($n$), the subgroup of diagonal matrices $T = \left\{ \mathrm{diag}(\,\mathrm{e}^{\mathrm{i}\theta_1}, \ldots,\, \mathrm{e}^{\mathrm{i}\theta_n})\,|\,\theta_i \in \mathbb{R} \right\}$ is a maximal torus. In other words, $T$ is the product of $n$ circles, so that U($n$) has rank $n$. If we additionally require $\sum \theta_i = 0$, then $T$ is a $(n-1)$-dimensional maximal torus of SU($n$). For SO($n$), the subgroup of block diagonal matrices with $2 \times 2$ blocks from SO(2) is a maximal torus. ◄

---

The theorem thus holds for all unitary representations of a compact (connected) Lie group. For abstract Lie groups, the theorem will not be proven here. An elegant proof uses the degree of the mapping

$$ F : G/T \times T \longmapsto G, \quad F(gT, t) = gtg^{-1}, \tag{14.58} $$

which is equal to the order of the Weyl group introduced below, as well as results from integration theory [60].

The normalizer $N_G(T)$ of a subgroup $T \leq G$ was already introduced in Definition 3.10 in Sect. 3.4. It is the largest subgroup in $G$ that contains $T$ as a normal subgroup, $T \trianglelefteq N_G(T) \leq G$.

**Definition 14.3 (Weyl Group of $G$)** Let $N_G(T)$ be the normalizer of a maximal torus $T \leq G$. Then $\mathcal{W} = N_G(T)/T$ is called the Weyl group of $G$.

The Weyl group $\mathcal{W}$ will play an important role in the discussion of the root system of a Lie algebra in Sect. 15.2 and the weights of its representations in Chap. 16. Before we continue with the general discussion, we consider the instructive example SU($n$).

### Example: Weyl Group of SU($n$)

Let $n$ be in the normalizer $N = N_{\mathrm{SU}(n)}(T)$ of the maximal torus $T = \{\mathrm{diag}(\mathrm{e}^{\mathrm{i}\theta_1}, \ldots, \mathrm{e}^{\mathrm{i}\theta_n})\}$, i.e., $ntn^{-1} = t'$ with diagonal $t, t' \in T$. Since $t$ and $t'$ are conjugate to each other, they have the same eigenvalues. However, the (along the diagonal) ordered eigenvalues of $t'$ can be a permutation $\pi$ of the ordered eigenvalues of $t$, and therefore

$$ntn^{-1} = t' \quad \text{or} \quad nt = t'n \implies n_{ij} \cdot \left(\mathrm{e}^{\mathrm{i}\theta_i} - \mathrm{e}^{\mathrm{i}\theta_{\pi(j)}}\right) = 0.$$

Every permutation is a product of transpositions, and it suffices to consider transpositions, e.g., the exchange of the angles $\theta_1$ and $\theta_2$. Clearly, $n \equiv n_{(1,2)}$ must have the following block form,

$$n_{(1,2)} = \begin{pmatrix} n_{\|} & 0 \\ 0 & n_{\perp} \end{pmatrix} \quad \text{with} \quad n_{\|} = \begin{pmatrix} 0 & \mathrm{e}^{\mathrm{i}s_{12}} \\ -\mathrm{e}^{\mathrm{i}s_{21}} & 0 \end{pmatrix}, \quad n_{\perp} = \mathrm{diag}\left(\mathrm{e}^{\mathrm{i}s_{33}}, \ldots, \mathrm{e}^{\mathrm{i}s_{nn}}\right).$$

If the non-vanishing phases $s_{ij}$ add up to zero, then $n_{(1,2)} \in \mathrm{SU}(n)$. Multiplying the matrix from the left with $t = \mathrm{diag}(\mathrm{e}^{\mathrm{i}\varphi_1}, \mathrm{e}^{\mathrm{i}\varphi_2}, \ldots, \mathrm{e}^{\mathrm{i}\varphi_n})$, the phases in $n_{(1,2)}$ shift according to $s_{ij} \mapsto s_{ij} + \varphi_i$. By appropriately choosing the angles $\varphi_i$, we can find an element $n_{(1,2)}$ with $s_{ij} = 0$ in each conjugacy class. For example, for $\mathrm{SU}(3)$, we can choose the following representatives $n_{(i,j)}$ in $N/T$, which swap the angles $\vartheta_i$ and $\vartheta_j$:

$$n_{(1,2)} = \begin{pmatrix} 0 & 1 & 0 \\ -1 & 0 & 0 \\ 0 & 0 & 1 \end{pmatrix}, \quad n_{(2,3)} = \begin{pmatrix} 1 & 0 & 0 \\ 0 & 0 & 1 \\ 0 & -1 & 0 \end{pmatrix}, \quad n_{(1,3)} = \begin{pmatrix} 0 & 0 & 1 \\ 0 & 1 & 0 \\ -1 & 0 & 0 \end{pmatrix}.$$

$$(14.59)$$

In the factor group $\mathcal{W} = N_{\mathrm{SU}(n)}(T)/T$, for example, $n_{(1,2)}^2$ and $\mathbb{1}$ are identified. Further representatives of $N/T$ are $n_{(1,2)}n_{(2,3)}$ and $n_{(1,2)}n_{(1,3)}$. With $n = \mathbb{1}$, $\mathcal{W} = N/T$ thus has 6 elements and is isomorphic to the permutation group $\mathcal{S}_3$. In Problem 14.3, we show that the Weyl group of $\mathrm{SU}(n)$ is isomorphic to the permutation group $\mathcal{S}_n$. ◀

For general compact Lie groups, the following holds:

**Lemma 14.7** *The Weyl group $\mathcal{W} \leq G$ is finite.*

**Proof** Let $T \leq G$ be a maximal torus with normalizer $N_G(T)$ and $N_0 \subseteq N_G(T)$ the connected component of the identity in the normalizer. First, we prove $N_0 = T$. Since $T$ is connected and contains the identity, it follows immediately that $T \subseteq N_0$. To prove $N_0 \subseteq T$, we consider the differential of the conjugation map $ntn^{-1}$ on the

Lie algebra $\mathfrak{h}$ of $T$ (this is the Cartan subalgebra $\mathfrak{h}$ introduced later in Sect. 15.2),

$$N_G(T) \mapsto \mathrm{Aut}(T) \cong \mathrm{GL}(r, \mathbb{Z}), \quad n \mapsto \mathrm{Ad}(n)\big|_{\mathfrak{h}} . \tag{14.60}$$

Here, $\mathrm{Ad}(n)H = nHn^{-1}$ for an element $H \in \mathfrak{h}$.

---

**Question**

The representatives in (14.59) can be identified as elements of $\mathrm{Aut}(T) \cong$ $\mathrm{GL}(3, \mathbb{Z})$ for U(3). How do you derive from this the elements of $\mathrm{Aut}(T) \cong$ $\mathrm{GL}(2, \mathbb{Z})$ for SU(3)? What does this have to do with the isomorphism $\mathrm{GL}(2, \mathbb{Z}_2) \cong \mathcal{S}_3$, see Sect. 2.2.

---

Because $\mathrm{GL}(r, \mathbb{Z})$ is discrete and $N_0$ is connected, the image of $N_0$ under this mapping is only the identity matrix (the identity in $N_0$ is mapped to the identity matrix). However, this means that all elements from $N_0$ commute with the elements from $\mathfrak{h}$. Then they also commute with all elements in a neighborhood of the identity of $T$, because the exponential map is a local diffeomorphism. Due to Theorem 8.3, they then commute with all elements of the connected Lie group $N_0$. Since $T$ is a maximal torus, we conclude $N_0 \subseteq T$, which then proves $N_0 = T$.

Thus, $\mathcal{W} = N_G(T)/N_0$ is isomorphic to the set of connected components of $N_G(T)$, which implies that $\mathcal{W}$ is compact ($N$ is compact) and discrete ($N_0$ is open, and thus also the cosets $gN_0$). This means, however, that $\mathcal{W}$ is finite.

---

## 14.5   Lorentz Group and Lorentz Algebra

The Lorentz transformations were discussed in Sect. 5.5. After choosing standard coordinates, they are parameterized by four-dimensional real matrices $\Lambda = (\Lambda^{\alpha}{}_{\beta})$ which satisfy the condition

$$\Lambda^{\alpha}{}_{\mu}\, \eta_{\alpha\beta}\, \Lambda^{\beta}{}_{\nu} = \eta_{\mu\nu} \tag{14.61}$$

with Lorentz metric $(\eta_{\mu\nu}) = \mathrm{diag}(1, -1, -1, -1)$. The Lie algebra $\mathfrak{so}(1, 3)$ of the Lorentz group SO(1, 3) is similar to the Lie algebra $\mathfrak{so}(4)$ of the orthogonal group SO(4) and plays a prominent role in relativistic physics. It is therefore worthwhile to study it in more detail.

As in Sect. 14.2, we consider a curve $\Lambda(t)$ with $\Lambda(0) = \mathbb{1}_4$ in the connected component $L^{\uparrow}_+$ and define the infinitesimal generator

$$\omega^{\alpha}{}_{\beta} = \frac{\mathrm{d}}{\mathrm{d}t}\Lambda^{\alpha}{}_{\beta}\Big|_{t=0} . \tag{14.62}$$

The derivative of the relation (14.61) at $t = 0$ yields the following linear condition for the matrices $\omega = (\omega^\mu_{\ \nu})$ in the Lorentz algebra $\mathfrak{so}(1, 3)$:

$$\omega^\alpha_{\ \mu} \eta_{\alpha\nu} + \eta_{\mu\beta} \omega^\beta_{\ \nu} \equiv \omega_{\nu\mu} + \omega_{\mu\nu} = 0 \,. \tag{14.63}$$

This means that the matrices in $\mathfrak{so}(1, 3)$, after lowering the contravariant index, are antisymmetric real $4 \times 4$ matrices. They form a six-dimensional vector space, whose elements can be parameterized as follows:

$$\omega(\boldsymbol{\alpha}, \boldsymbol{\theta}) = (\omega^\mu_{\ \nu}) = \begin{pmatrix} 0 & \alpha_1 & \alpha_2 & \alpha_3 \\ \alpha_1 & 0 & -\theta_3 & \theta_2 \\ \alpha_2 & \theta_3 & 0 & -\theta_1 \\ \alpha_3 & -\theta_2 & \theta_1 & 0 \end{pmatrix} \,. \tag{14.64}$$

Setting $\boldsymbol{\alpha} = 0$ and $\boldsymbol{\theta} = \theta \boldsymbol{e}$ with $\boldsymbol{e} \cdot \boldsymbol{e} = 1$, then the 1-parameter subgroup

$$\Lambda(\theta) = e^{\omega(\boldsymbol{0}, \boldsymbol{e})\theta}$$

describes rotations in Euclidean space around the fixed axis $\boldsymbol{e}$ with the angle $\theta$. It therefore contains neither a length contraction nor a time dilation and has the form

$$\Lambda = \begin{pmatrix} 1 & 0^T \\ 0 & R \end{pmatrix}, \quad R \in \mathrm{SO}(3) \,. \tag{14.65}$$

This SO(3) subgroup of the Lorentz group was thoroughly examined in Sect. 5.2.

We now focus on the other limiting case with $\boldsymbol{\theta} = 0$ and set $\boldsymbol{\alpha} = \alpha \boldsymbol{e}$ with $\boldsymbol{e} \cdot \boldsymbol{e} = 1$. Then for each fixed unit vector $\boldsymbol{e}$, the 1-parameter subgroup

$$\Lambda(\alpha) = e^{\omega(\boldsymbol{e}, \boldsymbol{0})\alpha} = \begin{pmatrix} \cosh(\alpha) & \sinh(\alpha) \cdot \boldsymbol{e}^T \\ \sinh(\alpha)\boldsymbol{e} & \mathbb{1}_3 + \big(\cosh(\alpha) - 1\big)\boldsymbol{e}\,\boldsymbol{e}^T \end{pmatrix} \tag{14.66}$$

describes Lorentz boosts in the direction of $\boldsymbol{e}$.

---

Task

Try to prove the last statement. Use the series representation of the exponential function.

As can be easily seen (e.g., by transforming into the rest frame of the particle), $\alpha$ and $\boldsymbol{e}$ are related to the relative velocity $\boldsymbol{v}$ of the two inertial frames as follows:

$$\cosh(\alpha) = \frac{1}{\sqrt{1 - \beta^2}} \equiv \gamma, \qquad \sinh(\alpha) \cdot \boldsymbol{e} = \gamma\,\boldsymbol{\beta}, \quad \boldsymbol{\beta} = \frac{\boldsymbol{v}}{c}, \quad \beta = |\boldsymbol{\beta}| \,. \tag{14.67}$$

For example, if the inertial frame $I$ moves relative to the inertial frame $I'$ with the velocity $v = v e_1$, then the Lorentz transformation is

$$\Lambda = \begin{pmatrix} \gamma & \gamma\beta & 0 & 0 \\ \gamma\beta & \gamma & 0 & 0 \\ 0 & 0 & 1 & 0 \\ 0 & 0 & 0 & 1 \end{pmatrix} , \tag{14.68}$$

and the coordinates transform according to

$$x'^0 = \gamma\left(x^0 + \beta x^1\right) \quad , \quad x'^2 = x^2$$

$$x'^1 = \gamma\left(\beta x^0 + x^1\right) \quad , \quad x'^3 = x^3. \tag{14.69}$$

If a particle is at rest at the origin of the inertial frame $I$, then it moves with the velocity $v$ in the 1-direction relative to the inertial frame $I'$.

## 14.5.1  Commutation Relations of the Lorentz Algebra $\mathfrak{so}(1, 3)$

We begin with the commutator of two generators, which are parameterized according to (14.64):

$$\left[\omega(\alpha, \theta), \omega(\alpha', \theta')\right] = \omega\left(\alpha \wedge \theta' - \alpha' \wedge \theta, \ \theta \wedge \theta' - \alpha \wedge \alpha'\right). \tag{14.70}$$

As a basis of the Lorentz algebra, we choose the matrices $K_i$ and $\Omega_i$ in the expansion

$$\omega(\alpha, \theta) = \sum_{i=1}^{3} \alpha_i K_i + \sum_{i=1}^{3} \theta_i \Omega_i . \tag{14.71}$$

The comparison with (14.64) leads to the following explicit form of the infinitesimal Lorentz boosts

$$K_1 = \begin{pmatrix} 0 & 1 & 0 & 0 \\ 1 & 0 & 0 & 0 \\ 0 & 0 & 0 & 0 \\ 0 & 0 & 0 & 0 \end{pmatrix} , \quad K_2 = \begin{pmatrix} 0 & 0 & 1 & 0 \\ 0 & 0 & 0 & 0 \\ 1 & 0 & 0 & 0 \\ 0 & 0 & 0 & 0 \end{pmatrix} , \quad K_3 = \begin{pmatrix} 0 & 0 & 0 & 1 \\ 0 & 0 & 0 & 0 \\ 0 & 0 & 0 & 0 \\ 1 & 0 & 0 & 0 \end{pmatrix} , \tag{14.72}$$

and the infinitesimal rotations in space,

$$\Omega_1 = \begin{pmatrix} 0 & 0 & 0 & 0 \\ 0 & 0 & 0 & 0 \\ 0 & 0 & 0 & -1 \\ 0 & 0 & 1 & 0 \end{pmatrix} , \quad \Omega_2 = \begin{pmatrix} 0 & 0 & 0 & 0 \\ 0 & 0 & 0 & 1 \\ 0 & 0 & 0 & 0 \\ 0 & -1 & 0 & 0 \end{pmatrix} , \quad \Omega_3 = \begin{pmatrix} 0 & 0 & 0 & 0 \\ 0 & 0 & -1 & 0 \\ 0 & 1 & 0 & 0 \\ 0 & 0 & 0 & 0 \end{pmatrix} . \tag{14.73}$$

Substituting the expansion (14.71) into the general commutation rule (14.70), we obtain the commutators of the basis elements:

**Lorentz Algebra**
The infinitesimal boosts and rotations in (14.72,14.73) satisfy the commutation rules

$$[K_i, K_j] = -\epsilon_{ijk}\Omega_k, \quad [\Omega_i, \Omega_j] = \epsilon_{ijk}\Omega_k, \quad [K_i, \Omega_j] = \epsilon_{ijk}K_k.$$
$$(14.74)$$

The Lorentz group is not compact. Its Lie algebra has rank 2 and possesses two linearly independent and commuting generators that can be simultaneously diagonalized, for example, $K_3$ and $\Omega_3$.

### 14.5.2 Another Basis of Generators

The above basis $\{K_i, \Omega_i\}$ emphasizes the different roles of spatial rotations on one hand and Lorentz boosts on the other. For the following structural investigation of the Lorentz algebra, we choose another basis and denote the 6 basis elements with $M_{\mu\nu} = -M_{\nu\mu}$. Note: here, $\mu$ and $\nu$ are not the indices of a matrix $M$, but each $M_{\mu\nu}$ is a matrix. An arbitrary element of the Lorentz algebra is then a linear combination of these matrices,

$$\frac{i}{2}\omega^{\mu\nu}M_{\mu\nu}, \quad M_{\mu\nu} = \left((M_{\mu\nu})^\rho{}_\sigma\right), \tag{14.75}$$

with real expansion coefficients $\omega^{\nu\mu} = -\omega^{\mu\nu}$. The factor i was introduced so that the (now imaginary) matrices $M_{\mu\nu}$ with spatial indices are Hermitian. In the linear combination, due to $M_{\nu\mu} = -M_{\mu\nu}$, each basis element appears twice. The condition (14.63) then reads

$$(M_{\mu\nu})_{\rho\sigma} = -(M_{\mu\nu})_{\sigma\rho}, \quad (M_{\mu\nu})_{\rho\sigma} = \eta_{\sigma\delta}(M_{\mu\nu})^\delta{}_\rho. \tag{14.76}$$

This means that the matrix $M_{\mu\nu}$ with covariant indices is antisymmetric. We choose the antisymmetric matrices

$$\left(M_{\mu\nu}\right)_{\rho\sigma} = -i\left(\eta_{\mu\rho}\eta_{\nu\sigma} - \eta_{\nu\rho}\eta_{\mu\sigma}\right). \tag{14.77}$$

Then $M_{0i} = iK_i$ and $M_{ij} = -i\epsilon_{ijk}\Omega_k$ hold. For example,

$$M_{01} = \begin{pmatrix} 0 & i & 0 & 0 \\ i & 0 & 0 & 0 \\ 0 & 0 & 0 & 0 \\ 0 & 0 & 0 & 0 \end{pmatrix} \quad \text{and} \quad M_{12} = \begin{pmatrix} 0 & 0 & 0 & 0 \\ 0 & 0 & i & 0 \\ 0 & -i & 0 & 0 \\ 0 & 0 & 0 & 0 \end{pmatrix} = M_{12}^\dagger \tag{14.78}$$

are two generators. The generators (14.77) satisfy the simple commutation relations

$$[M_{\mu\nu}, M_{\rho\sigma}] = i\left(\eta_{\mu\rho}M_{\nu\sigma} + \eta_{\nu\sigma}M_{\mu\rho} - \eta_{\mu\sigma}M_{\nu\rho} - \eta_{\nu\rho}M_{\mu\sigma}\right). \tag{14.79}$$

In the matrix product, a covariant index is contracted with a contravariant index, for example, $(M_{01}M_{12})^\rho{}_\sigma = (M_{01})^\rho{}_\delta (M_{12})^\delta{}_\sigma$. In the form (14.79), the commutation relations hold in any spacetime dimensions, and the matrices $M_{\mu\nu}$ in (14.77) form a basis of the Lorentz algebra $\mathfrak{so}(1, d-1)$.

---

**Task**

Verify that for the choice (14.77) of the basis elements $M_{\mu\nu}$, the following identification holds:

$$\frac{i}{2}\left(\omega^{\mu\nu}M_{\mu\nu}\right)^\rho{}_\sigma = \omega^\rho{}_\sigma \quad \text{or} \quad \frac{i}{2}(\omega, M) = \omega, \tag{14.80}$$

in agreement with (14.62). It may be somewhat confusing that on one hand, the $\omega^{\mu\nu}$ appear as expansion coefficients in $(\omega, M)$, and on the other hand, the matrix $(\omega^\rho{}_\sigma)$ is equal to the element $\frac{i}{2}(\omega, M)$ of the Lorentz algebra.

In 4 dimensions, the commutation relations (14.79) are equivalent to (14.74) and generalize the known commutation relations for the components of angular momentum. The latter are (up to an i) the generators of the rotation group, $J_i = i\Omega_i$. From (14.74), one obtains the following commutation rules for these Hermitian generators,

$$[J_i, J_j] = i\epsilon_{ijk}J_k. \tag{14.81}$$

The remaining generators $M_{0i}$ generate Lorentz boosts. An arbitrary Lorentz transformation in $L_+^\uparrow$ has the form

$$\Lambda = e^{\frac{i}{2}\omega^{\mu\nu}M_{\mu\nu}} = \mathbb{1} + \frac{i}{2}\omega^{\mu\nu}M_{\mu\nu} + O(\omega^2). \tag{14.82}$$

With the identification (14.80), this is consistent with (14.62).

---

**Task**

Determine the Killing form of the Lorentz algebra.

## 14.6  The Poincaré -Algebra

After the Lorentz transformations, we now consider the more general Poincaré transformations discussed in in Sect. 5.5. As in (5.48) we denote them by $(\Lambda, a)$. The normal subgroup of *translations* is generated by the infinitesimal translations $P_\mu$,

$$(\mathbb{1}, a) = \mathrm{e}^{\mathrm{i}\, a^\mu P_\mu} = \mathbb{1} + \mathrm{i}\, a^\mu P_\mu + O(a^2) \,. \tag{14.83}$$

The order of two translations does not matter so that their generators commute

$$[P_\mu, P_\nu] = 0, \qquad \mu, \nu = 0, \ldots, 3 \,. \tag{14.84}$$

The non-trivial commutation relations between infinitesimal Lorentz transformations and translations follow from

$$(\Lambda, 0)(\mathbb{1}, a)(\Lambda, 0)^{-1} = (\mathbb{1}, \Lambda a) \quad \text{bzw.} \quad \mathrm{e}^{\frac{\mathrm{i}}{2}\omega^{\mu\nu} M_{\mu\nu}} \, \mathrm{e}^{\mathrm{i}\, a^\rho P_\rho} \, \mathrm{e}^{-\frac{\mathrm{i}}{2}\omega^{\mu\nu} M_{\mu\nu}} = \mathrm{e}^{\mathrm{i}\,(\Lambda a)^\mu P_\mu} \,. \tag{14.85}$$

Here we set in the last exponent $\Lambda = \mathbb{1} + \frac{1}{2}\,\omega^{\mu\nu} M_{\mu\nu} + \ldots$ and compare terms linearly in $\omega$ and $a$. This yields

$$[\omega^{\mu\nu} M_{\mu\nu}, a^\rho P_\rho] = (\omega^{\mu\nu} M_{\mu\nu}\, a)^\sigma P_\sigma \,. \tag{14.86}$$

This applies to any $\omega^{\mu\nu}$ and $a^\rho$ and a comparison of coefficients leads to

$$[M_{\mu\nu}, P_\rho] = (M_{\mu\nu})^\sigma{}_\rho P_\sigma = \mathrm{i}\big(\eta_{\mu\rho} P_\nu - \eta_{\nu\rho} P_\mu\big). \tag{14.87}$$

In summary, we obtain the following commutation relations of the Poincaré algebra:

**Poincaré Algebra**
The generators of the Poincaré group $\{M_{\mu\nu}, P_\mu\}$ fulfill the commutation rules

$$[M_{\mu\nu}, M_{\rho\sigma}] = \mathrm{i}\big(\eta_{\mu\rho} M_{\nu\sigma} + \eta_{\nu\sigma} M_{\mu\rho} - \eta_{\mu\sigma} M_{\nu\rho} - \eta_{\nu\rho} M_{\mu\sigma}\big),$$

$$[M_{\mu\nu}, P_\rho] = \mathrm{i}\big(\eta_{\mu\rho} P_\nu - \eta_{\nu\rho} P_\mu\big), \tag{14.88}$$

$$[P_\mu, P_\nu] = 0 \,.$$

Because the translations form a normal subgroup of the Poincaré group, the infinitesimal translations define an ideal of the Poincaré algebra. The generators $P_\mu$ are the components of the 4-momentum.

### 14.6.1  Casimir Invariants of the Poincaré Algebra

The non-simple Poincaré algebra has the semi-simple Lorentz algebra as a sub-algebra and the infinitesimal translations as an abelian ideal. It can now be seen immediately that $P^2 = P_\mu P^\mu$ is a Casimir invariant of the Poincaré algebra:

**Lemma 14.8** $P^2 = P_\mu P^\mu$ *is a Casimir invariant of the Poincaré algebra.*

**Proof** Obviously, $[P^2, P_\mu] = 0$. In addition, with the Leibniz rule and the permutation relations (14.88) follows

$$[P^2, M_{\rho\sigma}] = P_\mu [P^\mu, M_{\rho\sigma}] + [P_\mu, M_{\rho\sigma}] P^\mu = 0. \tag{14.89}$$

More generally, the square $V^2 = V_\mu V^\mu$ of an arbitrary vector commutes with the $M_{\mu\nu}$, since the components of a vector $V$ have the same commutation relations with the generators of the Lorentz group as the components $P_\mu$ of the 4-momentum $P$. In other words, the square of every vector is a Lorentz invariant.

A second Casimir invariant can be defined using the vector $W$ from PAULI and LUBANSKI with components

$$W_\mu = \tfrac{1}{2}\epsilon_{\mu\nu\rho\sigma} P^\nu M^{\rho\sigma}, \quad (\epsilon_{0123} = 1). \tag{14.90}$$

This is orthogonal to the 4-momentum,

$$W_\mu P^\mu = 0, \tag{14.91}$$

and its time and space components have the form

$$W_0 = -P^i J_i, \quad W_i = P^0 J_i - \epsilon_{ijk} P^j K_k. \tag{14.92}$$

For a particle at rest, $P = (m, 0)$ and correspondingly $W = (0, mJ)$. Now follows the

**Lemma 14.9** *The square* $W^2 = W_\mu W^\mu$ *is a Casimir invariant of the Poincaré algebra.*

**Proof** Because of

$$[W_\mu, P_\xi] = \tfrac{1}{2}\epsilon_{\mu\nu\rho\sigma} P^\nu [M^{\rho\sigma}, P_\xi] = \mathrm{i}\epsilon_{\mu\nu\rho\xi} P^\nu P^\rho = 0 \tag{14.93}$$

commutes $W^2$ with the 4-momentum $P$. Moreover, since $W$ is a 4-vector, its square $W^2$ commutes with the infinitesimal Lorentz transformations $M_{\mu\nu}$. This then proves the lemma.

Apart from the quadratic Casimir invariants $P^2$ and $W^2$ there is no other independent invariant of the Poincaré algebra. Since the physical states in a relativistic theory are classified according to their transformation behaviour under the Poincaré symmetry, these invariants play a prominent role in the characterization of particles. They allow the classification of states according to their mass and spin. The invariant $P^2$ corresponds to the square of the mass, while $W^2$ is related to the spin of the particle.

## 14.7  General Lie Groups

For general Lie groups, you can prove analogous results as for matrix groups. The explicit exponential function is then replaced by the exponential mapping on differentiable manifolds. The one-parameter subgroups introduced in Definition 14.2 also exist for general Lie groups. The proofs of the following properties require knowledge of differential geometry and are therefore only sketched.

We start with the important result about the existence of a Lie group for a given Lie algebra $\mathfrak{g}$. First we recall that the mapping $X \mapsto \mathrm{ad}_X$ is a Lie algebra homomorphism from $\mathfrak{g}$ to $\mathfrak{gl}(\mathfrak{g})$. If the center $\mathfrak{z}_\mathfrak{g}$ of $\mathfrak{g}$ vanishes, then $\mathfrak{g}$ is isomorphic to the image $\mathrm{ad}_\mathfrak{g} \leq \mathfrak{gl}(\mathfrak{g})$ and the matrix subgroup $L_\mathfrak{g} \leq \mathrm{GL}(\mathfrak{g})$ generated by $\mathrm{ad}_\mathfrak{g}$— one speaks of the linear group belonging to $\mathfrak{g}$—in turn has $\mathrm{ad}_\mathfrak{g} \cong \mathfrak{g}$ as Lie algebra.

**Lemma 14.10**  *Let $\mathfrak{g}$ be a semi-simple Lie algebra with $\mathfrak{z} = 0$. Then $\mathfrak{g}$ is the Lie algebra of a Lie group, namely the linear group $L_\mathfrak{g}$ belonging to $\mathfrak{g}$.*

An extension of the theorem also applies to Lie algebras with non-trivial centers. This relates the image of the adjoint representation $Ad : G \mapsto \mathrm{GL}(n, \mathbb{R})$ with the linear group $L_\mathfrak{g}$ belonging to $\mathfrak{g}$. For matrix groups we had established this connection in Lemma 14.1. The same statement applies to general Lie groups:

**Theorem 14.2**  *Let $G$ be a connected Lie group with Lie algebra $\mathfrak{g}$. Then*

$$\mathrm{Ad}(G) = L_\mathfrak{g} \leq \mathrm{GL}(n, \mathbb{R}) \quad and \quad L_\mathfrak{g} \cong G/\mathcal{Z}. \tag{14.94}$$

*In particular, the Lie group $G/\mathcal{Z}$ is determined solely by the Lie algebra $\mathfrak{g}$ of $G$.*

Let $H \leq G$ be a Lie subgroup and $H \mapsto G$ the injective inclusion which defines an induced mapping of their Lie algebras. Consider the tangent space $T_e H$ as a linear subspace of $T_e G$. Since $H$ is a Lie group, $T_e H$ is invariant under the Lie bracket. This means that the theorem already known for matrix groups applies more generally:

**Theorem 14.3**  *The Lie algebra $\mathfrak{h}$ of a Lie subgroup $H < G$ is a Lie subalgebra of the Lie algebra $\mathfrak{g}$ of $G$.*

Similarly, one argues that a Lie normal subgroup leads to an ideal. The following theorem establishes the connection in the opposite direction

**Theorem 14.4** *Let $G$ be a connected Lie group with Lie algebra $\mathfrak{g}$ and $N < G$ a connected Lie subgroup with Lie algebra $\mathfrak{n} \leq \mathfrak{g}$. If $\mathfrak{n}$ is an ideal in $\mathfrak{g}$, then $N$ is a normal subgroup of $G$.*

## 14.8  Appendix A: Left-Invariant Vector Fields

When dealing with general Lie groups, one uses terms from differential geometry such as vector fields and differential forms on manifolds, pullback of differential forms, pushforward of vector fields as well as invariant forms and invariant vector fields. Some of these objects are outlined here. We assume that the underlying manifolds are smooth.

### 14.8.1  Vectors and Tangential Space

Let $X$ be a vector in $\mathbb{R}^n$ and $f : \mathbb{R}^n \to \mathbb{R}$ a differentiable mapping in the neighborhood of a point $x_0$. Then the directional derivative of $f$ in the direction of $X$ at the point $x_0$ is given by

$$D_X f = \sum_{\mu=1}^{n} \left. \frac{\partial f}{\partial x^\mu} \right|_{x_0} X^\mu \in \mathbb{R}. \tag{A.1}$$

The directional derivative is $\mathbb{R}$-linear and fulfills the *product rule*

$$D_X(f \cdot g) = D_X(f) \cdot g + f \cdot D_X(g). \tag{A.2}$$

Conversely, if $D_X = D_Y$ then $X = Y$ follows. To prove this, apply the derivatives to the coordinate functions $x^\mu$. The following property holds:

**Lemma 14.11** *For every $\mathbb{R}-$linear operator $D$ that fulfills the product rule, there is a vector $X$ (at the point $x_0$) with $D = D_X$.*

***Proof*** For a constant function $F = 1$ in a neighborhood of $x_0$ the product rule implies

$$D(1) = D(1 \cdot 1) = D(1) + D(1) \implies D(1) = 0.$$

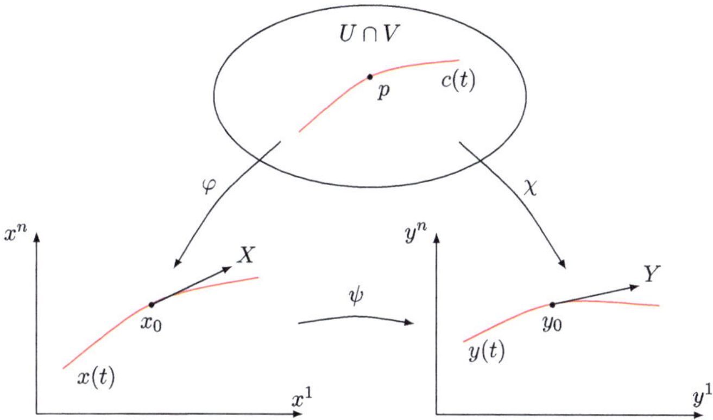


**Fig. 14.3**  A curve on the manifold and its images in two different coordinates

Due to linearity, $D$ then vanishes for all constant functions. Let $f$ now be a $C^\infty$-function in an neighborhood of $x_0 = 0$. Then we have

$$f(x) = f(0) + \sum_\mu \frac{\partial f}{\partial x^\mu}\Big|_0 x^\mu + \sum_{\mu\nu} g_{\mu\nu}(x)x^\mu x^\nu \quad \text{with} \quad g_{\mu\nu} \in C^\infty,$$

and due to linearity and the product rule

$$(Df)(0) = \sum_\mu \frac{\partial f}{\partial x^\mu}\Big|_0 D(x^\mu).$$

The choice $X^\mu = D(x^\mu)$ leads to the desired vector $X$ in $x_0$, for which $D_X f = Df$ holds.

After this preparation, we consider curves in a (differentiable) manifold $M$. Let $(U, \varphi)$ be a chart with coordinate domain $U$ containing $p$, see Fig. 14.3. Then a differentiable curve $c(t) \in U$, which passes through $p$ at $t = 0$, has the coordinate representation

$$x(t) = \varphi\,(c(t)) \in \mathbb{R}^n, \quad x(0) = \varphi(p) = x_0. \tag{A.3}$$

The coordinates $x^\mu(t)$ are differentiable functions of the curve parameter $t$. Then $\dot{x}(0) = X$ is the tangent vector of the curve at the point $x_0$. Conversely, for each $X$ there is a curve in $\mathbb{R}^n$ through $x_0$ with the tangent vector $X$. One speaks of a vector represented by the curve $x(t)$ or $c(t)$. Let $\bar{f}$ be a differentiable function in $U$ and $f = \bar{f} \circ \varphi^{-1}$ the same function expressed in local coordinates. Then

$$\bar{f}\,(c(t)) = f\,(x(t)) \implies \frac{\mathrm{d}}{\mathrm{d}t}\bar{f}(c(t))\Big|_0 = \frac{\mathrm{d}}{\mathrm{d}t}f(x(t))\Big|_0 = \frac{\partial f}{\partial x^\mu}\Big|_{x_0} X^\mu,$$

or

$$\frac{\mathrm{d}}{\mathrm{d}t}\bar{f}\left(c(t)\right)\Big|_0 = (D_X f)(x_0)\,. \tag{A.4}$$

The left side is independent of the chosen coordinates and leads to the definition

$$D_{\bar{X}}\bar{f} \equiv D_X f\,. \tag{A.5}$$

The derivative of the curve depends only on the tangent vector and is therefore independent of the chart. If $\psi = \chi \circ \varphi^{-1}$ is the coordinate transformation from $\varphi(U)$ to $\chi(V)$, then the following relationship holds between the vector components in the respective charts,

$$y(t) = \psi(x(t)) \implies Y^\mu = \frac{\partial \psi^\mu}{\partial x^\nu}X^\nu\,. \tag{A.6}$$

The derivative $D_{\bar{X}}$ is $\mathbb{R}$-linear with respect to $\bar{f}$ and satisfies the product rule. Two functions that agree in an arbitrarily small neighborhood of $p \in U$ have the same derivative. This leads to the concept of *germs of functions*. Two functions $\bar{f}$ and $\bar{g}$ have the same germ if they agree in a sufficiently small neighborhood of $p$. The set of all vectors $\bar{X}$ at point $p$ is denoted by $T_p M$. It is the *tangent space* at point $p$.

There exist equivalent definitions of $T_p M$.

1. In the geometric definition, one identifies an equivalence class of differentiable curves $c(t)$ with $c(0) = p$ as a tangent vector. Two curves are equivalent if they (in a chart) have the same tangent vector $(\varphi \circ c)'(0)$ at $p$.
2. In an algebraic definition, one identifies the set of derivations at point $p$ as the tangent space $T_p$. Derivations at $p$ are the linear mappings $\{D\}$, which satisfy the property

$$D(f \cdot g)\big|_p = (Df) \cdot g\big|_p + f \cdot (Dg)\big|_p \tag{A.7}$$

Derivations naturally form a real vector space.

The relationship between the geometric and algebraic definition is the following: if $c(t)$ is a curve with tangent vector $(\varphi \circ c)'(0) = X$ in a chart, then the associated derivation is the directional derivative $D_X$ in this chart, often denoted simply by $X$, so that $X(f)|_p = X^\mu \partial_\mu f|_p$ holds. In this sense, the partial derivatives $\partial_\mu$ at a point $p$ form a basis of $T_p M$ (in the chart with coordinates $x^\mu$).

A vector field is a (smooth) mapping that assigns to each $p \in M$ a vector $\bar{X}(p)$ in the vector space $T_p M$ or in a chart a derivation $X(x)$ to an $x \in \mathbb{R}^n$:

$$p \mapsto \bar{X}(p) \in T_p(M), \quad \text{or} \quad x \mapsto X^\mu(x)\partial_\mu \quad \text{with smooth} \quad X^\mu\,. \tag{A.8}$$

The infinite-dimensional linear space of vector fields is denoted by $\mathcal{T}M$.

### Lie Algebra of Vector Fields

The vector fields on $M$ form an infinite-dimensional Lie algebra. Here, the Lie bracket of $X, Y \in \mathcal{T}M$ is the commutator of their derivations:

$$[X, Y] : f \longmapsto X(Y(f)) - Y(X(f)) = [X, Y](f) . \tag{A.9}$$

The explicit calculation leads to the expression

$$[X, Y] = \left( X^\nu \frac{\partial X^\mu}{\partial x^\nu} - Y^\nu \frac{\partial X^\mu}{\partial x^\nu} \right) \partial_\mu . \tag{A.10}$$

**Lemma 14.12 (Lie Algebra of Vector Fields)** *The Lie bracket $[., .] : \mathcal{T}M \times \mathcal{T}M \to \mathcal{T}M$ is bilinear, antisymmetric, satisfies the Jacobi identity and the relation*

$$[fX, gY] = fg[X, Y] + f(Xg)Y - g(Yf)X . \tag{A.11}$$

### 14.8.2  The Mappings Pullback and Pushforward

Let $c(t)$ be a curve through $p \in M$ and $F : M \to N$ a diffeomorphism between manifolds. Then the image curve $F(c(t))$ describes a curve through $q = F(p)$ and the relation of the vector components $X^\mu$ and $Y^\mu$ is as in a coordinate change (A.6). The corresponding *linear mapping* from tangent vectors at $p$ to tangent vectors at $q$ is denoted by $F_* : T_pM \mapsto T_qN$. Now we want to establish the relationship between the derivatives in $M$ and $N$. For this, we introduce the so-called *pullback* of functions on $N$:

**Definition 14.4** Let $F : M \to N$ be a diffeomorphism and $g : N \to \mathbb{R}$ a smooth function. Then the pullback $F^*g$ of $g$ is defined by $(F^*g)(p) = g(F(p)) = g(q)$.

Non-injective functions can only be "pulled back" and not "pushed forward". The pullback $F^*$ is $\mathbb{R}$-linear. Now the following holds

**Lemma 14.13** *The derivative of $F^*g$ in the direction $X$ is equal to the derivative of $g$ in the direction $F_*X$, i.e. $X(F^*g) = (F_*X)(g)$.*

---

**Task**

Prove this lemma. You only need the definition of the pullback and the directional derivative.

**Left-Invariant Vector Fields of a Lie Group**

Let $L_g$ be the left translation on a Lie group $G$. The corresponding pushforward of vector fields of $G$ is then done with $(L_g)_*$. A vector field $X$ is called left-invariant if

$$(L_g)_* X = X \quad \text{for all} \quad g \in G. \tag{A.12}$$

In other words, for all real $C^\infty$ functions on $G$, the relation $X(L_g^* f) = X(f)$ holds. If a left-invariant vector field is known for only one $g_0 \in G$, then the vector $X(g_0)$ determines the entire vector field. Since the pushforwards are linear operations, the left-invariant vector fields form a linear subspace in the space of vector fields. Since the pushforwards also commute with the Lie bracket, the Lie bracket of left-invariant vector fields is again left-invariant.

With the help of the pushforward of vector fields, one defines the (related by duality) pullback of $p$-forms. The pullback $F^*$ of a map $F : M \mapsto N$, given by

$$(F^*\omega)(X_1, \ldots, X_p) = \omega(F_* X_1, \ldots, F_* X_p), \tag{A.13}$$

maps $p$-forms on $N$ to $p$-forms on $M$.

A $p$-form on a Lie group with the property $L_g^* \omega = \omega$ is called left-invariant. Obviously, the function $\omega(X_1, \ldots, X_p)$ is constant when a left-invariant $p$-form is evaluated on left-invariant vector fields.

## 14.9  Exercises for Chap. 14

**Problem 14.1 (Variation of the Determinant)**  Let $A(t)$ be a curve in the space of invertible matrices. Prove the identity

$$\frac{\mathrm{d}}{\mathrm{d}t} \log \det A = \mathrm{Tr}\left(A^{-1}\dot{A}\right).$$

**Problem 14.2 (Exponential Mapping)**  The exponential mapping $X \mapsto \mathrm{e}^X$ (for matrices or also general Lie groups) is differentiable and its differential at $0$ is the identity. Why does it follow that the exponential mapping is a local diffeomorphism? Hint: Recall the theorem about inverse functions.

**Problem 14.3 (Weyl Group of SU($n$))**  Prove that SU($n$) has the Weyl group $\mathcal{S}_n$. For SU(3) the mappings in Aut($T$) can be identified with matrices in GL(2, $\mathbb{Z}$), see (14.60). Which matrices are these?

**Problem 14.4 (Weyl Group of SO($n$))**  Determine the Weyl groups of the orthogonal groups SO($n$).

**Problem 14.5 (Lorentz Group in** $2$ **Spacetime Dimensions)** We investigate proper Lorentz transformations $x^\mu \mapsto \Lambda^\mu{}_\nu x^\nu$ in the two-dimensional Minkowski space with metric $(\eta_{\mu\nu}) = \mathrm{diag}(1, -1)$.

1. Show that the proper Lorentz transformations have the form

$$\Lambda = \begin{pmatrix} \cosh\alpha & -\sinh\alpha \\ -\sinh\alpha & \cosh\alpha \end{pmatrix}$$

   and form an abelian group.
2. Characterize the Lorentz algebra $\mathfrak{so}(1, 1)$.
3. The irreducible representations of an abelian group are one-dimensional. Therefore, the above representation decomposes into two irreducible one-dimensional representations. Find these.
4. Which linear combinations of $x^0$ and $x^1$ transform according to the irreducible representations? What is the significance of these coordinates in Minkowski space?
5. Are these representations unitary?

**Problem 14.6 (Lorentz Algebra)** Prove the commutation relations (14.76) for the generators (14.77).

**Problem 14.7 (Left-invariant Vector Fields of the Affine Group on** $\mathbb{R}$**)** We consider the group $G$ of affine transformations on $\mathbb{R}$. This has the matrix representation

$$A = \begin{pmatrix} \alpha_1 & \alpha_2 \\ 0 & 1 \end{pmatrix},$$

see (9.23). The left-invariant vector fields satisfy

$$X_A = (L_A)_* V = \frac{\mathrm{d}}{\mathrm{d}t}\big(AC(t)\big),$$

where $C(t)$ is a curve with $C(0) = \mathbb{1}$ and $\dot{C}(0) \in T_e G$. Choose in $T_e G$ the basis

$$X_1 = \begin{pmatrix} 1 & 0 \\ 0 & 0 \end{pmatrix}, \quad X_2 = \begin{pmatrix} 0 & 1 \\ 0 & 0 \end{pmatrix}.$$

What are the corresponding left-invariant vector fields?

# Root Systems and Cartan Classification 15



In this chapter, we analyze the commutation relations of semi-simple Lie algebras and their consequences more closely. According to Theorem 13.1 in Sect. 13.4, every semi-simple Lie algebra is a direct sum of simple ideals, so we only need to consider simple Lie algebras. First, we diagonalize a complete set of commuting basis vectors—whose linear span defines a Cartan subalgebra—in the adjoint representation. This leads to the roots of the Lie algebra $\mathfrak{g}$ and their quantization. For a fixed rank, there are only a few root systems that satisfy the quantization rules. Their analysis leads to the complete classification of all finite-dimensional simple Lie algebras—the so-called *Cartan classification*. Over the complex numbers, there are 4 infinite series (the classical Lie algebras) and 5 exceptional Lie algebras.[1]

In the investigation, we make use of the fact that $\mathfrak{g}$ always carries representations of certain $\mathfrak{su}(2)$ subalgebras. Therefore, a solid knowledge of the irreducible representations of $\mathfrak{su}(2)$ is indispensable for the classification of simple Lie algebras. The commutation rules in $\mathfrak{su}(2)$ are those of angular momentum. Therefore, $\mathfrak{su}(2)$ appears in the treatment of spherically symmetric systems within the framework of quantum mechanics, see e.g. [61]. Introductions to the classification of root systems of simple Lie algebras and the Weyl group operating on them can be found in [57, 58, 62].

---

[1] Over the real numbers, there are 12 infinite series and 23 exceptional Lie algebras.

© The Author(s), under exclusive license to Springer-Verlag GmbH, DE, part of Springer Nature 2026

A. Wipf, *Symmetries in Physics*, https://doi.org/10.1007/978-3-662-72675-4_15

## 15.1   Raising and Lowering Operators in $\mathfrak{su}(2)$

In Sect. 10.8.2, the irreducible representations of the unitary group SU(2) on the space of homogeneous polynomials in two complex variables were constructed. These representations induce representations of the Lie algebra $\mathfrak{su}(2)$, and we will analyze these using the methods of linear algebra.

The Lie algebra $\mathfrak{su}(2)$ contains the traceless anti-Hermitian $2 \times 2$ matrices with the commutator of matrices as the Lie product (see example box in Sect. 14.1.2). We multiply the matrices by i and consider traceless Hermitian matrices. As a basis, we choose the spin matrices (we set $\hbar = 1$)

$$S_i = \tfrac{1}{2}\sigma_i, \quad S_i = S_i^\dagger, \tag{15.1}$$

which were associated with spatial rotations in Problem 13.3. However, we will not only investigate this two-dimensional representation but also arbitrary irreducible representations of the $\mathfrak{su}(2)$ algebra. We denote the corresponding basis vectors by $J_i = \mathfrak{D}(S_i)$. These satisfy the commutation rules of the components of an angular momentum (spin, orbital angular momentum, total angular momentum)

$$[J_i, J_j] = i\epsilon_{ijk} J_k. \tag{15.2}$$

They are represented on a linear space with a scalar product, and we therefore often speak of operators instead of infinitesimal generators or vectors of the Lie algebra.

We will now construct "eigenvectors and eigenvalues of the angular momenta", using only the commutation relations and the hermiticity of the $J_i$. The results obtained will then apply to all irreducible representations. To do this, we first diagonalize $J_3$. Then $J_1, J_2$ will necessarily be non-diagonal, as they do not commute with $J_3$. Therefore, instead of $J_1$ and $J_2$, the ladder operators (raising and lowering operators) are introduced:

> **Ladder Operators for $\mathfrak{su}(2)$**
> The third component of the angular momentum $J_3$ and the ladder operators
> $J_\pm = J_1 \pm iJ_2$ satisfy the commutation relations
>
> $$[J_3, J_\pm] = \pm J_\pm, \quad [J_+, J_-] = 2J_3 \quad \text{with} \quad J_3^\dagger = J_3, \ J_-^\dagger = J_+. \tag{15.3}$$

---

**Task**

Show that (15.2) implies the commutation relations (15.3).

Now we make use of the fact that the quadratic Casimir invariant $\boldsymbol{J}^2$ commutes with the angular momentum (see Sect. 13.5.1),

$$[\boldsymbol{J}^2, J_i] = 0, \quad \text{where} \quad \boldsymbol{J}^2 = \sum J_i^2 . \tag{15.4}$$

We can therefore diagonalize $\boldsymbol{J}^2$ simultaneously with the component $J_3$. An eigenvalue $m$ of $J_3$ is referred to as magnetic quantum number.

The Casimir invariant is a positive operator, and we parameterize its eigenvalues according to $j(j+1)$ with $j \geq 0$. The parameter $j$ is called the angular momentum quantum number. We denote the common (normalized to one) eigenvectors of $\boldsymbol{J}^2$ and $J_3$ by $|jm\rangle$:

$$\boldsymbol{J}^2|jm\rangle = j(j+1)|jm\rangle \quad \text{and} \quad J_3|jm\rangle = m|jm\rangle \quad m, j \in \mathbb{R}, \ j \geq 0 . \tag{15.5}$$

To find the allowed values of $j$ and $m$, we rewrite the product of the ladder operators,

$$J_\pm J_\mp = J_1^2 + J_2^2 \mp i(J_1 J_2 - J_2 J_1) = \boldsymbol{J}^2 - J_3^2 \pm J_3 . \tag{15.6}$$

We first note that because $J_-^\dagger = J_+$, the matrix elements

$$\langle jm|J_\pm J_\mp|jm\rangle = \langle J_\mp jm|J_\mp jm\rangle = \left\| J_\mp|jm\rangle \right\|^2$$

are non-negative, and we rewrite the left side using (15.6),

$$\langle jm|J_\mp J_\pm|jm\rangle = \langle jm|(\boldsymbol{J}^2 - J_3^2 \mp J_3)|jm\rangle = c_\pm^2(j, m)\langle jm|jm\rangle . \tag{15.7}$$

It follows that the two constants

$$c_\pm^2(j, m) = j(j+1) - m(m \pm 1) = (j + \tfrac{1}{2})^2 - (m \pm \tfrac{1}{2})^2 \tag{15.8}$$

are non-negative. Both inequalities together restrict the values of the magnetic quantum number from above and below:

$$-j \leq m \leq j . \tag{15.9}$$

In the next step, we use the fact that the ladder operators $J_+$ and $J_-$ ascend and descend in the spectrum of $J_3$. Indeed, with the commutation relations (15.3), it follows

$$J_3 J_\pm|jm\rangle = J_\pm(J_3 \pm 1)|jm\rangle = J_\pm(m \pm 1)|jm\rangle = (m \pm 1)J_\pm|jm\rangle . \tag{15.10}$$

This means that $J_\pm$ maps the eigenvector $|jm\rangle$ either to the zero vector or to an eigenvector with magnetic quantum number $m \pm 1$. However, the eigenstates generated by the ladder operators will no longer be normalized to one. The relation

(15.7) determines the relative normalization of the states, and after a suitable choice of the phases of the eigenfunction, one obtains

$$J_{\pm}|jm\rangle = c_{\pm}(j, m)|j, m \pm 1\rangle. \tag{15.11}$$

In particular, the difference between two magnetic quantum numbers in any irreducible representation of $\mathfrak{su}(2)$ is an integer.

---

**Question**

Why would the representation be reducible if the difference between two $m$ were not an integer?

---

**The Action of the Ladder Operators**

If $m$ is an eigenvalue of $J_3$, then $(m \pm 1)$ is also an eigenvalue, unless the constant $c_{\pm}(j, m)$ in (15.8) vanishes. In the latter case, $J_{\pm}|jm\rangle = 0$.

---

We can apply $J_+$ multiple times to an eigenvector $|jm\rangle$ to generate eigenvectors with increasing magnetic quantum numbers $m + 1, m + 2, \ldots$. However, the inequalities (15.9) require that $m \leq j$, so at some point, the repeated action of $J_+$ on $|jm\rangle$ must produce the zero vector. Due to (15.7), this is only the case for

$$c_+(j, m) = 0 \quad \text{or} \quad m = \pm\left(j + \tfrac{1}{2}\right) - \tfrac{1}{2} \quad \text{for a} \quad m \leq j. \tag{15.12}$$

Of the two solutions to this equation, only the solution $m = j$ is possible due to (15.9). It follows that $J_3$ has the highest eigenvalue $j$. The analogous argument for the lowering operator $J_-$ shows that $J_3$ has the smallest eigenvalue $-j$. Since in an irreducible representation the difference between two magnetic quantum numbers must be an integer, we have $2j \in \mathbb{N}_0$.

**Lemma 15.1** *The commuting operators $J_3$ and $\mathbf{J}^2$ have eigenvalues $m$ and $j(j + 1)$, where*

$$j \in \frac{1}{2}\mathbb{N}_0 \quad \text{and} \quad m = -j, -j + 1, \ldots, j - 1, j. \tag{15.13}$$

For each angular momentum $j$ there are $2j+1$ different magnetic quantum numbers, so that the dimension of the representation with angular momentum $j$ is $2j + 1$. Since the normalized $|jm\rangle$ are simultaneous eigenvectors of the Hermitian operators $\mathbf{J}^2$ and $J_3$ with different pairs of eigenvalues they form an orthonormal system,

$$\langle jm|j'm'\rangle = \delta_{jj'}\,\delta_{mm'}. \tag{15.14}$$

In the eigenstate $|jm\rangle$, the third component of the spins is equal to $m$ and the "length" of the angular momentum vector is $\sqrt{j(j+1)}$.

---

**Example: Matrix Representation**

For a fixed $j$, there exist $2j+1$ eigenvectors $|jm\rangle$. In the standard basis

$$|jj\rangle = \begin{pmatrix} 1 \\ 0 \\ \vdots \\ 0 \end{pmatrix}, \quad |j,j-1\rangle = \begin{pmatrix} 0 \\ 1 \\ \vdots \\ 0 \end{pmatrix}, \quad |j,-j\rangle = \begin{pmatrix} 0 \\ 0 \\ \vdots \\ 1 \end{pmatrix} \tag{15.15}$$

the angular momentum operators for fixed $j$ are $(2j+1)$-dimensional square matrices. The Casimir invariant is $\boldsymbol{J}^2 = j(j+1)\mathbb{1}$ and

$$\langle jm|J_3|jm'\rangle = m\,\delta_{mm'}, \quad \langle jm|J_+|jm'\rangle = c_+(j,m')\,\delta_{m,m'+1}, \quad J_- = J_+^{\dagger}.$$

◀

---

## 15.2   Roots of a Simple Lie Algebra

We now generalize the method of raising and lowering operators to arbitrary simple Lie algebras. For this, we choose a basis orthogonal with respect to the Killing form. The first $r =\mathrm{rank}(\mathfrak{g})$ basis vectors $H_1, \ldots, H_r$ should commute with each other, so that

$$[\mathrm{ad}_{H_i}, \mathrm{ad}_{H_j}] = \mathrm{ad}_{[H_i,H_j]} = 0. \tag{15.16}$$

The maximal abelian subalgebra spanned by the $H_i$ is called a *Cartan subalgebra* and denoted by $\mathfrak{h}$. There are many Cartan subalgebras: if one conjugates a Cartan subalgebra with any group element, one obtains another isomorphic Cartan subalgebra. However, the following structural results are independent of the choice of $\mathfrak{h}$.

Due to (15.16), the mappings $\mathrm{ad}_{H_i}$ are simultaneously diagonalizable. In general, the mappings $\mathrm{ad}_{H_i}$, which are skew-symmetric with respect to the Killing form, can only be diagonalized in the complex domain. Therefore, we introduce the complexified Lie algebra $\mathfrak{g}_c$ of $\mathfrak{g}$ and extend the Killing form to a bilinear form on $\mathfrak{g}_c$. If the Lie algebra is simple, the elements of the Cartan subalgebra can be chosen orthonormal in the complexified Lie algebra,

$$K(H_i, H_j) = \mathrm{Tr}\,(\mathrm{ad}_{H_i}\mathrm{ad}_{H_j}) = -\delta_{ij}. \tag{15.17}$$

We now simultaneously diagonalize the commuting $\mathrm{ad}_{H_i}$ in $\mathfrak{g}_c$:

$$\mathrm{ad}_{H_i}(E_\alpha) = [H_i, E_\alpha] = \alpha_i E_\alpha, \quad 1 \le i \le r. \tag{15.18}$$

> **Root Vectors**
> The vector $\alpha \in \mathbb{R}^r$ is called a *root vector* of $\mathfrak{g}$ and $E_\alpha$ is the step operator
> associated with $\alpha$. We denote the set of root vectors by $\Phi$. Their number is
> $|\Phi| = \mathrm{Dim}(\mathfrak{g}) - \mathrm{Rank}(\mathfrak{g})$.

In the following, $(X, Y) = -K(X, Y)$ denotes the Killing form up to the sign.
For compact Lie algebras, $(., .)$ defines a scalar product. Two step operators $E_\alpha$, $E_\beta$
are orthogonal to each other with respect to this scalar product for $\alpha + \beta \neq 0$. This
follows from the invariance of the Killing form,

$$0 = \left(\mathrm{ad}_{H_i} E_\alpha, E_\beta\right) + \left(E_\alpha, \mathrm{ad}_{H_i} E_\beta\right) = (\alpha_i + \beta_i)\left(E_\alpha, E_\beta\right) \overset{\alpha \neq -\beta}{\Longrightarrow} \left(E_\alpha, E_\beta\right) = 0 . \tag{15.19}$$

Similarly, one proves that the elements of the Cartan subalgebra are orthogonal to
the step operators,

$$0 = \left(\mathrm{ad}_{H_j} H_i, E_\alpha\right) + \left(H_i, \mathrm{ad}_{H_j} E_\alpha\right) = \alpha_j \left(H_i, E_\alpha\right) \Longrightarrow (H_i, E_\alpha) = 0 , \tag{15.20}$$

since at least one $\alpha_j$ is not zero. With $\alpha$, $-\alpha$ must also be a root. If this were not the
case, $E_\alpha$ would be orthogonal to $\mathfrak{g}$, which is not possible for a simple Lie algebra.

Due to the Jacobi identity, we have

$$[H_i, [E_\alpha, E_\beta]] = -[E_\beta, [H_i, E_\alpha]] - [E_\alpha, [E_\beta, H_i]] = (\alpha_i + \beta_i)[E_\alpha, E_\beta] ,$$

which again means that with $\alpha$, $\beta$, $\alpha + \beta$ is also a root if $E_{\alpha+\beta} \propto [E_\alpha, E_\beta]$ does
not vanish. Finally, $[E_\alpha, E_{-\alpha}]$ commutes with all $H_i$ and must therefore lie in $\mathfrak{h}$,
$[E_\alpha, E_{-\alpha}] = \beta_j H_j$, where a sum over $j$ is implied. The left side in

$$(H_i, [E_\alpha, E_{-\alpha}]) = \beta_j \left(H_i, H_j\right) = \beta_i$$

can be written as follows due to the ad-invariance of $(X, Y)$,

$$([H_i, E_\alpha], E_{-\alpha}) = \alpha_i (E_\alpha, E_{-\alpha}) .$$

If we normalize the step operators according to $(E_\alpha, E_{-\alpha}) = 1$, then it follows
$\beta_i = \alpha_i$.

**Cartan-Weyl Basis**

The commutation rules for the elements $\{H_i, E_\alpha\}$ of the *Cartan-Weyl Basis* are:

$$[H_i, H_j] = 0, \quad [H_i, E_\alpha] = \alpha_i E_\alpha, \quad [E_\alpha, E_\beta] = \begin{cases} \alpha_i H_i & \text{if } \beta = -\alpha \\ N_{\alpha\beta} E_{\alpha+\beta} & \text{otherwise.} \end{cases} \tag{15.21}$$

The elements $H_i$, $E_\alpha$ are normalized as follows:

$$(H_i, H_j) = \delta_{ij}, \quad (H_i, E_\alpha) = 0, \quad (E_\alpha, E_\beta) = \delta_{\alpha+\beta,0}. \tag{15.22}$$

If $\alpha + \beta$ is not a root, then $N_{\alpha\beta}$ vanishes.

## 15.2.1 Explicit Basis for $\mathfrak{su}(3)$

As a basis for the Lie algebra $\mathfrak{su}(3)$ of Hermitian traceless $3 \times 3$ matrices we choose the matrices named after Murray Gell-Mann,

$$X_1 = \begin{pmatrix} 0 & 1 & 0 \\ 1 & 0 & 0 \\ 0 & 0 & 0 \end{pmatrix}, \quad X_2 = \begin{pmatrix} 0 & -i & 0 \\ i & 0 & 0 \\ 0 & 0 & 0 \end{pmatrix}, \quad X_3 = \begin{pmatrix} 1 & 0 & 0 \\ 0 & -1 & 0 \\ 0 & 0 & 0 \end{pmatrix},$$

$$X_4 = \begin{pmatrix} 0 & 0 & 1 \\ 0 & 0 & 0 \\ 1 & 0 & 0 \end{pmatrix}, \quad X_5 = \begin{pmatrix} 0 & 0 & -i \\ 0 & 0 & 0 \\ i & 0 & 0 \end{pmatrix}, \quad X_6 = \begin{pmatrix} 0 & 0 & 0 \\ 0 & 0 & 1 \\ 0 & 1 & 0 \end{pmatrix}, \tag{15.23}$$

$$X_7 = \begin{pmatrix} 0 & 0 & 0 \\ 0 & 0 & -i \\ 0 & i & 0 \end{pmatrix}, \quad X_8 = \frac{1}{\sqrt{3}} \begin{pmatrix} 1 & 0 & 0 \\ 0 & 1 & 0 \\ 0 & 0 & -2 \end{pmatrix}.$$

These are orthogonal with respect to the invariant scalar product defined by the trace,

$$(X_i, X_j) = \operatorname{Tr} X_i X_j = 2\,\delta_{ij}. \tag{15.24}$$

**Question**

What is the value of the constant $\lambda$ in $\operatorname{Tr}(X_i X_j) = \lambda\, K(X_i, X_j)$?

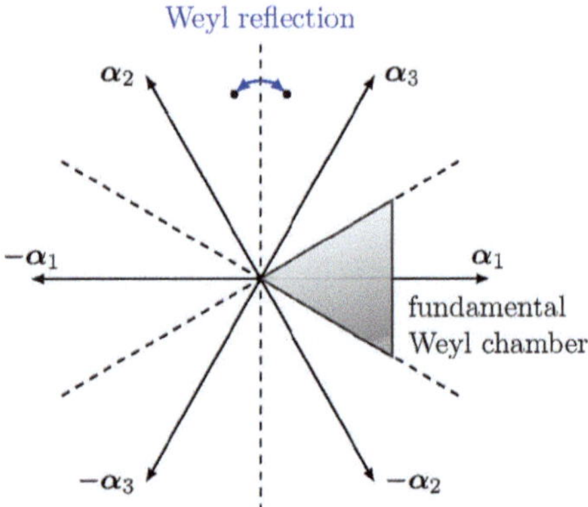



**Fig. 15.1** The roots and a fundamental or first Weyl chamber of $\mathfrak{su}(3)$ (see Sect. 15.5)

The non-vanishing structure constants $f_{ab}{}^{c}$ with $a < b < c$ are

| $abc$ | 123 | 147 | 156 | 246 | 257 | 345 | 367 | 458 | 678 |
|---|---|---|---|---|---|---|---|---|---|
| $f_{ab}{}^{c}$ | 2 | 1 | $-1$ | 1 | 1 | 1 | $-1$ | $\sqrt{3}$ | $\sqrt{3}$ |

They are completely antisymmetric in the three indices, for example, $f_{13}{}^{2} = -2$, and this property determines the remaining structure constants.

As a Cartan subalgebra, we choose the linear span of the diagonal matrices $H_1 = X_3$ and $H_2 = X_8$. We find the following three step operators and roots:

$$E_{\alpha_1} = \frac{1}{2}(X_1 + iX_2) : \quad \boldsymbol{\alpha}_1 = (2, 0)$$

$$E_{\alpha_2} = \frac{1}{2}(X_6 + iX_7) : \quad \boldsymbol{\alpha}_2 = (-1, \sqrt{3}), \tag{15.25}$$

$$E_{\alpha_3} = \frac{1}{2}(X_4 + iX_5) : \quad \boldsymbol{\alpha}_3 = (1, \sqrt{3}).$$

The lowering operators $E_{-\alpha}$ are transposed to the raising operators $E_{\alpha}$. The roots $\pm \boldsymbol{\alpha}_i$ all have the length 2 and define an equilateral hexagon, as shown in Fig. 15.1. Also shown are the reflection planes and a Weyl reflection, see Sect. 15.3.1.

## 15.3   Quantization of the Roots

We define the scalar products $\boldsymbol{\alpha} \cdot \boldsymbol{\beta} = \alpha_i \beta_i$ and $\boldsymbol{\alpha} \cdot H = \alpha_i H_i$, where summation over $i$ is implied. In the following, it is advantageous to choose a slightly different basis of the Cartan subalgebra:

$$H_\alpha = \frac{2}{\alpha^2}\, \alpha \cdot H \,. \tag{15.26}$$

At the same time, the step operators scale according to

$$E_\alpha \mapsto \sqrt{\frac{\alpha^2}{2}}\, E_\alpha \,. \tag{15.27}$$

From the second equation in (15.21), the commutation rules of the $H_\alpha$ with the step operators follow,

$$[H_\alpha, E_\beta] = \frac{2\alpha \cdot \beta}{\alpha^2}\, E_\beta \overset{\beta=\alpha}{\Longrightarrow} [H_\alpha, E_{\pm\alpha}] = \pm 2 E_{\pm\alpha} \,, \tag{15.28}$$

while the third equation in (15.21) for $\beta = -\alpha$ takes the following form,

$$[E_\alpha, E_{-\alpha}] = H_\alpha \,. \tag{15.29}$$

Thus, for each root $\alpha$, the three elements

$$\left\{ E_\alpha, E_{-\alpha}, \frac{H_\alpha}{2} \right\} \propto \{J_+, J_-, J_3\} \tag{15.30}$$

define an $\mathfrak{su}(2)$ Lie subalgebra in $\mathfrak{g}_c$ with the commutation relations of the angular momentum (15.3). This $\mathfrak{su}(2)$ acts on the linear space $\mathfrak{g}$ with the adjoint representation, which will generally be reducible. In the previous section, we showed that the eigenvalues of $J_3$ are half-integers. Thus, the eigenvalues of $H_\alpha$ in (15.28) are integers,

> **Quantization of the Roots**
> The roots of a Lie algebra satisfy the quantization rules
>
> $$n_{\alpha\beta} \equiv \frac{2\alpha \cdot \beta}{\alpha^2} \in \mathbb{Z} \quad \text{for all} \quad \alpha, \beta \in \Phi \,. \tag{15.31}$$

The commutation relations of $H_\alpha$ and $E_{\pm\alpha}$ in (15.28) then mean that the step operators $E_\alpha$ and $E_{-\alpha}$ increase or decrease the eigenvalues of $H_\alpha$ by 2, respectively.

Let $\beta = \eta\alpha$ be parallel to the root $\alpha$ with a scaling factor $\eta \neq 0$. Then

$$n_{\alpha\beta} = 2\eta \in \mathbb{Z} \quad \text{and} \quad n_{\beta\alpha} = \frac{2}{\eta} \in \mathbb{Z} \,,$$

from which it immediately follows that $|\eta|$ can only take the values $1/2$, $1$, and $2$.

**Definition 15.1** A root $\alpha$ is called indivisible if $\alpha/2$ is not a root. A system of roots is called reduced if it has no divisible roots.

In the following, it is always assumed that the considered root systems are reduced. Then only the root $-\alpha$ is parallel to $\alpha$.

The integers $n_{\alpha\beta}$ and $n_{\beta\alpha}$ defined in (15.31) have the same sign or both vanish. However, they cannot become arbitrarily large. To see this, we introduce the angle $\theta$ between two roots,

$$\cos\theta = \frac{\alpha\cdot\beta}{|\alpha||\beta|}.\tag{15.32}$$

The Schwartz inequality implies

$$0 \le n_{\alpha\beta}n_{\beta\alpha} = 4\frac{(\alpha\cdot\beta)^2}{\alpha^2\beta^2} \le 4,\tag{15.33}$$

which restricts the allowed values of $\cos\theta$,

$$n_{\alpha\beta}\,n_{\beta\alpha} = 4\cos^2\theta \in \{0, 1, 2, 3, 4\}.\tag{15.34}$$

If $\alpha\cdot\beta \neq 0$, then for the ratio of the integers $n_{\alpha\beta}$ and $n_{\beta\alpha}$ we obtain

$$\frac{n_{\alpha\beta}}{n_{\beta\alpha}} = \frac{\beta^2}{\alpha^2}.\tag{15.35}$$

The product of the integers in (15.34) is equal to 4 only for $\theta \in \{0, \pi\}$, i.e., only when $\beta$ and $\alpha$ are linearly dependent, which in a reduced root system means $\beta = \pm\alpha$. If $\alpha$ and $\beta$ are linearly independent, then the value 4 does not occur in (15.34).

For each allowed choice of $n_{\alpha\beta}$ and $n_{\beta\alpha}$, besides the angle $\theta$, the angle $\pi - \theta$ also appears, as can be seen with the help of a Weyl reflection on the plane perpendicular to $\alpha$ (these reflections map roots into roots and will be discussed in the next section). Therefore, it suffices to consider angles $\theta \in [0, \pi/2]$. The four possible configurations of two roots $\alpha$ and $\beta$ are summarized (up to an exchange of the roots) in Table 15.1 and illustrated in Fig. 15.2.

**Table 15.1** The possible values of the angle between two roots and their relative length

| $(n_{\alpha\beta}, n_{\beta\alpha})$ | $\cos^2\theta$ | $\theta$ | $\cos\theta$ | Length ratio |
|---|---|---|---|---|
| $(0, 0)$ | $0$ | $\pi/2 = 90°$ | $0$ | Arbitrary |
| $\pm(1, 1)$ | $1/4$ | $\pi/3 = 60°$ | $1/2$ | $\alpha^2 = \beta^2$ |
| $\pm(2, 1)$ | $1/2$ | $\pi/4 = 45°$ | $1/\sqrt{2}$ | $2\alpha^2 = \beta^2$ |
| $\pm(3, 1)$ | $3/4$ | $\pi/6 = 30°$ | $\sqrt{3}/2$ | $3\alpha^2 = \beta^2$ |

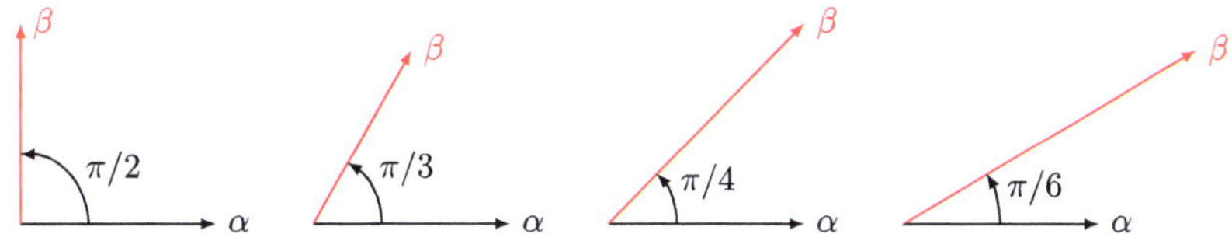

**Fig. 15.2**  Possible angles and relative lengths of two roots

## 15.3.1 Weyl Reflections

The *Weyl reflections* are automorphisms of the set $\Phi$ of roots. They originate from the adjoint action of the $\mathfrak{su}(2)$ subalgebras $(E_\alpha, E_{-\alpha}, H_\alpha)$ on $\mathfrak{g}_c$. To see this consider an eigenvector $E_\beta$ of $H_\alpha$, see (15.28). The step operator $E_{-\alpha}$ with $\alpha \neq \beta$ maps it to another eigenvector and lowers the eigenvalue by 2,

$$[H_\alpha, [E_{-\alpha}, E_\beta]] = \left( \frac{2\alpha \cdot \beta}{\alpha^2} - 2 \right) [E_{-\alpha}, E_\beta] . \tag{15.36}$$

By applying $\mathrm{ad}_{E_{-\alpha}}$ several times (say $m$ times) to $E_\beta$, one generates in the multiplet the eigenvector with the magnetic quantum number opposite to $E_\beta$, so that

$$\frac{2\alpha \cdot \beta}{\alpha^2} - 2m = -\frac{2\alpha \cdot \beta}{\alpha^2} , \tag{15.37}$$

from which $m$ is obtained. At the same time, the root $\beta$ changes into

$$\beta - m\alpha = \beta - \frac{2\alpha \cdot \beta}{\alpha^2} \alpha \equiv \sigma_\alpha(\beta) . \tag{15.38}$$

The linear mapping $\sigma_\alpha$ reflects on the plane perpendicular to the root $\alpha$, as sketched in Fig. 15.3. If $\alpha$ and $\beta$ are perpendicular to each other, then $\sigma_\alpha(\beta) = \beta$. Each Weyl reflection maps roots bijectively to roots. For each pair of roots $\alpha, \beta$,

$$\sigma_\alpha(\beta) = \beta - \frac{2\alpha \cdot \beta}{\alpha^2} \alpha \tag{15.39}$$

is again a root. These reflections are called *Weyl reflections*. They generate the Weyl group $\mathcal{W}$, which contains reflections and rotations.

## 15.3.2 Weyl Group

In Sect. 14.4 we introduced the Weyl group as the quotient of the normalizer $N_G(T)$ of a maximal torus and the torus $T \trianglelefteq N_G(T)$, $\mathcal{W} = N_G(T)/T$. Later we will see

**Fig. 15.3**  A Weyl reflection $\sigma_\alpha$ reflects a root $\beta$ on the plane perpendicular to $\alpha$

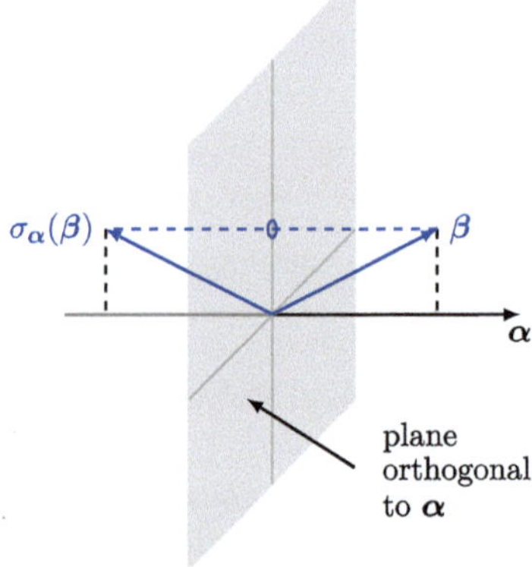


that this finite factor group can also be directly identified as the group generated by the Weyl reflections.

> **Weyl Group** $\mathcal{W}$
>
> The Weyl reflections generate the finite Weyl group $\mathcal{W}$ of the root system $\Phi = \{\alpha\}$, which contains not only the reflections but also the rotations generated by the reflections. The number of reflections is equal to the number of proper rotations in $\mathcal{W}$.

The earlier result (6.13) about the product of two reflections now tells us that $\sigma_\alpha \sigma_\beta$ is a rotation in the plane spanned by the two roots $\alpha$ and $\beta$, where the rotation angle $\varphi$ is twice the angle between the two roots. With the allowed angles between two roots in Table 15.1, it follows:

**Corollary 15.1**  *The rotations built from two Weyl reflections satisfy* $(\sigma_\alpha \sigma_\beta)^{m_{\alpha\beta}} = 1$ *with* $m_{\alpha\beta} = m_{\beta\alpha} \in \{2, 3, 4, 6\}$ *for the allowed angles* $\{\pi/2, \pi/3, \pi/4, \pi/6\}$ *between the roots. For* $\alpha = \beta$, $m_{\alpha\alpha} = 1$.

The corollary provides a presentation for each Weyl group and makes it clear that Weyl groups are special Coxeter groups. Their conjugacy classes are analyzed in [63].

**Corollary 15.2 (Weyl Group)**  *The root system* $\Phi$ *is invariant under the action of* $\mathcal{W}$, *and* $\mathcal{W}$ *is a finite subgroup of the orthogonal group* $O(r)$. *For each* $w \in \mathcal{W}$ *and each* $\alpha \in \Phi$, *it holds that* $\sigma_{w(\alpha)} = w \sigma_\alpha w^{-1}$.

The last property is proved as follows:

$$(w\sigma_\alpha w^{-1})(\beta) = (w\sigma_\alpha)(w^{-1}(\beta)) = w\left(w^{-1}(\beta) - \frac{2\alpha \cdot w^{-1}(\beta)}{\alpha^2}\,\alpha\right)$$

$$= \beta - \frac{2w(\alpha) \cdot \beta}{(w(\alpha))^2}\,w\alpha = \sigma_{w(\alpha)}(\beta)\,.$$

We made use of the invariance of the scalar product under reflections and rotations in $\mathcal{W}$.

---

**Reduced Root Systems**

A (reduced) root system $\Phi$ is a finite set of vectors in the real Euclidean vector space $\mathcal{V}$ with

1. $\Phi$ spans $\mathcal{V}$,
2. if $\alpha \in \Phi$, then the only multiples of $\alpha$ in $\Phi$ are the vectors $\pm\alpha$,
3. if $\alpha \in \Phi$, then $\sigma_\alpha \Phi = \Phi$.

---

In the classification or root system, one distinguishes between reducible and irreducible systems:

**Definition 15.2 (Irreducible Root Systems)** $\Phi$ is called reducible if it is the disjoint union of two root systems, $\Phi = \Phi_1 \cup \Phi_2$, and every element of $\Phi_1$ is orthogonal to every element of $\Phi_2$. $\Phi$ is called irreducible if there is no such decomposition.

A semi-simple Lie algebra is simple if and only if its associated root system $\Phi$ is irreducible. In the classification of root systems, one can therefore restrict to irreducible systems—reducible systems are composed in a simple way from irreducible systems.

### Action of $\mathcal{W}$ on $T$ and on $\Phi$

The group generated by the Weyl reflections acting on $\Phi$ is the same as the Weyl group $\mathcal{W}$ acting on $T$ as introduced in Sect. 14.4. We will examine this situation more closely here. For that purpose we consider a representative $w \in N_G(T)$ of the Weyl group $\mathcal{W} = N_G(T)/T$ and a $t \in T$. Since $w$ is in the normalizer of $T$, $\mathrm{Ad}_w(t) = wtw^{-1} = t' \in T$. Writing $t = \exp(iH)$ with $H \in \mathfrak{h}$, it follows that

$$\mathrm{Ad}_w\left(e^{iH}\right) = e^{i\mathrm{Ad}(w)H} \quad \text{with} \quad \mathrm{Ad}(w)H = wHw^{-1} \in \mathfrak{h}\,. \tag{15.40}$$

The adjoint mapping $\mathrm{Ad}_w : T \mapsto T$ induces, as expected, the mapping $\mathrm{Ad}(w) : \mathfrak{h} \mapsto \mathfrak{h}$. If $w$ represents a non-trivial element of the Weyl group, then $w \notin T$ and accordingly $\mathrm{Ad}(w)H \neq H$. Since $\mathrm{Ad}(w)$ is a Lie algebra homomorphism, it follows that

$$[H_i, E_\alpha] = \alpha_i E_\alpha \implies [\mathrm{Ad}(w)H_i, \mathrm{Ad}(w)E_\alpha] = \alpha_i \mathrm{Ad}(w)E_\alpha . \tag{15.41}$$

On the other hand, $\mathrm{Ad}(w)H_i = O_{ji}H_j$ with an orthogonal matrix $O$ from $\mathrm{O}(r)$. Substituting into (15.41) yields

$$[H_k, \mathrm{Ad}(w)E_\alpha] = (O\alpha)_k \mathrm{Ad}(w)E_\alpha . \tag{15.42}$$

This means that for each root $\alpha$, the image $O\alpha$ is also a root, or that the linear mapping $O$ permutes the elements of $\Phi$. An explicit realization for elements $w \in N_G(T)$, belonging to Weyl reflections, is given for matrix groups in Problem 15.2.

## 15.4   Roots of Lie Algebras with Rank $\leq 2$

### 15.4.1  Root System for Rank 1

The roots lie in $\mathbb{R}$, and besides $\alpha$, only $-\alpha$ is a root. We obtain the root system of the three-dimensional Lie algebra $\mathfrak{su}(2)$. The Weyl group $\mathbb{Z}_2$ consists of the identity and the reflection $\alpha \to -\alpha$, see Fig. 15.4.

### 15.4.2  Root Systems for Rank 2

The roots lie in $\mathbb{R}^2$ and there are at least two roots $\alpha$ and $\beta$. With each root $\alpha$, it is known that $-\alpha$ is also a root, so there must be at least four roots.

1. We begin with the possible angle $\theta = \pi/2$ in Table 15.1. Then $\alpha$ and $\beta$ are orthogonal to each other. The Weyl reflections do not generate any new roots, and we obtain the reducible root system in Fig. 15.5. The two roots $\alpha$ and $\beta$ do

**Fig. 15.4**  Root system of $A1$

**Fig. 15.5**  Root system of
$A_1 \oplus A_1$

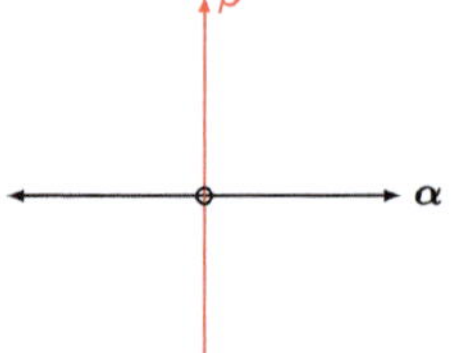

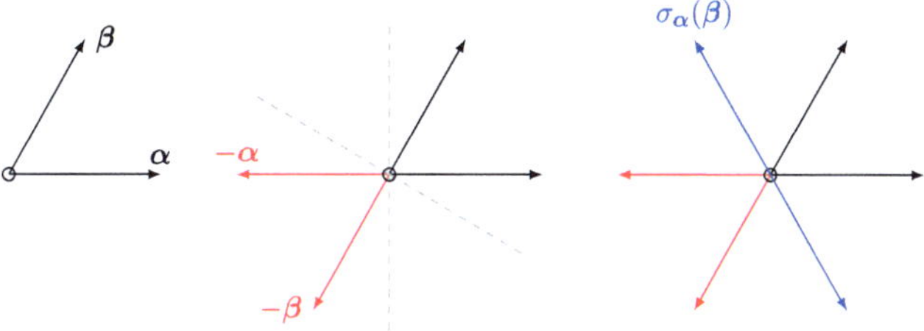

**Fig. 15.6**  Root system of $A_2$

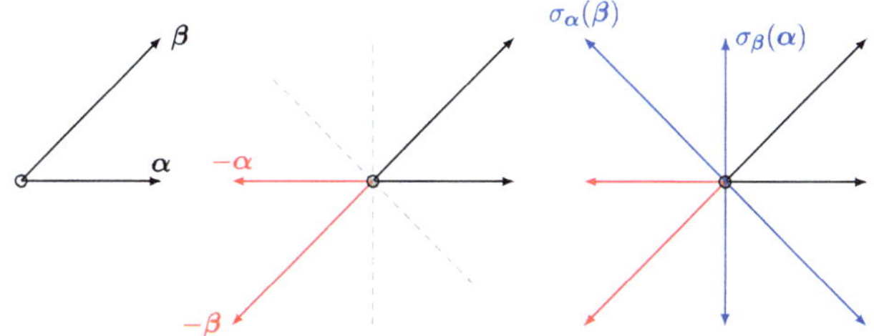

**Fig. 15.7**  Root system of $B_2$

not need to have the same length. Since

$$[H_\alpha, E_\beta] = [H_\beta, E_\alpha] = 0,  \tag{15.43}$$

the two sets or generators

$$\{H_\alpha, E_\alpha, E_{-\alpha}\} \quad \text{and} \quad \{H_\beta, E_\beta, E_{-\beta}\}  \tag{15.44}$$

form two three-dimensional subalgebras that commute with each other, $\mathfrak{g} = \mathfrak{su}(2) \oplus \mathfrak{su}(2)$. The Weyl group $\mathcal{W} = \mathbb{Z}_2 \times \mathbb{Z}_2$ acts reducibly on $\mathbb{R}^2$.

2. Now we consider the case $\theta = 60°$ in Fig. 15.6. We start with two roots $\boldsymbol{\alpha}$, $\boldsymbol{\beta}$ of the same length and an angle of $60°$ between them. Then $-\boldsymbol{\alpha}$, $-\boldsymbol{\beta}$ are also roots. The Weyl reflections then generate two additional roots. The equilateral hexagon spanned by the resulting roots is closed under all Weyl reflections. It defines the eight-dimensional Lie algebra $\mathfrak{su}(3)$.

   Its *Weyl group* is generated by the reflections in the three planes perpendicular to the roots $\boldsymbol{\alpha}$ and $\boldsymbol{\beta}$ and also contains the rotations by $2\pi/3$ and $4\pi/3$. It is isomorphic to the dihedral group $\mathcal{D}_3 \cong \mathcal{S}_3$ and acts transitively on $\Phi$.

3. Now we consider the case $\theta = 45°$ in Fig. 15.7. and start with two roots $\boldsymbol{\alpha}$, $\boldsymbol{\beta}$ with $\boldsymbol{\beta}^2 = 2\boldsymbol{\alpha}^2$ and an angle of $45°$ between them. We include $-\boldsymbol{\alpha}$, $-\boldsymbol{\beta}$ and the root vectors obtained using Weyl reflections. The resulting system of 8 roots is closed under all reflections and defines the ten-dimensional $\mathfrak{so}(5)$ Lie algebra.

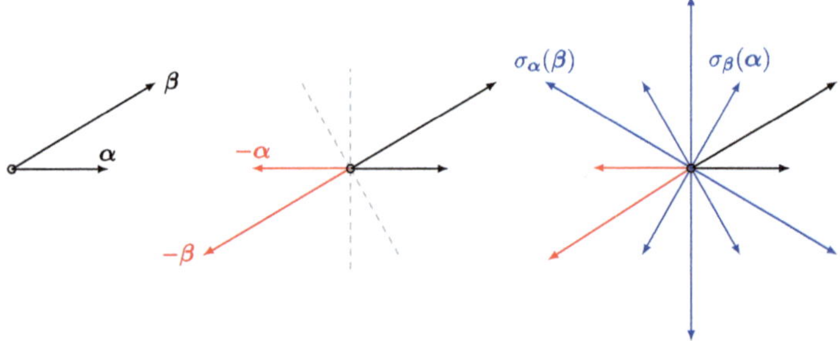


**Fig. 15.8** Root system of $G_2$

The *Weyl group* contains, in addition to the identity, the reflections corresponding to the four roots and the rotations by $\pi/2$, $\pi$ and $3\pi/2$ and is isomorphic to the dihedral group $\mathcal{D}_4$. It acts transitively on the set of short (or long) roots.

4. Finally, we examine the case $\theta = 30°$ in Table 15.1 and start with roots $\boldsymbol{\alpha}$, $\boldsymbol{\beta}$ with an angle of $30°$ between them and $\boldsymbol{\beta}^2 = 3\boldsymbol{\alpha}^2$. The resulting root system in Fig. 15.8, closed under Weyl reflections, defines the 14-dimensional exceptional Lie algebra $G_2$.

---

**Task**

Show that the *Weyl group* of the root system of $G_2$ is isomorphic to the dihedral group $\mathcal{D}_6$ and acts transitively on the set of short (or long) roots.

With this, we have constructed all simple (and reduced) Lie algebras of rank 2.

---

**Lie Algebras with Rank** 2
There exist three simple Lie algebras of rank 2. These are the Lie algebras $\mathfrak{su}(3)$, $\mathfrak{so}(5)$, and $G_2$ of dimensions 8, 10, and 14.

---

In the Cartan classification, these Lie algebras are called $A_2$, $B_2$, and $G_2$. The order of a Weyl group can be obtained in LiE with the command `W_order`:

```
LiE> W_order(A1) -> 2    W_order(A2) -> 6
LiE> W_order(B2) -> 8    W_order(G2) -> 12
```

The proper and improper rotations of the Weyl group $\mathcal{W}$ preserve the length of roots and the angles between roots. If two roots have different lengths, they cannot lie in the same orbit of the Weyl group. In LiE, one obtains the orbit of a root $\boldsymbol{\alpha}$ under the group action with the command `W_rt_orbit(`$\boldsymbol{\alpha}$`,g)`. For the Lie algebra

$B_2$, there exist long roots and short roots, and the ratio of their squared lengths is 2.
There are two orbits:

```
LiE> norm([1,0],B2) -> 4     /* [1,0] is the long simple root
LiE> norm([0,1],B2) -> 2     /* [0,1] is the short simple root
LiE> W_rt_orbit([1,0],B2) -> [[ 1, 2],[ 1, 0],[-1, 0],[-1,-2]]
LiE> W_rt_orbit([0,1],B2) -> [[ 1, 1],[ 0, 1],[ 0,-1],[-1,-1]]
```

The orbits of the long root $[1, 0]$ and the short root $[0, 1]$ are of equal length.

## 15.5   Properties of Root Systems

In general, the number of roots is greater than the rank $r$ of the Lie algebra, so they
are linearly dependent. Thus, one chooses a basis of roots

$$\{\gamma_1, \ldots, \gamma_r\}, \tag{15.45}$$

so that each root is a linear combination of these basis elements. The simple roots
provide a particular suitable basis:

**Theorem 15.1 (Simple Roots)** *One can choose a basis* $\Delta = \{\gamma_1, \ldots, \gamma_r\}$ *of
simple roots such that for each root* $\alpha$:

$$\alpha = \sum_{i=1}^{r} n_i \gamma_i, \tag{15.46}$$

*where the coefficients* $n_i$ *are integers and have the same sign (*$n_i = 0$ *is allowed).*

To prove this, we choose a vector $t \in \mathbb{R}^r$ that is not orthogonal to any root $\alpha \in \Phi$.
This allows us to polarize the root system,

$$\Phi = \Phi^+ \cup \Phi^-, \tag{15.47}$$

into positive and negative roots,

$$\Phi^+ = \{\alpha \in \Phi \,|\, (t, \alpha) > 0\} \quad \text{and} \quad \Phi^- = \{\alpha \in \Phi \,|\, (t, \alpha) < 0\}. \tag{15.48}$$

The roots appear in pairs $\{\alpha, -\alpha\}$ with one positive and one negative root. It follows
that $\Phi^+$ and $\Phi^-$ contain the same number of roots. We say $\alpha$ is greater than $\beta$ if $\alpha -
\beta$ is a sum of positive roots or $\alpha = \beta$. The definition of positive roots depends on the
choice of the Cartan subalgebra and $t$. However, many of the following properties
are independent of this choice.

**Definition 15.3 (Simple Roots)**  A root $\gamma$ is called simple if it is positive and cannot be written as a sum of positive roots. The set of simple roots is denoted by $\Delta$.

---

Task

Determine simple roots for the root system of $A_2$ or $\mathfrak{su}(3)$ in Fig. 15.6.

From Definition 15.3, it immediately follows that

**Lemma 15.2** *Every positive root is the sum of simple roots,* $\alpha = \sum n_i \gamma_i$ *with* $n_i \geq 0$.

***Proof*** If $\alpha \in \Phi^+$ is simple, then there is nothing to show. If not, then $\alpha = \alpha_1 + \alpha_2$ is the sum of positive roots with $\alpha_1, \alpha_2 < \alpha$. If $\alpha_1$ or $\alpha_2$ is not simple, then it can be written as a sum of smaller positive roots. This iterative process ends with simple roots. The *highest root* is the root for which the sum of the coefficients $n_i$—this sum is called the height of the root—is maximal. If the root $\alpha$ is negative, then $-\alpha$ is positive and therefore the sum of simple roots. It follows that every negative root has the expansion $\alpha = \sum n_i \gamma_i$ with $n_i \leq 0$. This completes the proof of Theorem 15.1.

The number of positive roots can be obtained in LiE with the command `n_pos_roots`. The positive roots are the rows of the matrix `pos_roots`, where the top $r$ rows contain the simple roots and the last row contains the highest root $\alpha \in \Phi^+$.

```
LiE> n_pos_roots(A2) -> 3
LiE> pos_roots(A2) -> [[1,0],[0,1]],[1,1]]
LiE> high_root(A2) -> [1,1]
LiE> n_pos_roots(G2) -> 6
LiE> high_root(G2) -> [3,2]
```

**Lemma 15.3** *If* $\gamma \neq \gamma'$ *are simple roots, then* $\gamma \cdot \gamma' \leq 0$ *and neither* $\gamma - \gamma'$ *nor* $\gamma' - \gamma$ *are roots. It follows for two simple roots* $[E_\gamma, E_{-\gamma'}] = [E_{\gamma'}, E_{-\gamma}] = 0$.

***Proof*** We first prove that the difference of two simple roots is not a root. Since roots come in pairs, with $\gamma - \gamma'$ also $\gamma' - \gamma$ would be a root (and vice versa) and one of them would be positive, e.g., $\gamma - \gamma'$. Then $\gamma$ would be the sum of the two positive roots $\gamma'$ and $\gamma - \gamma'$. This contradicts the assumption that $\gamma$ is simple. It follows that the step operators for $\gamma$ and $-\gamma'$ or for $-\gamma$ and $\gamma'$ commute. Next we show $\gamma \cdot \gamma' \leq 0$. We recall that $E_{\gamma'}$ is an eigenvector of $\mathrm{ad}_{H_\gamma}$ with eigenvalue $2\gamma \cdot \gamma'/\gamma^2$. Since $[E_{-\gamma}, E_{\gamma'}] = 0$, it is annihilated by the lowering operator $E_{-\gamma}$ and has in the $\mathfrak{su}(2)$-multiplet

$$E_{\gamma'}, \quad \mathrm{ad}_{E_\gamma}(E_{\gamma'}), \ldots \ldots, \mathrm{ad}_{E_\gamma}^{p-1}(E_{\gamma'}), \quad \text{with} \quad \mathrm{ad}_{E_\gamma}^{p}(E_{\gamma'}) = 0 \qquad (15.49)$$

the smallest possible magnetic quantum number. However, the minimum magnetic quantum number in each $\mathfrak{su}(2)$ multiplet is non-positive, and this proves the first statement in the lemma.

**Task**

Verify the statements in Lemma 15.3 for $\mathfrak{su}(3)$ in Fig. 15.6.

**Lemma 15.4** *The simple roots are linearly independent and form a basis in $\mathbb{R}^r$, where $r = rank\,(\mathfrak{g})$.*

**Proof** To prove the linear independence of the roots in $\Delta$, we assume they are linearly dependent. Then there would be a linear combination of the simple roots with

$$\sum c_i \boldsymbol{\gamma}_i = 0 . \tag{15.50}$$

We decompose the sum into two parts,

$$P = \sum_{i,c_i>0} c_i \boldsymbol{\gamma}_i \quad \text{and} \quad N = \sum_{j,c_j<0} c_j \boldsymbol{\gamma}_j , \tag{15.51}$$

so that by assumption $P + N = 0$. From $P = -N$ it follows

$$P^2 = N^2 = -N \cdot P = - \sum_{i,c_i>0} \sum_{j,c_j<0} c_i c_j \, \boldsymbol{\gamma}_i \cdot \boldsymbol{\gamma}_j .$$

The left side is non-negative, while the right side is non-positive, since for simple roots $\boldsymbol{\gamma}_i \cdot \boldsymbol{\gamma}_j \leq 0$. Therefore, $P$ and $N$ are equal to the zero vector, contradicting the assumption. To show that the simple roots span $\mathbb{R}^r$, we again assume the opposite. Then there would be a vector $\boldsymbol{v} \in \mathbb{R}^r$, which is perpendicular to all simple roots. Since every root is a linear combination of simple roots, $\boldsymbol{v}$ is perpendicular to all roots. We consider the element $h_* = v_i H_i$ in $\mathfrak{h}$, which commutes with all step operators, $[h_*, E_\alpha] = (\boldsymbol{v} \cdot \boldsymbol{\alpha})E_\alpha = 0$. Since $\mathfrak{g}_c$ is spanned by the $H_i$ and $E_\alpha$, the linear span of $h_*$ is a non-trivial ideal in $\mathfrak{g}$, which must not exist for a simple Lie algebra.

**Example: Simple Roots of $\mathfrak{su}(3)$**

With the choice $\boldsymbol{t} = \boldsymbol{\alpha}_{(1)}$ in (15.48), the three roots in the shaded area in Fig. 15.9 are positive and the remaining three roots are negative. $\boldsymbol{\alpha}_1$ is the sum of the positive roots $-\boldsymbol{\alpha}_2$ and $\boldsymbol{\alpha}_3$ and thus not simple. Therefore, $-\boldsymbol{\alpha}_2$ and $\boldsymbol{\alpha}_3$ are the simple roots. Their scalar product is $-2$, and thus negative. The four black-marked roots are not simple.

◄

**Fig. 15.9**  Roots of $\mathfrak{su}(3)$

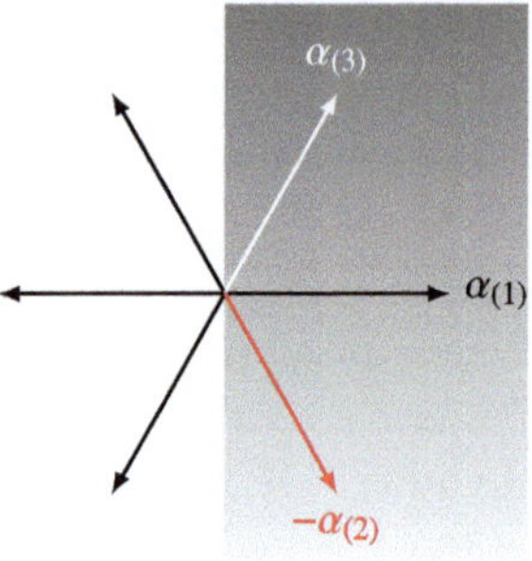


**Lemma 15.5**  *Let* $\Phi$ *be an irreducible root system. Then the Weyl group* $\mathcal{W}$ *acts irreducibly in* $\mathbb{R}^r$ *and the* $\mathcal{W}$-*orbit of each root spans* $\mathbb{R}^r$.

***Proof***  Irreducible root systems were introduced in Definition 15.2. If the statement in the lemma were not true, then there would be a proper invariant subspace $E \subset \mathbb{R}^r$ under the action of $\mathcal{W}$, and its orthogonal complement $E^\perp$ would also be a proper invariant subspace, see the proof of Theorem 10.1 in Sect. 10.3. Now we choose an $\alpha \notin E$ and act with the Weyl reflection $\sigma_\alpha$ on a $v \in E$. The image vector $\sigma_\alpha(v)$ lies also in $E$, since $E$ is invariant, as well as the difference vector $v - \sigma_\alpha(v)$. If $v$ is not perpendicular to $\alpha$, then this difference vector is non-zero and parallel to $\alpha$. This contradicts our assumption $\alpha \notin E$. Therefore, in the case $\alpha \notin E$, all vectors in $E$ must be perpendicular to $\alpha$. This means that each root must either be in $E$ or in $E^\perp$. However, this would split the root system into two orthogonal subsystems. For an irreducible root system, this is not possible and we conclude $E^\perp = \emptyset$. This proves the first statement in the lemma. Since the linear span of the $\mathcal{W}$-orbit of a root defines a non-trivial invariant subspace in $\mathbb{R}^r$, the second statement follows from the first.

**Lemma 15.6**  *In an irreducible root system* $\Phi$ *there are at most two different root lengths and* $\mathcal{W}$ *acts transitively on the long roots and on the short roots.*



Verify the second statement for the roots of $G_2$ in Fig. 15.8.

***Proof***  Let $\alpha$ and $\beta$ be arbitrary roots. In the Weyl orbit of $\alpha$, there is a root $\alpha'$, which is not perpendicular to $\beta$, since the linear span of the orbit is equal to $\mathbb{R}^r$. All roots in the orbit have the same length as $\alpha$. Therefore, the ratio of the squared lengths of $\alpha$ and $\beta$ is the same as that of $\alpha'$ and $\beta$, i.e., equal to 3, 2, 1, 1/2, or 1/3. If there were three root vectors with different lengths, then in addition to the ratios 1/2 and 1/3, the ratio 2/3 would also appear, which is not allowed. Let $\alpha$ and $\beta$ be any two roots of the same length. We may additionally assume $(\alpha, \beta) > 0$, since $\alpha$ and $-\alpha$ lie in the same Weyl orbit. Then $n_{\alpha\beta} = n_{\beta\alpha} = 1$, see Table 15.1, so that

$$\sigma_\alpha(\beta) = \beta - n_{\alpha\beta}\alpha = \beta - \alpha, \quad \sigma_\beta(\alpha) = \alpha - n_{\beta\alpha}\beta = \alpha - \beta,$$

from which it follows that

$$(\sigma_\alpha \sigma_\beta \sigma_\alpha)(\beta) = (\sigma_\alpha \sigma_\beta)(\beta - \alpha) = \sigma_\alpha(-\alpha) = \alpha. \tag{15.52}$$

This means that any two roots of the same length lie in the same orbit of the Weyl group.

**Lemma 15.7**  *Let $\gamma \in \Delta$ and $\alpha \neq \gamma$ be a positive root. Then $\sigma_\gamma(\alpha)$ is also positive.*

The lemma states that the Weyl reflection on a plane orthogonal to a simple root $\gamma$ permutes the positive roots $\alpha \neq \gamma$.

---

**Question**

Why does the lemma not hold for $\alpha = \gamma$? Why would it contradict Lemma 15.5 if we did not exclude $\alpha = \gamma$?

**Proof**  Let us first assume that $\gamma$ is one of the simple roots, say $\gamma_1$. We expand the positive root $\alpha = \sum n_i \gamma_i$ with $n_i \geq 0$ and find

$$\sigma_{\gamma_1}(\alpha) = \sum n_i \gamma_i - \left( \sum_{i=1}^{r} 2n_i \frac{\gamma_1 \cdot \gamma_i}{\gamma_1^2} \right) \gamma_1.$$

Since $\alpha \neq \gamma_1$, at least one of the coefficients $n_2, n_3, \ldots, n_r$ is positive. Therefore, according to Theorem 15.1, all coefficients are zero or positive, so that $\sigma_{\gamma_1}(\alpha)$ must be a positive root.

**Lemma 15.8 (Simple Weyl Reflections)**  *For each $\alpha \in \Phi$, there exists a $\gamma \in \Delta$ and a sequence of simple Weyl reflections whose composition maps $\gamma$ to $\alpha$. As a consequence, the simple Weyl reflections generate the Weyl group.*

**Proof**  Simple Weyl reflections are reflections in planes perpendicular to simple roots. It suffices to prove the statement for $\alpha \in \Phi^+$. Then it also holds for $\alpha \in \Phi^-$ because $\sigma_\gamma \gamma = -\gamma$. Let $\alpha = \sum_i n_i \gamma_i$ with $n_i \in \mathbb{N}_0$ and height $h_\alpha = \sum_i n_i$. For $h_\alpha = 1$, there is nothing to prove. For $h_\alpha \geq 2$, at least two coefficients $n_i$ are positive, and because of

$$\alpha^2 = \sum n_i \, \alpha \cdot \gamma_i > 0$$

we conclude $\alpha \cdot \gamma_i > 0$ for at least one $\gamma_i \in \Delta$ with $n_i > 0$. Without loss of generality, let this be $\gamma_1$. Reflecting $\alpha$ in the plane perpendicular to $\gamma_1$,

$$\sigma_{\gamma_1}(\alpha) = \alpha - \frac{2\gamma_1 \cdot \alpha}{\gamma_i^2}\, \gamma_1 ,$$

we obtain, according to Lemma 15.7, a positive root, but with reduced height, since $\gamma_1 \cdot \alpha$ is positive. Induction on the height of the roots proves the first part of Lemma 15.8. To prove the second part of the lemma, choose for $\alpha \in \Phi$ a product $w$ of simple reflections with $w(\gamma) = \alpha$. Then $\sigma_\alpha = \sigma_{w(\gamma)} = w\sigma_\gamma w^{-1}$, so that $\sigma_\alpha$ is also a product of simple reflections.

The action of the finite Weyl group $\mathcal{W} < O(r)$ on $\mathbb{R}^r$ or on the root system $\Phi$ is characterized by Weyl chambers, which are also relevant for representation theory. A special Weyl chamber is called fundamental:

**Definition 15.4 (Fundamental Weyl Chamber)** The fundamental Weyl chamber $\mathcal{W}_\Delta$ contains all $x \in \mathbb{R}^r$, that have a positive scalar product with all simple roots $\gamma_i$.

The fundamental Weyl chamber contains all $x \in \mathbb{R}^r$ that satisfy the inequalities $x \cdot \gamma_i > 0$ for all simple roots $\gamma_i$. As an intersection of open half-planes, $\mathcal{W}_\Delta$ is also open and convex. Due to Lemma 15.2, it also holds that

$$\mathcal{W}_\Delta = \{x \in \mathbb{R}^r | x \cdot \alpha > 0, \alpha \in \Phi^+\}. \tag{15.53}$$

In particular, the vector $t$ in (15.48), which was used for the polarization of the root system, always lies in the fundamental Weyl chamber $\mathcal{W}_\Delta$. Conversely, two vectors $t$ and $t'$ in the same fundamental Weyl chamber define the same positive roots. Let $w$ be any element in $\mathcal{W}$ and $\Phi = \Phi^+ \cup \Phi^-$ a polarization of the root system with a vector $t \in \mathbb{R}^r$ and $\Delta$ the corresponding simple roots. Then $w(\Phi) = \Phi = w(\Phi^+) \cup w(\Phi^-)$ is a polarization of the root system with respect to the image vector $w(t)$ and $w(\Delta) = \{w(\gamma_1), \ldots, w(\gamma_r)\}$ is a basis of simple roots with respect to the new polarization.

---

**Question**

How can these statements be justified without a calculation?

Fundamental Weyl chambers of $\mathfrak{su}(3)$ and $\mathfrak{so}(5)$ are the shaded areas in Figs. 15.1 and 15.10. The Weyl chambers of $G_2$ are determined in Problem 15.3.

The closure $\overline{\mathcal{W}}_\Delta$ of $\mathcal{W}_\Delta$ appearing in the following lemma is also characterized by (15.53) if the inequality is replaced by $\alpha \cdot x \geq 0$.

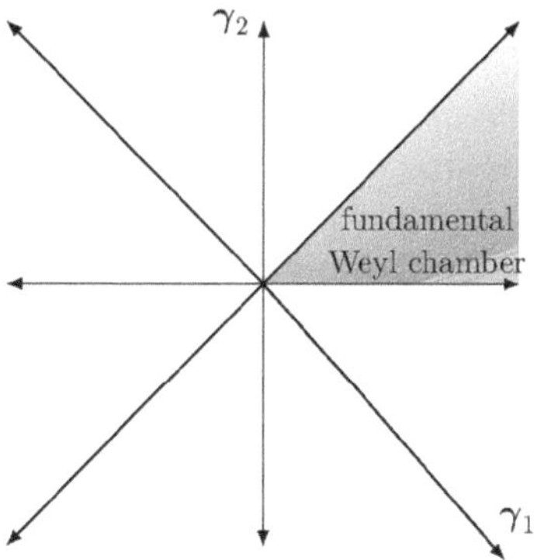

**Fig. 15.10** The roots and simple roots of $B_2$ with the corresponding fundamental Weyl chamber



**Lemma 15.9** *The closure $\overline{\mathcal{W}}_\Delta$ of the fundamental Weyl chamber $\mathcal{W}_\Delta$ is a fundamental domain for the action of the Weyl group $\mathcal{W}$.*

A subset $\overline{\mathcal{W}}_\Delta \subseteq \mathbb{R}^r$ is a fundamental domain of the action of the Weyl group on $\mathbb{R}^r$ if for every $x \in \mathbb{R}^r$ there exists exactly one $w \in \mathcal{W}$ with $w(x) \in \overline{\mathcal{W}}_\Delta$. Hence, in the orbit of $x_0 \in \overline{\mathcal{W}}_\Delta$ there is no other element in $\overline{\mathcal{W}}_\Delta$.

***Proof*** In $\mathbb{R}^r$ we introduce a partial order: $y \geq x$ if $y - x$ is a linear combination of simple roots with non-negative *real* coefficients. This means that $x$ increases when proceeding in the direction of a positive root. We first prove that every orbit of the Weyl group contains an element $x_0$ in $\overline{W}_\Delta$. If $x_0$ is the largest element in its orbit with respect to the partial order, then

$$x_0 - \sigma_i(x_0) = \frac{2x_0 \cdot \gamma_i}{\gamma_i^2}\, \gamma_i, \quad \text{with} \quad \sigma_i = \sigma_{\gamma_i}, \tag{15.54}$$

must be a linear combination of simple roots with non-negative coefficients. This proves $x_0 \cdot \gamma_i \geq 0$, i.e., $x_0 \in \overline{W}_\Delta$.

To prove that not several elements of an orbit lie in the fundamental domain, we introduce a subset $N(w)$ of the positive roots and the length $\ell(w)$ of $w \in \mathcal{W}$.

**Definition 15.5** For each element $w$ of $\mathcal{W}$, we define $N(w) = \{\alpha \in \Phi^+ \mid w(\alpha) \in \Phi^-\}$. Under $w$, $n(w) \equiv |N(w)|$ positive roots are mapped to negative roots.

Thus, according to Lemma 15.7, for a simple root $N(\sigma_\gamma) = \{\gamma\}$ and accordingly $n(\sigma_\gamma) = 1$.

**Definition 15.6** The minimal number of simple reflections needed for the factorization of $w$ is the length $\ell(w)$ of $w$.

**Lemma 15.10** *For each element of the Weyl group, $\ell(w) = n(w)$.*

***Proof*** A simple root maps exactly one positive root to a negative root and the other positive roots to positive roots. Therefore, a product $\sigma_1 \cdots \sigma_k$ of $k$ *simple* reflections can map at most $k$ positive roots to negative roots, and $n(w) \leq \ell(w)$.

We prove the lemma by induction on $k$. It is sufficient to show that a product $w = \sigma_\gamma \sigma_1 \cdots \sigma_k$ of $(k+1) > n(w)$ factors is not reduced. i.e., does not have the minimal length $\ell(w)$. We assume conversely that it is reduced. Then the product $w' = \sigma_1 \cdots \sigma_k$ of length $k$ is also reduced. By induction on $k$, it follows that $\ell(w') = n(w') = k$ and $w'$ maps $k$ positive roots to negative roots. We conclude that $n(w) < k + 1$ and thus $n(w) \leq n(w')$, and because $w' = \sigma_\gamma w$, the simple reflection $\sigma_\gamma$ transforms one of the negative roots back into a positive root. This negative root must be $-\gamma$, and $\gamma = w(\alpha)$ for some $\alpha \in \Phi^+$. This means that the simple reflections $\sigma_i$ in the product $w' = \sigma_1 \cdots \sigma_k$ leave the simple root $\alpha$ in $\Phi^+$ for a while, then reverse the sign, and then leave it in $\Phi^-$. Thus, we have

$$w_1 = \sigma_1 \sigma_2 \cdots \sigma_{\gamma'} \cdots \sigma_{k-1} \sigma_k = w' \sigma_{\gamma'} w'', \quad w''(\alpha) = \gamma', \quad w'(\gamma') = \gamma ,$$

and because $\sigma_{\gamma'}(\gamma') = -\gamma'$, it follows that $w_1(\alpha) = w' \sigma_{\gamma'} w''(\alpha) = -\gamma$. The property $w'(\gamma') = \gamma$ and Corollary 15.2 imply $w' \sigma_{\gamma'} w'^{-1} = \sigma_{w'(\gamma')} = \sigma_\gamma$. From this, it follows that

$$w = \sigma_\gamma w_1 = \sigma_\gamma w' \sigma_{\gamma'} w'' = w' w'' . \tag{15.55}$$

Thus, the original product for $w$ was not reduced, and the lemma is proven.

---

**Task**

Use Lemma 15.10 to prove that no more than one element of an orbit can lie in the fundamental region $\bar{\mathcal{W}}_\Delta$.

---

The Weyl chambers are defined as follows: To each root $\alpha \in \Phi$, we assign the hyperplane $L_\alpha$, which is perpendicular to the root and passes through the origin, $L_\alpha = \{x \in \mathbb{R}^r | (x, \alpha) = 0\}$. The connected components of $\mathbb{R}^r \setminus \bigcup_\alpha L_\alpha$ are the Weyl chambers. The fundamental Weyl chamber is the component that contains the vector $t$ used in the polarization $\Phi = \Phi^+ \cup \Phi^-$.

Using Lemma 15.10, one can prove further useful corollaries, e.g.

1. The only element in $\mathcal{W}$ that leaves $\Delta$ invariant is $w = 1$.
2. The Weyl group acts transitively on the set of Weyl chambers.
3. The number of Weyl chambers is equal to the cardinality of the Weyl group.

**Corollary 15.3 (Irreducible Root System)** *For an irreducible $\Phi$, $\Delta$ is not the disjoint union of two subsets $\Delta_1$ and $\Delta_2$, where each element of $\Delta_1$ is perpendicular to each element of $\Delta_2$.*

If the simple roots in $\Delta_1$ were perpendicular to the simple roots in $\Delta_2$, then both subsets would be invariant under simple Weyl reflections. According to Lemma 15.8, all roots would then lie either in $\Delta_1$ or in $\Delta_2$. However, this is impossible for an irreducible root system.

## 15.6  Cartan Matrix and Dynkin Diagrams

According to Lemma 15.8, by repeatedly applying simple Weyl reflections, one can reconstruct all roots in $\Phi$ from the simple roots. Therefore, all the information about the Lie algebra is encoded in the normalized scalar products of the simple roots. These form the integer matrix elements of the *Cartan matrix*,

$$K_{ij} = \frac{2\gamma_i \cdot \gamma_j}{\gamma_j^{\,2}}, \qquad K = (K_{ij}) \in \mathrm{GL}(r, \mathbb{Z}). \tag{15.56}$$

The matrix is regular because the $\gamma_i$ are linearly independent. All diagonal elements are 2, and thus $K$ contains redundant information. The off-diagonal elements are negative or zero. If all simple roots have the same length, then $K_{ij} = 2\,\hat{\gamma}_i \cdot \hat{\gamma}_j = K_{ji}$. Here, $\hat{\gamma} = \gamma/|\gamma|$ is the unit vector in the direction of $\gamma$. If $K$ is symmetric, then the Lie algebra is called *simply laced*.

---

**Example: Cartan Matrices of Simple Lie Algebras with Rank $2$**

The quantization of the roots in Table 15.1 allow for only three $2 \times 2$ Cartan matrices:

$$\mathfrak{su}(3): \; K = \begin{pmatrix} 2 & -1 \\ -1 & 2 \end{pmatrix}, \quad \mathfrak{so}(5): \; K = \begin{pmatrix} 2 & -1 \\ -2 & 2 \end{pmatrix}, \quad G_2: \; K = \begin{pmatrix} 2 & -1 \\ -3 & 2 \end{pmatrix}. \tag{15.57}$$

◀

The Cartan matrices are obtained in LiE using the command `Cartan`:

```
LiE> Cartan(G2)-> [[2,-1],[-3,2]]
LiE> Cartan(B3)-> [[ 2,-1, 0],[-1, 2,-2],[ 0,-1, 2]]
LiE> Cartan(F4)-> [[ 2,-1, 0, 0],[-1, 2,-2, 0],[ 0,-1, 2,-1],[ 0,
    0,-1, 2]]
```

The matrix $K$ is diagonalizable and has positive eigenvalues. This follows from

**Corollary 15.4** *The Cartan matrix $K$ is conjugate to the symmetric and positive matrix $\hat{K}$ with matrix elements $\hat{K}_{ij} = 2\,\hat{\gamma}_i \cdot \hat{\gamma}_j$. The matrix $D$ in $K = D\hat{K}D^{-1}$ is diagonal and positive, $D = diag\,(|\gamma_1|, \ldots, |\gamma_r|)$.*

The matrix $\hat{K}$ is positive because the quadratic form $\frac{1}{2}(x, \hat{K}x) = (y, y)$ with $y = \sum x_i \hat{y}_i$ is non-negative and only vanishes for $x = 0$, since the $\hat{y}_i$ are linearly independent. The eigenvalues of the symmetric and positive matrix $\hat{K}$, and thus those of $K$, are positive.

---

**Question**

A non-symmetric Cartan matrix does not need to be positive, although it has positive eigenvalues. Which of the Cartan matrices in (15.57) is not positive?

With increasing rank, it becomes more advantageous to encode the information contained in $K$ in a *Dynkin diagram* as follows:

---

**Dynkin Diagrams**

1. To each simple root $y_i$, associate a vertex $V_i$.
2. Two vertices $V_i$ and $V_j$ are connected with $K_{ij}K_{ji} \in \{0, 1, 2, 3\}$ lines.
3. If $V_i$ and $V_j$ are connected with more than one line, then $|y_i| \neq |y_j|$ and an arrow is drawn from $V_i$ to $V_j$ if $|y_i| > |y_j|$, and vice versa. Alternatively, vertices corresponding to short roots are marked with a filled circle, and those corresponding to long roots with an open circle.

---

A Dynkin diagram cannot decompose into two disconnected sub-diagrams, as the simple roots of one sub-diagram would then be perpendicular to the simple roots of the other sub-diagram. However, this is not possible due to Corollary 15.3.

If $x \neq 0$, then $\frac{1}{2}(x, \hat{K}x) > 0$. Since $\hat{K}_{ii} = 2$ and two different simple roots have a non-positive scalar product, the inequality implies

$$x^2 > \sum_{i<j} x_i x_j |\hat{K}_{ij}| \quad \text{for} \quad x \neq 0, \tag{15.58}$$

where the symmetric matrix $\hat{K}$ was introduced in Corollary 15.4. To apply this inequality, we consider a subset $\mathcal{A}$ of the index set $\{1, 2, \ldots, r\}$, and an $r$-tuple $x$ with components $x_i = 1$ for $i \in \mathcal{A}$ and $x_i = 0$ for $i \notin \mathcal{A}$. Then $x^2$ is the number of elements in $\mathcal{A}$ and the inequality (15.58) implies

**Corollary 15.5** *For any subset $\mathcal{A}$ of the index set $\{1, 2, \ldots, r\}$, the inequality holds*

$$|\mathcal{A}| > \sum_{i<j \in \mathcal{A}} |\hat{K}_{ij}|. \tag{15.59}$$

We identify the indices with the vertices $V_1, V_2, \ldots, V_r$ of a Dynkin diagram and recall the quantization of the roots, according to which $\hat{K}_{ij}^2$—this is the number of connecting lines between vertices $i$ and $j$ in the Dynkin diagram—only takes values in $\{0, 1, 2, 3\}$. Thus, for connected vertices, the number $|\hat{K}_{ij}|$ is greater than or equal to 1, and it follows

**Corollary 15.6** *For any subset $\mathcal{A}$ of vertices in the Dynkin diagram, the number of vertices in $\mathcal{A}$ is greater than the number of connected pairs within $\mathcal{A}$.*

With these inequalities one proves:

**Property 15.1** *Dynkin diagrams have no loops.*

If there were a loop we could choose for $\mathcal{A}$ the vertices traversed by the loop, and the number of connected pairs in $\mathcal{A}$ would be at least as large as the number of vertices in $\mathcal{A}$, contradicting Corollary 15.6. Similarly, one proves the statement

**Property 15.2** *In Dynkin diagrams, at most three lines are attached to a vertex.*

To see this, we choose for $\mathcal{A}$ a vertex $i$ and the vertices connected to it. Now, let $\mathcal{A}' \subset \mathcal{A}$ be the subset without the distinguished vertex, then the right side of the inequality (15.58) is less than or equal to $x_i \sum_{j \in \mathcal{A}'} x_j$. If we choose $x_i = 2$ and $x_j = 1$, then the inequality implies

$$4 + |\mathcal{A}'| > 2\,|\mathcal{A}'|. \tag{15.60}$$

This implies $|\mathcal{A}'| < 4$ and thus vertex $i$ can be connected to a maximum of 3 vertices.

Another piece of useful information is the following: a linear chain of vertices connected successively with one line each can be replaced by a single vertex.

$$\gamma_1, \ldots, \gamma_n \longrightarrow \bar{\gamma} = \gamma_1 + \gamma_2 + \cdots + \gamma_n, \tag{15.61}$$

This is sketched in Fig. 15.11.

To prove this statement, we divide the set of vertices into three subsets:

1. the vertices of the linear chain,
2. the vertices in $\mathcal{A}$, which are connected to the chain via vertex 1,
3. the vertices in $\mathcal{B}$, which are connected via vertex $n$ with the chain.

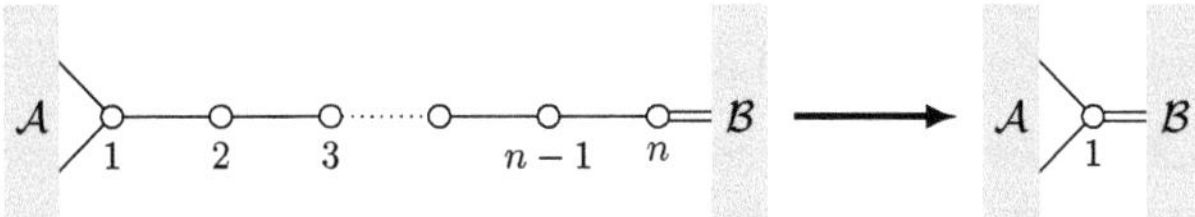

**Fig. 15.11** A linear chain of vertices $1, \ldots, n$ with one connection each can be replaced by a single vertex

---

**Question**

Why does $\bar{\boldsymbol{\gamma}}$ have the same length as the simple roots $\boldsymbol{\gamma}_1, \boldsymbol{\gamma}_2, \ldots, \boldsymbol{\gamma}_n$? Why do the $\mathcal{A}$-roots have the same scalar products with $\bar{\boldsymbol{\gamma}}$ as with $\boldsymbol{\gamma}_1$?

The only root in the linear chain that is not perpendicular to the roots of $\mathcal{A}$ is $\boldsymbol{\gamma}_1$. It follows that $\bar{\boldsymbol{\gamma}} \cdot \boldsymbol{\gamma}_a = \boldsymbol{\gamma}_1 \cdot \boldsymbol{\gamma}_a$ for all $\boldsymbol{\gamma}_a$ with $a \in \mathcal{A}$. A corresponding statement holds for the roots $\boldsymbol{\gamma}_b$ of $\mathcal{B}$ and $\boldsymbol{\gamma}_n$. Since the $\mathcal{A}$-roots are also perpendicular to the $\mathcal{B}$-roots, the Cartan matrix $K$ of the diagram with linear chain in Fig. 15.11 has the following block form:

$$
K = \begin{pmatrix} & & & u_{\mathcal{A}}^T & 0 \\ & K_n & & 0 & 0 \\ & & & 0 & u_{\mathcal{B}}^T \\ \hline v_{\mathcal{A}} & 0 & 0 & K_{\mathcal{A}} & 0 \\ 0 & 0 & v_{\mathcal{B}} & 0 & K_{\mathcal{B}} \end{pmatrix} \longrightarrow K_c = \begin{pmatrix} 2 & u_{\mathcal{A}}^T & u_{\mathcal{B}}^T \\ \hline v_{\mathcal{A}} & K_{\mathcal{A}} & 0 \\ v_{\mathcal{B}} & 0 & K_{\mathcal{B}} \end{pmatrix} \qquad (15.62)
$$

Here, the $n$-dimensional Cartan matrix $K_n$ of the linear chain has 2 on the diagonal and $-1$ on the two off-diagonals. The components of the vectors $u_{\mathcal{A}}$ and $v_{\mathcal{A}}$ contain the scalar products $\boldsymbol{\gamma}_1 \cdot \boldsymbol{\gamma}_a$, and those of $u_{\mathcal{B}}$ and $v_{\mathcal{B}}$ the scalar products $\boldsymbol{\gamma}_n \cdot \boldsymbol{\gamma}_b$. If we now replace all roots of the linear chain with the root $\bar{\boldsymbol{\gamma}}$, then the matrix $K_n$ is replaced by the number 2. The components of $u_{\mathcal{A}}$ and $v_{\mathcal{A}}$ now contain the scalar products of $\bar{\boldsymbol{\gamma}}$ with the $\mathcal{A}$-roots, and those of $u_{\mathcal{B}}$ and $v_{\mathcal{B}}$ the scalar products of $\bar{\boldsymbol{\gamma}}$ with the $\mathcal{B}$-roots. These now appear in the first row and first column of the Cartan matrix $K_c$ belonging to the contracted Dynkin diagram on the right in Fig. 15.11, in which the linear chain has been replaced by a single vertex. The lines connected to the vertices 1 and $n$ connect after contraction to the single vertex with root vector $\bar{\boldsymbol{\gamma}}$.

The Cartan matrices $K$ and $K_c$ of the original and contracted diagram give rise to the symmetric and positive matrices $\hat{K}$ and $\hat{K}_c$ in Corollary 15.4. These have the same block structure as the Cartan matrices. Specifically, the symmetric Cartan matrix $K_n$ of the linear chain appears unchanged in $\hat{K}$. If $\boldsymbol{p}$ is a column vector with $n$ identical components $p$, then $(\boldsymbol{p}, K_n \boldsymbol{p}) = 2p^2$. If $\boldsymbol{q}$ is another vector with $r - n$ arbitrary components, then

$$
(\boldsymbol{n}, \hat{K}\boldsymbol{n}) = (\boldsymbol{n}_c, \hat{K}_c \boldsymbol{n}_c), \quad \boldsymbol{n} = \begin{pmatrix} \boldsymbol{p} \\ \boldsymbol{q} \end{pmatrix}, \quad \boldsymbol{n}_c = \begin{pmatrix} p \\ \boldsymbol{q} \end{pmatrix} . \qquad (15.63)
$$

It follows that $\hat{K}$ cannot be positive if $\hat{K}_c$ is not. In other words: if we find a contraction with non-positive $\hat{K}_c$, then $\hat{K}$ is non-positive and $K$ cannot possibly be the Cartan matrix of a simple Lie algebra. From this follows:

**Property 15.3** *Dynkin diagrams contain at most one double line or one vertex with three connected lines, but not both simultaneously.*

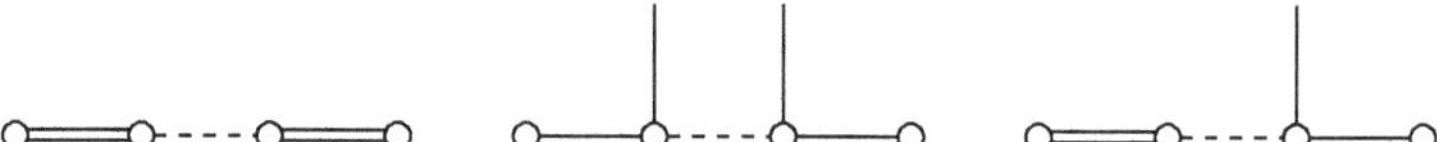

**Fig. 15.12** These diagrams are not Dynkin diagrams

**Fig. 15.13** Only the Dynkin diagram of $G_2$ has a triple line

**Fig. 15.14** Diagrams with $\det K = 0$

**Fig. 15.15** Only allowed diagrams with a double line

Three diagrams forbidden by this rule are shown in Fig. 15.12. These diagrams are forbidden because by contracting the linear chain (the dashed connecting line) a vertex with four or more attached lines would be created.

**Property 15.4** *The triple line only appears in the diagram with two vertices in Fig. 15.13.*

---

### Task

Determine the $K$-matrices of the diagrams in Fig. 15.14 and show that they have the eigenvectors $x^T = (1, 2, 3, 2, 1)$ and $x^T = (1, 2, 3, 4, 2)$ with eigenvalue 0.

This proves the

**Property 15.5** *The matrices $K$ for the diagrams (15.14) have* $\det K = 0$ *and are not allowed.*

**Property 15.6** *The only Dynkin diagrams with a double line are those in Fig. 15.15.*

To justify this, one notes that additional diagrams with a double line can be contracted to one of the forbidden diagrams in Fig. 15.14.

If a diagram is branched, then according to Property 15.3, it may only have one branching and all roots must be of equal length. This means that a branched diagram has the form sketched in Fig. 15.16.

However, not all branched diagrams of this form are permitted. For example, the diagram in Fig. 15.17, whose $K$-matrix is discussed in Problem 15.6, is forbidden.

**Property 15.7** *The matrix $K$ of the Diagram 15.17 has* $\det K = 0$ *and is not allowed.*

**Fig. 15.16** Initially admitted
diagrams with branching

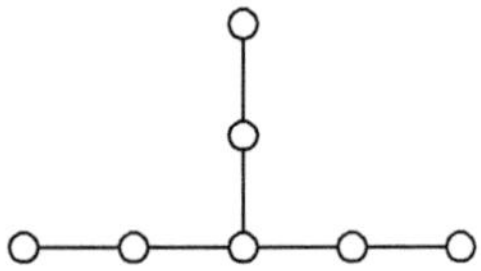

**Fig. 15.17** Forbidden
diagram with $\det K = 0$

**Fig. 15.18** Two forbidden
diagrams with $\det K = 0$

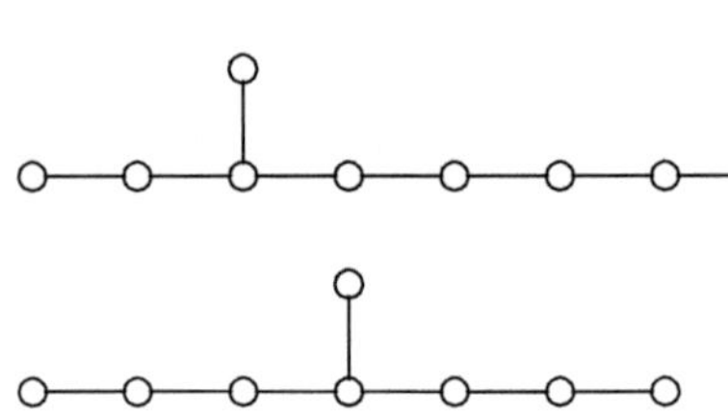

This means that of the diagrams in Fig. 15.16, only those with $n = 2$ occur. Finally, the Cartan matrices of the diagrams in Fig. 15.18 have the eigenvalue 0 and are also not allowed:

---

**Task**

Show that the $K$-matrices of the two diagrams in Fig. 15.18 have the following 8 eigenvalues:
$$\left\{0, 1, 2, 3, 4, \tfrac{1}{2}(5 \pm \sqrt{5}), \tfrac{1}{2}(3 \pm \sqrt{5})\right\} \qquad \text{and} \qquad \left\{0, 1, 2, 2, 3, 4, 2 \pm \sqrt{2}\right\}.$$

Finally, there is an interesting connection between the Dynkin diagrams of a Lie algebra $\mathfrak{g}$ and the Weyl group of its root system. For this, we recall the relations (15.1) for the reflections $\sigma_i$ at the simple roots $\boldsymbol{\gamma}_i$:

$$(\sigma_i \sigma_j)^{m_{ij}} = 1. \tag{15.64}$$

The exponents $m_{ij}$ can be easily read from the Dynkin diagram: if the vertices $V_i$ and $V_j$ are not connected, then $m_{ij} = 2$. For a single line, $m_{ij} = 3$, for a double line, $m_{ij} = 4$, and for a triple line, $m_{ij} = 6$.

---

## 15.7   Cartan Classification

The conditions derived in the previous section greatly restrict the form of the Dynkin diagrams. For each allowed diagram, there exists a simple Lie algebra with this diagram, as can be seen through an explicit construction of the root systems. These systems are given in Appendix 15.8. In this way, one arrives at Cartan's classification of simple complex Lie algebras.

> **All Simple Complex Lie Algebras**
> There exist four infinite families of Lie algebras and five exceptional Lie algebras. They are denoted by
>
> $$A_r, \quad B_r, \quad C_r, \quad D_r, \quad \text{and} \quad G_2, \quad F_4, \quad E_6, \quad E_7, \quad E_8 \,. \qquad (15.65)$$

Here, the index is equal to the rank of the Lie algebra or the number of vertices in the Dynkin diagram. The four families are the complexified Lie algebras of the classical groups,

$$A_r = \mathfrak{sl}(r+1, \mathbb{C}), \quad B_r = \mathfrak{so}(2r+1, \mathbb{C}), \quad C_r = \mathfrak{sp}(2r, \mathbb{C}),$$

$$D_r = \mathfrak{so}(2r, \mathbb{C}) \,, \qquad (15.66)$$

or their unique compact forms

$$A_r = \mathfrak{su}(r+1), \quad B_r = \mathfrak{so}(2r+1), \quad C_r = \mathfrak{usp}(2r), \quad D_r = \mathfrak{so}(2r) \,. \qquad (15.67)$$

The exceptional Lie algebras $G_2, F_4, E_6, E_7$, and $E_8$ have the dimensions $14, 52, 78, 133$, and $248$. They are the Lie algebras of the exceptional Lie groups. The Dynkin diagrams of the simple Lie algebras are shown in Fig. 15.19. The roots of $A_r, D_r, E_6, E_7$, and $E_8$ are of equal length—they are the simply laced Lie algebras.

The Dynkin diagrams can be obtained in LiE using the command `diagram`:

```
LiE> diagram(B4) -> 0---0---0=>=0
LiE>                1   2   3   4
LiE> diagram(F4) -> 0---0=>=0---0
LiE>                1   2   3   4
LiE>                        0 2
LiE>                        |
LiE>                        |
LiE> diagram(E8) -> 0---0---0---0---0---0---0
LiE>                1   3   4   5   6   7   8
```

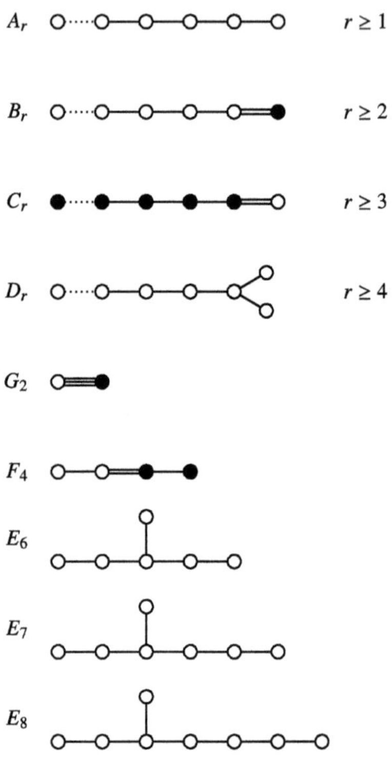

**Fig. 15.19** The Dynkin
diagrams of the simple Lie
algebras

$A_r$   o····o—o—o—o—o    $r \geq 1$

$B_r$   o····o—o—o—o⟹●    $r \geq 2$

$C_r$   ●····●—●—●—●⟸o    $r \geq 3$

$D_r$   o····o—o—o—o<    $r \geq 4$

$G_2$   o⟹●

$F_4$   o—o⟹●—●

$E_6$

$E_7$

$E_8$

---

## 15.8   Appendix A: Explicit Irreducible Root Systems

In this appendix, we provide explicit roots for all semi-simple Lie algebras in the
Cartan classification. Let $\{e_1, \cdots, e_n\}$ be a Cartesian basis in $\mathbb{R}^n$. Moreover, $\pm e_i \pm e_j$ stands for the 4 vectors $e_i + e_j$, $e_i - e_j$, $-e_i + e_j$ and $-e_i - e_j$.

**Root System of $A_r$** Let $E$ be the hyperplane in $\mathbb{R}^{r+1}$ orthogonal to the vector
$e_1 + e_2 + \cdots + e_{r+1}$. Then the root system contains the $r^2 + 2r$ roots

$$\Phi = \left\{ e_i - e_j \,|\, i \neq j \right\} . \tag{A.1}$$

The simple roots are chosen as the vectors

$$\Delta = \left\{ \gamma_i = e_i - e_{i+1} \,|\, i = 1, \ldots, r \right\} . \tag{A.2}$$

They all have the same length, so that the angle between any two simple roots must
be 120° or 90°. In particular, it holds that

$$\gamma_i \cdot \gamma_{i+1} = -1, \quad i = 1, 2, \ldots, r - 1, \tag{A.3}$$

and this means that the off-diagonal elements of the Cartan matrix $K$ are equal to $-1$. Furthermore, $\det(K) = r + 1$. The highest root is $e_1 - e_r$.

The Weyl group is the symmetric group $S_{r+1}$, and it contains the permutations of the vectors $\{e_1, \ldots, e_{r+1}\}$. For example, the Weyl group of $A_2$ is equal to $S_3 \cong \mathcal{D}_3$.

**Root System of $B_r$**   Here, $E = \mathbb{R}^r$ and the root system contains $2r^2$ roots,

$$\Phi = \{\pm e_i\} \cup \left\{\pm e_i \pm e_j \,|\, i \neq j\right\}. \tag{A.4}$$

The simple roots are chosen as

$$\Delta = \left\{\gamma_i = e_i - e_{i+1} \,|\, i = 1, \ldots, r-1\right\} \cup \left\{\gamma_r = e_r\right\}. \tag{A.5}$$

The highest root is $e_1 + e_2$ and the Weyl group is the group of permutations and sign changes of the set $\{e_1, \ldots, e_r\}$ and therefore contains $2^r r! = (2r)!!$ elements.

**Root System of $C_r$**   Again, $E = \mathbb{R}^r$. The root system is dual to that of $B_r$:

$$\Phi = \left\{\pm 2e_i\right\} \cup \left\{\pm e_i \pm e_j \,|\, i \neq j\right\}. \tag{A.6}$$

The simple roots are chosen as the vectors

$$\Delta = \left\{\gamma_i = e_i - e_{i+1} \,|\, i = 1, \ldots, r-1\right\} \cup \left\{\gamma_r = 2e_r\right\}. \tag{A.7}$$

The highest root is $2e_1$ and the Weyl group is identical to that of $B_r$.

**Root System of $D_r$**   Here too, $E = \mathbb{R}^r$ and the $2r(r-1)$ roots are

$$\Phi = \left\{\pm e_i \pm e_j \,|\, i \neq j\right\}. \tag{A.8}$$

The simple roots are chosen as

$$\Delta = \left\{\gamma_i = e_i - e_{i+1} \,|\, i = 1, \ldots, r-1\right\} \cup \left\{\gamma_r = e_{r-1} + e_r\right\}. \tag{A.9}$$

The highest root is $e_1 + e_2$ and the Weyl group is the group of permutations of $\{e_1, \ldots, e_r\}$ as well as the even sign changes of these basis vectors and contains $(2r)!!/2$ elements.

**Root System of $G_2$**   Here, $E$ is the plane hyperplane $\mathbb{R}^3$ perpendicular to $e_1 + e_2 + e_3$ and

$$\Phi = \left\{e_i - e_j \,|\, i \neq j\right\} \cup \left\{\pm (2e_i - e_j - e_k) \,|\, i \neq j \neq k\right\}. \tag{A.10}$$

The 2 simple roots are chosen as

$$\Delta = \left\{e_1 - e_2, -2e_1 + e_2 + e_3\right\}. \tag{A.11}$$

The highest root is $2e_3 - e_2 - e_1$ and the Weyl group is the group of permutations of $\{e_1, e_2, e_3\}$ as well as the even sign changes of these basis vectors and contains 12 elements. It is isomorphic to the dihedral group $D_6$.

**Root System of $F_4$**   For this root system $E = \mathbb{R}^4$ and

$$\Phi = \left\{ \pm e_i \right\} \cup \left\{ \pm e_i \pm e_j | i \neq j \right\} \cup \left\{ \tfrac{1}{2}(\pm e_1 \pm e_2 \pm e_3 \pm e_4) \right\}. \tag{A.12}$$

As 4 simple roots, one chooses

$$\Delta = \left\{ e_1 - e_2, e_2 - e_3, e_3, \tfrac{1}{2}(e_4 - e_1 - e_2 - e_3) \right\}. \tag{A.13}$$

The highest root is $e_1 + e_4$ and the Weyl group has 1152 elements.

**Root System of $E_8$**   The roots are from $E = \mathbb{R}^8$ and are

$$\Phi = \left\{ \pm e_i \pm e_j | i \neq j \right\} \cup \left\{ \tfrac{1}{2} \sum_i (-1)^{m(i)} e_i \ \Big| \ \sum_i m(i) \ \text{even} \right\}. \tag{A.14}$$

As the 8 simple roots, one chooses

$$\Delta = \left\{ \tfrac{1}{2}(e_1 + e_8 - \sum_{i=2}^{7} (-1)^{m(i)} e_i \right\} \cup \left\{ e_1 + e_2, e_2 - e_1, e_3 - e_2, \ldots, e_7 - e_6 \right\}. \tag{A.15}$$

The highest root is $e_7 + e_8$ and the Weyl group has 696 729 600 elements.

**Root System of $E_7$**   The root system is the intersection of the root system of $E_8$ with the vector space orthogonal to $e_7 + e_8$. The highest root is $e_8 - e_7$ and the Weyl group has 2 903 040 elements.

**Root System of $E_6$**   The root system is the intersection of the root system of $E_8$ with the vector space orthogonal to $e_6 - e_8$ and $e_7 - e_8$. The highest root is $\tfrac{1}{2}(e_1 + e_2 + e_3 + e_4 + e_5 - e_6 - e_7 - e_8)$ and the Weyl group has 51 840 elements.

To conclude, we list the number of positive roots and the order of the Weyl group for each simple Lie algebra in the Cartan classification in Table 15.2,

**Table 15.2**  The number of positive roots and the order of the Weyl group for the simple Lie algebras

| Type | $A_r$ | $B_r, C_r$ | $D_r$ | $E_6$ | $E_7$ | $E_8$ | $F_4$ | $G_2$ |
|---|---|---|---|---|---|---|---|---|
| $\lvert \Phi^+ \rvert$ | $\binom{l+1}{2}$ | $r^2$ | $r^2 - 1$ | 36 | 63 | 120 | 24 | 6 |
| $\lvert \mathcal{W} \rvert$ | $(r+1)!$ | $2^r r!$ | $2^{r-1} r!$ | $2^7 3^4 5$ | $2^{10} 3^4 5 \cdot 7$ | $2^{14} 3^5 5^2 7$ | $2^7 3^2$ | $2^2 3$ |

## 15.9  Exercises for Chap. 15

**Problem 15.1 (Example for Roots)**  Consider the generators

$$
H = \frac{1}{2}\begin{pmatrix} 0 & 0 & 0 \\ 0 & -1 & 0 \\ 0 & 0 & 1 \end{pmatrix}, \quad
X_1 = \frac{1}{2\sqrt{2}}\begin{pmatrix} 0 & -1 & 1 \\ -1 & 1 & 0 \\ 1 & 0 & 0 \end{pmatrix}, \quad
X_2 = \frac{i}{2\sqrt{2}}\begin{pmatrix} 0 & 1 & 1 \\ -1 & 0 & 0 \\ -1 & 0 & 0 \end{pmatrix}.
$$

1. Find a linear combination $E = c_1 X_1 + c_2 X_2$, such that $[H, E] = \alpha E$. What is the corresponding root?
2. Show that $[H, E^{\dagger}] = -\alpha E^{\dagger}$.
3. What is the complete list of roots?

**Problem 15.2 (Weyl Reflections)**  Show that a Weyl reflection of the step operators is implemented as follows:

$$
E_{\beta} \mapsto E_{\sigma_{\alpha}(\beta)} = \tilde{\sigma}_{\alpha} E_{\beta} \tilde{\sigma}_{\alpha}^{-1} \quad \text{with} \quad \tilde{\sigma}_{\alpha} = e^{-i\pi(E_{\alpha}+E_{-\alpha})/2} .
$$

**Problem 15.3 (Weyl Chamber of $G_2$)**  Find a fundamental Weyl chamber for $G_2$ and represent it graphically.

**Problem 15.4 (Positive Roots of $B_3$)**  Determine all positive roots of $B_3$.

**Problem 15.5 (Positive Roots of $D_4$)**  Determine all positive roots of $D_4$.

**Problem 15.6 (Forbidden Diagram)**  Verify that the $K$-matrix for the Diagram 15.17 has the following form

$$
K = \begin{pmatrix}
2 & -1 & 0 & 0 & 0 & 0 & 0 \\
-1 & 2 & -1 & 0 & 0 & 0 & 0 \\
0 & -1 & 2 & -1 & 0 & -1 & 0 \\
0 & 0 & -1 & 2 & -1 & 0 & 0 \\
0 & 0 & 0 & -1 & 2 & 0 & 0 \\
0 & 0 & -1 & 0 & 0 & 2 & -1 \\
0 & 0 & 0 & 0 & 0 & -1 & 2
\end{pmatrix} .
$$

It has the eigenvalues 0, 1, 1, 2, 3, 3, 4. What is the eigenvector with eigenvalue 0? Why is $K$ not a Cartan matrix of a simple Lie algebra?

**Problem 15.7 (Roots of $A_r$)**  Show that the Cartain matrix belonging to the (simple) roots in (A.1) is that of $A_r$.

**Problem 15.8 (Weyl Group of $A_r$)** Prove that the Weyl group of $A_r$ is the symmetric group $S_{r+1}$.

**Problem 15.9 (An Unique Element of Maximum Length in $\mathcal{W}$)** Argue that a Weyl group has a unique element $w_0$ of maximum length.

# Representations of Lie Algebras 16

*It is impossible to explain honestly the beauties of the laws of nature in a way that people can feel, without their having some deep understanding of mathematics. if someone does not understand mathematics. I am sorry, but this seems to be the case.*

—Richard Feynman

In this chapter, we will examine in more detail the representations of simple Lie algebras introduced in Sect. 13.3 and characterize all irreducible representations. In Chap. 14, we saw that every representation $D$ of a Lie group $G$ induces a representation $\mathfrak{D}$ of its Lie algebra $T_eG$, and the reverse is achieved through the exponential mapping.

We will see that an irreducible representation $\mathfrak{D}$ of a Lie algebra $\mathfrak{g}$ is characterized by its highest weight, similar to how an irreducible representation of $\mathfrak{su}(2)$ is characterized by the angular momentum $j$, the maximum possible value of the magnetic quantum number $m$. As in the analysis of the roots—these are the weights of the adjoint representation—we use the results on the irreducible representations of $\mathfrak{su}(2)$ in Sect. 15.1. In particular, we use the fact that for every root $\boldsymbol{\alpha}$, the elements $(H_\alpha, E_\alpha, E_{-\alpha})$ in (15.30) define a Lie subalgebra $\mathfrak{su}(2) \leq \mathfrak{g}$ acting on the representation space $\mathcal{V}$.

The components $\alpha_i$ of $\boldsymbol{\alpha}$ are the simultaneous eigenvalues of the commuting mappings $\mathrm{ad}_{H_i}$—the Cartan elements $H_i$ in the adjoint representation. In this chapter, we examine the action of the $H_i$ and $E_\alpha$ in an arbitrary irreducible representation $\mathfrak{D}$ of $\mathfrak{g}$. The simultaneous eigenvalues of the commuting $\mathfrak{D}(H_i)$ define the weights $\boldsymbol{\mu}$ of the representation.

In the second part of the chapter, the results are applied to derive Weyl's formula for the reduced Haar measure of connected compact Lie groups and to extend the results in Chap. 9. This leads to the important character and dimension formulas of Hermann Weyl.



A. Wipf, *Symmetries in Physics*, https://doi.org/10.1007/978-3-662-72675-4_16

337

The representation theory of simple Lie algebras is of great importance in fundamental physics, and accordingly, it is presented in many textbooks on group and representation theory and their applications in atomic, molecular, and particle physics, e.g., in the works [13, 14, 17, 58, 62, 64].

## 16.1   Weights of a Representation

Let $\{H_i, E_\alpha, E_{-\alpha}\}$ with $\alpha \in \Phi$ be the Cartan-Weyl basis of a simple Lie algebra $\mathfrak{g}$ introduced in Chap. 15, and $\mathfrak{D} : \mathfrak{g} \mapsto L(\mathcal{V})$ a unitary representation of $\mathfrak{g}$ on a linear space $\mathcal{V}$. We examine how their images $\{\mathfrak{D}(H_i), \mathfrak{D}(E_\alpha), \mathfrak{D}(E_{-\alpha})\}$ act on $\mathcal{V}$. Just like the $H_i$, the linear mappings $\mathfrak{D}(H_i)$ commute with each other and can be diagonalized simultaneously,

$$\mathfrak{D}(H_i)|\boldsymbol{\mu}\rangle = \mu_i|\boldsymbol{\mu}\rangle . \tag{16.1}$$

We use the Dirac notation and characterize the eigenvector $|\boldsymbol{\mu}\rangle$ by its eigenvalues. The $r$-tuple $\boldsymbol{\mu} = (\mu_1, \mu_2, \ldots, \mu_r)^t$ is called a weight of the representation $\mathfrak{D}$, and the simultaneous eigenvector $|\boldsymbol{\mu}\rangle$ a weight vector. We proceed similarly to the treatment of the roots. Due to the homomorphism property, for each root $\boldsymbol{\alpha}$, the linear mappings

$$\mathfrak{D}(H_\alpha), \mathfrak{D}(E_\alpha), \mathfrak{D}(E_{-\alpha}) \tag{16.2}$$

form an $\mathfrak{su}(2)$ subalgebra of $\mathfrak{gl}(\mathcal{V})$. Similarly as in (15.26), we conclude that for all roots $\boldsymbol{\alpha}$, the eigenvalues in

$$\mathfrak{D}(H_\alpha)|\boldsymbol{\mu}\rangle = \frac{2\,\boldsymbol{\alpha} \cdot \boldsymbol{\mu}}{\alpha^2}\,|\boldsymbol{\mu}\rangle \tag{16.3}$$

must be integers,

$$\frac{2\,\boldsymbol{\alpha} \cdot \boldsymbol{\mu}}{\alpha^2} \in \mathbb{Z} \quad \text{for all} \quad \boldsymbol{\alpha} \in \Phi . \tag{16.4}$$

This is the only quantization condition that a weight must satisfy and it defines the *weight lattice* $\Lambda_{\mathrm{W}}$ of $\mathfrak{g}$. Note that $2\boldsymbol{\alpha} \cdot \boldsymbol{\mu}/\mu^2$ does not need to be an integer, in contrast to $2\boldsymbol{\alpha} \cdot \boldsymbol{\mu}/\alpha^2$. Thus, every root is a weight, but the reverse is generally not true. This means that the root lattice $\Lambda_{\mathrm{R}}$, that by definition is the integral span of the elements of $\Phi$, is generally a genuine sublattice of the weight lattice.

---

**Question**

What are the possible weights for $\mathfrak{su}(2)$ (as multiples of $\boldsymbol{\alpha}$)?

For each $\alpha \in \Phi$, the linear space $\mathcal{V}$ carries a representation of the Lie subalgebra $\mathfrak{su}(2) < \mathfrak{g}$ spanned by the three generators (16.2). This representation is generally reducible, i.e. $\mathcal{V}$ will contain several irreducible $\mathfrak{su}(2)$ multiplets. To keep the notation simple, in the following we will again write $H_\alpha$ and $E_\alpha$ instead of $\mathfrak{D}(H_\alpha)$ and $\mathfrak{D}(E_\alpha)$.

Because of its definition, a root lattice is invariant under Weyl reflections. The same applies to the weight lattice:

**Corollary 16.1** *The Weyl reflections, which are known to map the root lattice onto itself, also map the weight lattice onto itself.*

To prove this, we consider the Weyl reflection $\sigma_\beta$ to a root $\beta \in \Phi$. Since reflections preserve the lengths and square to 1, we have for every root $\alpha$ and every weight $\mu$

$$\frac{2\,\alpha \cdot \sigma_\beta(\mu)}{\alpha \cdot \alpha} = \frac{2\,\sigma_\beta(\alpha) \cdot \mu}{\sigma_\beta^2(\alpha)} \in \mathbb{Z}\,, \tag{16.5}$$

because with $\alpha$, $\sigma_\beta(\alpha)$ is also a root. Therefore, with $\mu$, $\sigma_\beta(\mu)$ is also a weight. The Weyl reflections generate the Weyl group $\mathcal{W}$ and we conclude

> **Weyl Group Acts on Weights**
> The set of weights is invariant under the Weyl group.

## 16.1.1 Step Operators

If $|\mu\rangle$ is a weight vector, then (16.1) and the commutation relations (15.18) imply

$$H_i E_\alpha |\mu\rangle = \big(E_\alpha H_i + [H_i, E_\alpha]\big)|\mu\rangle = (\mu_i + \alpha_i) E_\alpha |\mu\rangle\,, \tag{16.6}$$

where, as previously announced, we write $H_i$ instead of $\mathfrak{D}(H_i)$. This relation means that $E_\alpha$ maps $|\mu\rangle$ either into the zero vector or into a weight vector $|\mu + \alpha\rangle$ with shifted weight,

$$E_\alpha |\mu\rangle = 0 \quad \text{oder} \quad E_\alpha |\mu\rangle \propto |\mu + \alpha\rangle\,. \tag{16.7}$$

It may be possible to shift the weight in the direction of other roots, for example

$$E_\beta E_\alpha |\mu\rangle = 0 \quad \text{oder} \quad E_\beta E_\alpha |\mu\rangle \propto |\mu + \alpha + \beta\rangle\,. \tag{16.8}$$

---

**Task**

Check that with $\mu \in \Lambda_W$ also $\mu + \beta$ fulfills the quantization condition (16.4).

By applying the step operators several times, all weight vectors of an *irreducible* representation of $\mathfrak{g}$ are obtained. If $E_\alpha$ annihilates the weight vector $|\mu\rangle$, then $\mu + \alpha$ need not be a weight of the representation. However, because each root is a linear combination of simple roots with integer coefficients, we conclude:

**Difference of Two Weights**
In every irreducible representation of $\mathfrak{g}$, the difference of two weights is a linear combination of simple roots with integer coefficients.

In Sect. (15.3.1) we analyzed the effect of the lowering operator $E_{-\alpha}$ in a $\mathfrak{su}(2)$ multiplet and concluded that with $\beta$ also $\sigma_\alpha(\beta)$ must be a root. The arguments are easily transferable to the multiple action of $E_{-\alpha}$ on a weight vector $|\mu\rangle$. One concludes that with $\mu$ also $\sigma_\alpha(\mu)$ must be a weight of the representation $\mathfrak{D}$.

---

**Question**

Why can we conclude that with $\mu$ *all* elements in the Weyl orbit of $\mu$ must be weights of the same irreducible representation?

## 16.1.2  Weyl Group, Weight Lattice and Fundamental Weights

We now consider not only the weights of a fixed irreducible representation of $\mathfrak{g}$, but all possible weights of the weight lattice $\Lambda_W$, i.e. all $r$-tuples, which satisfy the quantization condition (16.4). First, we note that it suffices to require the condition for the simple roots. To see this, we assume that it is fulfilled for a simple root $\gamma$ and a root $\alpha$. Then follows

$$2\frac{\sigma_\gamma(\alpha) \cdot \mu}{\sigma_\gamma^2(\alpha)} = 2\left(\alpha - \frac{2\gamma \cdot \alpha}{\gamma^2}\gamma\right) \cdot \frac{\mu}{\alpha^2} = \frac{2\alpha \cdot \mu}{\alpha^2} - \frac{2\gamma \cdot \alpha}{\alpha^2}\frac{2\gamma \cdot \mu}{\gamma^2} \in \mathbb{Z}, \qquad (16.9)$$

because all fractions on the right-hand side are integers. Therefore, in addition to $\gamma$ and $\alpha$, $\sigma_\gamma(\alpha)$ also fulfills the quantization condition. Since all roots are obtained by simple Weyl reflections from the simple roots (see Lemma 15.8, in Sect. 15.5) it is therefore sufficient to require the quantization condition (16.4) for the simple roots.

Corresponding to the simple roots, a basis of *fundamental weights* is introduced:

**Fundamental Weights**

There is a basis of *fundamental weights* $\boldsymbol{\mu}_i$, defined by

$$\frac{2\,\boldsymbol{\mu}_i \cdot \boldsymbol{\gamma}_j}{\boldsymbol{\gamma}_j^2} = \delta_{ij}, \qquad 1 \le i, j \le r. \tag{16.10}$$

Each weight is an integer linear combination of the fundamental weights,

$$\boldsymbol{\mu} = \sum n_i\,\boldsymbol{\mu}_i, \qquad n_i \in \mathbb{Z}. \tag{16.11}$$

The integer expansion coefficients $n_i$ are called Dynkin labels. Therefore, the weight lattice $\Lambda_W$ of $\mathfrak{g}$ is the integral span of the set of fundamental weights.

**Definition 16.1 (Dominant Weights)**   If all Dynkin labels $n_i$ in the expansion (16.11) are non-negative, then $\boldsymbol{\mu}$ is called a dominant weight.

---

**Question**

Why are dominant weights also characterized by $(\boldsymbol{\mu}, \boldsymbol{\alpha}) \ge 0$ for all $\boldsymbol{\alpha} \in \Phi^+$?

---

This means that a dominant weight lies in the fundamental domain $\bar{\mathcal{W}}_\Delta$ of the Weyl group, and with Lemma 15.9 it follows

**Lemma 16.1 (Dominant Weights and Weyl Orbits)** *Each weight $\boldsymbol{\mu}$ can be mapped to a unique dominant weight with a $w \in \mathcal{W}$, or in each Weyl orbit there is a unique dominant weight.*

We illustrate the situation for the Lie algebra $\mathfrak{su}(3)$.

---

**Example: Fundamental Weights of $\mathfrak{su}(3)$**

If we choose $t = \boldsymbol{\alpha}_1$ in (15.48), then $\boldsymbol{\alpha}_3$ and $-\boldsymbol{\alpha}_2$ are simple roots in Fig. 15.1,

$$\boldsymbol{\gamma}_1 = \boldsymbol{\alpha}_3 = 2 \begin{pmatrix} \cos\frac{\pi}{3} \\ \sin\frac{\pi}{3} \end{pmatrix} \quad \text{and} \quad \boldsymbol{\gamma}_2 = -\boldsymbol{\alpha}_2 = 2 \begin{pmatrix} \cos\frac{\pi}{3} \\ -\sin\frac{\pi}{3} \end{pmatrix}. \tag{16.12}$$

The corresponding fundamental weights

$$\boldsymbol{\mu}_1 = \frac{2}{\sqrt{3}} \begin{pmatrix} \cos\frac{\pi}{6} \\ \sin\frac{\pi}{6} \end{pmatrix} \quad \text{and} \quad \boldsymbol{\mu}_2 = \frac{2}{\sqrt{3}} \begin{pmatrix} \cos\frac{\pi}{6} \\ -\sin\frac{\pi}{6} \end{pmatrix} \tag{16.13}$$

are shown in Fig. 16.1.

◄

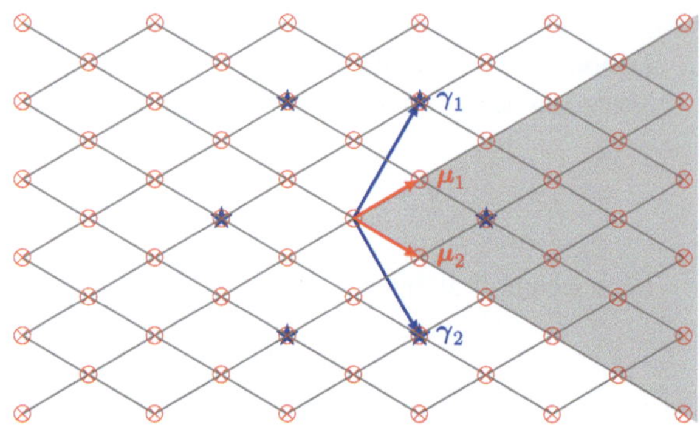

**Fig. 16.1** The roots and weight lattice of $\mathfrak{su}(3)$. The roots are marked with a star (*blue filled star*) and the weights in the weight lattice with (*crossed circle*). The simple roots $\gamma_1$, $\gamma_2$ and the fundamental weights $\mu_1$, $\mu_2$ are marked. All weights in the shaded area are dominant



**Corollary 16.2 (Simple Roots and Fundamental Weights)** *The Cartan matrix maps the fundamental weights to the simple roots,*

$$\gamma_i = \sum_{j=1}^{r} K_{ij}\mu_j \,. \tag{16.14}$$

To see this, take the scalar product of this relation with $\gamma_k$ and use (16.10), and the definition of the Cartan matrix.

With the help of the simple roots $\gamma_i$ and fundamental weights $\mu_i$, the coroots and coweights are defined as follows,

$$\gamma_i^\vee = \frac{2\gamma_i}{\gamma_i^2}, \quad \mu_i^\vee = \frac{2\mu_i}{\gamma_i^2} \,. \tag{16.15}$$

These have the following scalar products with the $\gamma_i$ and the $\mu_i$:

$$\gamma_i \cdot \gamma_j^\vee = K_{ij}, \quad \gamma_i^\vee \cdot \mu_j = \gamma_i \cdot \mu_j^\vee = \delta_{ij}, \quad \mu_i \cdot \mu_j^\vee = (K^{-1})_{ij} \,. \tag{16.16}$$

---

**Task**

Prove these identities using (16.14).

The lattice spanned by the coroots is reciprocal to the weight lattice $\Lambda_W$, and the root lattice $\Lambda_R$ spanned by the roots is reciprocal to the lattice of coweights. The relation (16.14) then implies that the unit cell of $\Lambda_R$ has a volume $\det(K)$ times larger than the unit cell of $\Lambda_W$.

### 16.1.3 Highest Weight

In every finite-dimensional irreducible representation of a Lie algebra, we can find a unique weight vector $|\lambda\rangle$ with the highest weight $\lambda$, characterized by

$$(\delta, \lambda) = \max\{(\delta, \mu)|\mu \text{ weight of the representation}\}. \tag{16.17}$$

Here, $\delta$ is a distinguished dominant weight: It is the weight of the Weyl vector $|\delta\rangle$, and can be expressed as a sum over all fundamental weights or as half the sum of the positive roots (positive roots $\alpha \in \Phi^+$ are denoted with $\alpha > 0$):

**Lemma 16.2** *The Weyl vector has the representations*

$$\delta = \frac{1}{2} \sum_{\alpha > 0} \alpha = \sum_i \mu_i \tag{16.18}$$

*and lies in the fundamental Weyl chamber $\mathcal{W}_\Delta$.*

**Proof** We assume $\delta$ is the half-sum of the positive roots and $\sigma_i$ the Weyl reflection on the plane perpendicular to the simple root $\gamma_i$. Then

$$\sigma_i(\delta) \cdot \gamma_i = \delta \cdot \sigma_i(\gamma_i) = -\delta \cdot \gamma_i. \tag{16.19}$$

The simple reflection $\sigma_i$ changes the sign of $\gamma_i$ and permutes all other positive roots according to Lemma 15.7. Therefore, $\sigma_i(\delta) = \delta - \gamma_i$, and with (16.19) this leads to the relation $2\delta \cdot \gamma_i = \gamma_i^2$. Due to the orthogonality relations between the simple roots and fundamental weights in (16.10), we also have

$$2 \sum_j \mu_j \cdot \gamma_i = \gamma_i^2.$$

Thus, the half-sum of the positive roots has the same scalar product with the simple roots as the sum of the fundamental weights. This proves the identity in the lemma. Furthermore, $\gamma_i \cdot \delta = \frac{1}{2}\gamma_i^2 > 0$ for all $\gamma_i \in \Delta$, and therefore the Weyl vector $\delta$ is in the fundamental Weyl chamber $\mathcal{W}_\Delta$. Its closure $\bar{\mathcal{W}}_\Delta$ is a fundamental domain of the Weyl group, and thus for every irreducible representation, there exists a *unique* highest weight.

A root $\alpha \in \Phi^+$ has a positive scalar product with $\delta$. Therefore, a weight vector $|\lambda\rangle$ with the highest weight is annihilated by all $E_\alpha$ with $\alpha \in \Phi^+$,

$$E_\alpha|\lambda\rangle = 0, \quad \alpha \in \Phi^+. \tag{16.20}$$

Otherwise, there would be a vector with weight $\lambda + \alpha$, for which $(\lambda + \alpha, \delta) > (\lambda, \delta)$ would hold. This is not possible for a highest weight.

**Question**

Why, conversely, does a weight vector $|\lambda\rangle$, which is annihilated by all $E_\alpha$ with $\alpha \in \Phi^+$, have a highest weight?

For each highest weight $\lambda$, there is thus a unique irreducible representation $\mathfrak{D}_\lambda$ on the vector space $\mathcal{V}_\lambda$, spanned by the weight vectors of the form

$$E_{-\alpha_1} E_{-\alpha_2} \cdots E_{-\alpha_m} |\lambda\rangle, \qquad \alpha_i > 0. \tag{16.21}$$

In the Weyl orbit of each weight $\mu$, there is always a dominant weight. The highest of the dominant weights of an irreducible representation—these lie in different Weyl orbits—is its highest weight. Conversely, every dominant weight is the highest weight of an irreducible representation [51,57]:

**Theorem 16.1 (Dominant Weights and Irreducible Representations)** *Let $\mathfrak{g}$ be a simple Lie algebra over $\mathbb{C}$ and $\lambda \in \Lambda_W$ a dominant weight. Then there is a unique finite-dimensional irreducible representation $\mathfrak{D}_\lambda$ of $\mathfrak{g}$ with $\lambda$ as the highest weight.*

Every weight is a linear combination of the fundamental weights with integer Dynkin labels $n_i$ as coefficients, see (16.11). The relations (16.10) in the form $2(\mu, \gamma_j) = n_j \gamma_j^2$ imply

$$\lambda = \sum_i n_i \mu_i \equiv [n_1, n_2, \ldots, n_r] \text{ dominant} \iff n_i \in \mathbb{N}_0. \tag{16.22}$$

**Irreducible Representations**
The irreducible representation $\mathfrak{D}_\lambda$ with hightest weights $\lambda$ is uniquely characterized by the Dynkin labels $[n_1, \ldots, n_r]$ with $n_i \in \mathbb{N}_0$ in the expansion $\lambda = \sum n_i \mu_i$. The $r$ representations with highest weights $\mu_i$ are called fundamental representations. $[0, \ldots, 0]$ is the one-dimensional trivial representation.

The program LiE uses this notation to characterize an irreducible representation. For example, for the Lie algebra $\mathfrak{su}(5)$, which has been used in the construction of Grand Unified Theories (GUTs):

```
LiE> setdefault A4
LiE> dim([0,0,0,0]) -> 1
LiE> dim([1,0,0,0]) -> 5
LiE> dim([0,1,0,0]) -> 10
LiE> dim([0,0,1,0]) -> 10
LiE> dim([0,0,0,1]) -> 5
LiE> dim([1,0,0,1]) -> 24
```

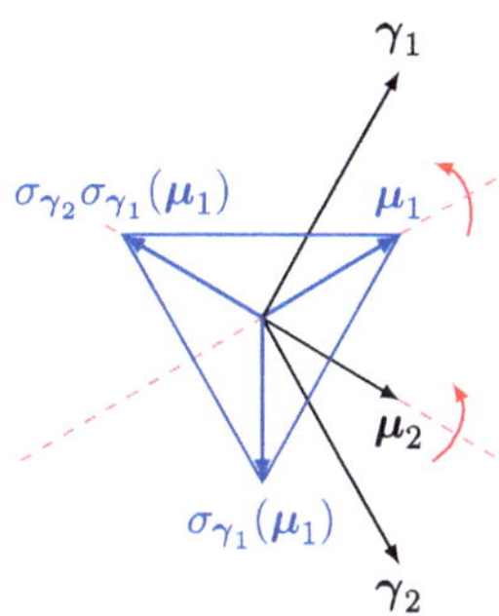

**Fig. 16.2**  Weyl reflections acting on the weights of the representation 3 of $\mathfrak{su}(3)$

The fundamental representations have the dimensions $5, 10, 10$, and $5$. As another example, we consider the Lie algebra of SU(3) in more detail.

---

**Example: The Weights of $\mathfrak{su}(3)$**

The weights of the three-dimensional fundamental representation with highest weight $\mu_1$ are obtained by acting with simple reflections on $\mu_1$ in Fig. 16.2. The reflection $\sigma_1$ leads to $\mu_1 - \gamma_1$ and the subsequent reflection $\sigma_2$ to $\mu_1 - \gamma_1 - \gamma_2$. Besides the weights

$$\mu_1, \quad \mu_1 - \gamma_1, \quad \mu_1 - \gamma_1 - \gamma_2$$

no further weights can occur, otherwise $\mu_1$ would not be the highest weight.

---

**Question**

Why can $0$ not be a weight of this fundamental representation?

The second fundamental representation with highest weight $\mu_2$ is also three-dimensional and has the weights:

$$\mu_2, \quad \sigma_2(\mu_2) = \mu_2 - \gamma_2,$$

$$\sigma_1\sigma_2(\mu_2) = \mu_2 - \gamma_1 - \gamma_2,$$

see Fig. 16.3. The fundamental representation $\mathfrak{D}_{\mu_1}$ is often denoted by 3 in the physics literature, and the representation $\mathfrak{D}_{\mu_2}$ by $\bar{3}$. These representations are complex conjugates of each other.

◀

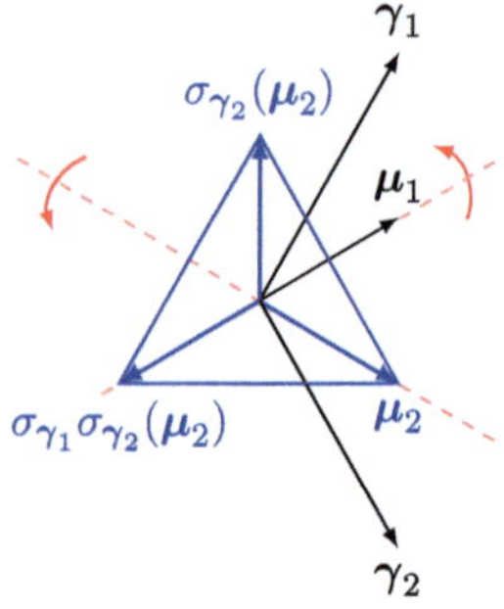



**Fig. 16.3** Weyl reflections acting on the weights of the representation $\bar{3}$ of $\mathfrak{su}(3)$

## 16.1.4  Tensor Product of Representations

In Sect. 14.1.4, we convinced ourselves that with $\mathcal{D} : \mathfrak{g} \mapsto \mathfrak{gl}(V)$ and $\mathcal{D}' : \mathfrak{g} \mapsto \mathfrak{gl}(V')$, the tensor product $\mathcal{D} \otimes \mathcal{D}' : \mathfrak{g} \mapsto \mathfrak{gl}(V \otimes V')$ is also a representation of $\mathfrak{g}$. If the representations $\mathcal{D}$ and $\mathcal{D}'$ assign the matrices $\mathcal{D}(X)$ and $\mathcal{D}'(X)$ to the element $X \in \mathfrak{g}$, then their tensor product assigns the matrix

$$(\mathcal{D} \otimes \mathcal{D}')(X) = \mathcal{D}(X) \otimes \mathbb{1} + \mathbb{1} \otimes \mathcal{D}'(X) \tag{16.23}$$

to the same element. As is common in the literature, we will no longer write the identities from now on.

If $|\mu\rangle$ is a weight vector of the representation $\mathcal{D}_\lambda$ with highest weight $\lambda$ and $|\mu'\rangle$ is a weight vector of the representation $\mathcal{D}_{\lambda'}$ with highest weight $\lambda'$, then their tensor product is a weight vector of the tensor product representation,

$$(\mathcal{D}_\lambda \otimes \mathcal{D}_{\lambda'})(H_i)\left(|\mu\rangle \otimes |\mu'\rangle\right) = (\mu_i + \mu_i')\left(|\mu\rangle \otimes |\mu'\rangle\right). \tag{16.24}$$

Thus, $|\mu\rangle \otimes |\mu'\rangle$ is a weight vector of the tensor product with weight $\mu + \mu'$. Using the representation property $(\mathcal{D}_\lambda \otimes \mathcal{D}_{\lambda'})(E_\alpha) = \mathcal{D}_\lambda(E_\alpha) + \mathcal{D}_{\lambda'}(E_\alpha)$, we obtain

**Corollary 16.3 (Highest Weight of Tensor Product)** *If $\mathcal{D}_\lambda$ and $\mathcal{D}_{\lambda'}$ are irreducible representations with highest weights $\lambda$ and $\lambda'$, then $\mathcal{D}_\lambda \otimes \mathcal{D}_{\lambda'}$ has the highest weight $\lambda + \lambda'$.*

In general, the tensor product is reducible, and the corollary means that

$$\mathcal{D}_\lambda \otimes \mathcal{D}_{\lambda'} = \mathcal{D}_{\lambda+\lambda'} \oplus \dots , \tag{16.25}$$

where the dots stand for further irreducible representations with smaller highest weights. If we denote representations with their Dynkin labels, then (16.25) reads

$$[n_1, \dots, n_r] \otimes [m_1, \dots, m_r] = [n_1 + m_1, \dots, n_r + m_r] \oplus \dots . \tag{16.26}$$

Corollary 16.3 implies that all dominant weights are highest weights of irreducible representations if only the fundamental weights are hightest weights, in accordance with Theorem 16.1.

---

**Example: Tensor Product $3 \times \bar{3}$ in su(3)**

Let's take a closer look at the product representation $\mathfrak{D}_{\mu_1} \otimes \mathfrak{D}_{\mu_2} = 3 \otimes \bar{3}$ of $\mathfrak{su}(3)$. Its 9 weights are the sums of the weights of the individual representations,

$$\mu_1 + \mu_2 - m_1 \gamma_1 - m_2 \gamma_2 = (1 - m_1)\gamma_1 + (1 - m_2)\gamma_2$$

with $(m_1, m_2)$ from the set

$$\{(0, 0), (1, 0), (0, 1), (1, 1), (1, 1), (1, 1), (2, 1), (1, 2), (2, 2)\}.$$

These weights are plotted in Fig. 16.4. The highest weight $\mu_1 + \mu_2 = \gamma_1 + \gamma_2$ is the unique positive root in the fundamental Weyl chamber.
It is the highest weight of the adjoint representation with the roots as weights, i.e., the representation $\mathfrak{D}_{\mu_1+\mu_1}$ is the eight-dimensional adjoint representation. The roots are of equal length, and the Weyl orbit of a root contains all 6 roots of $\mathfrak{su}(3)$. In addition to the 6 roots, the adjoint representation also has the weight $0$ twice.
In the above list of weights, the weight $0$ appears three times. Thus, the nine-dimensional product representation decomposes into the eight-dimensional adjoint representation and the one-dimensional trivial representation, $3 \otimes \bar{3} = 8 \oplus 1$, a result known to us, see (12.33). If we characterize the irreducible representations with their Dynkin labels, the decomposition reads $[1, 0] \otimes [0, 1] = [1, 1] + [0, 0]$. ◄

In LiE, the reduction of the tensor product is obtained as follows:

```
LiE> setdefault A2
LiE> dim [1,0] -> 3
LiE> dim [0,1] -> 3
LiE> dim [1,1] -> 8
LiE> tensor([1,0],[0,1]) -> [1,1] + [0,0]
```

**Fig. 16.4** Weights of the adjoint representation of $\mathfrak{su}(3)$

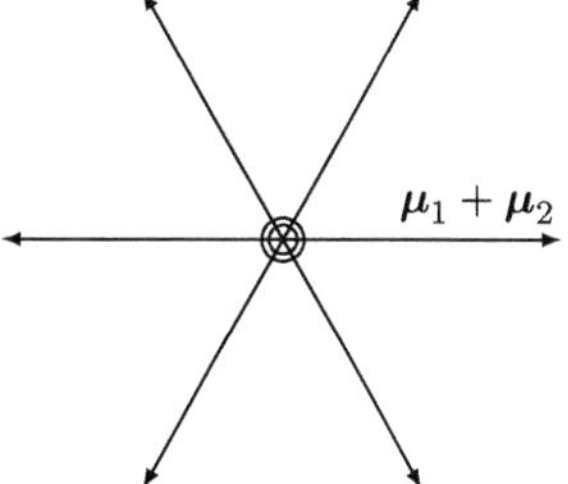

**Table 16.1** Dynkin labels and dimensions of the irreducible representations of $\mathfrak{su}(3)$ with the smallest dimensions

| $(n_1, n_2)$ | (0,0) | (0,1) | (1,0) | (0,2) | (2,0) | (1,1) | (0,3) | (3,0) | (1,2) | (2,1) |
|---|---|---|---|---|---|---|---|---|---|---|
| $D(n_1, n_2)$ | 1 | 3 | $\bar{3}$ | 6 | $\bar{6}$ | $8 = \bar{8}$ | 10 | $\overline{10}$ | 15 | $\overline{15}$ |

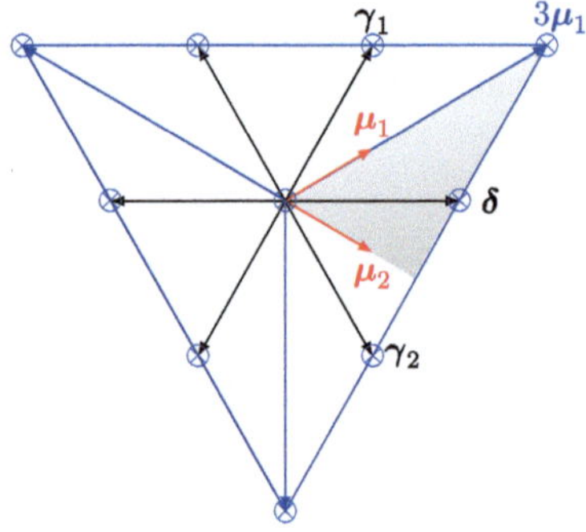


**Fig. 16.5** Weights of the representation $10 = [3, 0]$ of $\mathfrak{su}(3)$. The Weyl orbit of the highest weight $3\mu_1$ contains the three vertices of the triangle

The smallest irreducible representations of $\mathfrak{su}(3)$ are given in Table 16.1, The Weyl orbit of a weight $\mu = [n_1, \ldots, n_r]$ is found via

```
LiE> W_orbit([1,0],A2) -> [1,0],[-1,1],[0,-1]
LiE> W_orbit([1,1],A2) ->
     [1,1],[-1,2],[2,-1],[1,-2],[-2,1],[-1,-1]
LiE> W_orbit([1,0,0,0],A4) ->
     [1,0,0,0],[-1,1,0,0],[0,-1,1,0],[0,0,-1,1],[0,0,0,-1]
```

An important tensor product for the theory of strong interaction is

$$[1, 0] \otimes [1, 0] \otimes [1, 0] = [0, 0] \oplus [1, 1] \oplus [1, 1] \oplus [3, 0] \tag{16.27}$$

$$\text{or} \quad 3 \otimes 3 \otimes 3 = 1 \oplus 8 \oplus 8 \oplus 10 \tag{16.28}$$

The ten weights of the representation $10 = [3, 0]$ are the nine points on the sides of the equilateral triangle in Fig. 16.5 as well as the point at the origin. The vertex of the triangle at the top right belongs to the highest weight of the representation.

The weights of the representation $[3, 0]$ are non-degenerate. In contrast, the weight 0 of the adjoint representation is doubly degenerate. In general, the eigenspace spanned by the weight vectors with the same weight is denoted by $V_\mu$. The dimension of $V_\mu$, i.e., the degeneracy or multiplicity of the weight, is denoted by $m_\mu$.

## 16.2   Characters and Dimensions of Representations

In this section, we determine the characters $\chi_D$ of the irreducible representations of a simple compact Lie group. As class functions, characters are already determined by their restriction to a maximal torus. From the invariance under conjugations, only the invariance under the Weyl group $W = N_G(T)/T$ remains. Class functions can therefore be identified with Weyl-invariant functions on a maximal torus. To find their inner product, we need the reduced Haar measure, which was introduced in Chap. 9 and calculated for unitary groups. For arbitrary compact Lie groups, it was determined by Hermann Weyl.

### 16.2.1 Weyl's Integral Formula

We would like to express the integral of class functions for all compact and connected Lie groups as an integral over the maximal torus $T \leq G$. The homogeneous $G$-manifold $G/T$ has at the identity element $eT$ the tangent space

$$T_{eT}(G/T) = \mathfrak{g}/\mathfrak{h} = \mathfrak{p}\,, \tag{16.29}$$

and with respect to an Ad-invariant scalar product on $\mathfrak{g}$, $\mathfrak{p}$ is the orthogonal complement of the Cartan subalgebra, i.e., $\mathfrak{g} = \mathfrak{h} \oplus \mathfrak{p}$. The mapping

$$\phi : G/T \times T \mapsto G, \quad \phi(gT, t) = gtg^{-1} \tag{16.30}$$

is surjective, but generally not injective. In fact, if $w \in N_G(T)$ represents an element of the Weyl group, then

$$\phi(gw^{-1}, wtw^{-1}) = (gw^{-1})(wtw^{-1})(wg^{-1}) = gtg^{-1} = \phi(g, t)\,.$$

For regular points in $T$—these are not fixed points of a non-trivial Weyl transformation—$\phi$ is a covering of $G$ with fiber $W$ (coverings were discussed in Sect. 8.4.2). An $H \in \mathfrak{h}$ in the Cartan subalgebra transforms under the adjoint mapping according to

$$H \mapsto gHg^{-1} \qquad (t = \mathrm{e}^{\mathrm{i}H})\,. \tag{16.31}$$

Here, $gHg^{-1} = H$ only for $g$ in the coset $eT \in G/T$.

**Theorem 16.2 (Weyl's Integral Formula)**   *Let $f$ be a continuous class function on a compact connected Lie group with maximal torus $T$. Then*

$$\int_G f(g)\, \mathrm{d}\mu_G(g) = \frac{1}{|W|} \int_T (\det J)\, f(t)\, \mathrm{d}\mu_T(t) \quad with \quad J = \left(\mathbb{1} - \mathrm{Ad}(t^{-1})\right)_{\mathfrak{g}/\mathfrak{h}}, \tag{16.32}$$

*where $\mu_G$ and $\mu_T$ denote the normalized Haar measures of $G$ and $T$.*

Later it will be shown that the determinant of the Jacobian matrix $J$ is positive, so no absolute value signs are necessary.

Idea for proof: Due to the uniqueness of the Haar measure, we may assume that the group elements $g$ are (faithfully) represented by unitary matrices. We decompose the group elements as in (9.31) and write $g = vtv^{-1}$ with $t \in T$. As in (9.32), it follows that

$$g^{-1}\delta g = v\left(\delta t + \left(\mathrm{Ad}(t^{-1}) - \mathbb{1}\right)\delta v\right) v^{-1}. \tag{16.33}$$

The Haar measure and the invariant length element are invariant under conjugations, and we can omit the conjugation with $v$. The two terms $\delta t$ and $(\mathrm{Ad}(t^{-1}) - \mathbb{1})\delta v$ in this decomposition are orthogonal to each other with respect to an Ad-invariant scalar product,

$$\left(\delta t, \left(\mathrm{Ad}(t^{-1}) - 1\right)\delta v\right) = \left(\left(\mathrm{Ad}(t) - 1\right)\delta t, \delta v\right) = 0, \tag{16.34}$$

and the Jacobian matrix of the linear mapping $(\delta t, \delta v) \mapsto \left(\delta t, \left(\mathrm{Ad}(t^{-1}) - \mathbb{1}\right)\delta v\right)$ is $\mathbb{1}$ on the Cartan subalgebra $\mathfrak{h}$ and is $\mathrm{Ad}(t^{-1}) - \mathbb{1}$ on the orthogonal complement $\mathfrak{g}/\mathfrak{h}$. Therefore, the Haar measure factorizes as follows

$$\mathrm{d}\mu_G = \det J \, \mathrm{d}\mu_T \, \mathrm{d}\mu_{G/T}, \quad \text{with} \quad J = \left(\mathbb{1} - \mathrm{Ad}(t^{-1})\right)_{\mathfrak{g}/\mathfrak{h}}, \tag{16.35}$$

and a normalized measure $\mathrm{d}\mu_{G/T} \propto \mathrm{Tr}_{\mathfrak{g}/\mathfrak{h}}(\delta v \wedge \cdots \wedge \delta v)$ on the quotient space $G/T$.

> **Factorization of the Haar Measure**
>
> If $T$ is a maximal torus and $\mathcal{W}$ is the Weyl group of a connected compact Lie group $G$, then the Haar measure factors according to
>
> $$\int_G \mathrm{d}\mu_G(g)\, f(g) = \frac{1}{|\mathcal{W}|} \int_T \mathrm{d}\mu_T(t)\, (\det J) \int_{G/T} f(vtv^{-1})\, \mathrm{d}\mu_{G/T}, \tag{16.36}$$
>
> with Jacobian matrix in (16.35) and normalized Haar measures $\mathrm{d}\mu_G$ and $\mathrm{d}\mu_T$ on $G$ and $T$.

For regular points $t \in T$, $G/T \times T$ is the $|\mathcal{W}|$-fold covering of $G$, and this explains the factor $1/|\mathcal{W}|$. The singular points are a set of codimension 1 and do not need to be treated separately for continuous functions. The integration over the quotient space $G/T$ now leads to the reduced Haar measure

$$d\mu_{\mathrm{red}} = \frac{1}{|\mathcal{W}|} \, (\det J) \, d\mu_T(t)\,, \tag{16.37}$$

which appears in Weyl's integral formula. With this general formula, one easily recovers the reduced Haar measure of SU$(n)$ in (9.47), see Problem 16.5.

---

**Example: Reduced Haar Measure of SO$(2n)$**

As maximal torus of the orthogonal group SO$(2n)$, we choose the block-diagonal matrices $D(\boldsymbol{\varphi}) = \mathrm{diag}(R_2(\varphi_1), \dots, R_2(\varphi_n))$. The two-dimensional rotation matrices $R_2(\varphi)$ were introduced in (5.16). As in Appendix 5.6, we denote the matrices in the maximal torus with $D$ and not with $t$ as in the general analysis. Furthermore, let $E_{ij}$ be the $n$-dimensional matrix with 1 in row $i$ and column $j$ and otherwise only zeros. We form the tensor products $X_{ija} = E_{ij} \otimes e_a - E_{ji} \otimes e_a^T$ with the basis elements of $\mathfrak{gl}(2, \mathbb{R})$,

$$e_1 = \begin{pmatrix} 1 & 0 \\ 0 & 0 \end{pmatrix}, \quad e_2 = \begin{pmatrix} 0 & 1 \\ 0 & 0 \end{pmatrix}, \quad e_3 = \begin{pmatrix} 0 & 0 \\ 1 & 0 \end{pmatrix}, \quad e_4 = \begin{pmatrix} 0 & 0 \\ 0 & 1 \end{pmatrix}.$$

The elements with $i < j$ form a basis of $\mathfrak{so}(2n)/\mathfrak{h}$. Under the action of the linear mapping $\mathbb{1} - \mathrm{Ad}(D^{-1})$ the 4 elements with fixed $i, j$ form an invariant subspace,

$$\left(\mathbb{1} - \mathrm{Ad}(D^{-1})\right) X_{ij,a} = \sum_{b=1}^{4} B_{ba}(\varphi_i, \varphi_j) X_{ij,b}\,. \tag{16.38}$$

The transformation matrix $B$ has determinant $\det(B) = (2\cos(\varphi_i) - 2\cos(\varphi_j))^2$. Since the Weyl group has $(2n)!!/2$ elements, see Appendix 15.8, we conclude

$$d\mu_{\mathrm{red}} = \frac{2}{2^n n!} \prod_{i<j} (2\cos(\varphi_i) - 2\cos(\varphi_j))^2 \prod_k \frac{d\varphi_k}{2\pi} \qquad \text{for SO}(2n)\,. \tag{16.39}$$

The integration domain can be restricted to $[0, \pi]^n$ if simultaneously the $2\pi$ in the denominators are replaced by $\pi$. ◄

---

Actually, the determinant in the reduced Haar measure (16.37) is the absolute square of a simpler expression and thus always positive. For the proof, we recall that the root vectors $E_\alpha$ are eigenvectors of $\mathrm{Ad}(t^{-1})$ on $\mathfrak{g}/\mathfrak{h}$,

$$t = \mathrm{e}^{\mathrm{i}H} \;\Rightarrow\; \mathrm{Ad}(t^{-1})(E_\alpha) = \mathrm{e}^{-\mathrm{i}(\alpha, h)} E_\alpha \quad \text{for} \quad H = \sum h_i H_i\,, \tag{16.40}$$

where, as is common in the physics literature, we choose the $H_i$ to be Hermitian in a unitary representation. In many books on group and representation theory this is not the case and instead one writes $t = e^H$. Then the i's are missing in the following exponents.

Due to (16.40), the determinant can be directly expressed through the roots,

$$\det J = \prod_{\alpha \in \Phi} \left( 1 - e^{-i(\alpha, h)} \right) = \prod_{\alpha > 0} \left( 1 - e^{-i(\alpha, h)} \right) \left( 1 - e^{i(\alpha, h)} \right), \qquad (16.41)$$

and this leads to the following factorization of the determinant:

$$\det J = |\Theta|^2 \quad \text{with} \quad \Theta = \prod_{\alpha > 0} \left( 1 - e^{-i(\alpha, h)} \right). \qquad (16.42)$$

We see that the determinant of $J$ is always positive. The half-sum of the positive roots is equal to the Weyl vector $\delta$, so that

$$\Theta = e^{-i(\delta, h)} \prod_{\alpha > 0} \left( e^{\frac{i}{2}(\alpha, h)} - e^{-\frac{i}{2}(\alpha, h)} \right). \qquad (16.43)$$

Since $\bar{\Theta}$ is proportional to $e^{i(\delta, h)}$, $\det J$ has the equivalent factorization

$$\det J = A_\delta \bar{A}_\delta \quad \text{with} \quad A_\delta = \prod_{\alpha > 0} \left( e^{\frac{i}{2}(\alpha, h)} - e^{-\frac{i}{2}(\alpha, h)} \right). \qquad (16.44)$$

This ultimately leads to the following alternative and explicit form of the integral formula (16.36):

**Weyl's Integral Formula**
If the elements of the maximal torus are parameterized according to $t = e^{iH}$ with $H = \sum h_i H_i$, then the integral of a class function over the group is given by

$$\int_G f(g)\, d\mu_G(g) = \frac{1}{|\mathcal{W}|} \int_T f(h)\, A_\delta(h) \bar{A}_\delta(h)\, d\mu_T(h). \qquad (16.45)$$

The normalized Haar measure on the maximal torus is $d\mu_T(h) = \text{const}\, dh_1 \cdots dh_r$.

An alternative formula for the Haar measure of matrix groups is discussed in Problem 16.4.

## Functions on the Maximal Torus

The formula (16.44) for the function $A_\delta$ contains the product over all positive roots. Now there is an alternative representation that includes an averaging over the Weyl group. More generally, we now consider functions $\phi : T \mapsto \mathbb{C}$, i.e., functions $\phi(\mathrm{e}^{\mathrm{i}H}) = \phi(\mathrm{e}^{\mathrm{i}h_i H_i}) \equiv \psi(h)$. The elements of the Weyl group act on $\psi$ according to

$$(\Gamma_w \psi)(h) = \psi(w^{-1}h), \quad w \in \mathcal{W}. \tag{16.46}$$

The Weyl group $\mathcal{W}$ is a finite subgroup of $\mathrm{O}(r)$ and (16.46) is the restriction of the action (10.48)—in which $\mathbb{R}^3$ is replaced by $\mathbb{R}^r$ and $\mathrm{O}(3)$ by $\mathrm{O}(r)$—on $\mathcal{W}$. It follows that the mapping $w \mapsto \Gamma_w$ is a unitary representation of the Weyl group on the space of $L_2$ functions $\psi : T \mapsto \mathbb{C}$.

Now we consider the above function

$$A_\delta(h) = \prod_{\alpha>0} \left( \mathrm{e}^{\frac{\mathrm{i}}{2}(\alpha,h)} - \mathrm{e}^{-\frac{\mathrm{i}}{2}(\alpha,h)} \right) = \mathrm{e}^{\mathrm{i}(\delta,h)} \prod_{\alpha>0} \left( 1 - \mathrm{e}^{-\mathrm{i}(\alpha,h)} \right), \tag{16.47}$$

and note that for a Weyl reflection $\sigma_i$ on the plane orthogonal to the simple root $\boldsymbol{\gamma}_i$, the mapping $\Gamma_{\sigma_i}$ acts on this function as follows,

$$(\Gamma_{\sigma_i} A_\delta)(h) = A_\delta(\sigma_i^{-1}h) = \prod_{\alpha>0} \left( \mathrm{e}^{\frac{\mathrm{i}}{2}(\sigma_i\alpha,h)} - \mathrm{e}^{-\frac{\mathrm{i}}{2}(\sigma_i\alpha,h)} \right). \tag{16.48}$$

The simple reflection $\sigma_i$ changes the sign of $\boldsymbol{\gamma}_i$ and permutes the remaining positive roots, so that

$$(\Gamma_{\sigma_i} A_\delta)(h) = -A_\delta(h). \tag{16.49}$$

If $w \in \mathcal{W}$ is the product of an even number of reflections, then $w$ is a proper rotation with $\det(w) = 1$, otherwise an improper rotation with $\det(w) = -1$. Obviously,

$$(\Gamma_w A_\delta)(h) = \det(w)\, A_\delta(h) = (-1)^{\ell(w)} A_\delta(h), \tag{16.50}$$

where $\ell(w)$ denotes the length of $w$ introduced in Definition 15.6 of Sect. 15.5. Functions with this property, namely $\Gamma_w \psi = (-1)^{\ell(w)}\psi$, are called *Weyl-antisymmetric*. On the other hand, functions with $\Gamma_w \psi = \psi$ are called *Weyl-symmetric* or Weyl-invariant. Weyl-symmetric and Weyl-antisymmetric functions on maximal abelian subgroups have been extensively studied in the work [65], where they are called symmetric and antisymmetric orbit functions.

We follow the discussions on projection operators in Sect. 11.2.4 and define the projectors onto the trivial representation $S : w \mapsto 1$ and alternating sign-representation $A : w \mapsto \det(w)$ of the Weyl group.

**Corollary 16.4 (Projectors onto Trivial and Alternate Representations)** *The projectors*

$$P_S = \frac{1}{|\mathcal{W}|} \sum_{w \in \mathcal{W}} \Gamma_w \quad and \quad P_A = \frac{1}{|\mathcal{W}|} \sum_{w \in \mathcal{W}} \det(w)\,\Gamma_w \,, \tag{16.51}$$

*with the $\Gamma_w$ in (16.46), project onto the subspaces of Weyl-invariant and Weyl-antisymmetric functions on the maximal abelian torus.*

---

**Task**

Prove the properties

$$\Gamma_w P_S = P_S \Gamma_w = P_S \quad and \quad \Gamma_w P_A = P_A \Gamma_w = \det(w) P_A \tag{16.52}$$

for the projectors $P_S$, $P_A$ and linear mappings $\Gamma_w$ in (16.46).

---

To arrive at another representation for the Weyl-antisymmetric function $A_\delta = P_A A_\delta$, we multiply out the product in (16.47) and obtain

$$A_\delta(h) = \sum_\beta (-1)^{|\beta|}\, e^{i(\delta,h)-i(\beta,h)} = (P_A A_\delta)(h) = P_A \sum_\beta (-1)^{|\beta|}\, e^{i(\delta-\beta,h)} \,,$$

$$\tag{16.53}$$

with a sum over roots $\beta$ that can be written as a sum of *different* positive roots $\alpha$. The number of ways to write $\beta$ as a sum of positive roots is the value of the partition function $P_K(\beta)$ of Bertram Kostant, and is here briefly denoted by $|\beta|$. The sum also includes a term for $\beta = 0$.

The Weyl vector $\delta$ is element of the weight lattice $\Lambda_W$, and this is also true for every difference $\delta - \beta$ in the last exponent of (16.53). Due to $P_A \Gamma_w = \det(w) P_A$, all $\beta$ for which $\delta - \beta$ lie in the same Weyl orbit contribute, up to a possible sign, the same to the last sum in (16.53). Thus, we can restrict the sum over $\beta$ to those $\beta$ for which $\delta - \beta$ is dominant, i.e., lies in the fundamental Weyl chamber,

$$A_\delta(h) = P_A \sum_{\beta:\delta-\beta \in \mathcal{W}_\Delta} c_\beta\, e^{i(\delta-\beta,h)} \,, \quad with \quad c_0 = |\mathcal{W}| \,. \tag{16.54}$$

The value for $c_0$ follows from a direct comparison of the expressions on the left in (16.53) and in (16.54). The Dynkin labels $n_i$ of the dominant weights $\delta - \beta$ are non-negative. Now we show that only $\beta$ contribute to the sum (16.54) when the Dynkin labels of $\delta - \beta$ are strictly positive. For example, if $n_i = 0$, then $\sigma_i(\delta - \beta) = \delta - \beta$ and accordingly $\Gamma_{\sigma_i}\, e^{i(\delta-\beta,h)} = e^{i(\delta-\beta,h)}$. Acting with $P_A$ on this relation, it follows

$$P_A \Gamma_{\sigma_i}\, e^{i(\delta-\beta,h)} = P_A\, e^{i(\delta-\beta,h)} \,. \tag{16.55}$$

But with $P_A \Gamma_{\sigma_i} = \det(\sigma_i) P_A = -P_A$, it also follows that the left side is equal to the right side with the opposite sign. This proves that dominant $\delta - \beta$ with non-strictly positive Dynkin labels do not contribute to the sum (16.54). The Weyl vector has Dynkin labels $[1, 1, \ldots, 1]$, and as a sum of positive roots, every $\beta \neq 0$ has at least one positive Dynkin label. This means that except for $\beta = 0$, no root $\beta$ contributes to the sum. With the value of $c_0$ in (16.54), it follows

**Lemma 16.3 (Weyl's Denominator Formula)** *The Weyl-antisymmetric function $A_\delta$ in (16.44) is proportional to the projection of $e^{i(\delta, h)}$ onto the subspace of antisymmetric functions,*

$$A_\delta(h) = \prod_{\alpha > 0} \left( e^{\frac{i}{2}(\alpha, h)} - e^{-\frac{i}{2}(\alpha, h)} \right) = \sum_{w \in \mathcal{W}} \det(w)\, e^{i(w\delta, h)} . \qquad (16.56)$$

While the left side yields a factorization into $\frac{1}{2}(\dim \mathfrak{g} - \dim \mathfrak{h})$ factors, the sum on the right side contains $|\mathcal{W}|$ terms. For large groups, the order of the Weyl group grows exponentially with the rank of the group.

## 16.2.2 Characters

Class functions are Weyl-invariant functions on a maximal torus or on the Cartan subalgebra $\mathfrak{h} \leq \mathfrak{g}$. Let $\mathfrak{D}_\lambda$ be the irreducible representation of the Lie algebra with highest weight $\lambda$, which exponentiates to a representation $D_\lambda$ of the Lie group. Its character is

$$\chi_\lambda(h) = \mathrm{Tr}\left( e^{i\mathfrak{D}(H)} \right) = \sum_\mu \langle \mu | e^{i\mathfrak{D}(H)} | \mu \rangle = \sum_\mu m_\mu\, e^{i(h, \mu)}, \quad H = \sum h_i H_i ,$$

$$(16.57)$$

where the sum is over all weights $\mu$ of the representation with highest weight $\lambda$, and $m_\mu$ denotes the multiplicity of $\mu$. For semi-simple compact Lie groups, the following theorem by Hermann Weyl holds:

**Theorem 16.3 (Weyl's Character Formula)** *The character of the irreducible representation $\mathfrak{D}_\lambda$ with highest weight $\lambda$ is*

$$\chi_\lambda(h) = \frac{\displaystyle\sum_{w \in \mathcal{W}} \det(w)\, e^{i(w(\lambda + \delta), h)}}{\displaystyle\sum_{w \in \mathcal{W}} \det(w)\, e^{i(w(\delta), h)}} . \qquad (16.58)$$

*The Weyl vector $\delta$ has been defined in (16.18).*

Before we proceed with the proof, we will once again calculate the characters of SU(2):

**Example: Characters of SU(2)**

The Weyl group of SU(2) has two elements, namely the identity and a reflection. If $n$ is the dimension of the representation, then its highest weight is $n$. The Weyl vector is equal to the fundamental weight $n = 1$. Therefore, the Weyl character formula gives

$$\chi_n(h) = \frac{e^{i(n+1)h} - e^{-i(n+1)h}}{e^{ih} - e^{-ih}} = \frac{\sin(n+1)h}{\sin h}, \tag{16.59}$$

which reproduces the result (12.18). ◄

***Proof*** We relate the characters of $G$ with those of the maximal abelian subgroup $T \le G$. The exponential functions $e_\mu(h) = e^{i(\mu, h)}$ form an orthonormal basis in the function space $L_2(T)$ with Haar measure of the maximal abelian group,

$$\left(e_\mu, e_{\mu'}\right)_{L_2(T)} = \int_T d\mu_T(h)\, e^{-i(\mu, h)}\, e^{i(\mu', h)} = \delta_{\mu\mu'}. \tag{16.60}$$

If we antisymmetrize the exponential functions, we obtain the Weyl-antisymmetric function

$$A_\mu(h) = \sum_{w \in \mathcal{W}} \det(w)\, e^{i(w\mu, h)} \tag{16.61}$$

i.e. the function (16.56) with $\delta$ replaced by $\mu$. Since $A_{w\mu}(h) = \det(w)A_\mu(h)$, it suffices again to study the functions with dominant $\mu$. If two dominant $\mu$ and $\mu'$ are different, then their Weyl orbits are disjoint and

$$\int_T d\mu_T(h)\, A_\mu^*(h)A_{\mu'}(h) = |\mathcal{W}|\,\delta_{\mu\mu'}, \quad \mu, \mu' \text{ dominant}. \tag{16.62}$$

The $A_\mu$ with dominant $\mu$ define an orthogonal basis for the Weyl-antisymmetric functions.

**Task**

Justify the orthogonality relation (16.62) and the subsequent claim.

The characters of $G$ are (in contrast to those of $T$) Weyl-symmetric, such that for the character $\chi_\lambda$ of a highest-weight representation $\mathfrak{D}_\lambda$, the function $A_\delta(h)\chi_\lambda(h)$ is Weyl-antisymmetric. Thus it is a linear combination of the basis functions in (16.61), so that

$$A_\delta(h)\chi_\lambda(h) = \sum_{\mu \text{ dominant}} c_\mu A_\mu(h). \tag{16.63}$$

The characters $\chi_\lambda$ of the irreducible representations of $G$ are orthonormal with respect to the reduced Haar measure, so that with the integration formula (16.45) it follows

$$1 = \int_T \mathrm{d}\mu_{\mathrm{red}}\, \chi_\lambda^* \chi_\lambda = \frac{1}{|\mathcal{W}|} \sum_{\mu,\mu'} c_\mu^* c_{\mu'} \int \mathrm{d}\mu_T\, A_\mu^* A_{\mu'} \overset{(16.62)}{=} \sum_\mu |c_\mu|^2 . \qquad (16.64)$$

In the second step, we used that the factor $A_\delta$ in the reduced Haar measure in (16.45) and $A_\delta$ in (16.63) cancel each other. If we substitute the defining sums (16.57) and (16.61) for the factors on the left side of (16.63) and order the exponential functions by the size of the weights, it follows

$$A_\delta(h)\chi_\lambda(h) = 1 \cdot e^{i(\delta+\lambda,h)} + \dots , \qquad (16.65)$$

where the dots represent exponential functions with $\mu < \delta + \lambda$. The coefficient of the leading term is one because the highest weight $\lambda$ has the multiplicity $m_\lambda = 1$. It follows that $c_{\delta+\lambda} = 1$, and due to (16.64), all other coefficients must then be zero. This proves the celebrated character formula of Weyl.

### 16.2.3 Weyl's Dimension Formula

The dimension of an irreducible representation with highest weight $\lambda$ is $\dim \mathfrak{D}_\lambda = \chi_\lambda(e)$. If one uses the character formula (16.58) here, then one deals with the quotient of two functions, both of which vanish at $h = 0$. We apply the rule of l'Hopital to the quotient (16.58) with argument $h = \varepsilon\delta$ and then let $\varepsilon$ approach 0. For the analysis of the numerator at $h = \varepsilon\delta$, we use $\det(w) = \det(w^{-1})$ in

$$A_{\lambda+\delta}(\varepsilon\delta) = \sum_w \det(w)\, e^{i(\delta+\lambda, w^{-1}(\varepsilon\delta))} = \sum_w \det\left(w^{-1}\right) e^{i(\delta, w(\varepsilon\delta+\varepsilon\lambda))}$$

$$= A_\delta(\varepsilon\delta + \varepsilon\lambda) . \qquad (16.66)$$

We have used the fact that the argument is proportional to $\delta$. The product representation of $A_\delta$ now leads to the following expansion for small $\varepsilon$,

$$A_\delta(\varepsilon\delta + \varepsilon\lambda) = \prod_{\alpha>0}\left(e^{\frac{i}{2}\varepsilon(\alpha,\delta+\lambda)} - e^{-\frac{i}{2}\varepsilon(\alpha,\delta+\lambda)}\right) = (i\varepsilon)^n \prod_{\alpha>0}(\alpha, \delta+\lambda) + O\left(\varepsilon^{n+1}\right),$$

$$\qquad (16.67)$$

where $n$ is the number of positive roots. Thus, it follows that

$$\dim \mathfrak{D}_\lambda = \lim_{\varepsilon\to 0} \chi_\lambda(\varepsilon\delta) = \lim_{\varepsilon\to 0} \frac{A_\delta(\varepsilon\delta + \varepsilon\lambda)}{A_\delta(\varepsilon\delta)} = \frac{\prod_{\alpha>1}(\alpha, \delta+\lambda)}{\prod_{\alpha>0}(\alpha, \delta)} . \qquad (16.68)$$

This proves the following formula by Weyl:

**Lemma 16.4 (Weyl's Dimension Formula)**   *The irreducible representation $\mathfrak{D}_\lambda$ with highest weight $\lambda$ has the dimension*

$$\dim \mathfrak{D}_\lambda = \prod_{\alpha > 0} \frac{(\delta + \lambda, \alpha)}{(\delta, \alpha)} \,. \tag{16.69}$$

To evaluate this, we write the positive roots as sums of simple roots

$$\alpha = \sum_i k_\alpha^i \gamma_i \in \Phi^+, \quad k_\alpha^i \in \mathbb{N}_0 \,. \tag{16.70}$$

---

**Dimension Formula**

If the highest weight of the representation $\lambda$ has the Dynkin labels $[n_1, \ldots, n_r]$, then

$$\dim \mathfrak{D}_\lambda = \prod_{\alpha > 0} \frac{\sum_{i,j} k_\alpha^i (1 + n_j)(\mu_j, \gamma_i)}{\sum_{i,j} k_\alpha^i (\mu_j, \gamma_i)} \stackrel{(16.10)}{=} \prod_{\alpha > 0} \frac{\sum_i k_\alpha^i (1 + n_i) \gamma_i^2}{\sum_i k_\alpha^i \gamma_i^2} \,. \tag{16.71}$$

Here, $k_\alpha^i \in \mathbb{N}_0$ are the expansion coefficients in (16.70).

---

The dimension is thus the product of factors, one for each positive root $\alpha$. For a given $\alpha$, the numerator and denominator contain sums over those simple roots that add up to $\alpha$. Each simple root contributes the term $k_\alpha^i (1 + n_i) \gamma_i^2$ to the numerator and the term $k_\alpha^i \gamma_i^2$ to the denominator, where $n_i$ is the $i$-th Dynkin label of the highest weight $\lambda$.

For the "simple-laced" Lie algebras $A_\ell$, $D_\ell$, $E_6$, $E_7$, and $E_8$, the roots have the same length, and the formula simplifies to

$$\dim \mathfrak{D}_\lambda = \prod_{\alpha > 0} \frac{\sum_i k_\alpha^i (1 + n_i)}{\sum_i k_\alpha^i} \,. \tag{16.72}$$

---

**Example: Representations of $\mathfrak{su}(3)$**

The positive roots are $\gamma_1$, $\gamma_2$ and $\gamma_1 + \gamma_2$. The representation with Dynkin labels $[n_1, n_2]$ then has the dimension $\dim[n_1, n_2] = f(1 + n_1, 1 + n_2)$ with

$$f(x, y = x \cdot y \cdot \frac{x + y}{2} \,. \tag{16.73}$$

For example, one finds the known dimensions $\dim[1, 0] = \dim[0, 1] = 3$ and $\dim[1, 1] = 8$. The corresponding formula for arbitrary $\mathfrak{su}(n)$ is calculated in Problem 16.7. ◀

---

**Example: Representations of $G_2$**

This exceptional Lie algebra has 6 positive roots, of which 2 are simple roots. Let $\boldsymbol{\gamma}_1$ be the shorter of the simple roots. Then $\boldsymbol{\gamma}_2^2 = 3\boldsymbol{\gamma}_1^2$. The non-simple short positive roots are $\boldsymbol{\gamma}_1 + \boldsymbol{\gamma}_2$ and $2\boldsymbol{\gamma}_1 + \boldsymbol{\gamma}_2$. The non-simple long positive roots are $3\boldsymbol{\gamma}_1 + \boldsymbol{\gamma}_2$ and $3\boldsymbol{\gamma}_1 + 2\boldsymbol{\gamma}_2$. The representation with Dynkin labels $[n_1, n_2]$ has the dimension $\dim[n_1, n_2] = f(1 + n_1, 1 + n_2)$ with

$$f(x, y) = x \cdot y \cdot \frac{x + 3y}{1 + 3} \cdot \frac{2x + 3y}{2 + 3} \cdot \frac{3x + 3y}{3 + 3} \cdot \frac{3x + 6y}{3 + 6}. \tag{16.74}$$

For example, one finds $\dim[1, 0] = 7$, $\dim[0, 1] = 14$ and $\dim[1, 1] = 64$. ◀

This can be confirmed with LiE:

```
LiE> setdefault G2
LiE> norm([1,0]) -> 2
LiE> norm([0,1]) -> 6
LiE> pos_roots ->   [[1,0],[0,1],[1,1],[2,1],[3,1],[3,2]]
LiE> dim([1,1]) -> 64
```

## 16.2.4  Multiplicity of Weights

Each root $\boldsymbol{\alpha}$ defines an $\mathfrak{su}(2)$ subalgebra, and the weight vectors appear in multiplets of this subalgebra. The multiplet with the weight vector $|\boldsymbol{\mu}\rangle$ contains the weights

$$\boldsymbol{\mu} + k\boldsymbol{\alpha}, \quad -m \leq k \leq n, \quad m, n \in \mathbb{N}_0, \tag{16.75}$$

and is called the $\boldsymbol{\alpha}$-string of weights through the weight $\boldsymbol{\mu}$. A weight in the string is mapped to another weight in the string by the Weyl reflection $\sigma_\alpha$,

$$\sigma_\alpha(\boldsymbol{\mu} + k\boldsymbol{\alpha}) = \boldsymbol{\mu} - \left(k + \frac{2(\boldsymbol{\alpha}, \boldsymbol{\mu})}{\alpha^2}\right)\boldsymbol{\alpha}. \tag{16.76}$$

The weight $\boldsymbol{\mu} + n\boldsymbol{\alpha}$ with maximum $k$ is mapped by $\sigma_\alpha$ to the weight $\boldsymbol{\mu} - m\boldsymbol{\alpha}$ with minimum $k$, so that

$$m - n = \frac{2(\boldsymbol{\alpha}, \boldsymbol{\mu})}{\alpha^2} \tag{16.77}$$

must hold. From this, the length of the $\boldsymbol{\alpha}$-string through $\boldsymbol{\mu}$ immediately follows:

> **$\alpha$-strings**
>
> An $\alpha$-string through $\mu$ contains $|m - n| + 1 = 2|(\alpha, \mu)|/\alpha^2 + 1$ weights of the form $\mu + k\alpha$.

Weights $\mu$ can have different multiplicities $m_\mu$, i.e., they can characterize several linearly independent weight vectors. For example, in the adjoint representation of $\mathfrak{su}(3)$, the subspace spanned by the weight vectors with weight $\mu = 0$ has dimension $m_0 = 2$. The non-vanishing weights $\mu$ of the highest-weight representation $\mathfrak{D}_\lambda$ satisfy

$$\mu = \lambda - \sum_{\alpha>0} k_\alpha \alpha, \tag{16.78}$$

and the multiplicity $m_\mu = \dim V_\mu$ is equal to the number of ways to write the difference $\lambda - \mu$ as a sum of positive roots.

Hans Freudenthal succeeded in finding a recursive formula for the multiplicities $m_\mu$ [66]. A proof uses properties of Casimir invariants and is independent of Weyl's character formula, see for example in [51, 57, 62].

**Lemma 16.5 (Freudenthal's Recursive Formula)** *The multiplicity $m_\mu$ of a weight $\mu$ in an irreducible representation with highest weight $\lambda$ is*

$$m_\mu = \frac{2}{(\delta + \lambda)^2 - (\lambda + \mu)^2} \sum_{\alpha>0} \sum_{k\geq1} (\mu + k\alpha, \alpha)\, m_{\mu+k\alpha}, \quad \mu \neq \lambda. \tag{16.79}$$

Starting with the simple multiplicity $m_\lambda = 1$ of the highest weight $\lambda$, this formula can be used recursively to determine the multiplicity of those weights whose depth increases by $\Delta k = 1$. In calculating the scalar product in the numerator, one first considers the weights $\lambda + k\alpha$ after specifying a positive root $\alpha$. We illustrate the procedure using the adjoint representation of $\mathfrak{su}(3)$.

---

**Example: Multiplicities for Adjoint Representation of $\mathfrak{su}(3)$**

The weights of the adjoint representation are the roots shown in Fig. 16.4, and their highest weight is $\lambda = \gamma_1 + \gamma_2 = \delta$. To the weight $\gamma_1$, we can only add the positive weight $\gamma_2$ to obtain a weight of the adjoint representation again. Since $\gamma_2^2 = \gamma_1^2 = -2(\gamma_1, \gamma_2)$, it follows that

$$m_{\gamma_1} = \frac{2(\gamma_1 + \gamma_2, \gamma_2)}{4(\gamma_1 + \gamma_2)^2 - (2\gamma_1 + \gamma_2)^2} = \frac{2(\gamma_1, \gamma_2) + 2\gamma_2^2}{3\gamma_1^2 + 4(\gamma_1, \gamma_2)} = 1.$$

Similarly, one finds $m_{\gamma_2} = 1$. Now we can determine the multiplicity of $\mu = 0$,

$$m_0 = \frac{2}{4(\gamma_1 + \gamma_2)^2 - (\gamma_1 + \gamma_2)^2} \left( \gamma_1^2 m_{\gamma_1} + \gamma_2^2 m_{\gamma_2} + (\gamma_1 + \gamma_2)^2 m_{\gamma_1 + \gamma_2} \right) = 2 \,.$$

All weights of the adjoint representation except for $\mu = 0$ have the multiplicity 1. The weight $\mu = 0$ has the multiplicity 2. ◄

An alternative method for calculating the multiplicities goes back to B. Kostant [67]. His formula, which is not proven here, makes use of the partition function $P_K(\mu)$, which is equal to the number of ways to write the weight $\mu$ as a sum of positive roots. If $\mu$ is not the sum of positive roots, then $P_K(\mu) = 0$.

**Lemma 16.6 (Kostant's Formula)** *The multiplicity $m_\mu$ of a weight $\mu$ in a irreducible representation with highest weight $\lambda$ is*

$$m_\mu = \sum_{w \in \mathcal{W}} \det(w) \, P_K\left( w(\delta + \lambda) - (\delta + \mu) \right). \tag{16.80}$$

The proof of this interesting formula can be found in texts where Freudenthal's formula is also proven. Since the order of the Weyl group grows exponentially with the size of the Lie algebra, it is in most cases more effective to calculate the $m_\mu$ using Freudenthal's recursive formula.

## 16.3  Young Diagrams

An efficient method to determine the decomposition of a tensor product of irreducible representations makes use of Young diagrams and Young tableaux. These have already appeared in Sects. 4.2 and 11.4 during the analysis of symmetric groups and their representations. Let $V$ be a complex vector space with basis $|1\rangle, \ldots, |n\rangle$, which carries an $n$-dimensional representation $U$ of the matrix group $G$. We use the symbol $U$ for representations of $G$ because in the following the symbol $D$ will be used for representations of the occurring symmetric group. The tensor products $|i_1 \ldots i_m\rangle = |i_1\rangle \otimes \cdots \otimes |i_m\rangle$ form a basis in the product space $V^{\otimes m} = V \otimes \cdots \otimes V$, and a tensor of rank $m$ has the expansion

$$T = \sum_{i_1, \ldots, i_m} T_{i_1 i_2 \ldots i_m} |i_1, i_2, \ldots, i_m\rangle \,. \tag{16.81}$$

The product space $V^{\otimes m}$ carries two representations:

- The tensor product $U^{\otimes m} = U \otimes \cdots \otimes U$ of $U$ acts on the product basis according to

$$U^{\otimes m} : |i_1, i_2, \ldots, i_m\rangle \longmapsto \sum_{j_1,\ldots,j_m=1}^{n} U_{j_1 i_1} U_{j_2 i_2} \cdots U_{j_m i_m} |j_1, j_2, \ldots, j_m\rangle,$$

$$(16.82)$$

or equivalently on the components of the tensor as

$$U^{\otimes m} : T_{i_1 i_2 \ldots i_m} \longmapsto T'_{i_1, i_2, \ldots, i_m} = \sum_{j_1,\ldots,j_m=1}^{n} U_{i_1 j_1} \cdots U_{i_m j_m} T_{j_1 j_2 \ldots j_m} .$$

$$(16.83)$$

It acts only within the factors of the tensor product and is generally reducible. $U^{\otimes m}$ is the obvious generalization of the product $U \otimes U$, which was introduced in Sect. 10.5. In Corollary 14.2, the representation $U_*^{\otimes m}$ of the Lie algebra induced by $U^{\otimes m}$ was given. In components, it reads

$$T^{i_1 \ldots i_m} \mapsto X^{i_1}_{\ k} T^{k i_2 \ldots i_m} + \cdots + X^{i_m}_{\ k} T^{i_1 \ldots i_{p-1} k} , \tag{16.84}$$

where we briefly wrote $X$ for $U_*(X)$.

- A permutation $\pi \in \mathcal{S}_m$ permutes the basis vectors according to

$$D(\pi) : |i_1\rangle \otimes \cdots \otimes |i_m\rangle \longmapsto |\pi^{-1}(i_1)\rangle \otimes \cdots \otimes |\pi^{-1}(i_m)\rangle , \tag{16.85}$$

and is extended as a linear map to the entire space $\mathcal{V}^{\otimes m}$. It only acts between the factors $\mathcal{V}$ of the tensor product. On the components of a tensor, it acts according to

$$D(\pi) : T_{i_1 i_2 \ldots i_m} \longmapsto T_{\pi(i_1)\pi(i_2)\ldots\pi(i_m)} . \tag{16.86}$$

The representations act on different spaces: $U^{\otimes m}$ vertically within each factor $\mathcal{V}$ in the product $\mathcal{V} \otimes \cdots \otimes \mathcal{V}$, and $D$ horizontally between the factors. Therefore, they commute,

$$U^{\otimes m}(g)D(\pi) = D(\pi)U^{\otimes m}(g), \quad \text{for all} \quad g \in G, \ \pi \in \mathcal{S}_m . \tag{16.87}$$

The $U^{\otimes m}(g)$ also commute with the Young symmetrizers introduced in (11.38). Therefore, the reduction of $D$ leads to a (partial) reduction of $U^{\otimes m}$.

**Task**

Show that if $T_{i_1 i_2 \ldots}$ is symmetric in two indices, then $T'_{i_1 i_2 \ldots}$ in (16.83) is also symmetric in the corresponding two indices.

As in Sect. 11.4, let $\mathcal{A}(\mathcal{S}_m)$ denote the group algebra of the symmetric group $\mathcal{S}_m$. With the help of the Young symmetrizers, we define the linear subspaces

$$
\mathcal{V}_\lambda^{\otimes m} = \mathcal{A}(\mathcal{S}_m) Y_\lambda \mathcal{V}^{\otimes m} \subseteq \mathcal{V}^{\otimes m} \quad \text{and} \quad \mathcal{V}_\lambda^{\otimes m}(T) = \mathcal{A}(\mathcal{S}_m) Y_\lambda T \subseteq \mathcal{V}_\lambda^{\otimes m} ,
\tag{16.88}
$$

where $Y_\lambda$ is the symmetrizer of the (normal) tableau $T_\lambda$ that belongs to the partition $\lambda$ of $m$, and $T$ is an arbitrary tensor in $\mathcal{V}^{\otimes m}$. The $m$-fold tensor product of $\mathcal{V}$ decomposes into direct sums of these subspaces,

$$
\mathcal{V}^{\otimes m} = \bigoplus_\lambda \mathcal{V}_\lambda^{\otimes m} = \bigoplus_\lambda \bigoplus_T \mathcal{V}_\lambda^{\otimes m}(T) .
\tag{16.89}
$$

We know that $\mathcal{V}_\lambda^{\otimes m}$ carries the representation $D_\lambda$ of the symmetric group $\dim(D_\lambda)$ times.

**Definition 16.2** For a fixed $T_\lambda$, elements in $\mathcal{V}_\lambda^{\otimes m}$ are tensors of the symmetry class $\lambda$.

**Example: Totally Antisymmetric Tensors**

The matrix group $\mathrm{GL}(n)$ acts on totally antisymmetric tensors in $\mathcal{V}^{\otimes n}$ (note that $m = n$), which must be proportional to the totally antisymmetric $\varepsilon$-tensor, as follows

$$
T'_{i_1 \ldots i_n} = T' \varepsilon_{i_1 \ldots i_n} = A_{i_1 j_1} \cdots A_{i_n j_n} T \varepsilon_{j_1 \ldots j_n} = \det(A) T \, \varepsilon_{i_1 \ldots i_n} \quad \text{i.e.}
$$

$$
T' = \det(A) T .
$$

This is the determinant homomorphism of the matrix group. For $\mathrm{SL}(n)$ and $\mathrm{SU}(n)$, it is the trivial representation. ◄

**Corollary 16.5** *If the subspace* $\mathcal{V}_\lambda^{\otimes m}(T)$ *is not empty, then it carries an irreducible representation of* $\mathcal{S}_m$ *corresponding to the Young tableau* $T_\lambda$.

***Proof*** The corollary has already been proven in Sect. 11.4. It is repeated here in an adapted form due to its relevance for the representations of $G$. The subspace $\mathcal{V}^{\otimes m}(T)$ is empty if $Y_\lambda(T)$ vanishes. Otherwise, we consider a tensor $S$ in this subspace, which by assumption has the form $S = \rho T$ with $\rho \in \mathcal{A}(\mathcal{S}_m) Y_\lambda$. Then

$$
\pi S = \pi(\rho T) = (\pi \rho) T \in \mathcal{V}_\lambda(T), \quad \text{since} \quad \pi \rho \in \mathcal{A}(\mathcal{S}_m) Y_\lambda .
\tag{16.90}
$$

This means that $\mathcal{V}_\lambda(T)$ is invariant under $\mathcal{S}_m$. To show that it carries the irreducible representation corresponding to $T_\lambda$, we consider a basis $\rho_i$ of $\{\mathcal{A}(\mathcal{S}_m)Y_\lambda\}$ with the corresponding basis $\rho_i T$ of $\mathcal{V}_\lambda^{\otimes m}(T)$. The action of a permutation $\pi$ on a basis vector $\rho_i$ is

$$\pi \rho_i = \rho_j D_\lambda(\pi)_{ji}\,, \tag{16.91}$$

and on $\rho_i T$

$$\pi \rho_i T = \rho_j D_\lambda(\pi)_{ji} T = \rho_j T\, D_\lambda(\pi)_{ji}\,. \tag{16.92}$$

The representation matrices in (16.91) and (16.92) are (after choosing an adapted basis) identical and independent of $T$. According to the remarks in Sect. 11.4, they belong to irreducible representations of the symmetric group $\mathcal{S}_m$. We conclude that the representing matrices $D_\lambda(\pi)$ in $\pi|\lambda, T, a\rangle = |\lambda, T, b\rangle\, D_\lambda(\pi)_{ba}$ do not depend on $T$.

**Corollary 16.6** *The subspaces $\mathcal{V}_\lambda(a)$, spanned by $|\lambda, T, a\rangle$ with fixed $\lambda$ and $a$, are invariant and irreducible under the action of $G$, and the representation matrices do not depend on $a$.*

---

### Question

Why does this mean that different standard tableaux for the same $\lambda$ lead to equivalent representations?

***Proof*** The representation $U^{\otimes m}$ of $G$ does not change the symmetry class, and therefore

$$U^{\otimes m}(g)|\lambda, S, a\rangle = |\lambda, T, b\rangle\, U_\lambda(g)_{(Tb)(Sa)}\,, \tag{16.93}$$

where the sum is taken over $(T, b)$ for fixed $\lambda$. We show that the representation matrices are diagonal in the indices $a, b$. First, we act with $D_\lambda$ and then with $U^{\otimes m}$, obtaining

$$U^{\otimes m}(g)\, D_\lambda(\pi)|\lambda, S, a\rangle = |\lambda, T, b\rangle\, U_\lambda(g)_{(Tb)(Sc)}\, D_\lambda(\pi)_{ca}\,,$$

and if we act in the reverse order, then we find

$$D_\lambda(\pi)U^{\otimes m}(g)|\lambda, S, a\rangle = |\lambda, T, b\rangle\, D_\lambda(\pi)_{bc}\, U_\lambda(g)_{(Tc)(Sa)}\,.$$

The right-hand sides must match, and in matrix form this means

$$U_\lambda(g)_{TS} D_\lambda(\pi) = D_\lambda(\pi)U_\lambda(g)_{TS}\,. \tag{16.94}$$

The $D_\lambda$ form an irreducible representation of $\mathcal{S}_m$ and according to Schur's lemma, $U_\lambda$ must be a multiple of the identity, $U_\lambda(g)_{(Tc)(Sa)} = U_\lambda(g)_{TS}\,\delta_{ab}$. We conclude

$$U^{\otimes m}(g)\,|\lambda, S, a\rangle = |\lambda, T, a\rangle\, U_\lambda(g)_{TS}\,, \qquad (16.95)$$

which proves the corollary. We summarize:

---

**Reduction of the Product Representation $U^{\otimes m}$**
The decomposition (16.89) of the regular representation of $\mathcal{S}_m$ on $\mathcal{V}^{\otimes m}$ into invariant and irreducible subspaces $\mathcal{V}_\lambda^{\otimes m}(T)$ leads to a reduction of the product representation $U^{\otimes m}$ of $G$. The irreducible representations of $G$ on the subspaces $\mathcal{V}_\lambda(a)$ do not depend on $a$ and have the same Young diagrams with $m$ boxes. The latter have a maximum of $n = \dim\mathcal{V}$ rows because the antisymmetrization of a tensor with $n + 1$ indices results in zero.

---

## 16.3.1 Dimensions of Representations of GL($n$)

To each Young diagram $T_\lambda$ with $m$ boxes belongs an irreducible representation $D_\lambda$ of the symmetric group $\mathcal{S}_m$ and an irreducible representation $U_\lambda$ of the matrix group $G$ on $\mathcal{V}^{\otimes m}$. The dimension of $D_\lambda$ and its multiplicity in $\mathcal{V}_\lambda^{\otimes m}$ are both equal to $d_\lambda$, which is given by the hook length formula (11.40). We have also seen that the multiplicity of $D_\lambda$ is equal to the dimension of $U_\lambda$ and vice versa. To determine the multiplicity we must consider which Young symmetrizers, applied to the normal tableau, generate linearly independent elements. The result can be extracted from the Young diagram:

---

**Representations $U_\lambda$ of GL($n$)**
The tableaux with non-decreasing numbers from $\{1, \ldots, n\}$ from left to right in each row and increasing numbers from top to bottom in each column lead to linearly independent $|\lambda, T, a\rangle$.

---

**Example: Dimensions of Representations of GL($2$)**

The following diagrams give rise to irreducible representations with the given dimensions:

$$\dim(U_\lambda) = 2 \quad , \quad \dim(U_\lambda) = 4\,.$$

Justification:

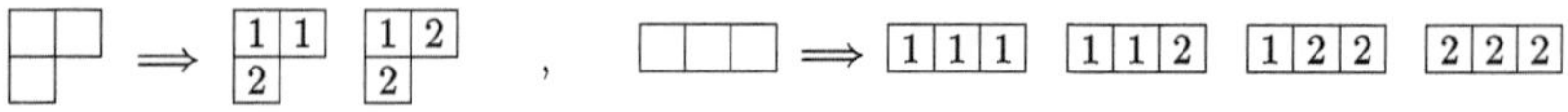

$$\young(\ \ ,\ ) \;\Rightarrow\; \young(11,2)\;\young(12,2)\;,\qquad\qquad \young(\ \ \ )\;\Rightarrow\; \young(111)\;\young(112)\;\young(122)\;\young(222)$$

◀


## 16.3.2  Irreduzible Representations of U($n$) and SU($n$)

The irreducible representations of $GL(n, \mathbb{C})$ are also representations of their subgroups, but as such are generally reducible. They remain irreducible for the subgroups $U(n)$ and $SU(n)$, but not for the subgroups $O(n)$ and $SO(n)$. This follows from how the corresponding Lie subalgebras are embedded in $\mathfrak{gl}(n)$ and will not be explained in detail here. However, it must be taken into account that a Young diagram has a maximum of $n$ rows and that a column of length $n$, whose entries are completely antisymmetrized by $Y_\lambda$, results in the determinant homomorphism, i.e. the trivial representation for $SU(n)$. Therefore, the diagrams for $SU(2)$

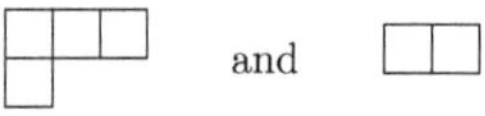

$$\young(\ \ ,\ )\qquad\text{and}\qquad\young(\ \ )\tag{16.96}$$

belong to the same representation of dimension 3. Besides the trivial representation with one column and two rows, only diagrams with one row need to be considered. For $SU(3)$, for example, the following two diagrams

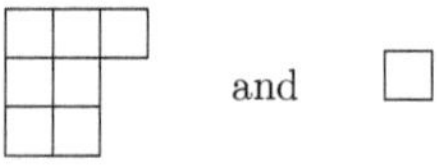

$$\young(\ \ \ ,\ \ )\qquad\text{and}\qquad\young(\ )\tag{16.97}$$

belong to the same irreducible representation.

---

**Task**

Show that for $SU(2)$ a diagram with $m$ boxes (in a row) gives rise to a representation of dimension $m + 1$.

---

**Example: Representations of SU(3)**

Only diagrams of the form

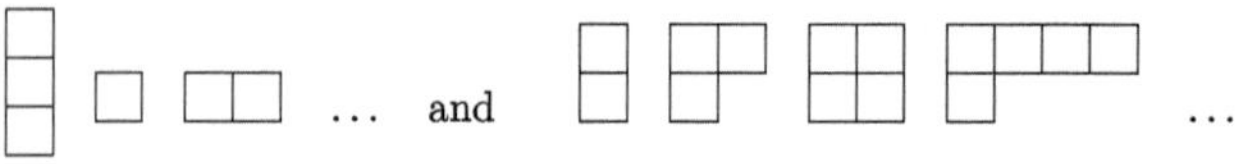

$$\young(\ ,\ ,\ )\quad\young(\ )\quad\young(\ \ )\;\ldots\;\text{and}\quad\young(\ ,\ )\;\young(\ \ ,\ )\;\young(\ \ ,\ \ )\;\young(\ \ \ ,\ )\;\ldots$$

contribute. According to the rules, we may fill the boxes with the numbers $1, 2$, and $3$. It follows

$$\square \quad \dim = 3, \qquad \square\!\!\square \quad \dim = 3, \qquad \square\!\!\square \quad \dim = 8 \ .$$

◀

Without proof, we note the formula for the dimension of the representation $U_\lambda$ of SU($n$):

**Dimension of Representations for SU($n$)**

For SU($n$), the irreducible representation $U_\lambda$ has the dimension

$$\dim U_\lambda = \prod_{ij} \frac{n + j - i}{h_{ij}}, \tag{16.98}$$

where $i$ is the row index, $j$ is the column index, and $h_{ij}$ is the hook length of cell $(ij)$, as defined in Sect. 11.4.1.

For example, one finds:

$$\dim\left(\square\right) = n, \qquad \dim\left(\square\!\square\right) = \frac{n^2 + n}{2}, \qquad \dim\left(\square\!\square\!\square\right) = \frac{1}{6}\prod_{i=0}^{2}(n + i)$$

$$\dim\left(\begin{smallmatrix}\square\\\square\end{smallmatrix}\right) = \frac{n^2 - n}{2}, \qquad \dim\left(\begin{smallmatrix}\square\square\\\square\end{smallmatrix}\right) = \frac{n^3 - n}{3}, \qquad \dim\left(\begin{smallmatrix}\square\\\square\\\square\end{smallmatrix}\right) = \frac{1}{6}\prod_{i=0}^{2}(n - i)$$

The simple formula (16.98) is equivalent to the formula (16.105) in Problem 16.7 for the dimensions of the irreducible SU($n$) representations. When comparing results of the graphical method, based on Young diagrams with rows of lengths $\ell_1 \geq \ell_2 \geq \dots$, with results of the analytical method, the following relationship between row lengths and Dynkin labels is used:

$$[n_1, n_2, \dots, n_{r-1}, n_r] = [\ell_1 - \ell_2, \ell_2 - \ell_3, \dots, \ell_{r-1} - \ell_r, \ell_r], \quad r = n - 1 . \tag{16.99}$$

Young diagrams with only one row of $n$ boxes have Dynkin labels $[n, 0, 0, \dots]$. For diagrams with two rows, only the first two Dynkin labels are non-zero, etc.

**Task**

Show that (16.98) and (16.105) yield the same results for SU(3) and SU(4).

### 16.3.3 Tensor Products with Young Diagrams

Given two irreducible representations $U_\lambda$ and $U_{\lambda'}$ of the Lie groups SU($n$), U($n$) or GL($n$) with Young diagrams $T_\lambda$ and $T_{\lambda'}$. In the reduction of the corresponding product representation $U_\lambda \otimes U_{\lambda'}$, the graphical method with Young diagrams is very efficient and superior to the analytical method using characters. Here we only provide the algorithm, without showing that it is correct. The graphical rules for $U_\lambda \otimes U_{\lambda'}$ are (see e.g. [13]):

1. Write the number 1 in the boxes of the first row of the diagram $T_{\lambda'}$, the number 2 in the boxes of the second row, etc.
2. Add the boxes of $T_{\lambda'}$ to $T_\lambda$. In the first step, the ones, in the second step, the twos, etc., while observing the following rules:
   (a) In each step, the resulting diagram must be a valid Young diagram and not have more than $n$ rows,
   (b) a number may appear at most once in a column,
   (c) reading the numbers row by row from right to left and from top to bottom, there must never be more $i$ than $(i - 1)$ in the sequence.
3. If a diagram is obtained multiple times in this way, only diagrams with different distributions of numbers are considered.
4. For SU($n$), columns with $n$ boxes can be omitted.

**Example: Tensor Products of SU(2) Representations**

We consider the tensor product of the representations with spin 3 and 2,

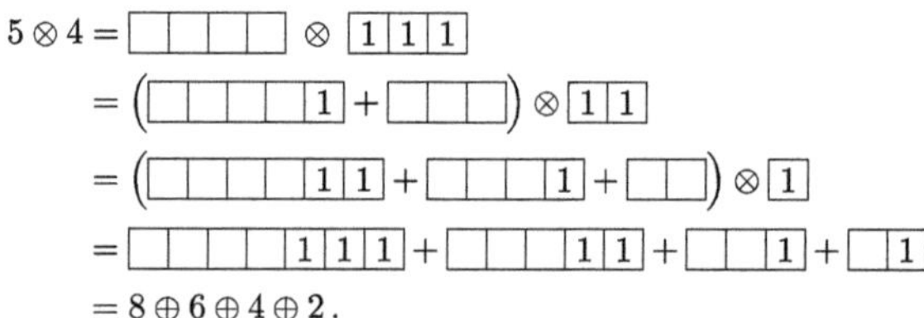

$$= 8 \oplus 6 \oplus 4 \oplus 2.$$

Many tableaux were omitted because they either had two rows or appeared multiple times.  ◄

**Example: Tensor Products of SU(3) Representations**

We consider the tensor product of the representations $3$ and $\bar{3}$:

$$3 \otimes 3 = \square \otimes \boxed{1} = \boxed{\phantom{1}1} \oplus \boxed{\begin{smallmatrix}\phantom{1}\\1\end{smallmatrix}} = 6 \oplus \bar{3}$$

$$\bar{3} \otimes 3 = \begin{smallmatrix}\square\\\square\end{smallmatrix} \otimes \boxed{1} = \begin{smallmatrix}\boxed{\phantom{1}1}\\\square\end{smallmatrix} \oplus \begin{smallmatrix}\phantom{1}\\1\end{smallmatrix} = 8 \oplus 1$$

◄

Thus, using simple Young tableaux, the results obtained using characters in Sect. 12.3 were reproduced.

## 16.4 Exercises for Chap. 16

**Problem 16.1 (Spherical Harmonics as Irreducible Representations of SO(3))**
Let $x^i$ be Cartesian coordinates in $\mathbb{R}^3$. The function space $L_2(S^2)$ is spanned by the polynomials $P_n(x) = \sum f_{i_1 \ldots i_n} x^{i_1} \cdots x^{i_n}$ (with complex coefficients), where the embedding of $S^2$ in $\mathbb{R}^3$ is given by $x^2 = 1$. See the explanations in Sect. 10.8.1. We also need the space $C_n(S^2) = P_0(x) \oplus P_1(x) \oplus \cdots \oplus P_n(x)$ of all polynomials of degree $\leq n$. The representation of SO(3) on $\mathbb{R}^3$—given by $x \to Rx$ with $R \in$ SO(3)—defines a representation on the continuous functions on $S^2$ by $\psi(x) \mapsto (\Gamma(R)\psi)(x) = \psi(R^{-1}x)$.

1. Determine the induced representation of $\mathfrak{su}(2)$ on $C_n(S^2)$, i.e., determine the action of the generators $J_i$ on a polynomial.
2. Find the polynomials with $J_+ P_n(x) = 0$, i.e., the highest weight vectors, including multiplicities of the corresponding representations.
3. Find explicitly the representation spaces for angular momentum 0, 1, and 2 as polynomials in the Cartesian coordinates.

**Problem 16.2 (Roots and Weights of $\mathfrak{so}(5)$)** In Fig. 16.6 the 4 short and 4 long roots, fundamental weights, the Weyl vector, and the fundamental Weyl chamber of $B_2$ are shown. The dashed line separates positive and negative roots. The weight $\mu_1$ is simultaneously one of the short positive roots. The (red) dots mark the weights of the representation $[1, 0]$ and the (blue) squares the weights of the representation $[0, 1]$. Try to answer the following questions:

1. The Weyl orbit of $\mu_1$ contains 4 weights. Why does $\mu = 0$ belong to the irreducible representation $\mathfrak{D}_{\mu_1} = [1, 0]$?
2. The Weyl orbit of $\mu_2$ also contains 4 weights. Why can $0$ not possibly be a weight of the representation $\mathfrak{D}_{\mu_2} = [0, 1]$?
3. What is the highest weight of the adjoint representation?

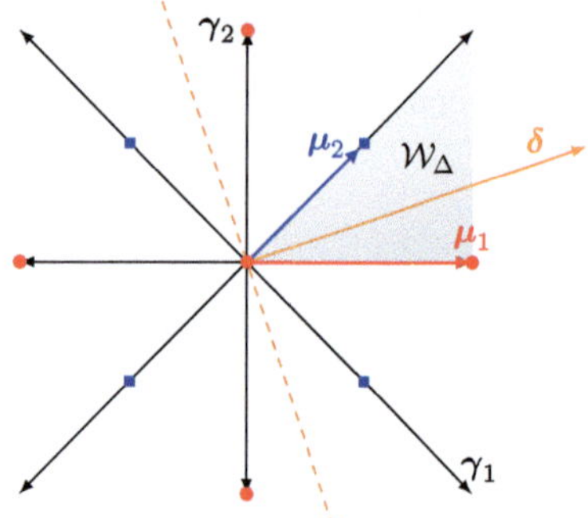

**Fig. 16.6** Simple roots, Weyl vector, fundamental weights, and fundamental Weyl chamber for the Lie algebra $B_2$

4. Which of the tensor products, $[1, 0] \otimes [1, 0]$, $[0, 1] \otimes [0, 1]$ or $[1, 0] \otimes [0, 1]$ contains the adjoint representation?

**Problem 16.3 (Weight and Root Lattice)** In Lemma 13.6, a connection was established between a Lie algebra $\mathfrak{g}$ and its image $\mathrm{ad}_{\mathfrak{g}}$ under the adjoint representation. Why does this imply the property $\Lambda_{\mathrm{W}}/\Lambda_{\mathrm{R}} \cong \mathfrak{z}_{\mathfrak{g}}$ for the weight lattice $\Lambda_{\mathrm{W}}$ and root lattice $\Lambda_{\mathrm{R}}$? Why is the ratio of the volumes of the elementary cells of $\Lambda_{\mathrm{R}}$ and $\Lambda_{\mathrm{W}}$ equal to $\det K$?

**Problem 16.4 (An Alternative Formula for the Haar Measure)** In this somewhat longer exercise, we calculate the Haar measure using methods familiar to physicists. We start with the invariant line element $\mathrm{d}s^2 = -\mathrm{Tr}\left(U^{-1}\mathrm{d}U\,U^{-1}\mathrm{d}U\right) = g_{ab}\mathrm{d}x^a\mathrm{d}x^b$ and derive from it the Haar measure $\sqrt{g}\,\mathrm{d}^n x$ as in Sect. 9.5. We choose coordinates $U(x) = \mathrm{e}^{\mathrm{i}X(x)}$ with $X = x^a X_a$ and accordingly $\mathrm{d}X = \mathrm{d}x^a X_a$ (we apply the summation convention)

1. First, prove the interesting formula

$$\omega = U^{-1}\mathrm{d}U = \mathrm{e}^{-\mathrm{i}X} \int_0^1 \mathrm{d}t\ \mathrm{e}^{\mathrm{i}(1-t)X}(\mathrm{id}X)\,\mathrm{e}^{\mathrm{i}tX} = \mathrm{i}\mathrm{d}x^a \int_0^1 \mathrm{e}^{-\mathrm{i}tX} X_a\, \mathrm{e}^{\mathrm{i}tX}\,.$$

$$(16.100)$$

Hint: For matrices, the exponential function is given by its Taylor series. You may need the integration formula

$$\int_0^1 \mathrm{d}t\ (1-t)^n t^m = \frac{n!\,m!}{(1+n+m)!}\,.$$

$$(16.101)$$

2. Use the formula (16.100) twice to prove

$$g_{ab} = \int_{-1}^1 \mathrm{d}t\ (1 - |t|)\, c_{ab}(t, x) \quad \text{with} \quad c_{ab}(t, x) = \mathrm{Tr}\left( \mathrm{e}^{\mathrm{i}tX} X_a\, \mathrm{e}^{-\mathrm{i}tX} X_b \right).$$

Hint: A change to new variables $t = t_2 - t_1$ and $2s = t_1 + t_2$ might be helpful.

3. Derive the differential equation for $c_{ab}$ as a function of $t$ and use it to prove

$$g_{ab} = \int_{-1}^{1} dt \, (1 - |t|) \, e^{t M(x)} , \quad \text{with} \quad M^a{}_b = x^c f_{cb}{}^a .\tag{16.102}$$

4. The eigenvalues of the antisymmetric matrix $M$ occur in complex-conjugate pairs, $M v_k = i\lambda_k v_k$ and $M v_k^* = -i\lambda_k v_k^*$, so that the diagonal matrix $D$ in $M = V D V^{-1}$ has the form $D = \mathrm{diag}(i\lambda_1, -i\lambda_1, i\lambda_2, \dots)$. Show that the density of the Haar measure is given by

$$\sqrt{g} = \Delta^2, \quad \Delta = \prod_{\lambda_k > 0} \left( \frac{\sin(\frac{1}{2}\lambda_k)}{\frac{1}{2}\lambda_k} \right) .\tag{16.103}$$

5. Show that $v$ is an eigenvector of $M$ with eigenvalue $i\lambda$ if and only if $v = v^a X_a$ is an eigenvector of $\mathrm{ad}_X$ with eigenvalue $-\lambda$.

   Hint: Compare the eigenvalue equations $[X, v] = -\lambda v$ and $M v = iv$ and recall the definition of $M$.

6. Finally, prove that

$$d\mu = C \cdot \frac{\det \left( \sin \left( \frac{1}{2}\mathrm{ad}_X \right) \right)}{\det \left( \frac{1}{2}\mathrm{ad}_X \right)} \, d^n x .\tag{16.104}$$

7. Why does $\mathrm{ad}_X$ have the $r$-fold eigenvalue 0? Why is this not a problem in (16.104)?

**Problem 16.5 (Reduced Haar Measure for SU($n$))**    Use the general formula (16.37) to confirm the result (9.47) for the reduced Haar measure of SU($n$).

**Problem 16.6 (Dimensions of the $\mathfrak{so}(5)$ Representations)**    The roots of $B_2 \cong \mathfrak{so}(5)$ were given in Appendix 15.8. Determine a Weyl vector, the corresponding positive roots, and fundamental weights. Let $\lambda = n_1 \mu_1 + n_2 \mu_2$ be the highest weight of an irreducible representation $[n_1, n_2]$. What is the dimension of this representation?

**Problem 16.7 (Dimensions of the $\mathfrak{su}(n)$ Representations)**    The roots and simple roots of $\mathfrak{su}(n)$ were given in Appendix 15.8. Show that:

1. The corresponding fundamental weights $\mu_1, \mu_2, \dots, \mu_r$ with $r = n - 1$ and the Weyl vector have the form

$$\mu_i = \sum_{p=1}^{i} e_p - \frac{i}{n} \sum_{p=1}^{n} e_p, \quad i = 1, \dots, r, \quad \text{and} \quad \delta = \sum_{p=1}^{n} \left( \frac{n+1}{2} - p \right) e_p .$$

2. The positive roots are $\alpha_{pq} = e_p - e_q$ with $q > p$ and $(\delta, \alpha_{pq}) = q - p$.
3. Let $\lambda = \sum n_i \mu_i$ be a weight. Prove the identity

$$(\mu, \alpha_{pq}) = \sum_{i=p}^{q-1} n_i, \quad 1 \le p < q \le n.$$

4. With this, you should obtain the following general formula for the dimension of the irreducible representation with highest weight $\lambda = \sum n_i \mu_i$:

$$\dim[n_1, \ldots, n_r] = \prod_{1 \le p < q \le n} \frac{\Delta_{pq} + q - p}{q - p}, \qquad \Delta_{pq} = n_p + \cdots + n_{q-1}.$$

$$(16.105)$$

**Problem 16.8 (Dimensions of the $\mathfrak{su}(5)$ Representations)** The irreducible representations of $\mathfrak{su}(5)$ are characterized by their Dynkin labels $[n_1, n_2, n_3, n_4]$. We define the numbers $x_{ij\ldots} = (1 + n_i) + (1 + n_j) + \cdots \in \mathbb{N}$. Show that

$$\dim[n_1, n_2, n_3, n_4] = x_1 x_2 x_3 x_4 \cdot \frac{x_{12}}{2} \frac{x_{23}}{2} \frac{x_{34}}{2} \cdot \frac{x_{123}}{3} \frac{x_{234}}{3} \cdot \frac{x_{1234}}{4}. \qquad (16.106)$$

# Symmetries in Quantum Mechanics 17

<br>

> *Money, machines, algebra - the three monsters of today's civilization.*
>
> —Simone Weil

In the past, it often happened that experimental data suggested the existence of conserved quantities. In most cases, such conserved quantities owe their existence to a symmetry of the system, and knowledge of the symmetry can be helpful in finding the correct equations of motion. In fundamental theories, symmetries are usually known and are implemented from the beginning. For example, the spectrum of an atom will not change if we measure it in Weimar instead of in Jena. We will therefore demand that the transition amplitudes or eigenvalues of observables do not change under a translation or rotation of the physical system. In this chapter, we use natural units with $\hbar = c = 1$. The role of symmetries in quantum mechanics is emphasized in many textbooks, e.g., in the monographs [61, 68–70].

In the formalism of quantum mechanics, two vectors in a Hilbert space $\mathcal{H}$ describe the same (pure) state if they are parallel. This property defines an equivalence relation,

$$|\psi\rangle \cong |\phi\rangle \Leftrightarrow |\psi\rangle = \lambda|\phi\rangle \quad \text{with} \quad \lambda \in \mathbb{C} \setminus \{0\}. \tag{17.1}$$

Equivalent vectors define a ray $\mathcal{R}_\psi = \{\lambda|\psi\rangle | \lambda \neq 0\}$, and the set of rays $\mathcal{R}_\mathcal{H}$ is identified with the set of pure states. The state space is therefore

$$\mathcal{R}_\mathcal{H} = \mathcal{H} \setminus \{0\}/ \cong . \tag{17.2}$$

A symmetry of a quantum system with Hilbert space $\mathcal{H}$ is now defined as follows:

**Symmetry**
A symmetry is a bijective mapping $\mathcal{S} : \mathcal{R}_{\mathcal{H}} \mapsto \mathcal{R}_{\mathcal{H}}$ that preserves transition probabilities

$$\frac{|\langle \phi|\psi\rangle|}{\|\phi\|\|\psi\|} = \frac{|\langle \mathcal{S}\phi|\mathcal{S}\psi\rangle|}{\|\mathcal{S}\phi\|\|\mathcal{S}\psi\|}. \tag{17.3}$$

An important theorem by Eugene Wigner states that:

**Theorem 17.1 (Wigner)**   *For a symmetry* $\mathcal{S} : \mathcal{R}_{\mathcal{H}} \mapsto \mathcal{R}_{\mathcal{H}}$, *there exists a unitary or anti-unitary mapping* $\Gamma_{\mathcal{S}} : \mathcal{H} \mapsto \mathcal{H}$, *which is compatible with* $\mathcal{S}$.

The mapping $\Gamma_{\mathcal{S}} : \mathcal{H} \mapsto \mathcal{H}$ is called compatible with $\mathcal{S}$ if $\Gamma_{\mathcal{S}}|\psi\rangle \in \mathcal{S}(\mathcal{R}_{\psi})$ holds for all elements in $\mathcal{H}$. An anti-unitary mapping $U : \mathcal{H} \mapsto \mathcal{H}$ satisfies $\langle U\psi|U\phi\rangle = \langle \phi|\psi\rangle$. It can be shown that unitary mappings are linear and anti-unitary mappings are anti-linear. It is said that symmetries are implemented by (anti)unitary operators. The exact assumptions of Wigner's theorem and its proof can be found, for example, in [68] or [69].

**Example: Spatial Reflections**

Under a spatial reflection of a system of $N$ point particles, the coordinates of all particles are reflected at the origin,

$$\mathcal{P} : \{\boldsymbol{x}_1, \ldots, \boldsymbol{x}_N\} \longrightarrow \{-\boldsymbol{x}_1, \ldots, -\boldsymbol{x}_N\}. \tag{17.4}$$

For mirror-symmetric systems, there exists according to Wigner's theorem a (in this case) unitary operator $\Gamma_{\mathcal{P}}$, that implements the reflection,

$$|\psi'\rangle = \Gamma_{\mathcal{P}}|\psi\rangle \quad , \quad A' = \Gamma_{\mathcal{P}} A \Gamma_{\mathcal{P}}^{-1}.$$

The reflection acts on position eigenstates according to $|\boldsymbol{x}_1, \ldots, \boldsymbol{x}_N\rangle \to |-\boldsymbol{x}, \ldots, -\boldsymbol{x}_N\rangle$. A wave function in position space thus transforms under a reflection into

$$\psi'(\boldsymbol{x}_1, \ldots, \boldsymbol{x}_N) = (\Gamma_{\mathcal{P}}\psi)(\boldsymbol{x}_1, \ldots, \boldsymbol{x}_N) = \psi(-\boldsymbol{x}_1, \ldots, -\boldsymbol{x}_N). \tag{17.5}$$

◀

**Question**

Why is $\Gamma_{\mathcal{P}}$ unitary on the Hilbert space $L_2(\mathbb{R}^{3N})$?

## 17.1   Many Particle Systems

Interactions propagate at a finite speed, and therefore even the classical interaction energy cannot depend solely on the particle positions at a fixed time. However, if the relative velocities of the particles are small compared to the speed of light, then their coordinates (for example, positions and spins) change only slightly during the interaction transfer between the particles. Then, up to terms of order $(v/c)^2$, the Hamiltonian can be defined as a function solely of the positions, momenta, and spins of the particles.

For non-relativistic many-particle systems the Hamiltonian operator takes the form

$$H = \sum_{i=1}^{N} \frac{p_i^2}{2m_i} + V + W. \tag{17.6}$$

The potential $V$ is of the order $(v/c)^0$ and describes an interaction that depends only on the particle positions. The correction $W$ is of the order $(v/c)^2$ and includes the spin-orbit coupling. It is a function of the positions, momenta, and spins of the particles and partially accounts for the retardation of the interaction.

The time evolution of the undisturbed many-body system follows from the time-dependent Schrödinger equation

$$i\hbar \frac{\partial}{\partial t} |\psi\rangle = H |\psi\rangle. \tag{17.7}$$

The entire system is composed of subsystems. In the formalism of quantum mechanics, the Hilbert space $\mathcal{H}$ is the tensor product of the Hilbert spaces $\mathcal{H}_1, \ldots, \mathcal{H}_N$ of the subsystems,

$$\mathcal{H} = \mathcal{H}_1 \otimes \mathcal{H}_2 \otimes \cdots \otimes \mathcal{H}_N. \tag{17.8}$$

For the tensor product of the $|\psi_i\rangle \in \mathcal{H}_i$ we write $|\psi_1\rangle \otimes |\psi_2\rangle \otimes \cdots \otimes |\psi_N\rangle \equiv |\psi_1 \psi_2 \cdots \psi_N\rangle$. The product is linear in each factor. It describes separable states. The scalar product of two tensor products is the product of the scalar products of its factors,

$$\langle \phi_1 \cdots \phi_N | \psi_1 \cdots \psi_N \rangle_{\mathcal{H}} = \langle \phi_1 | \psi_1 \rangle_{\mathcal{H}_1} \cdots \langle \phi_N | \psi_N \rangle_{\mathcal{H}_N}. \tag{17.9}$$

The tensor products of basis vectors of the individual Hilbert spaces,

$$|n_1 \ldots n_N\rangle \equiv |n_1\rangle \otimes \cdots \otimes |n_N\rangle, \tag{17.10}$$

form a basis of $\mathcal{H}$. If a vector $|\psi\rangle \in \mathcal{H}$ is not a tensor product, but only the sum of tensor products, then it describes an *entangled state*.

For observables $A_i : \mathcal{H}_i \mapsto \mathcal{H}_i$ of the subsystems the tensor product $A_1 \otimes \cdots \otimes A_N$ is an observable of the entire system and acts as follows on tensor products,

$$(A_1 \otimes \cdots \otimes A_N)(|\psi_1\rangle \otimes \cdots \otimes |\psi_N\rangle) = |A_1\psi_1\rangle \otimes \cdots \otimes |A_N\psi_N\rangle. \tag{17.11}$$

Tensor products of linear mappings are discussed in detail in Sect. 10.5.

The operators $A_i$ act in different Hilbert spaces and cannot be added. However, if it is tacitly assumed that $A_i$ acts trivially on the spaces $\mathcal{H}_j$ with $j \neq i$, then they can be added and one defines, for example,

$$A_1 + A_2 \equiv A_1 \otimes \mathbb{1} + \mathbb{1} \otimes A_2. \tag{17.12}$$

With this convention

$$(A_1 + A_2)(|\psi_1\rangle \otimes |\psi_2\rangle) = |A_1\psi_1\rangle \otimes |\psi_2\rangle + |\psi_1\rangle \otimes |A_2\psi_2\rangle. \tag{17.13}$$

This notation is used, for example, in the addition of angular momenta in Sect. 17.5.

**Tensor Products of Eigenvectors of the Subsystems**
Let $|a_i\rangle \in \mathcal{H}_i$ be an eigenvector of $A_i$ with eigenvalue $a_i$. Then the tensor product

$$|a_1\rangle \otimes \cdots \otimes |a_N\rangle \equiv |a_1 \ldots a_N\rangle \in \mathcal{H}_1 \otimes \cdots \otimes \mathcal{H}_N \tag{17.14}$$

is an eigenvector of the sum and the tensor product of the operators $A_i$,

$$(A_1 + \cdots + A_N)|a_1 \ldots a_N\rangle = (a_1 + \cdots + a_N)|a_1 \ldots a_N\rangle$$
$$(A_1 \otimes \cdots \otimes A_N)|a_1 \ldots a_N\rangle = (a_1 \cdots \cdots a_N)|a_1 \ldots a_N\rangle. \tag{17.15}$$

**Task**

Prove these useful properties.

To fully characterize state vectors in $\mathcal{H}$, one can choose a complete set of commuting observables in each subspace $\mathcal{H}_i$ and diagonalize them simultaneously. If $\xi_i$ represents the eigenvalues of a complete set of commuting observables in $\mathcal{H}_i$, and $|\xi_i\rangle$ represents the corresponding eigenvectors, then the products

$$|\xi_1 \ldots \xi_N\rangle \equiv |\xi_1\rangle \otimes \cdots \otimes |\xi_N\rangle \tag{17.16}$$

form a basis of $\mathcal{H}$. For example, we can choose the common (improper) eigenvectors of the commuting position and spin operators.

$$|x_1 s_1, \ldots, x_N s_N\rangle = |x_1 s_1\rangle \otimes \cdots \otimes |x_N s_N\rangle \,. \tag{17.17}$$

An arbitrary state is then a superposition of this basis,

$$|\psi\rangle = \sum_{s_1,\ldots,s_N} \int d^3x_1 \ldots d^3x_N \, \psi(x_1, s_1, \ldots, x_N, s_N) \, |x_1 s_1, \ldots, x_N s_N\rangle \,, \tag{17.18}$$

where the expansion coefficients are the $N$-particle wave functions,

$$\psi(x_1, s_1, \ldots, x_N, s_N) = \langle x_1 s_1, \ldots, x_N s_N|\psi\rangle \,. \tag{17.19}$$

The scalar product of two state vectors is

$$\langle\phi|\psi\rangle = \sum_{s_1,\ldots,s_N} \int d^3x_1 \ldots d^3x_N \, \bar{\phi}(x_1, s_1 \ldots, x_N, s_N)\psi(x_1, s_1, \ldots, x_N, s_N) \,. \tag{17.20}$$

The wave function (17.19) normalized to one is the probability amplitude for finding the first particle at position $x_1$ with third spin component $s_1$, the second particle at position $x_2$ with third spin component $s_2$, ... and the $N$th particle at position $x_N$ with third spin component $s_N$. For identical particles, the wave functions must show a prescribed behavior when the arguments $(x_i, s_i)$ are permuted.

## 17.2  Translations

The simplest continuous symmetries are the translations in space. Here, the Cartesian coordinates of all particles are shifted,

$$x_i \longrightarrow x_i + a, \quad i = 1, \ldots, N \,, \tag{17.21}$$

and the shifts form the three-dimensional Abelian Lie group $\mathbb{R}^3$. State vectors and operators transform according to

$$|\psi\rangle \longrightarrow |\psi'\rangle = \Gamma_a|\psi\rangle \quad, \quad A \longrightarrow A' = \Gamma_a A \Gamma_a^{-1} \,, \tag{17.22}$$

with unitary operator $\Gamma_a$ on the Hilbert space $\mathcal{H}$. In position space, it is

$$(\Gamma_a \psi)(x_1, \ldots, x_N) = \psi(x_1 + a, \ldots, x_N + a) \,. \tag{17.23}$$

**Question**

Why are the operators $\Gamma_a$ unitary?

The mapping $a \to \Gamma_a$ defines a unitary representation of the translation group $\mathbb{R}^3$ on $\mathcal{H}$,

$$\Gamma_{a+b} = \Gamma_a \Gamma_b \quad \text{and} \quad \Gamma_a^{-1} = \Gamma_{-a} \, . \tag{17.24}$$

The representation property implies that the operators $\Gamma_a$ commute,

$$[\Gamma_a, \Gamma_b] = 0, \qquad \forall a, b \, , \tag{17.25}$$

and therefore can be simultaneously diagonalized. Under a shift the Hamiltonian transforms according to

$$\Gamma_a H \left(p_1, \ldots, p_N, x_1, \ldots, x_N\right) \Gamma_a^{-1} = H \left(p_1, \ldots, p_N, x_1 + a, \ldots, x_N + a\right) \, . \tag{17.26}$$

The operator (17.6) with $W = 0$ is translation invariant, $\Gamma_a H \Gamma_a^{-1} = H$, if the potential is invariant,

$$V \left(x_1 + a, \ldots, x_N + a\right) = V \left(x_1, \ldots, x_N\right) \, . \tag{17.27}$$

For such potentials, $H$ and the operators $\Gamma_a$ can be simultaneously diagonalized. The translations are generated by the total momentum of the system,

$$\Gamma_a = e^{ia \cdot P}, \qquad P = p_1 + \cdots + p_N \, , \tag{17.28}$$

see Problem 17.1, and the translation invariance of the Hamiltonian operator implies that the total momentum commutes with $H$, $[P, H] = 0$. Thus, for homogeneous systems, the total momentum is conserved. The diagonalization of the unitary translation operators is equivalent to the diagonalization of the Hermitian momentum $P$.

An important class of homogeneous systems includes pair interactions that depend only on the difference vectors between the particle pairs,

$$H = \sum \frac{1}{2m_i} p_i^2 + \sum_{i<j} V_{ij}(x_i - x_j) \, . \tag{17.29}$$

The Coulomb interaction is of this type. For a particle in an *external* potential, the invariance condition is $V(x + a) = V(x)$ is only fulfilled for constant potentials. This means that only in the absence of external forces is the Hamiltonian operator translation invariant.

## 17.3   Periodic Potentials and Bloch Waves

Many physical properties and processes in ordered solids (crystals) are influenced or even caused by electrons. To understand a solid, one encounters the difficulty of solving a complex many-body problem. A significant simplification is achieved by restricting the discussion to the effective dynamics of a single electron. The electron-electron and electron-nucleus interactions are taken into account by an averaged effective lattice potential acting equally on each electron.

If we further simplify and assume the crystal lattice to be infinite, this approximation leads to a single-electron problem in the effective lattice potential

$$V(x) = V(x + a), \tag{17.30}$$

where $a$ is a lattice vector, connecting any two points of the lattice $\Lambda$. We consider the effective dynamics of the electron in a $d$-dimensional lattice spanned by the lattice vectors $a_1, \ldots, a_d$. For a hypercubic lattice $\Lambda$ we have

$$a_i = a\,e_i, \quad a = \sum_i n_i\,a_i, \quad n_i \in \mathbb{Z}, \tag{17.31}$$

where the $e_i$ form an orthonormal basis in $\mathbb{R}^d$ and $a$ denotes the lattice spacing and thus the translation period. Understanding the energy spectrum in the periodic potential leads to insight into why some crystals are insulators and others are conductors.

With a shift by a lattice vector $a$, the wave functions in real space change according to

$$\psi(x) \longrightarrow (\Gamma_a \psi)(x) = \psi(x + a), \quad a \in \Lambda, \tag{17.32}$$

and the effective single-particle Hamiltonian operator

$$H(p, x) = \frac{p^2}{2m} + V(x)$$

commutes with the shift operators when the effective lattice potential is periodic,

$$V(x) = V(x + a), \quad a \in \Lambda. \tag{17.33}$$

In this section, we consider translations that map the crystal lattice onto itself. These form a discrete Abelian group, and the $\Gamma_a$ represent this group in the Hilbert space $L_2(\mathbb{R}^3)$. For a lattice potential (17.33), the unitary $\Gamma_a$ commute with the Hamiltonian operator and can be diagonalized simultaneously with $H$. Their eigenvalues are pure phases.

### 17.3.1 Bloch Waves

The shifts of the lattice with $a \in \Lambda$ form a discrete Abelian group, whose irreducible representations, according to Lemma 11.2, are all one-dimensional. Each irreducible representation thus defines an eigenvector of the representation matrices $\Gamma_a$. What do the common eigenfunctions of the unitary shift operators look like? Due to (17.28), the plane waves are eigenfunctions with eigenvalues $e^{i k \cdot a}$,

$$\Gamma_a \, e^{i k \cdot x} = e^{i k \cdot a} \cdot e^{i k \cdot x} \, . \tag{17.34}$$

These eigenvalues are invariant under

$$k \longrightarrow k + 2\pi \sum_i m_i \, a_i^* \quad \text{with} \quad (a_i^*, a_j) = \delta_{ij} \, ,$$

where the $a_1^*, \ldots, a_d^*$ form a dual basis to the $a_1, \ldots, a_d$. We may therefore assume that the wave vector lies in the elementary cell of the dual lattice. Due to

$$\Gamma_a(\phi\psi) = (\Gamma_a\phi)(\Gamma_a\psi)$$

$\phi\psi$ and $\psi$ have the same eigenvalue if $\phi$ is periodic, $\Gamma_a\phi = \phi$.

**Bloch Waves**
The simultaneous eigenfunctions of the shift operators $\Gamma_a$ are the Bloch waves,

$$\psi_{n,k}(x) = e^{i k \cdot x} \phi_{n,k}(x) \quad \text{with periodic} \quad \phi_{n,k} \, . \tag{17.35}$$

The corresponding eigenvalue of $\Gamma_a$ is $\exp(i k \cdot a)$.

We may therefore assume that the energy eigenfunctions are Bloch waves, where $k$ lies in the elementary cell of the dual lattice. In fact, it suffices to consider the shifts in the direction of the basis vectors $a_i$, as these generate all translations,

$$a = \sum n_i \, a_i \implies \Gamma_a = \Gamma_{a_1}^{n_1} \ldots \Gamma_{a_d}^{n_d} \, .$$

The eigenvalues of the generators $\Gamma_{a_i}$ depend only on the wave vector $k$ in the elementary cell of the dual lattice,

$$\Gamma_{a_i} \psi_{n,k} = e^{i k \cdot a_i} \psi_{n,k} \, ,$$

and two Bloch waves $\psi_{n,k}$ and $\psi_{n',k}$ for the same $k$ differ only by a periodic function.

Note that the periodicity of the lattice potential only transfers to the probability density $|\psi_{n,k}|^2$ of the particle in the periodic potential—the wave function itself is not periodic.

**Example: Kronig-Penney Model**

As an instructive model, we consider the one-dimensional periodic and infinitely extended lattice with an effective single-particle potential

$$V(x) = \sum_{n=-\infty}^{\infty} V\delta(x - na), \quad a^* = \frac{1}{a}. \tag{17.36}$$

We seek solutions in the form of Bloch waves $\psi_{n,k} = e^{ikx}\phi_{n,k}$ with periodic $\phi_{n,k}$. In the one-dimensional system, $k$ lies in the elementary cell $[-\pi a^*, \pi a^*]$ of the dual lattice. In the interval $0 < x < a$, the potential vanishes, and the solutions are plane waves,

$$\psi_{n,k} = e^{ikx}\phi_{n,k}(x) = \alpha\, e^{iqx} + \beta\, e^{-iqx}, \quad \text{with} \quad q = \sqrt{2mE}. \tag{17.37}$$

The conditions that the Bloch wave

$$\phi_{n,k}(x) = \alpha\, e^{i(q-k)x} + \beta\, e^{-i(q+k)x}$$

is periodic and $\psi_{n,k}$ satisfies the Schrödinger equation lead to two homogeneous linear equations for the expansion coefficients $\alpha, \beta$. From this, one obtains by elimination an equation for $\alpha$. This is solvable if

$$\cos(ka) = \cos(qa) + \frac{mVa}{\hbar^2}\sin(qa)qa$$

with $|k| \leq \frac{\pi}{a}$ holds. The solutions of this transcendental equation are best discussed graphically [61]. The allowed energies lie in a band, and between the bands are energy gaps, as shown in Fig. 17.1. This band structure with gaps between the bands is typical for all periodic potentials, also in 3 dimensions.

◀

**Fig. 17.1** Spectrum of the Hamilton operator of the Kronig-Penney model

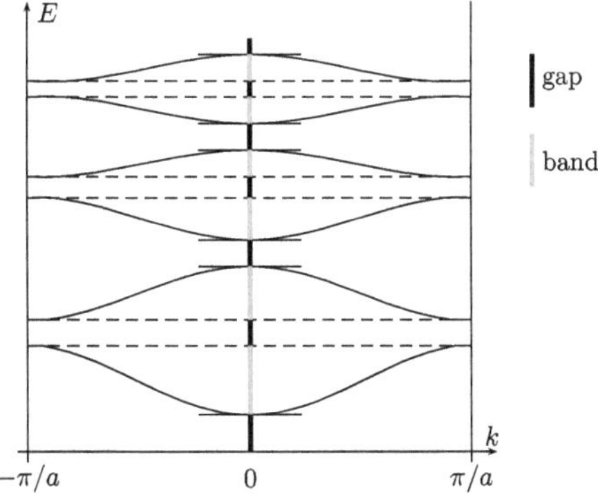

## 17.4   Angular Momentum in Quantum Mechanics

In quantum mechanics, a particle without spin is described by a complex wave function $\psi \in L_2(\mathbb{R}^3)$, which satisfies the Schrödinger equation. It transforms in position space under rotations according to

$$\psi(\boldsymbol{x}) \mapsto (\Gamma_R \psi)(\boldsymbol{x}) = \psi(R^{-1}\boldsymbol{x}) \qquad (17.38)$$

see (10.49) or Problem 10.4. A particle with spin $\frac{1}{2}$, on the other hand, is assigned a two-component wave function $\psi \in L_2(\mathbb{R}^3) \otimes \mathbb{C}^2$, which satisfies the Pauli equation and transforms under rotations according to

$$\psi(\boldsymbol{x}) \mapsto \big(\Gamma_U \psi\big)(\boldsymbol{x}) = U\psi(R^{-1}\boldsymbol{x}) \quad \text{with} \quad R = R(U) \qquad (17.39)$$

Here, $U \mapsto R$ is the double covering of SO(3) by SU(2), see (8.17).

---

**Task**

Show that $\Gamma_U$ defines a unitary representation of SU(2) on $L_2(\mathbb{R}^3) \otimes \mathbb{C}^2$.

The wave function of a many-particle system (17.19) transforms accordingly, e.g. the wave function for two particles with spin $\frac{1}{2}$

$$\big(\Gamma_U \psi\big)(\boldsymbol{x}_1, s_1, \boldsymbol{x}_2, s_2) = \sum_{s_1', s_2'} U_{s_1, s_1'} U_{s_2, s_2'}\, \psi\Big(R^{-1}\boldsymbol{x}_1, s_1', R^{-1}\boldsymbol{x}_2, s_2'\Big). \qquad (17.40)$$

Here, the spin degrees of freedom transform with the tensor product $U \otimes U$. For several particles, possibly with different spins, they transform with a unitary representation $\mathcal{S}(U)$ of SU(2).

The infinitesimal rotations of the wave functions define the angular momentum operators: If $R$ is a rotation around the axis $e$ with a small angle $\theta$, then

$$\Gamma_U \psi = e^{i\theta\, e \cdot J} \psi = \psi + i\theta e \cdot J\psi + \mathcal{O}(\theta^2). \qquad (17.41)$$

In other words, $\theta e \cdot J$ describes the change of the wave function during an infinitesimal rotation with $\theta \ll 1$ about the axis $e$. The angular momenta $J_i$ are Hermitian, since $\Gamma_U$ is unitary. They form a basis of the $\mathfrak{su}(2)$ representation induced by $\Gamma_U$.

### Example: Total Angular Momentum of a Particle with Spin

With the explicit form of the covering map SU(2) $\mapsto$ SO(3) (see Problem 17.5), the transformation (17.39) for a particle with spin $\frac{1}{2}$ has the form

$$
(\Gamma_U \psi)(x) = e^{i\theta e \cdot S} \psi\left(e^{\theta e \cdot \Omega} x\right)
$$

$$
= \left(1 + i\theta e \cdot S + i\theta(e \cdot \Omega x) \cdot p\right)\psi(x) + \mathcal{O}(\theta^2) \qquad (17.42)
$$

with momentum operator $p = \frac{1}{i}\nabla$. The comparison with the general result (17.41) shows that the total momentum is the sum of the orbital angular momentum and the spin,

$$
e \cdot J = e \cdot S + (e \times x) \cdot p \quad \text{or} \quad J = L + S \quad \text{with} \quad L = x \times p . \qquad (17.43)
$$

The representation induced by $U \mapsto \Gamma_U$ maps the generators $S_i$ to the $J_i$. $\blacktriangleleft$

### Task

Verify that for a many-particle system, the total angular momentum in (17.41) is the sum of all orbital angular momenta plus the sum of all spins of the particles, $J = L + S$ with $L = \sum L_i$ and $S = \sum s_i$.

## 17.5  Addition of Angular Momenta

The Hamilton operator $H$ of a many-body system commutes with the total orbital angular momentum $L$ only if its spin-dependent terms in (17.6) are neglected. If not, then $H$ only commutes with the total angular momentum $J = L + S$. We therefore want to address the question of how the eigenstates of the total angular momentum are composed of the eigenstates of the angular momenta of the subsystems. Let $J$ denote the total angular momentum and $J_1$, $J_2$ the commuting angular momenta of the subsystems. The components of all angular momenta satisfy the angular momentum algebra

$$
[J_x, J_y] = iJ_z \quad \text{and cyclic.} \qquad (17.44)
$$

In this section, $J_x$, $J_y$, $J_z$ (and not $J_i$) denote the projections of $J$ onto the axes of the coordinate system. In Sect. 15.1, we saw that $J^2$ and $J_z$ can be simultaneously diagonalized and how the ladder operators (15.3) act on the angular momentum eigenstates $|j, m\rangle$ with fixed $j$. The $(2j+1)$-dimensional invariant subspace spanned by the $|j, m\rangle$ is denoted by $\mathfrak{h}_j$.

## 17.5.1  Reduction of Tensor Products

A first subsystem has the angular momentum $j_1$ and a second one has the angular momentum $j_2$. What can we say about the angular momentum of the entire system in this situation? The state space of the entire system $\mathfrak{h}_{j_1 j_2} = \mathfrak{h}_{j_1} \otimes \mathfrak{h}_{j_2}$ is spanned by the products of the orthonormal eigenstates of the individual angular momenta,

$$|j_1 m_1 j_2 m_2\rangle = |j_1, m_1\rangle \otimes |j_2, m_2\rangle , \tag{17.45}$$

and has the dimension $(2j_1 + 1)(2j_2 + 1)$.

The product states are eigenstates of the mutually commuting operators $J_1^2, J_2^2, J_{1z}, J_{2z}$:

$$J_i^2 |j_1 m_1 j_2 m_2\rangle = j_i(j_i + 1)|j_1 m_1 j_2 m_2\rangle ,$$
$$J_{iz}|j_1 m_1 j_2 m_2\rangle = m_i|j_1 m_1 j_2 m_2\rangle , \quad i = 1, 2 . \tag{17.46}$$

The total angular momentum $\boldsymbol{J} = \boldsymbol{J}_1 + \boldsymbol{J}_2$ also satisfies the angular momentum algebra, and we can search for the common eigenstates of the commuting operators $\boldsymbol{J}^2$ and $J_z$. Since both commute with $J_1^2$ and $J_2^2$ we can simultaneously diagonalize the four operators $\boldsymbol{J}^2, J_z, J_1^2$, and $J_2^2$. We denote the corresponding orthonormal eigenvectors by $|j_1 j_2 jm\rangle$:

$$\boldsymbol{J}^2|j_1 j_2 jm\rangle = j(j + 1)|j_1 j_2 jm\rangle, \quad J_z|j_1 j_2 jm\rangle = m|j_1 j_2 jm\rangle ,$$
$$J_i^2|j_1 j_2 jm\rangle = j_i(j_i + 1)|j_1 j_2 jm\rangle, \quad i = 1, 2 . \tag{17.47}$$

The $|j_1 m_1 j_2 m_2\rangle$ and $|j_1 j_2 jm\rangle$ form two orthonormal basis systems of $\mathfrak{h}_{j_1 j_2}$. We write the $|j_1 j_2 jm\rangle$ as a linear combination of the basis $|j_1 m_1 j_2 m_2\rangle$,

$$|j_1 j_2 jm\rangle = \sum_{m_1, m_2} \langle j_1 m_1 j_2 m_2|j_1 j_2 jm\rangle \, |j_1 m_1 j_2 m_2\rangle . \tag{17.48}$$

The eigenvalues of $J_1^2$ and $J_2^2$ in the eigenstates on the left and right must be equal so that the scalar products do not vanish. The occurring matrix elements

$$\langle j_1 m_1 j_2 m_2|j_1 j_2 jm\rangle \tag{17.49}$$

are the *Clebsch-Gordan coefficients* (CG coefficients). They are, for example, tabulated in the book by Condon and Shortley [71]. It should be noted that the symbols used for the coefficients are not universal. Commonly used symbols are

$$\langle j_1 m_1 j_2 m_2|j_1 j_2 jm\rangle = \langle j_1 m_1 j_2 m_2|jm\rangle = C^{jm}_{j_1 m_1 j_2 m_2} . \tag{17.50}$$

We usually choose the first one. In explicit calculations, where the $j's$ and $m's$ are replaced by numbers, the second symbol is preferable.

Next we derive important properties of the CG coefficients. Since $J_z$ is Hermitian, we have

$$\langle j_1 m_1 j_2 m_2 | J_z | j_1 j_2 jm \rangle = m \langle j_1 m_1 j_2 m_2 | j_1 j_2 jm \rangle = (m_1 + m_2) \langle j_1 m_1 j_2 m_2 | j_1 j_2 jm \rangle,$$

and it follows a first

> **Selection Rule**
> The CG coefficients $\langle j_1 m_1 j_2 m_2 | j_1 j_2 jm \rangle$ can only be non-zero if
>
> $$m = m_1 + m_2. \tag{17.51}$$

It follows that

$$|j_1 j_2 jm\rangle = \sum_{m_1 + m_2 = m} \langle j_1 m_1 j_2 m_2 | j_1 j_2 jm \rangle \, |j_1 m_1 j_2 m_2 \rangle. \tag{17.52}$$

Due to the sum rule (17.51), the degeneracy of $m$ is equal to the number of pairs $(m_1, m_2)$ with $m_1 + m_2 = m$. For $j_1 = 4$ and $j_2 = 3$, the possible pairs are marked with a point in Fig. 17.2. There is exactly one eigenstate with $m = j_1 + j_2$,

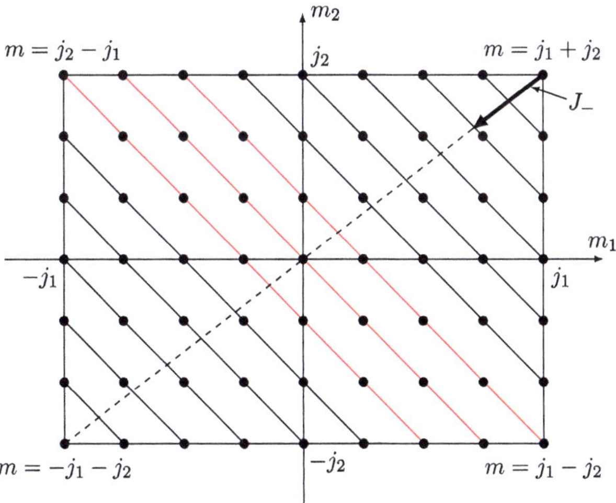


**Fig. 17.2** Possible values of $m_1, m_2$ and the action of $J_-$. The points characterize elements of the product basis (17.45). Elements on the same descending line have the same magnetic quantum number $m = m_1 + m_2$. In the figure, $j_1 = 4$ and $j_2 = 3$

namely the product state $|j_1 j_1\rangle \otimes |j_2 j_2\rangle$ with maximum $J_{iz}$ eigenvalues. This state is annihilated by the two raising operators $J_{i+}$. Due to

$$\boldsymbol{J}^2 = (\boldsymbol{J}_1 + \boldsymbol{J}_2)^2 = \boldsymbol{J}_1^2 + \boldsymbol{J}_2^2 + 2J_{1z}J_{2z} + J_{1+}J_{2-} + J_{1-}J_{2+} \qquad (17.53)$$

it is automatically an eigenvector of $\boldsymbol{J}^2$ with the eigenvalue

$$j_1(j_1 + 1) + j_2(j_2 + 1) + 2j_1 j_2 = (j_1 + j_2)(j_1 + j_2 + 1)\,.$$

Thus, we have found an eigenvector of the total angular momentum with $j = j_1 + j_2$ and magnetic quantum number $m = j$. By repeatedly applying the lowering operator $J_-$ to this vector, we generate the $2j + 1$ eigenvectors

$$|j_1 j_2 jm\rangle \quad \text{with} \quad j = j_1 + j_2, \quad -j \leq m \leq j\,. \qquad (17.54)$$

The total system thus contains a multiplet with angular momentum $j_1 + j_2$. What remains now? The largest remaining $m$ is $j_1 + j_2 - 1$. Therefore, a multiplet with $j = j_1 + j_2 - 1$ exists, as only this has such a maximum $m$. Now, one continues in this way and concludes that $j$ can take the values $j_1 + j_2, j_1 + j_2 - 1, \ldots, |j_1 - j_2|$.

**Triangle Rule**

The possible values of the total angular momentum are

$$|j_1 - j_2| \leq j \leq j_1 + j_2\,. \qquad (17.55)$$

**Question**

Why does it follow from the sum formula

$$\sum_{j=|j_1-j_2|}^{j_1+j_2} (2j + 1) = (2j_1 + 1)(2j_2 + 1)\,,$$

that the eigenvectors (17.54) form a basis of $\mathfrak{h}_{j_1 j_2}$?

Thus, the following reduction formula holds:

**Reduction of the Tensor Product**
The tensor product of two representations $\mathfrak{h}_{j_1}$ and $\mathfrak{h}_{j_2}$ of the individual angular momenta contains the following representations of the total angular momentum:

$$\mathfrak{h}_{j_1} \otimes \mathfrak{h}_{j_2} = \mathfrak{h}_{j_1+j_2} \oplus \mathfrak{h}_{j_1+j_2-1} \oplus \cdots \oplus \mathfrak{h}_{|j_1-j_2|} . \tag{17.56}$$

This is exactly the reduction formula (12.23) derived with the help of characters.

In atomic physics, multiplets are often denoted by their dimension, for example, the subspace $\mathfrak{h}_1$ spanned by the triplet is denoted by **3**. With this convention, for example, the nine-dimensional space $\mathbf{3} \otimes \mathbf{3}$ decomposes into states with total angular momentum 0, 1, and 2, i.e., a singlet, triplet, and quintet, $\mathbf{3} \otimes \mathbf{3} = \mathbf{1} \oplus \mathbf{3} \oplus \mathbf{5}$.

## 17.5.2 Clebsch-Gordan (CG) Coefficients

We now turn to the further properties of the expansion coefficients (17.50) and their calculation. Instead of expanding the $|j_1 j_2 j m\rangle$ in terms of the $|j_1 m_1 j_2 m_2\rangle$ as above, we can also expand the $|j_1 m_1 j_2 m_2\rangle$ in terms of the $|j_1 j_2 j m\rangle$,

$$|j_1 m_1 j_2 m_2\rangle = \sum_{j,m=m_1+m_2} \langle j_1 j_2 j m | j_1 m_1 j_2 m_2\rangle \, |j_1 j_2 j m\rangle . \tag{17.57}$$

The matrix elements

$$\langle j_1 j_2 j m | j_1 m_1 j_2 m_2\rangle \tag{17.58}$$

are called inverse CG coefficients and satisfy the orthogonality relations,

$$\sum_{j,m} \langle j_1 m_1' j_2 m_2' | j_1 j_2 j m\rangle \langle j_1 j_2 j m | j_1 m_1 j_2 m_2\rangle = \delta_{m_1 m_1'} \delta_{m_2 m_2'} . \tag{17.59}$$

Here, the sum is over all $j$ allowed by the triangle inequality (17.55) and the corresponding $2j + 1$ values of $m$. The CG coefficients satisfy the orthogonality relations

$$\sum_{m_1+m_2=m} \langle j_1 j_2 j m | j_1 m_1 j_2 m_2\rangle \langle j_1 m_1 j_2 m_2 | j_1 j_2 j' m'\rangle = \delta_{jj'} \delta_{mm'} . \tag{17.60}$$

The recursion relations derived below show that all coefficients can be chosen to be real. Then

$$\langle j_1 j_2 j m | j_1 m_1 j_2 m_2 \rangle = \langle j_1 m_1 j_2 m_2 | j_1 j_2 j m \rangle ,$$

and (17.60) becomes for $j = j'$ and $m = m'$

$$\sum_{m_1 + m_2 = m} \langle j_1 j_2 j m | j_1 m_1 j_2 m_2 \rangle^2 = 1 . \tag{17.61}$$

The coefficients can be calculated recursively. To do this, one acts with the raising and lowering operators $J_\pm = J_{1\pm} + J_{2\pm}$ in the following matrix element on the ket and bra vector

$$\langle j_1 m_1 j_2 m_2 | J_\pm | j_1 j_2 j m \rangle ,$$

and uses $J_\pm^\dagger = J_\mp$. With (15.11) this leads to the following

---

**Recursion Formulas for the CG Coefficients**

$$c_{jm}^\pm \langle j_1 m_1 j_2 m_2 | j_1 j_2 j, m \pm 1 \rangle$$

$$= c_{j_1 m_1}^\mp \langle j_1, m_1 \mp 1, j_2 m_2 | j_1 j_2 j m \rangle + c_{j_2 m_2}^\mp \langle j_1 m_1 j_2, m_2 \mp 1 | j_1 j_2 j m \rangle .$$

$$\tag{17.62}$$

---

Together with (17.61), these relations can be used to calculate the coefficients. The corresponding recursion relations for the inverse CG coefficients are

---

**Recursion Formulas for the Inverse CG Coefficients**

$$c_{jm}^\pm \langle j_1 j_2 j, m \pm 1 | j_1 m_1 j_2 m_2 \rangle$$

$$= c_{j_1 m_1}^\mp \langle j_1 j_2 j m | j_1, m_1 \mp 1, j_2 m_2 \rangle + c_{j_2 m_2}^\mp \langle j_1 j_2 j m | j_1 m_1 j_2, m_2 \mp 1 \rangle .$$

$$\tag{17.63}$$

---

**Task**

Prove the recursion relations (17.62) and (17.63).

### Example: Spin-Orbit Coupling

We add the spin and orbital angular momentum of an electron to the total angular momentum $J$ with eigenstates $|jm\rangle$. Let $j_1 = \ell$ be the orbital angular momentum and $j_2 = \frac{1}{2}$ the spin. Because

$$\mathfrak{H}_0 \otimes \mathfrak{H}_{\frac{1}{2}} = \mathfrak{H}_{\frac{1}{2}} \quad \text{and} \quad \mathfrak{H}_\ell \otimes \mathfrak{H}_{\frac{1}{2}} = \mathfrak{H}_{\ell-\frac{1}{2}} \oplus \mathfrak{H}_{\ell+\frac{1}{2}} \quad (\ell \geq 1) \tag{17.64}$$

the total angular momentum of an $s$-electron with orbital angular momentum $\ell = 0$ is equal to $\frac{1}{2}$, of a $p$-electron with orbital angular momentum $\ell = 1$ is equal to $\frac{1}{2}$ or $\frac{3}{2}$, etc. For an $s$-electron, we have

$$\langle 00\tfrac{1}{2}m_s | \tfrac{1}{2}m \rangle = \delta_{m,m_s} . \tag{17.65}$$

For $\ell \geq 1$, $j$ can take the values $\ell \pm \frac{1}{2}$. Only the product $|\ell\ell\rangle \otimes |\frac{1}{2}\frac{1}{2}\rangle$ has the maximum magnetic quantum number $m = \ell + \frac{1}{2}$, and therefore

$$\langle \ell\ell\tfrac{1}{2}\tfrac{1}{2} | \ell + \tfrac{1}{2}, \ell + \tfrac{1}{2} \rangle = 1 . \tag{17.66}$$

We use (17.62) with the lower sign and $m_2 = \frac{1}{2}$,

$$c^-_{jm} \langle \ell m_\ell \tfrac{1}{2}\tfrac{1}{2} | j, m-1 \rangle = c^+_{\ell m_\ell} \langle \ell, m_\ell+1, \tfrac{1}{2}\tfrac{1}{2} | jm \rangle , \tag{17.67}$$

where the rules (17.51) and (17.55) must be observed. We conclude with $m \to m + 1$ and $j = \ell + \frac{1}{2}$,

$$\sqrt{\ell + m + \tfrac{3}{2}} \, \langle \ell, m-\tfrac{1}{2}, \tfrac{1}{2}\tfrac{1}{2} | \ell+\tfrac{1}{2}, m \rangle = \sqrt{\ell + m + \tfrac{1}{2}} \, \langle \ell, m+\tfrac{1}{2}, \tfrac{1}{2}\tfrac{1}{2} | \ell+\tfrac{1}{2}, m+1 \rangle .$$

After iterating this relation using (17.66), one finds

$$\langle \ell, m - \tfrac{1}{2}, \tfrac{1}{2}\tfrac{1}{2} | \ell + \tfrac{1}{2}, m \rangle = \sqrt{\frac{\ell + m + \frac{1}{2}}{2\ell + 1}} . \tag{17.68}$$

◀

The remaining CG coefficients of the spin-orbit coupling are now calculated analogously, and Table 17.1 contains the coefficients $\langle \ell m_\ell \tfrac{1}{2} m_s | jm \rangle$ with $m_\ell =$

**Table 17.1** CG coefficients for the coupling of orbital angular momentum with spin

| $j \backslash m_s$ | $\frac{1}{2}$ | $-\frac{1}{2}$ |
|---|---|---|
| $\ell + \frac{1}{2}$ | $\sqrt{\dfrac{\ell+m+1/2}{2\ell+1}}$ | $\sqrt{\dfrac{\ell-m+1/2}{2\ell+1}}$ |
| $\ell - \frac{1}{2}$ | $-\sqrt{\dfrac{\ell-m+1/2}{2\ell+1}}$ | $\sqrt{\dfrac{\ell+m+1/2}{2\ell+1}}$ |

**Table 17.2**  CG coefficients for the coupling of angular momentum $\ell$ and angular momentum 1

| $j \backslash m_2$ | 1 | 0 | $-1$ |
|---|---|---|---|
| $\ell + 1$ | $\sqrt{\dfrac{(\ell+m)(\ell+m+1)}{(2\ell+1)(2\ell+2)}}$ | $\sqrt{\dfrac{(\ell-m+1)(\ell+m+1)}{(2\ell+1)(\ell+1)}}$ | $\sqrt{\dfrac{(\ell-m)(\ell-m+1)}{(2\ell+1)(2\ell+2)}}$ |
| $\ell$ | $-\sqrt{\dfrac{(\ell+m)(\ell-m+1)}{2\ell(\ell+1)}}$ | $\dfrac{m}{\sqrt{\ell(\ell+1)}}$ | $\sqrt{\dfrac{(\ell-m)(\ell+m+1)}{2\ell(\ell+1)}}$ |
| $\ell - 1$ | $\sqrt{\dfrac{(\ell-m)(\ell-m+1)}{2\ell(2\ell+1)}}$ | $-\sqrt{\dfrac{(\ell-m)(\ell+m)}{\ell(2\ell+1)}}$ | $\sqrt{\dfrac{(\ell+m)(\ell+m+1)}{2\ell(2\ell+1)}}$ |

$m - m_s$. The corresponding coefficients $(\ell m_\ell 1 m_2 | j m)$ for the coupling of two systems with angular momenta $\ell$ and 1 can be found in Table 17.2.

### 17.5.3  Wigner-Eckart Theorem

Matrix elements (ME) of tensor operators between eigenstates of angular momentum play an important role in applications of quantum mechanics. For example, the behavior of atoms and molecules during emission, absorption, and scattering of electromagnetic radiation is determined by the ME of the dipole vector operator. Other examples of vector operators are momentum, angular momentum, or spin. Examples of higher tensor operators are the atomic multipole moments. Many MEs of tensor operators vanish due to selection rules, and others are related to each other because of the isotropy of space. We begin with the discussion of scalar operators and then consider MEs of general tensor operators.

#### Scalar Operators

Scalar operators are rotation invariant and commute with angular momentum. Important representatives are the Hamiltonian operator, kinetic energy, or $\boldsymbol{L} \cdot \boldsymbol{S}$. Let $S$ be a scalar,

$$\Gamma_U S \Gamma_U^{-1} = S \iff [\boldsymbol{J}, S] = 0 . \qquad (17.69)$$

By letting the angular momenta act on the ket and the bra in the equations

$$\langle jm|[\boldsymbol{J}^2, S]|j'm'\rangle = 0 \quad \text{and} \quad \langle jm|[J_z, S]|j'm'\rangle = 0$$

be obtain the

**Selection Rule for Scalars**

$$\langle jm|S|j'm'\rangle = \delta_{jj'}\delta_{mm'}\langle jm|S|jm\rangle . \qquad (17.70)$$

In fact, the ME $\langle jm|S|jm\rangle$ do not depend on $m$ at all. To prove this, we calculate the ME of $J_+SJ_-$ in two ways. First, we act with $J_-$ on the ket and with $J_+$ on the bra,

$$\langle jm|J_+SJ_-|jm\rangle \overset{(15.11)}{=} (c^-_{jm})^2\langle j,m-1|S|j,m-1\rangle ,$$

and then we use the fact that $J_+$ commutes with $S$. With $c^+_{j,m-1}=c^-_{jm}$, we conclude that

$$\langle jm|J_+SJ_-|jm\rangle = \langle jm|SJ_+J_-|jm\rangle \overset{(15.11)}{=} (c^-_{jm})^2\langle jm|S|jm\rangle .$$

A comparison of the right-hand sides shows that the matrix elements $\langle jm|S|jm\rangle$ are independent of $m$,

$$\langle jm|S|j'm'\rangle = \delta_{jj'}\delta_{mm'}\langle j\|S\|j\rangle . \tag{17.71}$$

The reduced ME $\langle j\|S\|j\rangle$ is determined by calculating the left side, e.g., for $m=j$.

## Tensor Operators

We begin with the behavior of eigenstates of angular momentum $|jm\rangle$ under (quantum mechanical) rotations $U\in SU(2)$,

$$\Gamma_U|jm\rangle = \sum_{m'=-j}^{j} D^j_{m'm}(U)|jm'\rangle, \qquad D^j_{m'm}(U) = \langle jm'|\Gamma_U|jm\rangle . \tag{17.72}$$

The $D^j$ form a $(2j+1)$-dimensional unitary representation of $SU(2)$ on the subspace $\mathfrak{h}_j$.

---

**Task**

Show that from the representation property of $U\mapsto \Gamma_U$, it follows that $U\mapsto D^j(U)$ is also a representation of $SU(2)$.

The product states (17.45) transform according to

$$\Gamma_U|j_1m_1j_2m_2\rangle = \sum_{m_1',m_2'} D^{j_1}_{m_1'm_1}(U)D^{j_2}_{m_2'm_2}(U)|j_1m_1'j_2m_2'\rangle . \tag{17.73}$$

**Definition 17.1 (Spherical Tensor)** An irreducible spherical tensor operator of rank $J$ is an object $T^J$, whose $2J + 1$ components $T_M^J$ transform under rotations as follows,

$$\Gamma_U T_M^J \Gamma_U^{-1} = \sum_{M'=1}^{2J+1} D_{M'M}^J(U) T_{M'}^J, \quad U \in \mathrm{SU}(2).$$ (17.74)

The $T_M^J$ are called normal components or spherical components of $T^J$.

The states $T_M^J |jm\rangle$ transform under rotations just like the product states $|JMjm\rangle$, as can be easily shown,

$$\Gamma_U T_M^J |jm\rangle = \Gamma_U T_M^J \underbrace{\Gamma_U^{-1}\Gamma_U}_{\mathbb{1}} |jm\rangle = \sum_{M',m'} D_{M'M}^J(U) \, D_{m'm}^j(U) \, T_{M'}^J |jm'\rangle.$$

(17.75)

Thus they decompose into irreducible multiplets with angular momenta between $J + j$ and $|J - j|$.

To obtain the infinitesimal version of (17.74), we set $\Gamma_U = \exp(i\alpha J)$ on the left side of this equation and note that the normal components $T_M^J$ transform just like the eigenstates $|JM\rangle$. This results in the following commutators for the angular momenta and normal components of a tensor operator,

$$[J_\pm, T_M^J] = c_{JM}^\pm T_{M\pm1}^J \quad \text{and} \quad [J_3, T_M^J] = M T_M^J,$$ (17.76)

where the $c_{JM}^\pm$ were introduced in (15.8). Using the last commutator, we find

$$J_3 T_M^J |jm\rangle = \left( T_M^J J_3 + M T_M^J \right) |jm\rangle = (m + M) \, T_M^J |jm\rangle.$$ (17.77)

This means that the component $T_M^J$ of $T^J$ increases the $J_z$-eigenvalue by $M$, leading to the

**First Selection Rule for Matrix Elements of Tensor Operators**

$$m \neq M + m' \implies \langle jm|T_M^J|j'm'\rangle = 0.$$ (17.78)

Note that the states $T_M^J |jm\rangle$ are generally not eigenvectors of $\boldsymbol{J}^2$.

Now we consider matrix elements of the first commutation relation in (17.76)

$$\langle jm|J_\mp T_M^J - T_M^J J_\mp - c_{JM}^\mp T_{M\mp1}^J|j'm'\rangle = 0,$$

and find after applying the ladder operators to the state vectors the

**Recursion Relations for Matrix Elements of Tensor Operators**

$$c_{jm}^{\pm}\langle j, m \pm 1|T_M^J|j'm'\rangle = c_{JM}^{\mp}\langle jm|T_{M\mp1}^J|j'm'\rangle$$

$$+ c_{j'm'}^{\mp}\langle jm|T_M^J|j', m' \mp 1\rangle. \qquad (17.79)$$

These recursion relations are identical to those for the inverse CG coefficients in (17.63), and this proves the

**Theorem 17.2 (Wigner-Eckart Theorem)** *If $T_M^J$ are the normal components of a tensor operator of rank $J$ and $|jm\rangle$ the eigenvectors of the angular momentum operator, then*

$$\langle jm|T_M^J|j'm'\rangle = \langle jm|JMj'm'\rangle\,\langle j||T^J||j'\rangle \qquad (17.80)$$

*with a reduced matrix element $\langle j||T^J||j'\rangle$ that is independent of $m, m'$ and $M$.*

The Wigner-Eckart theorem implies a

**Second Selection Rule for Matrix Elements of Tensor Operators**
$$\langle jm|T_M^J|j'm'\rangle = 0 \quad \text{for} \quad j \notin \{j'+J, j'+J-1,\ldots,|j'-J|\}. \qquad (17.81)$$

In the study of atomic transitions, one often encounters matrix elements of the dipole operator, see Problem 17.6. The dipole operator, like the angular momentum, is a (pseudo)vector, for which further properties apply.

### Vector Operators

A vector operator is a tensor operator of rank $J = 1$. Since the spherical harmonics transform like the eigenvectors $|1m\rangle$, the relationship between Cartesian components $V_x, V_y, V_z$ and normal components $V_M$ is the same as that between Cartesian coordinates $(x, y, z)$ and spherical harmonics $Y_{1m}$,

$$Y_{11} = -\frac{c}{\sqrt{2}}(x + iy), \quad Y_{10} = c \cdot z, \quad Y_{1-1} = \frac{c}{\sqrt{2}}(x - iy). \qquad (17.82)$$

Thus, the normal components of a vector operator are

$$V_1 = -\frac{1}{\sqrt{2}}\left(V_x + iV_y\right), \qquad V_0 = V_z, \qquad V_{-1} = \frac{1}{\sqrt{2}}\left(V_x - iV_y\right). \qquad (17.83)$$

The commutation relations $[J_i, V_j] = i\epsilon_{ijk}V_k$ for the Cartesian components are equivalent to the commutation relations (17.76) for the normal components $V_M \equiv T_M^1$. For a vector operator, the Wigner-Eckart Theorem states

$$\langle jm|V_M|j'm'\rangle = \langle jm|1Mj'm'\rangle\,\langle j\|V\|j'\rangle. \qquad (17.84)$$

**Example: Matrix Elements of Angular Mmomentum**

The normal components of angular momentum are

$$J_1 = -\frac{1}{\sqrt{2}}J_+, \quad J_0 = J_z, \quad J_{-1} = \frac{1}{\sqrt{2}}J_-. \qquad (17.85)$$

Obviously, for $j \neq j'$, the left side of

$$\langle jm|J_M|j'm'\rangle = \langle jm|1Mj'm'\rangle\,\langle j\|J\|j'\rangle, \qquad (17.86)$$

vanishes, and we conclude that the reduced ME vanishes for $j \neq j'$. For $j' = j$, we can choose $M = 0$ and $m = m' = j$ and obtain

$$\langle jj|J_0|jj\rangle = \langle jj|10jj\rangle\,\langle j\|J\|j\rangle.$$

The ME on the left side is $j$ and the CG coefficient on the right is equal to the element in the second column and second row of Table 17.2. This results in the reduced ME

$$\langle j\|J\|j\rangle = \sqrt{j(j+1)},$$

and (17.86) provides the following expression for the CG coefficients,

$$\langle jm|1Mjm'\rangle = \frac{\langle jm|J_M|jm'\rangle}{\sqrt{j(j+1)}}. \qquad (17.87)$$

◄

We utilize the result (17.87) to relate the ME of any vector operator to the ME of angular momentum. Choosing in (17.84) $j = j'$ and using this result, we find

$$\langle jm|V_M|jm'\rangle = \langle jm|J_M|jm'\rangle\,\frac{\langle j\|V\|j\rangle}{\sqrt{j(j+1)}}. \qquad (17.88)$$

This relationship also holds for the Cartesian components $V_i$ and $J_i$,

$$\langle jm|V_i|jm'\rangle = \langle jm|J_i|jm'\rangle \frac{\langle j||V||j\rangle}{\sqrt{j(j+1)}}. \tag{17.89}$$

To find a simple formula for the reduced matrix element, we multiply by $\langle jm'|J_i|jm''\rangle$ and sum over $m'$ and $i$. This leads to

$$\langle jm|V\cdot J|jm''\rangle = \langle jm|J^2|jm''\rangle \frac{\langle j||V||j\rangle}{\sqrt{j(j+1)}} = \delta_{mm''}\sqrt{j(j+1)}\,\langle j||V||j\rangle. \tag{17.90}$$

Since $V\cdot J$ is a scalar, we can replace the left side here according to (17.71) with the reduced ME. Therefore, $\langle j||V||j\rangle$ is proportional to $\langle j||V\cdot J||j\rangle$, and substituting into (17.89) we obtain

$$\langle jm|V|jm'\rangle = \langle jm|J|jm'\rangle \frac{\langle j||V\cdot J||j\rangle}{j(j+1)}. \tag{17.91}$$

Apart from a $j$-dependent constant, $V$ and $J$ have the same ME on the subspace $\mathfrak{h}_j$. This does not mean that the operators are proportional. Unlike $J$, a general vector operator has nonzero matrix elements $\langle jm|V|j\pm 1, m'\rangle$.

---

## 17.6  Algebraic Solution of the Hydrogen Atom

For a closed system of two non-relativistic point masses interacting via a central force, the angular momentum $L = x \times p$ of the relative motion is conserved, and the orbit lies in the plane perpendicular to $L$. If the potential is a Coulomb or Newton potential,

$$V(r) = -\frac{\gamma}{r}, \quad r = |x| = |x_1 - x_1|, \tag{17.92}$$

then, in addition to the three components of the angular momentum $L$, there are three more conserved quantities. These are the components of the Laplace-Runge-Lenz vector[1]

$$Q = \frac{1}{\mu\gamma}\, p \times L - \frac{x}{r}, \tag{17.93}$$

where $\mu$ denotes the reduced mass of the particles. For the Kepler problem, $\gamma = Gm_1m_2$ and for the Coulomb problem $\gamma = q_1q_2$. The vector $Q$ lies in the orbital

---

[1] A more suitable name for the vector would be Hermann-Bernoulli-Laplace vector, see [72,73].

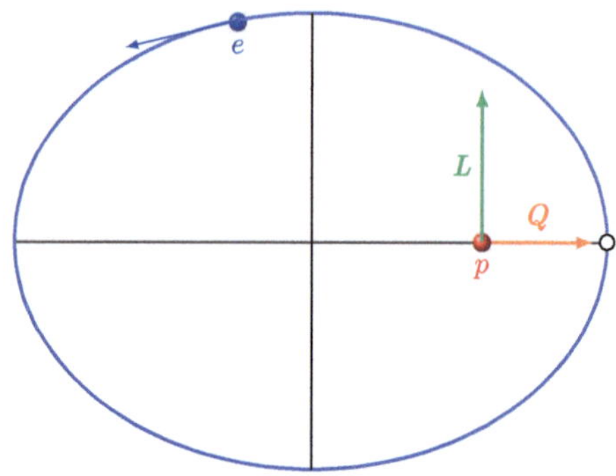

**Fig. 17.3** The angular momentum $L$ is perpendicular to the orbital plane, and the Laplace-Runge-Lenz vector $Q$ lies in the orbital plane and points towards the perihelion

plane, is thus orthogonal to the orbital angular momentum, and points from the focus to the perihelion of the elliptical orbit. The existence of the conserved quantity $Q$ explains why the Kepler ellipses in Newton's theory of gravitation do not precess. In quantum mechanics, we choose the order of the operators such that $Q$ is Hermitian,

$$Q = \frac{1}{2\mu\gamma}\,(p \times L - L \times p) - \frac{x}{r}\,, \tag{17.94}$$

where $x$, $p$, and $L$ are the operators corresponding to the classical quantities (see Fig. 17.3). These are represented in a Hilbert space, e.g., the position space $L_2(\mathbb{R}^3)$.

---

**Question**

Why are the components of the Laplace-Runge-Lenz vector $Q$ Hermitian?

The components of the vector operator $Q$ commute with the Hamiltonian operator

$$H = \frac{p^2}{2\mu} - \frac{\gamma}{r}\,. \tag{17.95}$$

The proof of this interesting property can be found, for example, in [61].

**Additional Conservation Laws in the Coulomb Field**
For the Coulomb potential, in addition to $L$, $Q$ also commutes with the Hamiltonian operator.

### 17.6.1  Lie Algebra of the Conserved Quantities

The conserved quantities of a physical system generate symmetries and therefore form a Lie algebra. In the Coulomb problem, besides the energy and the components of the angular momentum and Laplace-Runge-Lenz vector, there are no other conserved quantities. Since the operator $H$ commutes with all conserved quantities by definition, we can treat it as a constant and thus conclude that the six conserved operators $L_i$ and $Q_i$ form a Lie algebra.

To determine this symmetry algebra, in addition to the known commutation rules for the angular momenta, the commutators $[L_i, Q_j]$ and $[Q_i, Q_j]$ are needed. The components of the *vector operator* $Q$ have the following commutation relations with the angular momentum operators:

$$[L_i, Q_j] = i\varepsilon_{ijk}Q_k \, . \tag{17.96}$$

The calculation of the commutation relation of the $Q_i$ is more complex. One finds

$$[Q_i, Q_j] = \frac{2}{i\mu\gamma^2} H\varepsilon_{ijk}L_k \, , \tag{17.97}$$

see Problem 17.7 at the end of the chapter and [61]. To examine the action of the conserved charges on the bound states with negative energy we introduce

$$K_i = \sqrt{-\frac{\mu\gamma^2}{2H}}\, Q_i \, . \tag{17.98}$$

These rescaled operators have the dimension of an angular momentum and also define conserved quantities. They satisfy the simpler commutation relations:

$$[K_i, K_j] = i\,\varepsilon_{ijk}L_k \, . \tag{17.99}$$

We summarize the results:

**Lie Algebra of Conserved Quantities**
The conserved quantities $L$ and $K$ satisfy the $\mathfrak{so}(4)$ commutation relations

$$[L_i, L_j] = i\varepsilon_{ijk}L_k, \quad [L_i, K_j] = i\varepsilon_{ijk}K_k, \quad [K_i, K_j] = i\varepsilon_{ijk}L_k \, . \tag{17.100}$$

Additionally, the vector operators are orthogonal to each other, $K \cdot L = L \cdot K = 0$.

In Problem 17.7, the square of $Q$ is also calculated,

$$Q^2 = \frac{2H}{\mu\gamma^2}\left(L^2 + 1\right) + 1 \ . \tag{17.101}$$

If we replace $Q$ with $K$ in this relation, we can solve for $H$ and express $H$ in terms of the conserved quantities.

$$H = -\frac{\mu\gamma^2}{2}\frac{1}{L^2 + K^2 + 1} \ . \tag{17.102}$$

Finally, we define the linear combinations

$$U = \frac{1}{2}\left(L + K\right) \quad \text{and} \quad V = \frac{1}{2}\left(L - K\right) \ . \tag{17.103}$$

These generate two commuting $\mathfrak{su}(2)$ Lie subalgebras:

$$[U_i, U_j] = i\varepsilon_{ijk}U_k, \quad [V_i, V_j] = i\varepsilon_{ijk}V_k, \quad [U_i, V_j] = 0 \ . \tag{17.104}$$

We see explicitly here that the six conserved quantities generate the Lie algebra $\mathfrak{su}(2) \oplus \mathfrak{su}(2) \cong \mathfrak{so}(4)$. It follows immediately, that the operators $U^2$, $U_3$, $V^2$ and $V_3$ can be diagonalized simultaneously.

With the help of the vector operators $U$ and $V$, the Hamiltonian operator is written as

$$H = -\frac{\mu\gamma^2}{2}\frac{1}{2U^2 + 2V^2 + 1} \ , \tag{17.105}$$

and the orthogonality relation $L \cdot K = 0$ reads

$$U^2 = V^2 \ . \tag{17.106}$$

For three Hermitian operators $J_i$, which satisfy the $\mathfrak{su}(2)$ commutation rules $[J_i, J_j] = i\varepsilon_{ijk}J_k$, we determined the eigenvalues and eigenvectors of the commuting operators $(J^2, J_3)$ in Sect. 15.1, These results apply to the pairs $(U^2, U_3)$ and $(V^2, V_3)$. But due to the constraint (17.106), the operators $U^2$ and $V^2$ have the same eigenvalues $j(j + 1)$. Thus, the eigenvalues of $H$ are

$$E_j = -\frac{1}{2}\frac{\mu\gamma^2}{(2j + 1)^2}, \qquad j \in \tfrac{1}{2}\mathbb{N}_0 \ . \tag{17.107}$$

We further define the principal quantum number $n = 2j + 1$ and set for hydrogen-like ions $\gamma = q_1 q_2 = Z\,e^2$, then we obtain the well-known energies

$$E_n = -\frac{\mu c^2}{2}\frac{(Z\alpha)^2}{n^2}, \quad \alpha = \frac{e^2}{\hbar c} \tag{17.108}$$

for an electron in the field of a nucleus with atomic number $Z$. In the last expression, we have written out $\hbar$ again.

---

**Question**

Why is the orbital angular momentum $\ell$ integer, even though the quantum number $j$ of $U$ and $V$ can also be half-integer?

---

The large degeneracy of the energies is also explained with the help of the enhanced symmetry. For a fixed angular momentum $j$, the eigenvalues of the independent operators $U_3$ and $V_3$ each take $2j+1$ values. Therefore, the degeneracy of the energy $E_n$ is equal to $(2j+1)^2 = n^2$.

A generalization of the Runge-Lenz vector also exists in $d \neq 3$ spatial dimensions for Schrödinger operators with $1/r$ potential and their supersymmetric extensions, see [74].

---

## 17.7  Exercises for Chap. 17

**Problem 17.1 (Infinitesimal Translations)** In position space, the unitary shift operators have the form (17.23). Show for a particle that these are generated by the momentum, i.e., justify the representation

$$(\Gamma_{\boldsymbol{a}}\psi)(\boldsymbol{x}) = (\mathrm{e}^{\mathrm{i}\boldsymbol{a}\cdot\boldsymbol{P}}\psi)(\boldsymbol{x}).$$

in (17.28).

Note: For smooth wave functions, you can expand the exponential function

$$(\Gamma_{\boldsymbol{a}}\psi)(\boldsymbol{x}) = \psi(\boldsymbol{x}) + a^i \partial_i \psi(\boldsymbol{x}) + \dots,$$

and rewrite the right side as an exponentiated derivative operator. Repeat the argument for several particles.

**Problem 17.2 (Three Particles with Spin $\frac{1}{2}$)** For a system consisting of three particles with spin $\frac{1}{2}$, there are eight spin states. Classify these according to their total spin. Determine all eigenvectors (spin functions) $|ss_3\rangle$, described by the values $s$ and $s_3$ of the total spin $\boldsymbol{S} = \boldsymbol{s}^{(1)} + \boldsymbol{s}^{(2)} + \boldsymbol{s}^{(3)}$.

**Problem 17.3 (Clebsch-Gordan Coefficients)** Consider a system of two angular momenta $j_1 = 1$ and $j_2 = 1$. The eigenvalues of the z-components of the individual angular momenta are $m_1$ and $m_2$, respectively. The states of the entire system are labeled with the quantum numbers of the total angular momentum $j$ and $m$. Calculate the CG coefficients for the quantum numbers in the following tables (for the cases $m = 0$ and $m = 1$) and additionally for the case $m = 2$.

| $m = 0$ | $j = 2$ | $j = 1$ | $j = 0$ |
|---|---|---|---|
| $m_1 = 1, m_2 = -1$ | | | |
| $m_1 = 0, m_2 = 0$ | | | |
| $m_1 = -1, m_2 = 1$ | | | |

| $m = 1$ | $j = 2$ | $j = 1$ |
|---|---|---|
| $m_1 = 1, m_2 = 0$ | | |
| $m_1 = 0, m_2 = 1$ | | |

**Problem 17.4 (Quadrupole Tensor)**  The Cartesian components of the quadrupole tensor are

$$Q_{ik} = Q_{ki} = 3x_i x_k - r^2 \delta_{ik} \, .$$

The tensor is symmetric and traceless and therefore has 5 linearly independent components. Except for a factor of $r^2$, they can be written as a linear combination of the spherical harmonics with $\ell = 2$.

1. Write the components $Q_{ik}$ as linear combinations of the $Y_{2m}$.
   Hint: $Y_{\ell m} \propto e^{im\varphi}$.
2. What are the commutators $[L_k, Y_{\ell m}]$?
   Hint: Apply the commutator to a wave function in position space.
3. Determine the commutators of $L_3$ with the components $Q_{ik}$.
4. $-i[L_3, Q]$ can be written as the commutator of an antisymmetric matrix $\Omega_3$ with $Q$. Determine $\Omega_3$ and interpret the result.
5. You should, for example, obtain $[L_3, Q_{13}+iQ_{23}] = -Q_{12}+i(Q_{11}-Q_{33})$. Using the relationships between $Q_{ik}$ and $Y_{2m}$, calculate the following commutation relations:

$$[L_+, Q_{12}] \quad \text{and} \quad [L_-, Q_{12}] \, .$$

**Problem 17.5 (Double Covering of SO(3))**  Prove that the double covering (8.17) of SO(3) by SU(2) defines the following explicit homomorphism:

$$U = \exp(i\theta e \cdot S) \longmapsto R = \exp(-\theta e \cdot \Omega) \, .$$

The $S_i = \frac{1}{2}\sigma_i$ are the spin matrices and the matrices $\Omega_i$ with elements $(\Omega_i)_{jk} = -\varepsilon_{ijk}$ are the infinitesimal rotations about the axes $e_i$. The $\Omega_i$ are the skew-symmetric matrices in (14.8) and satisfy the $\mathfrak{so}(3)$ commutation relations $[\Omega_i, \Omega_j] = \epsilon_{ijk}\Omega_k$.

**Problem 17.6 (Dipole Operator)**  The rate at which an atom in state $|m\rangle$ absorbs (or emits) radiation of frequency $\omega = \Delta E$ and transitions to state $|n\rangle$ is, in the dipole approximation, proportional to

$$\Gamma^{\text{Abs.}}_{n \to m} \propto \left| \langle m | \boldsymbol{e} \cdot \boldsymbol{d} | n \rangle \right|^2 ,$$

where the atomic states are eigenvectors of $\boldsymbol{L}^2$ and $L_z$. The unit vector $\boldsymbol{e}$ points in the direction of polarization and $\boldsymbol{d} \propto \boldsymbol{x}$ is the dipole operator. Prove the selection rules

$$\Delta \ell = \pm 1 \quad \text{and} \quad \Delta m = 0, \pm 1 .$$

**Problem 17.7 (Laplace-Runge-Lenz Vector)**  In the algebraic solution of the hydrogen atom, the Hermitian Runge-Lenz vector operator $\boldsymbol{Q}$ in (17.94) played an important role. Determine

1. the square $\boldsymbol{Q}^2$ of the vector,
2. the commutation relations of its components.

Assistance can be found, for example, in Chapter 28 of [61].

# Symmetries in Relativistic Quantum Mechanics

18



The unification of the principles of quantum mechanics and special relativity explains unusual properties of matter and is the starting point for the relativistic field theories discussed in Chaps. 19 and 20. Already L. de Broglie and E. Schrödinger tried to formulate a covariant wave equation for the electron, and the equation they established for scalar particles is today called the Klein-Gordon equation. Schrödinger soon noticed that his relativistic equation could not explain the fine structure of the hydrogen spectrum. In addition, the Klein-Gordon equation had problems with the positivity of the probability amplitude, and this problem was only later solved by W. Pauli and V. Weisskopf.

The Dirac theory of the electron is the first *effective* attempt to make wave mechanics relativistically invariant. Inspired by his transformation theory, P. Dirac sought a first-order wave equation. He realized that such an equation can be found for a 4-component electron field. His Dirac equation then automatically implies that an electron has spin $\frac{1}{2}$ and a magnetic moment. The hydrogen spectrum with fine structure is also successfully explained.

After the Lorentz group and its representations, we discuss in the present chapter the representations of the Poincaré-algebra in 4 space-time dimensions. This leads to the characterization and classification of elementary particles. It follows an in-depth discussion of Clifford algebras, spin groups and spinor representations of these groups in arbitrary dimensions, as they play a prominent role in recent developments in fundamental physics. As an application, we discuss the covariant Dirac equation for charged fermions. A detailed presentation of relativistic quantum mechanics and its symmetries can be found, for example, in the textbooks [75–77].

© The Author(s), under exclusive license to Springer-Verlag GmbH, DE,     403
part of Springer Nature 2026
A. Wipf, *Symmetries in Physics*, https://doi.org/10.1007/978-3-662-72675-4_18

## 18.1   Representations of the Lorentz Group

A relativistic generalization of non-relativistic quantum mechanics must not distinguish any inertial frame—it must be covariant and have the same form in such frames. In particular, observables must not distinguish inertial frames. According to Wigner's theorem, in a quantum theory the Lorentz symmetry is implemented by unitary representations of the Lorentz group. Therefore, we study the representations of the Lorentz group $SO(1, 3)$ or its Lie algebra $\mathfrak{so}(1, 3)$ at the beginning of the chapter.

The starting point for the construction of the (irreducible) representations of the Lorentz group and its universal covering $SL(2, \mathbb{C})$ (see Problem 8.5) are the commutation relations (14.74) for the basis elements $\Omega_i$ and $K_i$ of the Lorentz algebra $\mathfrak{so}(1, 3)$. In the *complexified* Lie algebra $\mathfrak{so}(4, \mathbb{C})$, where the ladder operators lie, we choose the new basis

$$A_i = \tfrac{i}{2}(\Omega_i - iK_i) \quad \text{and} \quad B_i = \tfrac{i}{2}(\Omega_i + iK_i)\,. \tag{18.1}$$

The $\{A_i\}$ commute with the $\{B_i\}$ and both sets form bases of an $\mathfrak{su}(2)$,

$$[A_i, B_j] = 0, \quad [A_i, A_j] = i\epsilon_{ijk}A_k, \quad [B_i, B_j] = i\epsilon_{ijk}B_k\,. \tag{18.2}$$

This means that $\mathfrak{so}(4, \mathbb{C})$ is the direct sum of two Lie algebras,

$$\mathfrak{so}(4, \mathbb{C}) \cong \mathfrak{su}_c(2) \oplus \mathfrak{su}_c(2) \cong \mathfrak{sl}(2, \mathbb{C}) \oplus \mathfrak{sl}(2, \mathbb{C})\,. \tag{18.3}$$

Here the complexified $\mathfrak{su}_c(2)$ appear, because we allow in $\mathfrak{so}(4, \mathbb{C})$ complex linear combinations of the basis elements. The finite-dimensional irreducible complex representations of the Lie algebra $\mathfrak{su}(2)$ are well-known to us (these correspond to the finite-dimensional irreducible representation of $\mathfrak{su}_c(2)$) and they are characterized by the half-integer spin, see Sect. 15.1. It follows that each irreducible representation of the Lorentz algebra is characterized by two half-integer quantum numbers $(n, m)$, where $n$ and $m$ are the spins of the two $\mathfrak{su}(2)$ representations.

The representations of the two summands $\mathfrak{su}_c(2, \mathbb{C})$ in (18.3) are connected by reflections, which are not in $SO^\uparrow(1, 3)$. In order to see this, we note how a space reflection acts on the infinitesimal rotations and boosts,

$$\Omega_i \mapsto \Omega_i \quad \text{and} \quad K_i \mapsto -K_i\,. \tag{18.4}$$

This means that in a reflection, the generators $A_i$ and $B_i$ swap their roles. If the generators $\Omega_i$ and $K_i$ are represented by anti-Hermitian matrices or operators, then the parity operation is equivalent to a Hermitian conjugation.

**Question**

Why do the irreducible representations of the Lorentz algebra have dimensions $(2m + 1)(2n + 1)$ with $m, n \in \frac{1}{2}\mathbb{Z}$?

We consider the representations with the smallest spins:

- $(m, n) = (0, 0)$: this is a one-dimensional trivial representation.
- $(m, n) = (\frac{1}{2}, 0)$: 2-component vectors that transform according to this representation are called left-handed spinors. They transform according to the spin-$\frac{1}{2}$ representation of the first $\mathfrak{su}(2)$ in (18.3) and trivially under the second $\mathfrak{su}(2)$.
- $(m, n) = (0, \frac{1}{2})$: 2-component vectors that transform according to this representation are called right-handed spinors. These transform trivially under the first $\mathfrak{su}(2)$ in (18.3) and with the spin-$\frac{1}{2}$ representation of the second $\mathfrak{su}(2)$.
- A reflection swaps right- and left-handed spinors and therefore a mirror-symmetric theory contains left- and right-handed spinors. Thus, the field of a Dirac electron transforms according to the four-dimensional reducible representation

$$(\tfrac{1}{2}, 0) \oplus (0, \tfrac{1}{2}) \, . \tag{18.5}$$

- By repeatedly tensoring the left- and right-handed spinor representations, one generates every representation of the Lorentz algebra. In particular, the tensor product of the two spinor representations yields the (also four-dimensional) representation

$$(\tfrac{1}{2}, 0) \otimes (0, \tfrac{1}{2}) = (\tfrac{1}{2}, \tfrac{1}{2}) \, . \tag{18.6}$$

This is the irreducible vector representation, according to which 4-vectors transform.

- The tensor product of two left-handed spinor representations is

$$(\tfrac{1}{2}, 0) \otimes (\tfrac{1}{2}, 0) = (0, 0) \oplus (1, 0) \, , \tag{18.7}$$

and leads to the trivial and the spin-1 representation. The latter contains the antisymmetric (anti)self-dual tensors of rank 2.

**Example: Decomposition of Second Rank Tensors**

Second rank tensors transform with the tensor product of two vector representations,

$$T^{\mu\nu} \in (\tfrac{1}{2}, \tfrac{1}{2}) \otimes (\tfrac{1}{2}, \tfrac{1}{2}) \overset{(18.6)}{=} (\tfrac{1}{2}, 0) \otimes (0, \tfrac{1}{2}) \otimes (0, \tfrac{1}{2}) \otimes (\tfrac{1}{2}, 0)$$

$$= \left[(1, 0) \oplus (0, 0)\right] \otimes \left[(0, 1) \oplus (0, 0)\right]$$

$$= (1, 1) \oplus (1, 0) \oplus (0, 1) \oplus (0, 0) \, .$$

The tensor contains a spin-0 part $(0, 0)$, spin-1 part $(1, 0) \oplus (0, 1)$ and spin-2 part $(1, 1)$. This reduction corresponds to the decomposition of the tensor into its trace part, its antisymmetric part and its symmetric and trace-free part:

$$T^{\mu\nu} = \left(\tfrac{1}{4}T^{\rho}_{\ \rho}\right)\eta^{\mu\nu} + A^{\mu\nu} + S^{\mu\nu} \quad \text{with} \quad A^{\mu\nu} = -A^{\nu\mu}, \quad S^{\mu\nu} = S^{\nu\mu}, \quad S^{\rho}_{\ \rho} = 0.$$

The antisymmetric term can be further decomposed into its self-dual and anti-self-dual parts. Under a space reflection, these are interchanged, see Problem 18.1. ◄

## 18.2   Representations of the Poincaré-Algebra

As a basis for the Poincaré-Algebra in 4 dimensions, we choose the infinitesimal translations and Lorentz transformations $P_{\mu}$ and $J_{\mu\nu}$ in (14.88). In Sect. 14.6.1, we convinced ourselves that

$$P^2 = P_{\mu}P^{\mu} \quad \text{and} \quad W^2 = W_{\mu}W^{\mu} \tag{18.8}$$

are Casimir invariants. The polarization vector of Pauli and Lubanski is quadratic in the generators of the Poincaré-Algebra, $W_{\mu} = \tfrac{1}{2}\epsilon_{\mu\nu\rho\sigma}P^{\nu}J^{\rho\sigma}$, and is perpendicular to the 4-momentum, $W_{\mu}P^{\mu} = 0$. By definition, the two Casimir invariants commute with all elements of the Poincaré-Algebra and therefore have a fixed value in each irreducible representation. These values characterize the irreducible representations.

For a massive particle, there exists a rest frame, in which the spatial momentum vanishes and $P^0$ is the rest energy, $P = (m, 0)$. Therefore, the first Casimir invariant has the value $P^2 = m^2$. Massless particles moves at the speed of light and have no rest frame. This explains why the representations with $P^2 = 0$ and $P^2 > 0$ are physically different and a distinction is made between massive and massless representations of the Poincaré algebra.

The infinitesimal Lorentz transformations in a relativistic field theory are generalized angular momenta $J_{\mu\nu}$ and contain an orbital and a spin part,

$$J_{\mu\nu} = L_{\mu\nu} + S_{\mu\nu}, \qquad L_{\mu\nu} = x_{\mu}P_{\nu} - x_{\nu}P_{\mu}, \qquad [x^{\mu}, P_{\nu}] = i\delta^{\mu}_{\ \nu}. \tag{18.9}$$

This is elaborated in Sects. 18.3.3 and 19.5. Orbital and spin parts commute with each other and the $L_{\mu\nu}$, $S_{\mu\nu}$ and $J_{\mu\nu}$ all satisfy the same $\mathfrak{so}(1, 3)$-commutation rules as the $M_{\mu\nu}$.

**Task**

Convince yourself that $\epsilon_{\mu\nu\rho\sigma}P^{\nu}L^{\rho\sigma}$ vanishes and therefore

$$W_{\mu} = \tfrac{1}{2}\epsilon_{\mu\nu\rho\sigma}P^{\nu}S^{\rho\sigma}. \tag{18.10}$$

### 18.2.1 Massive Representations

Casimir invariants have the same value in all inertial reference frames and can be calculated in any (accessible) frame. For massive representations, we choose the rest frame, in which the Pauli-Lubanski pseudovector (18.10) is proportional to the particle spin,

$$W_0 = 0 \quad \text{and} \quad W_i = m S_i \, . \tag{18.11}$$

The component $S_i$ of the spin generates the spin rotations about the axis $i$,

$$S_i = -\tfrac{1}{2}\epsilon_{ijk} S^{jk}, \qquad [S_i, S_j] = \mathrm{i}\epsilon_{ijk} S_k \, . \tag{18.12}$$

The Pauli-Lubanski pseudovector can thus be interpreted as a relativistic generalization of spin. Because of

$$W^2 = -m^2 \boldsymbol{S}^2 \tag{18.13}$$

massive representations are uniquely characterized by mass and spin of the particle:

$$P^2 = m^2 \quad \text{and} \quad W^2 = -m^2 s(s+1), \quad s = 0, \tfrac{1}{2}, 1, \ldots \, . \tag{18.14}$$

Massive particles thus form $(2s + 1)$-dimensional irreducible multiplets, whose states are distinguished by the third component of the spin and the continuous eigenvalue of $\boldsymbol{P}$.

### 18.2.2 Massless Representations

For a massless particle, $P^2 = m^2 = 0$ and there is no rest frame. But we can always switch to an inertial frame, in which

$$P_\mu = (P_0, 0, 0, P_3) \quad \text{with} \quad P_0^2 - P_3^2 = 0 \, . \tag{18.15}$$

Because $P_\mu W^\mu = 0$, the Pauli-Lubanski pseudovector in this frame has the form

$$W_\mu = (\lambda P_0, W_1, W_2, \lambda P_3) \quad \text{with} \quad W^2 = -W_1^2 - W_2^2 \le 0 \, . \tag{18.16}$$

For $W^2 = 0$, $P_\mu$ and $W_\mu$ are obviously linearly dependent,

$$W_\mu = \lambda P_\mu \, . \tag{18.17}$$

This relationship holds in every inertial frame, and therefore $\lambda$ is a Lorentz scalar. But a scalar quantity commutes with all generators of the Poincaré -group,

$$[\lambda, P^\mu] = [\lambda, J^{\mu\nu}] = 0 \, , \tag{18.18}$$

and defines a Casimir invariant. From $W_0 = -P^i J_i$ it follows that $\lambda$ is proportional to the helicity of the massless particle,

$$\lambda = \frac{\boldsymbol{P} \cdot \boldsymbol{J}}{P_0} . \tag{18.19}$$

It can be shown that $\lambda$ can only take half-integer values (see [78], page 90). For example, the photon has the helicity $\pm 1$.

The remaining case $P^2 = 0$ and $W^2 < 0$ describes massless particles with an infinite number of polarization states. Such representations do not seem to be realized in nature.

## 18.3   Clifford Algebras in $d$ Dimensions

Both in applications and in theories beyond the standard model, field theories in $d \neq 4$ dimensions occur and therefore we now consider spinors in arbitrary space-time dimensions. With the help of a representation of the Clifford algebra in $d$ dimensions one can construct an explicit basis of the Lie algebra of the Spin group $\mathrm{Spin}(1, d - 1)$. The latter is the double covering group of the connected component $\mathrm{SO}_0(1, d - 1) \leq \mathrm{SO}(1, d - 1)$, which contains the identity.

The Clifford algebra is the free algebra of the $d$ elements $\gamma_0, \dots, \gamma_{d-1}$, modulo the quadratic relation

$$\gamma_\mu \gamma_\nu + \gamma_\nu \gamma_\mu \equiv \{\gamma_\mu, \gamma_\nu\} = 2\eta_{\mu\nu}, \quad (\eta_{\mu\nu}) = \mathrm{diag}(1, -1, \dots, -1) . \tag{18.20}$$

A simple realization of this algebra is achieved with matrices of dimension $d_s = 2^{[d/2]}$, where $[d/2]$ is the largest integer less than or equal to $d/2$:

$$\begin{aligned}
\gamma_0 &= \sigma_1 \otimes \sigma_0 \otimes \sigma_0 \otimes \dots & \gamma_2 &= i\sigma_3 \otimes \sigma_1 \otimes \sigma_0 \otimes \dots & \gamma_4 &= i\sigma_3 \otimes \sigma_3 \otimes \sigma_1 \otimes \dots \\
\gamma_1 &= i\sigma_2 \otimes \sigma_0 \otimes \sigma_0 \otimes \dots & \gamma_3 &= i\sigma_3 \otimes \sigma_2 \otimes \sigma_0 \otimes \dots & \dots &\; .
\end{aligned} \tag{18.21}$$

The proof of this statement will be dealt with in Problem 18.2.

The representation matrices in odd spacetime dimensions $d$ are the same as those in one dimension less, supplemented by the $\frac{1}{2}(d - 1)$-fold tensor product $\gamma_{d-1} = i\sigma_3 \otimes \dots \otimes \sigma_3$. In low dimensions:

$$d = 2: \quad \gamma_0 = \sigma_1, \quad \gamma_1 = i\sigma_2 ,$$

$$d = 3: \quad \gamma_0 = \sigma_1, \quad \gamma_1 = i\sigma_2, \quad \gamma_2 = i\sigma_3 , \tag{18.22}$$

$$d = 4: \quad \gamma_0 = \sigma_1 \otimes \sigma_0, \quad \gamma_1 = i\sigma_2 \otimes \sigma_0, \quad \gamma_2 = i\sigma_3 \otimes \sigma_1, \quad \gamma_3 = i\sigma_3 \otimes \sigma_2 .$$

The matrices are Hermitian or anti-Hermitian,

$$\gamma_0^\dagger = \gamma_0 \quad\text{and}\quad \gamma_i^\dagger = -\gamma_i \quad\text{or}\quad \gamma_\mu^\dagger = \gamma_0\gamma_\mu\gamma_0 = \gamma^\mu = \eta^{\mu\nu}\gamma_\nu . \tag{18.23}$$

Equivalent representations

$$\gamma_\mu' = U\gamma_\mu U^{-1} . \tag{18.24}$$

(with unitary $U$) also have the properties (18.23), which we assume in the following. We need the ordered products of $n$ $\gamma$-matrices,

$$\gamma_{\mu_1\mu_2\ldots\mu_n} = \gamma_{\mu_1}\gamma_{\mu_2}\cdots\gamma_{\mu_n} \quad\text{with}\quad \mu_1 < \mu_2 < \cdots < \mu_n . \tag{18.25}$$

Equivalent to these, one defines the anti-symmetrized products

$$\gamma_{\mu_1\mu_2\ldots\mu_n} = \frac{1}{n!} \sum_{\pi\in\mathcal{S}_n} \operatorname{sign}(\pi)\, \gamma_{\pi(\mu_1)}\gamma_{\pi(\mu_2)}\cdots\gamma_{\pi(\mu_n)} , \tag{18.26}$$

which for ordered indices are equal to the ordered products and for unordered and unequal indices coincide with the ordered product up to a sign. The matrix without index (with $n = 0$) is the identity matrix. There exist $\binom{d}{n}$ different anti-symmetrized products for each $n \in \{0, 1, \ldots, d\}$, so there are in total

$$\sum_{n=0}^{d} \binom{d}{n} = (1 + 1)^d = 2^d$$

of them. In fact, the ordered products

$$\frac{1}{\sqrt{d_s}}\big\{\mathbb{1},\, \gamma_\mu,\, \gamma_{\mu_1\mu_2},\, \ldots,\, \gamma_{\mu_1\ldots\mu_D}\big\}, \quad\text{with}\quad \begin{cases} D = d & \text{for even } d, \\ D = \tfrac{1}{2}(d - 1) & \text{for odd } d, \end{cases} \tag{18.27}$$

form an *orthonormal basis* in the vector space $\mathrm{Mat}(d_s, \mathbb{C})$ with scalar product $(M, N) = \mathrm{Tr}(M^\dagger N)$.

---

**Question**

In Problem 18.5 we show that different $\gamma_{\mu_1\ldots\mu_n}$ with $0 \le n \le D$ are orthogonal. Why does this imply that the matrices (18.27) form a basis of the Clifford algebra and the $\gamma^\mu$ are irreducibly represented?

From the last property in (18.23) it now follows

$$\gamma^{\dagger}_{\mu_1\mu_2\ldots\mu_n} = \left(\gamma_{\mu_1}\gamma_{\mu_2}\cdots\gamma_{\mu_n}\right)^{\dagger} = \gamma^{\mu_n}\cdots\gamma^{\mu_2}\gamma^{\mu_1} = (-)^{n(n-1)/2}\gamma^{\mu_1\mu_2\ldots\mu_n}\,,$$
$$(18.28)$$

so that every $d_s$-dimensional matrix $M$ can be written as a linear combination of the orthonormal basis (18.27),

$$M = \frac{1}{d_s}\sum_{n=0}^{D}\sum_{\mu_1,\ldots,\mu_n}\frac{1}{n!}(-)^{n(n-1)/2}\,\gamma_{\mu_1\ldots\mu_n}\,\mathrm{Tr}\left(\gamma^{\mu_1\ldots\mu_n}M\right)\,. \qquad (18.29)$$

The second sum is over all index sets $\mu_1, \mu_2, \ldots, \mu_n$, and not just the ordered ones. This explains the normalization factor $1/n!$. The expansion (18.29) is the starting point for the important Fierz identities in field theories with spinor fields.

Of particular importance is the product of all $\gamma$-matrices,

$$\gamma_* = \mathrm{i}^{[\frac{1}{2}(d-1)]}\gamma^0\ldots\gamma^{d-1}\quad\text{with}\quad \gamma_*^{\dagger} = \gamma_*,\quad \gamma_*\gamma_* = 1\,. \qquad (18.30)$$

This is the generalization of the matrix $\gamma_5$ in 4 space-time dimensions, for arbitrary dimensions. The prefactor was chosen such that the matrix is Hermitian.

---

**Task**

Convince yourself that the matrix $\gamma_*$ has the properties (18.30) and for even $d$ anti-commutes with all $\gamma_\mu$ and for odd $d$ commutes with all $\gamma_\mu$.

With the $\gamma^\mu$, the matrices $-\gamma^\mu$ also form an irreducible representation of the Clifford algebra. For even $d$, the two representations are equivalent due to $\gamma_*\gamma^\mu\gamma_* = -\gamma^\mu$. For odd $d$, however, they are inequivalent, since $\gamma_*$ commutes with all basis elements $\gamma_{\mu_1\ldots\mu_n}$ and therefore must be a multiple of the identity matrix. Because of the last property in (18.30), it then follows that $\gamma_* = \pm 1$. However, since $\gamma_*$ is a product of an *odd* number of $\gamma^\mu$, it has a different sign for the representations $\{\gamma^\mu\}$ and $\{-\gamma^\mu\}$. Hence the two representations cannot be equivalent, since $U\gamma_*U^{-1} \neq -\gamma_*$ for $\gamma_* = 1$.

## 18.3.1 Irreducible Representations of the Clifford Algebra

We want to convince ourselves that, up to equivalence, there are only the just mentioned irreducible representations of the Clifford algebra. For the proof, we consider the *Clifford group* of order $2\cdot 2^d$, consisting of the elements

$$1, -1, \gamma^\mu, -\gamma^\mu, \gamma^{\mu\nu}, -\gamma^{\mu\nu}, \ldots \qquad (18.31)$$

In even dimensions, the group has $2^d + 1$ conjugacy classes, namely the two classes $\{\mathbb{1}\}$ and $\{-\mathbb{1}\}$ of the center elements and the $2^d - 1$ classes consisting of the pairs

$$\{\gamma^{\mu_1\mu_2\cdots\mu_n}, -\gamma^{\mu_1\mu_2\cdots\mu_n}\}, \quad n = 1, \ldots, d. \tag{18.32}$$

According to Theorem 11.7 in Sect. 11.3 there then exist $2^d + 1$ inequivalent irreducible representations. Because of (18.22), there exists at least one irreducible representation of dimension $d_s = 2^{d/2}$, so that according to Burnside's Theorem 11.6, we have

$$2 \cdot 2^d = \sum_{n=1}^{2^d}(\dim D_n)^2 + d_s^2 \quad \text{or} \quad \sum_{n=1}^{2^d}(\dim D_n)^2 = 2^d. \tag{18.33}$$

Thus, for even $d$, the Clifford group has $2^d$ one-dimensional irreducible representations and one $d_s$-dimensional irreducible representation.

In odd dimensions, in addition to $\pm\,\mathbb{1}$, there are two further center elements $\pm\,\gamma^0\gamma^1\cdots\gamma^{d-1}$ proportional to $\gamma_*$ in (18.30), so that there are $2^d + 2$ conjugacy classes and the same number of irreducible representations. We have already encountered two irreducible representations of dimensions $d_s = 2^{(d-1)/2}$, so that according to Burnside's theorem

$$2 \cdot 2^d = \sum_{n=1}^{2^d}(\dim D_n)^2 + d_s^2 + d_s^2 \quad \text{or} \quad \sum_{n=1}^{2^d}(\dim D_n)^2 = 2^d. \tag{18.34}$$

Thus, for odd $d$, the Clifford group has $2^d$ one-dimensional and two $d_s$-dimensional irreducible representations.

However, the one-dimensional representations of the Clifford group are not representations of the Clifford algebra, since one-dimensional matrices cannot represent the relation $\gamma^0\gamma^1 = -\gamma^1\gamma^0$. This then proves the following theorem:

**Theorem 18.1 (Irreducible Representations of the Clifford Algebra)** *In even dimensions, there exists (up to equivalence) only one irreducible representation of the Clifford algebra. This has the dimension $d_s = 2^{d/2}$. In odd dimensions, there exist two irreducible representations. Both have the dimension $d_s = 2^{(d-1)/2}$.*

### 18.3.2 Spin Group as Covering of the Lorentz Group

Here we generalize the construction of the universal covering $SL(2, \mathbb{C})$ of the proper orthochronous Lorentz group $SO^{\uparrow}(1, 3)$ (see Problem 8.5) to arbitrary dimensions. The results will be useful in the formulation of the covariant Dirac equation for spinor fields in Sect. 18.4. The construction is done using the $d_s$-dimensional $\gamma^{\mu}$-

matrices and their antisymmetrized products

$$S^{\mu\nu} = -S^{\nu\mu} = \frac{1}{2i}\gamma^{\mu\nu}, \quad \gamma^{\mu\nu} = \frac{1}{2}[\gamma^{\mu}, \gamma^{\nu}] \in \mathrm{Mat}(d_s, \mathbb{C}). \tag{18.35}$$

These satisfy the following commutation relations with the $\gamma^{\mu}$ matrices and themselves,

$$[S^{\mu\nu}, \gamma^{\rho}] = i\left(\eta^{\mu\rho}\gamma^{\nu} - \eta^{\nu\rho}\gamma^{\mu}\right), \tag{18.36}$$

$$[S_{\mu\nu}, S_{\rho\sigma}] = i\left(\eta_{\mu\rho}S_{\nu\sigma} + \eta_{\nu\sigma}S_{\mu\rho} - \eta_{\mu\sigma}S_{\nu\rho} - \eta_{\nu\rho}S_{\mu\sigma}\right), \tag{18.37}$$

and we prove these relations in Problem 18.7. The matrices $S_{\mu\nu}$ thus define a $d_s$-dimensional representation of the Lorentz algebra (14.80). If $(\omega, S) = \omega_{\mu\nu}S^{\mu\nu}$ is an arbitrary representation matrix, then according to the general theory

$$\mathcal{S}(s) = e^{i\frac{s}{2}(\omega, S)} \tag{18.38}$$

is a one-parameter subgroup in a $d_s$-dimensional representation of the universal covering of $SO_0(1, d - 1)$. This universal covering is the spin group ($d = 3$ is an exception) and is denoted by $\mathrm{Spin}(1, d - 1)$.

> **Spin Groups and Signatures of the Metric**
> The double covering of $SO(d)$ constructed analogously with the Euclidean $\gamma^{\mu}$ matrices—these satisfy $\{\gamma^{\mu}, \gamma^{\nu}\} = 2\delta^{\mu\nu}$—is denoted by $\mathrm{Spin}(d)$. It is simply connected for $d > 2$ and therefore the universal covering of $SO(d)$. On the other hand, $\mathrm{Spin}(1, 2)$ is not the universal covering of $SO_0(1, 2)$, since $\pi_1(\mathrm{Spin}(1, 2)) = \mathbb{Z}$. The groups $\mathrm{Spin}(1, p)$ with $p \neq 2$ are again simply connected. For a space with an indefinite metric of signature $(p, q)$, the double covering $\mathrm{Spin}(p, q)$ is generally not simply connected and therefore not the universal covering.

To find a covariant field equation, we need the adjoint action of $\mathcal{S}(s)$ on the $\gamma$ matrices,

$$\Gamma^{\rho}(s) = \mathcal{S}^{-1}(s)\gamma^{\rho}\mathcal{S}(s). \tag{18.39}$$

With the help of (18.36), we obtain the ordinary differential equation

$$\frac{d}{ds}\Gamma^{\rho}(s) = -\frac{i}{2}\omega_{\mu\nu}\mathcal{S}^{-1}(s)\left[S^{\mu\nu}, \gamma^{\rho}\right]\mathcal{S}(s) = \omega^{\rho}_{\ \sigma}\Gamma^{\sigma}(s),$$

whose solution with the initial condition $\Gamma^\rho(0) = \gamma^\rho$ reads:

$$\Gamma^\rho(s) = \mathcal{S}^{-1}(s)\gamma^\rho \mathcal{S}(s) = \left(e^{s\omega}\right)^\rho{}_\sigma \gamma^\sigma. \tag{18.40}$$

We remind that for $\omega_{\mu\nu} = -\omega_{\nu\mu}$ the matrix $(e^\omega)^\rho{}_\sigma$ is a Lorentz transformation, see Sect. 14.5. This leads to the important relation

$$\mathcal{S}^{-1}\gamma^\rho \mathcal{S} = \Lambda^\rho{}_\sigma \gamma^\sigma, \quad \text{with} \quad \mathcal{S} = e^{\frac{i}{2}(\omega, S)} = e^{\frac{1}{4}\omega^{\mu\nu}\gamma_{\mu\nu}}, \quad \Lambda = e^\omega. \tag{18.41}$$

This quadratic relation, which assigns a Lorentz transformation to each spin rotation, defines a representation of the spin group by Lorentz transformations,

$$\mathrm{Spin}(1, d-1) \ni \mathcal{S} \longmapsto \Lambda(\mathcal{S}) \in \mathrm{SO}(1, d-1). \tag{18.42}$$

Spin transformations are generally not unitary, but satisfy the condition

$$\gamma^0 \mathcal{S}^\dagger \gamma^0 = \mathcal{S}^{-1}. \tag{18.43}$$

---

**Task**

Prove this property of spin transformations.

---

Each spin transformation is assigned a unique Lorentz transformation in (18.41). Conversely, the same Lorentz transformation is assigned to the two spin transformations $\pm \mathcal{S}$. Therefore, the group $\mathrm{Spin}(1, d-1)$ is the double covering of the connected component $\mathrm{SO}_0(1, d-1) \leq \mathrm{SO}(1, d-1)$, which contains the identity— similar to how the quantum mechanical rotation group $\mathrm{SU}(2)$ is the double covering of the rotation group $\mathrm{SO}(3)$.

In 4 dimensions, the representation splits into two irreducible representations, namely the left- and right-handed spinor representation, see (18.5).

---

**Example: Spin Transformations in Chiral Representation**

We supplement the Pauli matrices $\sigma_k$ with $\sigma_0 = \mathbb{1}_2$ and define the Hermitian matrices

$$\sigma_\mu = \tilde{\sigma}^\mu = \{\sigma_0, \sigma_k\} \quad \text{or} \quad \sigma^\mu = \tilde{\sigma}_\mu = \{\sigma_0, -\sigma_k\}. \tag{18.44}$$

In the chiral representation the matrices $\gamma^\mu$ and $S^{\mu\nu}$ have the form

$$\gamma^\mu = \begin{pmatrix} 0 & \sigma^\mu \\ \tilde{\sigma}^\mu & 0 \end{pmatrix} \quad \text{and} \quad S^{\mu\nu} = \frac{1}{2i}\begin{pmatrix} \sigma^\mu\tilde{\sigma}^\nu - \sigma^\nu\tilde{\sigma}^\mu & 0 \\ 0 & \tilde{\sigma}^\mu\sigma^\nu - \tilde{\sigma}^\nu\sigma^\mu \end{pmatrix}, \tag{18.45}$$

and $\mathcal{S}$ in (18.41) is block diagonal,

$$\mathcal{S} = \begin{pmatrix} A & 0 \\ 0 & B \end{pmatrix}. \tag{18.46}$$

The matrices $A$, $B$ are generated by traceless matrices and are therefore in the spin group $\mathrm{SL}(2, \mathbb{C})$. The reality condition (18.43) translates into $B = A^{-1\dagger}$. The mapping $A \mapsto B$ defines a representation of $\mathrm{SL}(2, \mathbb{C})$, so that $A \mapsto \mathcal{S}$ defines a reducible representation of the spin group. $\mathcal{S}$ thus decomposes into two irreducible $\mathrm{SL}(2, \mathbb{C})$ representations. These are the left- and right-handed spinor representations in the decomposition (18.5). With (18.41) it follows

$$A(x^\mu \sigma_\mu) A^\dagger = (\Lambda^\mu{}_\nu x^\nu)\, \sigma_\mu \quad \text{and} \quad B(x^\mu \tilde{\sigma}_\mu) B^\dagger = (\Lambda^\mu{}_\nu x^\nu)\, \tilde{\sigma}_\mu \,. \tag{18.47}$$

The first relation is the group homomorphism $\mathrm{SL}(2, \mathbb{C}) \mapsto \mathrm{SO}(1,3)^\uparrow_+$ in Problem 8.5. ◄

The group $\mathrm{Spin}(1, 1)$ contains two connected components, while the spin groups in $d > 2$ dimensions are connected. They are isomorphic to well-known Lie groups, as we will now explicitly prove in 3 and in 5 dimensions.

### Example: Spin Group Spin$(1, 2)$

In 3 spacetime dimensions, we choose a representation with imaginary $\gamma$-matrices,

$$\{\gamma_0, \gamma_1, \gamma_2\} = \{\sigma_2, \mathrm{i}\sigma_1, \mathrm{i}\sigma_3\}. \tag{18.48}$$

Then it is immediately apparent that the infinitesimal generators of the spin transformations $\mathcal{S}$ in (18.41)—these are the traceless and real matrices

$$\mathrm{i}(\omega, S) = \frac{1}{2} \begin{pmatrix} w^{01} & -w^{02} + \omega^{12} \\ -\omega^{02} - \omega^{12} & -\omega^{01} \end{pmatrix}, \tag{18.49}$$

span the three-dimensional Lie algebra $\mathfrak{sl}(2, \mathbb{R})$. This generates the connected Lie group $\mathrm{SL}(2, \mathbb{R})$. This proves the isomorphism of $\mathrm{Spin}(1, 2)$ and $\mathrm{SL}(2, \mathbb{R})$. The group $\mathrm{SL}(2, \mathbb{R})$ is not simply connected. Its maximal compact subgroup is $\mathrm{SO}(2)$, so that $\pi_1\big(\mathrm{SL}(2, \mathbb{R})\big) = \mathbb{Z}$. Thus, $\mathrm{Spin}(1, 2)$ is the double, but not the universal covering of $\mathrm{SO}_0(1, 2)$. ◄

In other dimensions, similar isomorphisms exist and one can show that the spin groups $\mathrm{Spin}(1, d - 1)$ with $d \neq 3$ are simply connected. As an instructive example, we consider the double covering of the ten-dimensional Lorentz group in the five-dimensional Minkowski space.

**Table 18.1** The double coverings Spin$(1, d-1)$ of the component containing the identity of the proper Lorentz groups $SO_0(1, d-1)$ and isomorphic Lie groups

| Spin$(1, d-1)$ | Spin$(1, 1)$ | Spin$(1, 2)$ | Spin$(1, 3)$ | Spin$(1, 4)$ | Spin$(1, 5)$ | ... |
|---|---|---|---|---|---|---|
| $\cong$ | GL$(1, \mathbb{R})$ | SL$(2, \mathbb{R})$ | SL$(2, \mathbb{C})$ | Sp$(1, 1)$ | SL$(2, \mathbb{H})$ | ... |

---

**Example: Spin Group Spin$(1, 4)$**

In $d = 5$ dimensions, there exists an antisymmetric charge conjugation matrix $C$ with $C^{-1}\gamma_\mu C = \gamma_\mu^T$, see Sect. 18.4.4. This relation and the hermiticity properties (18.23) imply the following relations for the infinitesimal spin transformations:

$$\gamma_{\mu\nu}^T C^{-1} + C^{-1}\gamma_{\mu\nu} = 0, \quad \gamma_{\mu\nu}^\dagger \gamma^0 + \gamma^0\gamma_{\mu\nu} = 0. \tag{18.50}$$

The corresponding group elements $S$ in (18.41) then satisfy the relations

$$S^T C^{-1} S = C^{-1}, \quad S^\dagger \gamma^0 S = \gamma^0. \tag{18.51}$$

The matrix $C$ is anti-symmetric and squares to $-\mathbb{1}$, whereas $\gamma^0$ is Hermitian and has the eigenvalues $\{1, 1, -1, -1\}$. Therefore, the relation (18.50) defines the Lie group Sp$(1, 1)$, see [3]. The maximal compact subgroup is the simply connected group SU$(2) \oplus$ SU$(2)$. This is generated by the elements $\gamma_{ij}$. ◀

Like the proper Lorentz group SO$(1, 3)$ in Sect. 5.5.1, the groups SO$(p, q)$ for $p, q \geq 1$ decompose into two connected components. The double coverings of the identity-components $SO_0(1, d-1)$ are isomorphic to the Lie groups in Table 18.1.

### 18.3.3 Transformations of Spinor Fields

We proceed similarly to the non-relativistic theory in Sect. 17.4 and examine the transformation of Dirac spinor fields when we change the inertial frame,

$$\psi(x) \mapsto (\Gamma_S \psi)(x) = S\,\psi(\Lambda^{-1}x), \quad \text{with} \quad S = e^{\frac{i}{2}(\omega, S)}, \quad \Lambda = e^\omega, \tag{18.52}$$

specifically for Lorentz transformations close to the identity,

$$\delta_\omega \psi = \left(\tfrac{i}{2}(\omega, S) - \omega^\alpha{}_\beta x^\beta \partial_\alpha\right)\psi(x) \equiv \frac{i}{2}\omega^{\mu\nu}\left(S_{\mu\nu} + L_{\mu\nu}\right)\psi(x) = \frac{i}{2}\omega^{\mu\nu} J_{\mu\nu}\psi(x). \tag{18.53}$$

The infinitesimal Lorentz transformations $J_{\mu\nu}$ are thus the sum of an orbital and spin part, as anticipated in (18.9). Like the $J_{\mu\nu}$ the $L_{\mu\nu}$ and $S_{\mu\nu}$ in the decomposition

$$J_{\mu\nu} = L_{\mu\nu} + S_{\mu\nu}, \quad L_{\mu\nu} = x_\mu p_\nu - x_\nu p_\mu, \quad (p_\mu = -i\partial_\mu) \tag{18.54}$$

satisfy the Lorentz algebra (14.80). The orbital part is universal and arises from the transformation of the coordinates and occurs for all fields, especially for scalar fields. The spin part originates from the representation $\mathcal{S}$ of the spin group in (18.52) and is absent for scalar fields.

## 18.4   Covariance of the Dirac Equation

The Dirac equation plays a similarly important role in relativistic quantum mechanics as the Schrödinger equation in non-relativistic quantum mechanics. We introduce and discuss it in 4 spacetime dimensions—the generalization to arbitrary dimensions is not difficult. In 4 dimensions, Dirac spinors transform according to the four-dimensional reducible spinor representation $(\frac{1}{2}, 0) \oplus (0, \frac{1}{2})$ of the spin group $SL(2, \mathbb{C})$. In the chiral representation for the $\gamma$-matrices the spin transformations $\mathcal{S}$ in (18.46) are block-diagonal and the reducibility of the representation is evident.

Under $SL(2, \mathbb{C})$ a 4-component Dirac spinor field $\psi(x)$ transforms as follows

$$\psi'(x') = \mathcal{S}\psi(x), \quad x' = \Lambda x + a . \tag{18.55}$$

Because of (18.43), the Dirac-conjugated spinor field $\bar{\psi} = \psi^{\dagger}\gamma^0$ transforms accordingly

$$\bar{\psi}'(x') = \bar{\psi}(x)\mathcal{S}^{-1} \tag{18.56}$$

In contrast to tensor fields, whose components transform with $(\Lambda^{\mu}{}_{\nu})$ when changing the inertial frame, the four components of a spinor field transform with the spin transformation $\mathcal{S}$. But the following 16 bilinears are (pseudo)tensor fields:

$$
\begin{aligned}
S(x) &= \bar{\psi}(x)\psi(x) && \text{scalar field} && 1 \\[4pt]
j^{\mu}(x) &= \bar{\psi}(x)\gamma^{\mu}\psi(x) && \text{vector field (current)} && 4 \\[4pt]
T^{\mu\nu}(x) &= \bar{\psi}(x)[\gamma^{\mu}, \gamma^{\nu}]\psi(x) && \text{antisymm. tensor field} && 6 \\[4pt]
A^{\mu}(x) &= \bar{\psi}(x)\gamma_5\gamma^{\mu}\psi(x) && \text{pseudovector field (axial current)} && 4 \\[4pt]
P(x) &= \bar{\psi}(x)\gamma_5\psi(x) && \text{pseudoscalar field} && 1
\end{aligned}
\tag{18.57}
$$

Here, the Hermitian matrix $\gamma_5 \equiv \gamma_* = i\gamma^0\gamma^1\gamma^2\gamma^3$ introduced in (18.30) appears, which anti-commutes with all $\gamma^{\mu}$.

---

**Question**

How do the $\gamma^{\mu\nu\cdots}$ transform under spin transformations, i.e. what is $\mathcal{S}^{-1}\gamma^{\mu\nu\cdots}\mathcal{S}$? Why does the transformation $T'^{\mu\nu}(x') = \Lambda^{\mu}{}_{\alpha}\Lambda^{\nu}{}_{\beta}T^{\alpha\beta}(x)$ follow from this?

### 18.4.1 Dirac Equation for Free Particles

The Dirac equation is a wave equation for the 4-component Dirac spinor field $\psi$. It is a first order differential equation in the time and the spatial coordinates. The first derivatives transform like the covariant components of a 4-vector,

$$\frac{\partial}{\partial x'^{\mu}} = \Lambda_{\mu}{}^{\nu}\frac{\partial}{\partial x^{\nu}}. \tag{18.58}$$

Thus, using (18.41), we obtain the identity

$$\gamma^{\mu}\mathcal{S}\frac{\partial}{\partial x'^{\mu}} = \mathcal{S}\gamma^{\mu}\frac{\partial}{\partial x^{\mu}}, \tag{18.59}$$

so that $\gamma^{\mu}\partial_{\mu}\psi$ transforms under spin transformations just like the Dirac spinor field $\psi$,

$$\gamma^{\mu}\frac{\partial}{\partial x'^{\mu}}\psi'(x') \stackrel{(18.55)}{=} \gamma^{\mu}\frac{\partial}{\partial x'^{\mu}}\mathcal{S}\psi(x) \stackrel{(18.59)}{=} \mathcal{S}\gamma^{\mu}\partial_{\mu}\psi(x). \tag{18.60}$$

Therefore, the Dirac equation for a Dirac spinor field,

$$(i\slashed{\partial} - m)\psi = 0, \qquad \slashed{\partial} = \gamma^{\mu}\partial_{\mu}, \tag{18.61}$$

is a covariant equation that does not distinguish any inertial frame. The constant $m$ appearing here has the unit of an inverse length and is identified with the inverse Compton wavelength of the spin-$\frac{1}{2}$ particle described by the equation.

**Covariance of the Dirac Equation**
If $\psi(x)$ satisfies the Dirac equation $(i\slashed{\partial} - m)\psi = 0$ in the inertial frame with coordinates $x^{\mu}$, then $\psi'(x') = \mathcal{S}\psi(x)$ satisfies the equation in the inertial frame with coordinates $x' = \Lambda x + a$.

Due to (18.23), the Dirac-conjugated field $\bar{\psi} = \psi^{\dagger}\gamma^{0}$ satisfies the equation

$$i\partial_{\mu}\bar{\psi}\gamma^{\mu} + m\bar{\psi} = 0. \tag{18.62}$$

**Question**

In arbitrary dimensions, the Dirac equation also has the form (18.61) with $d_s$-dimensional matrices $\gamma^{\mu}$, which satisfy the relation (18.20). Why is it a covariant equation?

### 18.4.2 Parity Symmetry

If the dimension $d$ of spacetime is even, then $x \mapsto -x$ is a space inversion. Let $\psi(x)$ be a solution of the free Dirac equation (18.61). If we reverse the directions of all space coordinates, this is called parity operation, we find the equation $(i\gamma^0 \partial_0 - i\gamma^k \partial_k - m)\psi(t, -x) = 0$. Multiplying from the left by $\gamma^0$ and using the relations (18.20), it follows that the parity-transformed field

$$(\Gamma_{\mathcal{P}}\psi)(x) = \gamma^0 \psi(Px), \quad P = \mathrm{diag}(1, -\mathbb{1}) \quad (d \text{ even}) \tag{18.63}$$

satisfies the Dirac equation as well. It is the space inverted wave function. In 4 dimensions and the chiral representation (18.45), parity swaps right- and left-handed Weyl spinors.

---

**Question**

Why is $x \mapsto -x$ not a space reflection, when the dimension of spacetime is odd?

If the dimension of spacetime is odd, then we only invert one spatial coordinate, e.g., the last coordinate $x^{d-1}$, and with this choice, parity acts on a spinor field as follows:

$$(\Gamma_{\mathcal{P}}\psi)(x) = \gamma^{d-1}\psi(Px), \quad P = \mathrm{diag}(\mathbb{1}, -1) \quad (d \text{ odd}). \tag{18.64}$$

It solves the Dirac equation with the opposite sign of $m$.

### 18.4.3 Time Reversal

We start again with a solution of the Dirac equation (18.61). If we reverse the direction of time and then complex conjugate the resulting equation, we get $(i\gamma^{0*}\partial_t - i\gamma^{*k}\partial_k - m)\psi^*(-t, x) = 0$. If $\mathcal{T}$ is an invertible matrix with

$$\mathcal{T}\gamma_0^*\mathcal{T}^{-1} = \gamma_0, \quad \mathcal{T}\gamma_k^*\mathcal{T}^{-1} = -\gamma_k, \tag{18.65}$$

then the transformed spinor field

$$(\Gamma_{\mathcal{T}}\psi)(x) = \mathcal{T}\psi^*(Tx), \quad T = \begin{pmatrix} -1 & 0 \\ 0 & \mathbb{1} \end{pmatrix}, \tag{18.66}$$

satisfies the Dirac equation in the inertial frame with reversed time. Note that time reversal acts anti-linearly on spinor fields.

---

**Example: $\mathcal{T}$-Matrix in Standard Representation in 4 Dimensions**

In the standard representation $\gamma^0 = \sigma_3 \otimes \sigma_0$ and $\gamma^i = i\sigma_2 \otimes \sigma_k$ only $\gamma^2$ is imaginary and a possible choice for $\mathcal{T}$ is $i\gamma^1\gamma^3 = \sigma_0 \otimes \sigma_2$. Because $\mathcal{T}$ is imaginary, we have

$$\left(\Gamma_{\mathcal{T}} \circ \Gamma_{\mathcal{T}} \psi\right)(x) = -\psi(x).$$

The property $\Gamma_{\mathcal{T}} \circ \Gamma_{\mathcal{T}} = -\mathbb{1}$ originates from the half-integer intrinsic spin of a Dirac particle[75]. ◄

## 18.4.4 Coupling to the Electromagnetic Field and Charge Conjugation

The coupling of a Dirac spinor field to the electromagnetic field is achieved through *minimal coupling*. The form of the coupling is determined by the required local gauge symmetry and is discussed in detail in Sect. 20.3.1. As a result, partial derivatives are replaced by covariant derivatives with vector potential $A_\mu$,

$$\partial_\mu \psi \longrightarrow D_\mu \psi = \left(\partial_\mu + ieA_\mu\right)\psi, \quad A^\mu = (\phi, \boldsymbol{A}). \tag{18.67}$$

Here, the charge $e$ of the particle described by $\psi$ appears. To see how a charge conjugation acts on a Dirac spinor field, we consider the Dirac equation in the external electromagnetic field,

$$i\gamma^\mu D_\mu(e)\psi - m\psi = 0. \tag{18.68}$$

If we multiply the equation from the left by $\gamma^0$, use the relation (18.23) and complex conjugate the obtained relation, we find

$$-i\gamma^{\mu T} D_\mu^*(e)\gamma^{0*}\psi^* - m^*\gamma^{0*}\psi^* = 0. \tag{18.69}$$

To continue, we need the transpose of the Dirac-conjugated spinor,

$$\bar{\psi}^T = \gamma^{0T}\psi^* = \gamma^{0*}\psi^*. \tag{18.70}$$

With this, we replace $\psi^*$ in (18.69) and obtain

$$i\gamma^{\mu T} D_\mu(-e)\bar{\psi}^T + m^*\bar{\psi}^T = 0. \tag{18.71}$$

Further below we will prove that in all dimensions a charge conjugation matrix $\mathcal{C}$ with the following property exists,

$$\mathcal{C}\gamma_\mu^T \mathcal{C}^{-1} = \eta\gamma_\mu \iff \gamma_\mu^* = \eta\mathcal{C}^{-1}\gamma^\mu\mathcal{C} \quad (\eta = \pm 1). \tag{18.72}$$

With this matrix, we define the charge-conjugated spinor

$$\psi_C = C\bar{\psi}^T = C\gamma_0^T \psi^* \,. \tag{18.73}$$

Finally, we multiply (18.71) from the left by $C$ and obtain

$$i\gamma^\mu D_\mu(-e)\psi_C + \eta m^* \psi_C = 0 \,, \tag{18.74}$$

which means that $\psi_C$ satisfies the Dirac equation with opposite charge and parameter $-\eta m$. This justifies the name charge conjugation for the mapping $\psi \mapsto \psi_C$ (the sign of $m$ in the Dirac equation is physically irrelevant).

If all $\gamma^\mu$ are real or all are imaginary, then we can choose $C = \gamma_0^T$. Then the following holds

$$\gamma^\mu \text{ real} \implies \gamma_0^T = \gamma_0 = C, \qquad \eta = 1, \qquad \psi_C = \psi^*$$

$$\gamma^\mu \text{ imaginary} \implies \gamma_0^T = -\gamma_0 = C, \quad \eta = -1, \quad \psi_C = \psi^* \,. \tag{18.75}$$

Such representations of the $\gamma$-matrices are called Majorana representations. In these representations, charge conjugation is just complex conjugation. A spinor that is invariant under charge conjugation, $\psi_C = \psi$, is called a Majorana spinor. Majorana spinors are real in a Majorana representation. They describe particles that are identical to their anti-particles.

Finally, we want to convince ourselves that a matrix $C$ exists in every dimension.

**Lemma 18.1** *In every dimension there exists a symmetric or antisymmetric matrix $C$ with the properties*

$$\gamma_\mu^T = \eta \, C^{-1}\gamma_\mu C, \qquad \eta = \pm 1 \,. \tag{18.76}$$

A formal proof can be found, for example, in [79]. Here we take the simpler route and give $C$ explicitly in the representations (18.21). If we change to another representation with (18.24), then $C$ in the new representation has the form

$$C' = UCU^T \,, \tag{18.77}$$

and is thus symmetric or antisymmetric in all equivalent representations.

***Proof*** *In even dimensions* we use

$$\sigma_1\sigma_1\sigma_1 = \sigma_1^T, \quad \sigma_1\sigma_2\sigma_1 = \sigma_2^T, \quad \sigma_1\sigma_3\sigma_1 = -\sigma_3^T, \quad \sigma_2\sigma_i\sigma_2 = -\sigma_i^T \tag{18.78}$$

**Table 18.2** In even dimensions, there exist two $\mathcal{C}$ matrices with $\eta = \pm 1$ and in odd dimensions one. They are either symmetric (S) or antisymmetric (A)

| $d$ | 1 | 2 | 3 | 4 | 5 | 6 | 7 | 8 |
|-----|---|---|---|---|---|---|---|---|
| $\mathcal{C}_+$ | S | S |   | A | A | A |   | S |
| $\mathcal{C}_-$ |   | A | A | A |   | S | S | S |

to show that in the representation (18.21) the symmetric and anti-symmetric matrices

$$\mathcal{C}_+ = \sigma_1 \otimes \sigma_2 \otimes \sigma_1 \otimes \sigma_2 \otimes \ldots \qquad (\eta = +1)$$

$$\mathcal{C}_- = \sigma_2 \otimes \sigma_1 \otimes \sigma_2 \otimes \sigma_1 \otimes \ldots \propto \mathcal{C}_+ \gamma_* \qquad (\eta = -1) \qquad (18.79)$$

fulfill the condition (18.76).

---

**Task**

Convince yourself that this statement is correct.

The matrices $\mathcal{C}$ are Hermitian, unitary, and symmetric or anti-symmetric,

$$\mathcal{C}_+^T = (-1)^{[d/4]}\,\mathcal{C}_+, \qquad \mathcal{C}_-^T = (-1)^{[(d+2)/4]}\,\mathcal{C}_- . \qquad (18.80)$$

*In odd dimensions* $d$, we use the fact that the $d-1$ matrices $\gamma_0, \ldots, \gamma_{d-2}$ can be chosen as $\gamma^\mu$ matrices in $d-1$ dimensions, for which the two matrices $\mathcal{C}_\pm$ in (18.79) exist. In the representation (18.21), the missing matrix is $\gamma_{d-1} = i\sigma_3 \otimes \cdots \otimes \sigma_3$, and with

$$\mathcal{C}_\pm^{-1}\gamma_{d-1}\mathcal{C}_\pm = (-1)^{[d/2]}\gamma_{d-1}^T ,$$

it follows that $\eta = (-)^{[d/2]}$, or that $\mathcal{C}_+$ is a charge conjugation matrix in $1 + 4n$ dimensions and $\mathcal{C}_-$ in $3 + 4n$ dimensions. Thus, there exist two $\mathcal{C}$ matrices in even dimensions and one $\mathcal{C}$ matrix in odd dimensions. Solutions of (18.76) in the dimensions $d = 1, 2, \ldots, 8$ are given in Table 18.2. The structure is periodic in the dimension with a period of 8 and thus it is sufficient to note the solutions for $d \leq 8$ in the table.

---

## 18.5  Exercises for Chap. 18

**Problem 18.1 (Self-Dual Tensors)**  An antisymmetric tensor $F_{\mu\nu}$ with $\mu, \nu \in \{1, 2, 3, 4\}$ is called self-dual if

$$F_{\mu\nu} = \frac{1}{2}\sum_{\alpha,\beta=1}^{4} \epsilon_{\mu\nu\alpha\beta} F_{\alpha\beta}$$

It is called anti-self-dual if there is a minus sign on the right-hand side. Show that every antisymmetric tensor is the sum of a self-dual and an anti-self-dual tensor. Is this decomposition invariant under rotations in SO(4)?

Remark: This is the decomposition of the field strength tensor $F_{\mu\nu}$ in Euclidean electrodynamics.

**Problem 18.2 (Representation of the Gamma Matrices)** Prove that the matrices in (18.21) represent the Clifford algebra. Use

$$\{A_1 \otimes \cdots \otimes A_p, B_1 \otimes \cdots \otimes B_p\} = A_1 B_1 \otimes \cdots \otimes A_p B_p + B_1 A_1 \otimes \cdots \otimes B_p A_p \,.$$

The anti-commutator of two tensor products simplifies if you use

$$\{A_1 \otimes \cdots \otimes A_p, B_1 \otimes \cdots \otimes B_p\} = A_1 B_1 \otimes \cdots \otimes \{A_q, B_q\} \otimes \cdots \otimes A_p B_p \,,$$

in case $[A_i, B_i] = 0$ for all $i \neq q$.

**Problem 18.3 (Chiral Representation)** Convince yourself that the matrices in (18.44) provide a representation of the Clifford algebra in 4 dimensions. What is $\gamma_5 \equiv \gamma_*$ in this representation?

**Problem 18.4 (Dirac Matrices in 4 Dimensions)** Show that the matrices $\{\mathbb{1}_4, \gamma^\mu, \gamma^{\mu\nu}, \gamma_5\gamma^\mu, \gamma_5\}$ form a basis for the complex $4 \times 4$ matrices, i.e., that they are linearly independent. Multiply the equation

$$a\mathbb{1}_4 + b_\mu\gamma^\mu + c_{\mu\nu}\gamma^{\mu\nu} + d_\mu\gamma_5\gamma^\mu + e\gamma_5 = 0$$

by the matrices $\{\mathbb{1}_4, \gamma^\mu, \gamma^{\mu\nu}, \gamma_5\gamma^\mu, \gamma_5\}$ and form the trace. Show that $a = b_\mu = c_{\mu\nu} = d_\mu = e = 0$ must hold.

**Problem 18.5 (Products of $\gamma$ Matrices in $d$ Dimensions)** Prove the following identities for the $\gamma^\mu$ matrices:

$$d \text{ even:} \quad \mathrm{Tr}(\gamma_{\mu_1\mu_2\ldots\mu_n}) = 0, \quad 1 \leq n \leq d \,, \tag{18.81}$$

$$d \text{ odd:} \quad \mathrm{Tr}(\gamma_{\mu_1\mu_2\ldots\mu_n}) = 0, \quad 1 \leq n < d \,. \tag{18.82}$$

Hint: To prove the trace properties, you need the invariance of the trace under cyclic permutation of the matrices and the existence of an anti-commuting $\gamma_*$ for even $d$.

The trace $\mathrm{Tr}(M^\dagger N)$ is a scalar product on the linear space of matrices. Prove

$$d \text{ even} \quad (\gamma_{\mu_1,\ldots,\mu_m}, \gamma_{\nu_1,\ldots,\nu_n}) = d_s \delta_{mn}\delta_{\mu\nu} \,, \tag{18.83}$$

$$d \text{ odd} \quad (\gamma_{\mu_1,\ldots,\mu_m}, \gamma_{\nu_1,\ldots,\nu_n}) = d_s \left(\delta_{mn}\delta_{\mu\nu} + \kappa\,\delta_{m+n,0}\,\varepsilon_{\mu\nu}\right) \,, \tag{18.84}$$

where $\mu$ and $\nu$ are multi-indices and correspondingly $\delta_{\mu\nu} = \prod_i \delta_{\mu_i \nu_i}$ and $\varepsilon_{\mu\nu} = \varepsilon_{\mu_1\mu_2\ldots\mu_m\nu_1\nu_2\ldots\nu_n}$ with $\varepsilon_{01\ldots(d-1)} = 1$. The constant $\kappa$ in the last term is a complex number of magnitude 1 and appears because in odd dimensions $\gamma_*$ in each irreducible representation is proportional to the identity.

**Problem 18.6 (Weyl Spinors and Weyl Equations)** Show that for $m = 0$ the Dirac equation (18.61) in the chiral representation (18.45) decomposes into two independent equations for the 2-component Weyl spinors $\chi$ and $\lambda$ in the decomposition

$$\psi = \begin{pmatrix} \chi \\ \lambda \end{pmatrix}.$$

These are the Weyl equations. How do the Weyl spinors transform when changing the inertial frame? What does the decomposition of $\psi$ look like in other representations of the $\gamma^\mu$?

**Problem 18.7 (Commutation Rules for $S_{\mu\nu}$)** Prove, starting from the anti-commutation relations (18.20), the commutation rules (18.37) for the generators $S_{\mu\nu}$ of the spin transformations.

**Problem 18.8 (Clifford Group in 2 Dimensions)** Consider the Clifford group for the $\gamma$-matrices $\gamma^0$, $\gamma^1$ in 2 spacetime dimensions. This has 8 elements. Convince yourself that the regular representation decomposes into 4 one-dimensional and 2 two-dimensional representations, and that the latter two are equivalent.

**Problem 18.9 (Matrix $T$ for Time Reversal)** Why does a time reversal matrix $T$ exist, which fulfills the condition (18.65), at least in $d \bmod 8 = 0, 1, 2, 4, 5, 6$?

# Relativistic Field Theories

*Things should be made as simple as possible, but not simpler.*

—Albert Einstein

Interactions compatible with the principles of special relativity are successfully described by relativistic field theories. If the associated field equations can be derived from the Hamiltonian principle of least action with Poincaré-invariant action, then they are covariant and do not single out any inertial frame. The most well-known *classical* relativistic field theory is Maxwell's electrodynamics, developed in the nineteenth century.

If the field of a classical field theory is "quantized" according to the rules of quantum mechanics—we then speak of a *quantized* field theory or a quantum field theory (QFT)—it describes particles with fixed mass and fixed spin. A scalar field assigns a scalar to each point in space and describes spinless particles, for example, Higgs bosons. In contrast, a vector field assigns a vector to each point in space and describes particles with spin 1, for example, photons. Fermions like electrons or quarks, on the other hand, are described by spinor fields, which assign a spinor to each point in space. Also, gravity within the framework of general relativity is a classical field theory invariant under local Lorentz transformations and local coordinate transformations.

An introduction to classical relativistic field theories can be found, for example, in [80–82]. Quantized field theories are presented in numerous works, for example, [78, 83–86].

## 19.1 Lagrangian Formalism

A field is a mapping that assigns an element $\phi(x)$ in a target space $\mathcal{T}$ to each point in spacetime with coordinates $x$. If the target space is linear, then the infinite-dimensional space of all fields $\mathcal{F} = \{\phi(x)\}$—referred to as field space or configuration space of the field theory—is also linear. In a local field theory, the action is the spacetime integral over a local Lagrange density $\mathcal{L}(x)$,

$$S = \int \mathrm{d}^d x \, \mathcal{L}(x) \equiv \int \mathrm{d}t \mathrm{d}\boldsymbol{x} \, \mathcal{L}(t, \boldsymbol{x}) \,, \tag{19.1}$$

where, for the majority of field theories, $\mathcal{L}$ is a function of the field and its first derivatives.[1] We will always assume this in the following. Also, to keep the notation simple, we will often write $\mathcal{L}(\phi)$ instead of $\mathcal{L}(\phi, \partial_\mu \phi)$. The field $\phi$ can have several components, e.g., be a collection of several scalar fields, like the Higgs field in the standard model with target space $\mathcal{T} = \mathbb{C}^2$, or a tensor or spinor field.

The space-time dimension $d$ is not fixed, as theories in $1{+}1$ and $1{+}2$ dimensions are also of interest in applications. Only when discussing electrodynamics will we assume that spacetime is four-dimensional. We denote the volume element $\mathrm{d}^{d-1}x$ of space with $\mathrm{d}\boldsymbol{x}$. We use natural units with $\hbar = c = 1$.

The (Frechet) derivative of the action $\delta S/\delta \phi(x)$ is defined by

$$S[\phi + \delta\phi] = S[\phi] + \delta S + \mathcal{O}(\delta\phi^2) = S[\phi] + \int \mathrm{d}^d x \, \frac{\delta S}{\delta \phi(x)} \delta\phi(x) + \mathcal{O}(\delta\phi^2) \,, \tag{19.2}$$

where the variation $\delta\phi$ is a test function that vanishes at the boundary of the considered region (or "at infinity").[2] The variation of an action with Lagrange density $\mathcal{L}(\phi, \partial_\mu \phi)$ is equal to

$$\begin{aligned}
\delta S &= \int \mathrm{d}^d x \left( \frac{\partial \mathcal{L}}{\partial \phi} \delta\phi + \frac{\partial \mathcal{L}}{\partial(\partial_\mu \phi)} \partial_\mu \delta\phi \right) \\
&= \int \mathrm{d}^d x \left( \frac{\partial \mathcal{L}}{\partial \phi} - \partial_\mu \frac{\partial \mathcal{L}}{\partial(\partial_\mu \phi)} \right) \delta\phi + \int \mathrm{d}^d x \, \partial_\mu \left( \frac{\partial \mathcal{L}}{\partial(\partial_\mu \phi)} \delta\phi \right) \,.
\end{aligned} \tag{19.3}$$

The last integral can be converted into a surface integral using Gauss's theorem. This is zero, because by assumption the variation $\delta\phi$ vanishes on the boundary of the integration region.

---

[1] An early investigation of field theories with higher derivatives can be found in the classic work by Pais and Uhlenbeck [87]. The corresponding quantized field theories often have problems with unitarity.

[2] Mathematical aspects of calculus of variations are presented clearly and comprehensively in [88].

**Euler-Lagrange Equation**

The Hamiltonian principle of least action $\delta S = 0$ leads to the equation of motion

$$\partial_\mu \frac{\partial \mathcal{L}}{\partial(\partial_\mu \phi)} - \frac{\partial \mathcal{L}}{\partial \phi} = 0 . \tag{19.4}$$

For a density $\mathcal{L}(\phi, \partial_\mu \phi)$, the equation of motion is a generalized wave equation of second order (for fermionic fields one finds first-order field equations).

A real scalar field, which describes an uncharged free particle with inverse Compton wavelength $\mu$, has the quadratic action

$$S_{\mathrm{M}}[\phi] = \int d^d x\, \mathcal{L}_{\mathrm{M}}(\phi) \quad \text{with} \quad \mathcal{L}_{\mathrm{M}}(\phi) = \frac{1}{2}\left(\partial_\mu \phi \partial^\mu \phi - \mu^2 \phi^2\right) . \tag{19.5}$$

The index on $\mathcal{L}_{\mathrm{M}}(\phi)$ is intended to indicate that the field $\phi$ describes matter. The associated Euler-Lagrange equation is the linear Klein-Gordon equation

$$\Box \phi + \mu^2 \phi = 0, \qquad \Box = \partial_\mu \partial^\mu . \tag{19.6}$$

The general solution is a superposition of plane waves $e^{ikx}$, where the components of the wave number vector satisfy the dispersion relation $k_\mu k^\mu = k_0^2 - \boldsymbol{k}^2 = \mu^2$.

The action should be invariant under Poincaré transformations (5.46) so that the equivalence principle is not violated. Under these transformations, a tensor field changes according to

$$\phi'(x') = D(\Lambda)\phi(x), \quad x' = \Lambda x + a \tag{19.7}$$

with invertible $D(\Lambda)$. For a scalar field, $D(\Lambda) = \mathbb{1}$ and accordingly $\phi'(x') = \phi(x)$.

**Task**

Show that the action $S_{\mathrm{M}}[\phi]$ in (19.5) is invariant under Poincaré transformations.

In contrast, the 4-potential of electrodynamics $A^\mu = (A^0, \boldsymbol{A})$ is a vector field with

$$A'_\mu(x') = \Lambda_\mu{}^\alpha A_\alpha(x) . \tag{19.8}$$

The target space is $\mathcal{T} = \mathbb{R}^4$ and $D(\Lambda) = \Lambda$. The field strength tensor $F_{\mu\nu} = \partial_\mu A_\nu - \partial_\nu A_\mu$ is a second-order anti-symmetric tensor field with

$$F'_{\mu\nu}(x') = \Lambda_\mu{}^\alpha \Lambda_\nu{}^\beta F_{\alpha\beta}(x)\,. \tag{19.9}$$

Its components are the components of the electric and magnetic fields,

$$\left(F_{\mu\nu}\right) = \begin{pmatrix} 0 & E_1 & E_2 & E_3 \\ -E_1 & 0 & -B_3 & B_2 \\ -E_2 & B_3 & 0 & -B_1 \\ -E_3 & -B_2 & B_1 & 0 \end{pmatrix} = (\boldsymbol{E}, \boldsymbol{B})\,. \tag{19.10}$$

More generally, a tensor field of rank $p$ transforms according to the tensor product representation,

$$\phi'_{\mu_1\ldots\mu_p}(x') = D(\Lambda)_{\mu_1\ldots\mu_p}^{\nu_1\ldots\nu_p}\, \phi_{\nu_1\ldots\nu_p}(x)\,, \quad \text{with} \quad D(\Lambda)_{\mu_1\ldots\mu_p}^{\nu_1\ldots\nu_p} = \Lambda_{\mu_1}^{\nu_1}\cdots\Lambda_{\mu_p}^{\nu_p}\,. \tag{19.11}$$

The transformations (19.7) define a group action (see Sect. 5.1) of the Poincaré group on the infinite-dimensional field space $\mathcal{F}$, provided the mapping $\Lambda \mapsto D(\Lambda)$ is a homomorphism of the Lorentz group,

$$D(\mathbb{1}) = \mathbb{1} \quad \text{and} \quad D(\Lambda'\Lambda) = D(\Lambda')D(\Lambda)\,. \tag{19.12}$$

However, this does not mean that the homomorphism is a representation of the Lorentz group. There exist interesting relativistic field theories with non-linear target space $\{\phi\}$, for which $D$ does not define a representation.

---

**Question**

Do you know of a field theory for which $\Lambda \mapsto D(\Lambda)$ is not a representation of the Lorentz group?

---

> **Matter Fields Transform with Representations of the Lorentz Group**
> If the target space is a vector space and $D$ is a linear mapping, then $\Lambda \mapsto D$ in (19.7) is a representation of the Lorentz group.

For a spinor field $D$ is a representation of the double covering $\mathrm{Spin}(1, d-1)$ of $\mathrm{SO}_0(1, d-1)$.

## 19.2  Hamiltonian Formalism

As in classical mechanics, in addition to the covariant Lagrange formalism, there is the Hamiltonian formulation. The transition between the two is accomplished with the help of a Legendre transformation, in which the velocity field $\dot{\phi} = \partial_t \phi$ is replaced by the canonically conjugate momentum field $\pi$. The latter is defined as

$$\pi(x) = \frac{\partial \mathcal{L}}{\partial(\partial_t \phi(x))}.\tag{19.13}$$

The canonically conjugate fields $\{\phi, \pi\}$ at a fixed time

$$\big(\phi(x), \pi(x)\big)\big|_{x^0} \equiv \big(\phi(\boldsymbol{x}), \pi(\boldsymbol{x})\big)\tag{19.14}$$

are coordinates in the phase space of the underlying field theory.[3] In contrast to the Lagrange formalism, the Hamilton formalism distinguishes a special time slice. It is therefore not covariant, as simultaneity is a property dependent on the inertial frame.

The *fundamental Poisson brackets* of the field and conjugate momentum field are

$$\{\phi(\boldsymbol{x}), \phi(\boldsymbol{y})\} = \{\pi(\boldsymbol{x}), \pi(\boldsymbol{y})\} = 0, \qquad \{\phi(\boldsymbol{x}), \pi(\boldsymbol{y})\} = \delta(\boldsymbol{x} - \boldsymbol{y}).\tag{19.15}$$

We would like to emphasize once again that the fields appear here at a fixed moment in time. More generally, one introduces the Poisson bracket of functions on phase space. Since the phase space itself is a function space, one speaks of functionals on phase space.

**Poisson Bracket of Functionals on Phase Space**
The Poisson bracket of two functionals $F[\phi, \pi]$ and $G[\phi, \pi]$ is

$$\{F, G\} = \int \mathrm{d}\boldsymbol{x} \left( \frac{\delta F}{\delta \phi(\boldsymbol{x})} \frac{\delta G}{\delta \pi(\boldsymbol{x})} - \frac{\delta F}{\delta \pi(\boldsymbol{x})} \frac{\delta G}{\delta \phi(\boldsymbol{x})} \right).\tag{19.16}$$

The Poisson bracket is obviously bilinear and antisymmetric. But it also satisfies the product rule (derivation rule)

$$\{F, GH\} = G\{F, H\} + \{F, G\}H \implies \{FG, H\} = F\{G, H\} + \{F, H\}G,\tag{19.17}$$

---

[3] Alternatively, one can choose the set of solutions $\phi(t, \boldsymbol{x})$ of the field equation as phase space.

and the Jacobi identity

$$\{F, \{G, H\}\} + \{H, \{F, G\}\} + \{G, \{H, F\}\} = 0 \,. \tag{19.18}$$

The proofs of these properties are the content of Problem 19.3. Alternative definitions of Poisson brackets are conceivable, and an example is given in Problem 19.4.

From the Definition (19.16) now follow for a phase space functional $F[\phi, \pi]$ the relations

$$\{\phi(\boldsymbol{x}), F\} = \frac{\delta F}{\delta \pi(\boldsymbol{x})}, \qquad \{\pi(\boldsymbol{x}), F\} = -\frac{\delta F}{\delta \phi(\boldsymbol{x})} \,. \tag{19.19}$$

With this, and with the identities

$$\frac{\delta \phi(\boldsymbol{y})}{\delta \phi(\boldsymbol{x})} = \delta(\boldsymbol{y} - \boldsymbol{x}), \qquad \frac{\delta \pi(\boldsymbol{y})}{\delta \pi(\boldsymbol{x})} = \delta(\boldsymbol{y} - \boldsymbol{x}) \,, \tag{19.20}$$

one finds the fundamental Poisson brackets (19.15) again. As known from mechanics, one now obtains the Hamilton function as the Legendre transform of the Lagrangian function,

$$H[\phi, \pi] = \int d\boldsymbol{x} \, \pi(\boldsymbol{x})\dot{\phi}(\boldsymbol{x}) - L, \qquad L[\phi, \dot{\phi}] = \int d\boldsymbol{x} \, \mathcal{L}(\dot{\phi}, \nabla\phi, \phi) \,. \tag{19.21}$$

In this, the velocity field $\dot{\phi}$ is expressed by the momentum field $\pi$ using (19.13). The field equations are the Hamiltonian equations of motion

$$\dot{\phi}(\boldsymbol{x}) = \{\phi(\boldsymbol{x}), H\} = \frac{\delta H}{\delta \pi(\boldsymbol{x})} \,, \tag{19.22}$$

$$\dot{\pi}(\boldsymbol{x}) = \{\pi(\boldsymbol{x}), H\} = -\frac{\delta H}{\delta \phi(\boldsymbol{x})} \,. \tag{19.23}$$

In contrast to the Euler-Lagrange equations, these are first order differential equations in time.

### Example: Hamilton Function for the Free Real Scalar Field

The Lagrangian function for the real scalar field with Lagrange density (19.5) is

$$\mathcal{L}_{\mathrm{M}}(\phi) = \frac{1}{2} \int d\boldsymbol{x} \left( \dot{\phi}^2 - (\nabla\phi)^2 - \mu^2\phi^2 \right) \,, \tag{19.24}$$

so that $\pi = \dot{\phi}$. If one substitutes $\dot{\phi} = \pi$ into (19.21), then one finds the Hamilton function

$$H = \frac{1}{2} \int d\boldsymbol{x} \left( \pi^2 + (\nabla\phi)^2 + \mu^2\phi^2 \right) . \tag{19.25}$$

It is a quadratic functional on phase space. The Hamiltonian equations of motion are

$$\dot{\phi}(\boldsymbol{x}) = \pi(\boldsymbol{x}) \quad , \quad \dot{\pi}(\boldsymbol{x}) = \Delta\phi(\boldsymbol{x}) - \mu^2\phi(\boldsymbol{x}) . \tag{19.26}$$

They again yield the Klein-Gordon equation:

$$\ddot{\phi}(\boldsymbol{x}) = \dot{\pi}(\boldsymbol{x}) = \Delta\phi(\boldsymbol{x}) - \mu^2\phi(\boldsymbol{x}) \implies \Box\phi + \mu^2\phi = 0 . \tag{19.27}$$

◀

After this introduction to basic properties of classical field theories, we turn to their possible symmetries.

## 19.3 Noether's Theorem for Internal Symmetries

Translations, rotations and Lorentz boosts are space-time symmetries which also act on the argument $x$ of fields. In addition to these, there are the so-called internal symmetries, which leave $x$ fixed. For example, the Lagrange density for a complex scalar field

$$\mathcal{L}_M(\phi) = \partial_\mu\phi^* \partial^\mu\phi - V(\phi^*\phi) \tag{19.28}$$

is invariant under phase transformations,

$$\phi(x) \mapsto e^{i\alpha}\phi(x), \quad \phi^*(x) \mapsto e^{-i\alpha}\phi^*(x) \tag{19.29}$$

with real and constant phase $\alpha$.

More generally, we consider fields with values in a vector space $\mathcal{V}$ with scalar product. This means that for each $x$ the value $\phi(x)$ lies in $\mathcal{V}$ and the scalar product $\big(\phi(x), \phi(x)\big)$ does not change, when $\phi(x)$ is transformed with a unitary matrix,

$$\big(U\phi(x), U\phi(x)\big) = \big(\phi(x), \phi(x)\big) . \tag{19.30}$$

We first assume that $U$ is a global transformation, which means that $U$ does not depend on $x$. Then $U\partial_\mu\phi = \partial_\mu(U\phi)$, and the Lagrange density

$$\mathcal{L}_M(\phi) = (\partial_\mu\phi, \partial^\mu\phi) - V\big((\phi, \phi)\big) \tag{19.31}$$

is invariant under the global transformations

$$\phi(x) \longmapsto U\phi(x). \tag{19.32}$$

A unitary matrix has the representation $U = \exp(\mathrm{i}X)$ with a Hermitian matrix $X$. The infinitesimal symmetries are therefore

$$\phi \longmapsto U\phi \approx \phi + \mathrm{i}X\phi \equiv \phi + \delta_X\phi, \qquad X^\dagger = X. \tag{19.33}$$

According to the (first) theorem of Emmy Noether, every real parameter of a continuous symmetry group leads to a covariantly conserved current and thus to a conserved Noether charge. We will now prove this for internal symmetries.

If a Lagrange density does not change under transformations (19.33), then

$$0 = \delta_X \mathcal{L} = \frac{\partial \mathcal{L}}{\partial(\partial_\mu\phi)}\, \partial_\mu\,(\delta_X\phi) + \frac{\partial \mathcal{L}}{\partial\phi}\,\delta_X\phi. \tag{19.34}$$

Using the Euler-Lagrange equations in the form

$$\frac{\partial \mathcal{L}}{\partial\phi} = \partial_\mu\left(\frac{\partial \mathcal{L}}{\partial(\partial_\mu\phi)}\right), \tag{19.35}$$

in the last term, we immediately obtain the covariantly conserved Noether current.

**Noether Current for Internal Symmetries**
If the Lagrange density $\mathcal{L}$ is invariant under the transformations (19.33), then

$$\partial_\mu J_X^\mu = \partial_0 J_X^0 + \operatorname{div} \boldsymbol{J}_X = 0, \qquad J_X^\mu = \frac{\partial \mathcal{L}}{\partial(\partial_\mu\phi)}\delta_X\phi. \tag{19.36}$$

From this continuity equation, the conservation of the associated Noether charge immediately follows. To this end, we integrate the continuity equation over space at a fixed time and convert the volume integral into a surface integral,

$$0 = \int \mathrm{d}\boldsymbol{x}\,\partial_\mu J_X^\mu = \frac{\partial}{\partial x^0}\int \mathrm{d}\boldsymbol{x}\,J_X^0 + \int \mathrm{d}\boldsymbol{x}\,\operatorname{div}\boldsymbol{J}_X = \frac{\partial}{\partial x^0}\int \mathrm{d}\boldsymbol{x}\,J_X^0 + \oint \mathrm{d}\boldsymbol{f}\,\boldsymbol{J}_X. \tag{19.37}$$

If the spatial current density $\boldsymbol{J}_X(\boldsymbol{x})$ vanishes at spatial infinity quickly enough (which we want to assume), then the conservation of the Noether charge follows,

$$\frac{\mathrm{d}}{\mathrm{d}t}Q_X = 0, \quad Q_X = \int_{x^0} \mathrm{d}\boldsymbol{x}\,J_X^0(x) = \int_{x^0} \mathrm{d}\boldsymbol{x}\,\pi(x)\,\delta_X\phi(x). \tag{19.38}$$

In the last step, we made use of the fact that the derivative of $\mathcal{L}$ with respect to $\partial_0\phi$ in (19.36) is equal to the momentum field $\pi$ canonically conjugated to $\phi$.

We have seen that for every generator of a symmetry $X$, there is a time-conserved Noether charge. The number of linearly independent conserved charges is thus equal to the dimension of the continuous symmetry group. The question now arises as to how these charges act on the fields of the theory. With the help of the product rule and the fundamental Poisson brackets, one finds

$$\{\phi(\boldsymbol{x}),\, Q_X\} = \int \mathrm{d}\boldsymbol{y}\, \{\phi(\boldsymbol{x}),\, \pi(\boldsymbol{y})\, \delta_X\phi(\boldsymbol{y})\} = \delta_X\phi(\boldsymbol{x})$$

$$\{\pi(\boldsymbol{x}),\, Q_X\} = \int \mathrm{d}\boldsymbol{y}\, \{\pi(\boldsymbol{x}),\, \pi(\boldsymbol{y})\, \delta_X\phi(\boldsymbol{y})\} = \delta_X\pi(\boldsymbol{x})\,. \tag{19.39}$$

It was assumed here that $\delta_X\phi$ does not contain time derivatives of the field. On the right-hand sides are the phase space variables transformed with the infinitesimal symmetries.

**Noether Charges Generate Symmetries**
A conserved Noether charge $Q_X$ generates via the Poisson bracket the infinitesimal symmetry transformation $\phi \mapsto \phi + \delta_X\phi$, from which it originally arose.

To illustrate these results we return to the complex scalar field with Lagrange density (19.28).

**Example: Global U$(1)$ Gauge Transformations**

The infinitesimal form of the phase transformation (19.29) is

$$\delta_\alpha\phi = \mathrm{i}\alpha\phi, \quad \delta_\alpha\phi^* = -\mathrm{i}\alpha\phi^*, \tag{19.40}$$

and the corresponding Noether currents are

$$J_\alpha^\mu = \alpha J^\mu, \qquad J_\mu = \mathrm{i}\big(\partial_\mu\phi^\dagger\phi - \phi^\dagger\partial_\mu\phi\big). \tag{19.41}$$

Using the field equation $\Box\phi + V'(\phi^*\phi)\phi$, where the prime denotes the derivative with respect to the argument, one quickly verifies that $J^\mu$ is covariantly conserved. The conserved Noether charge is the electric charge carried by the field,

$$Q = \mathrm{i}\int_{x^0} \mathrm{d}\boldsymbol{x}\,\big(\pi_\phi\phi - \pi_{\phi^*}\phi^*\big)\,. \tag{19.42}$$

This generates the phase transformations $\{\phi,\, Q\} = \mathrm{i}\phi$ and $\{\phi^*,\, Q\} = -\mathrm{i}\phi^*$.  ◀

### 19.3.1 Phase Transformations for Dirac Spinor Fields

A charged fermion with spin $\frac{1}{2}$ is described by a Dirac spinor field $\psi(x)$. Without interaction this satisfies the free Dirac equation (18.61) which is the Euler-Lagrange equation for the action $S_{\mathrm{M}}[\psi] = \int \mathrm{d}^d x \, \mathcal{L}_{\mathrm{M}}(\psi)$ with Lagrange density

$$\mathcal{L}_{\mathrm{M}}(\psi) = \bar{\psi}(\mathrm{i}\slashed{\partial} - m)\psi, \qquad \slashed{\partial} = \gamma^\mu \partial_\mu \, . \tag{19.43}$$

Here, $\bar{\psi} = \psi^\dagger \gamma^0$ is the Dirac-conjugate spinor field.

**Question**

Why does $\mathcal{L}_{\mathrm{M}}$ transform like a scalar field, $\mathcal{L}'_{\mathrm{M}}(x') = \mathcal{L}_{\mathrm{M}}(x)$, under Lorentz transformations?

The Lagrange density for the spinor field is invariant under global U(1) phase transformations

$$\psi(x) \longrightarrow \mathrm{e}^{\mathrm{i}\lambda}\psi(x), \quad \bar{\psi}(x) \longrightarrow \mathrm{e}^{-\mathrm{i}\lambda}\bar{\psi}(x) \, , \tag{19.44}$$

and this internal symmetry leads to the covariantly conserved current density

$$\partial_\mu j^\mu(x) = 0, \quad j^\mu(x) = \bar{\psi}(x)\gamma^\mu \psi(x) \, , \tag{19.45}$$

which is identified with the electric current density in spacetime. The corresponding conserved Noether charge

$$Q = \int \mathrm{d}\boldsymbol{x} \, \psi^\dagger \psi \, , \tag{19.46}$$

is the electric charge of the system described by $\psi$.

## 19.4  Noether's Theorem for Translations

Under spacetime translations, a field transforms as follows:

$$\phi(x) \mapsto \phi(x - a) = \phi(x) - a^\mu \partial_\mu \phi(x) + \mathcal{O}(a^2) = \phi(x) + \delta_a \phi(x) + \mathcal{O}(a^2) \, . \tag{19.47}$$

The Lagrange density is a scalar field and therefore transforms just like $\phi$ under infinitesimal translations,

$$\delta_a \mathcal{L} = \partial_\mu V_a^\mu, \qquad V_a^\mu = -a^\mu \mathcal{L} \, . \tag{19.48}$$

Translations are elements of the Poincaré group, and in a relativistic theory, the action must be invariant under translations. This is the case because the change in the Lagrange density $\delta_a \mathcal{L}$ is a total divergence which integrates to zero in the spacetime integral in (19.1).

We can also determine the change of $\mathcal{L}$ for small translations in another way, since we know the change of its arguments,

$$\delta_a \mathcal{L} = \frac{\partial \mathcal{L}}{\partial(\partial_\mu \phi)} \, \partial_\mu\big(\delta_a \phi\big) + \frac{\partial \mathcal{L}}{\partial \phi}\, \delta_a \phi = \partial_\mu\left( \frac{\partial \mathcal{L}}{\partial(\partial_\mu \phi)}\, \delta_a \phi \right). \qquad (19.49)$$

In the last step, we used the field equation (19.35), to eliminate $\partial \mathcal{L}/\partial \phi$. If we equate the right-hand sides of (19.48) and (19.49), this leads to the celebrated theorem of Noether. Only property (19.48) and the Euler-Lagrange equations were used in the derivation. Therefore, the theorem applies not only to translations, but to all spacetime symmetries for which $\delta_a \mathcal{L}$ is a total divergence and thus the action is invariant.

**Noether's Theorem for Spacetime Symmetries**
For all spacetime symmetries, $\delta \mathcal{L} = \partial_\mu V^\mu$ is a divergence and the Noether theorem holds

$$\partial_\mu J^\mu = 0 \quad \text{with} \quad J^\mu = \frac{\partial \mathcal{L}}{\partial(\partial_\mu \phi)}\, \delta\phi - V^\mu. \qquad (19.50)$$

The theorem applies to translations and Lorentz transformations.

Since $\mathcal{L}$ is preserved under infinitesimal spacetime symmetries only up to a total divergence, the corresponding Noether currents, compared to those of an internal symmetry in (19.36), have the additional term $-V^\mu$.

In particular, for translations, $\delta\phi = \delta_a \phi$, as given in (19.47), and correspondingly, $V^\mu = V_a^\mu$ in (19.48). Then the Noether theorem reads

$$\partial_\mu J_a^\mu = 0, \qquad J_a^\mu = -a^\nu \Theta^\mu{}_\nu, \qquad \Theta_{\mu\nu} = \frac{\partial \mathcal{L}}{\partial(\partial^\mu \phi)}\, \partial_\nu \phi - \eta_{\mu\nu} \mathcal{L}. \qquad (19.51)$$

The translations form a $d$-parameter symmetry group, and therefore there exist $d$ covariantly conserved Noether currents, encoded in the *canonical energy-momentum tensor* $\Theta^\mu{}_\nu$.

**Task**

Prove directly that the canonical energy-momentum tensor is covariantly conserved, $\partial_\mu \Theta^\mu{}_\nu = 0$. In the proof, you need the field equations.

The Noether charges are the energy $H = P^0$ and the momentum $\boldsymbol{P}$ stored in the field,

$$(P^\mu) = \begin{pmatrix} P^0 \\ \boldsymbol{P} \end{pmatrix}, \qquad P^\mu = \int_{x^0} \mathrm{d}\boldsymbol{x}\, \Theta^{0\mu}, \qquad \dot{P}^\mu = 0. \tag{19.52}$$

For illustration, we consider a real scalar field.

### Example: Energy and Momentum of the Real Scalar Field

The scalar field has the Lagrange density and the symmetric energy-momentum tensor

$$\mathcal{L}(\phi) = \frac{1}{2}\partial_\mu\phi\partial^\mu\phi - V(\phi) \quad \text{and} \quad \Theta_{\mu\nu} = \partial_\mu\phi\partial_\nu\phi - \eta_{\mu\nu}\mathcal{L}_\mathrm{M}(\phi). \tag{19.53}$$

With $\dot{\phi} = \pi$, energy and momentum are given by the integrals

$$H = \int_{x^0} \mathrm{d}\boldsymbol{x}\, \mathcal{H}(\boldsymbol{x}) \quad , \quad \boldsymbol{P} = \int_{x^0} \mathrm{d}\boldsymbol{x}\, \mathcal{P}(\boldsymbol{x}) \tag{19.54}$$

with the Hamiltonian density and the momentum density as integrands,

$$\mathcal{H} = \frac{1}{2}\pi^2 + \frac{1}{2}(\nabla\phi)^2 + V(\phi) \quad \text{and} \quad \mathcal{P} = \pi\,\nabla\phi. \tag{19.55}$$

Conversely, the Noether charges (19.52) generate the infinitesimal translations,

$$\{P_\mu, \phi\} = -\partial_\mu\phi. \tag{19.56}$$

For $\mu = 0$, we recognize the Hamilton equation of motion (19.22). ◀

If we contract the relation (19.56) with the constant vector $a^\mu$, we get

$$\delta_a\phi \equiv -a^\mu\partial_\mu\phi(x) = \{aP, \phi\}, \quad aP = a^\mu P_\mu. \tag{19.57}$$

As expected, energy and momentum generate the infinitesimal translations. To integrate these infinitesimal translations into finite ones, we first assign a linear operator $\mathrm{ad}_F$ to a function $F$ in phase space, defined by

$$\mathrm{ad}_F G = \{F, G\}, \tag{19.58}$$

such that the infinitesimal translation (19.57) takes the form

$$\delta_a\phi = \mathrm{ad}_{aP}\,\phi. \tag{19.59}$$

For smooth fields, a finite displacement of the arguments is given by the Taylor series,

$$\phi(x - a) = \phi(x) - a^\mu \partial_\mu \phi(x) + \frac{1}{2}(a^\mu \partial_\mu)^2 \phi(x) + \cdots = e^{-a^\mu \partial_\mu}\phi(x)\,, \qquad (19.60)$$

i.e., by repeatedly applying $\mathrm{ad}_{aP}$,

$$\phi(x - a) = \left(e^{\mathrm{ad}_{aP}}\,\phi\right)(x)\,. \qquad (19.61)$$

While $\mathrm{ad}_{aP}$ generates an infinitesimal translation, $\exp(\mathrm{ad}_{aP})$ maps $\phi(x)$ to $\phi(x - a)$.

---

**Task**

Prove the identities $\mathrm{ad}_{\alpha F + \beta G} = \alpha\,\mathrm{ad}_F + \beta\,\mathrm{ad}_G$ and $\mathrm{ad}_{\{F,G\}} = \mathrm{ad}_F\,\mathrm{ad}_G - \mathrm{ad}_G\,\mathrm{ad}_F$.

### 19.4.1 Energy-Momentum of the Electromagnetic Field

Together, the electric and magnetic field form the field strength tensor $(F_{\mu\nu}) = (E, B)$ in (19.10). Raising the indices is equivalent to the sign change of the electric field, $(F^{\mu\nu}) = (-E, B)$. The Lagrange density of electrodynamics is

$$\mathcal{L}_{\mathrm{ED}} = \frac{1}{2}\left(E^2 - B^2\right) = -\frac{1}{4}F^{\mu\nu}F_{\mu\nu} = -\frac{1}{4}F^{\mu\nu}\left(\partial_\mu A_\nu - \partial_\nu A_\mu\right)\,, \qquad (19.62)$$

and from the last representation it directly follows

$$\frac{\partial \mathcal{L}_{\mathrm{ED}}}{\partial(\partial_\mu A_\rho)} = -F^{\mu\rho}\,. \qquad (19.63)$$

This result is needed for the calculation of the energy-momentum tensor (19.51),

$$\Theta_{\mu\nu} = -F_\mu{}^\rho \partial_\nu A_\rho + \frac{1}{4}\eta_{\mu\nu}F^{\rho\sigma}F_{\rho\sigma}\,. \qquad (19.64)$$

The Noether charges energy and momentum of the electromagnetic field have the form

$$H = \int_{x^0} d\boldsymbol{x}\left(E_i \dot{A}_i - \mathcal{L}_{\mathrm{ED}}\right) \quad \text{and} \quad P_i = \int_{x^0} d\boldsymbol{x}\left(E_j \partial_i A_j\right)\,. \qquad (19.65)$$

Although we were able to construct a covariantly conserved energy-momentum tensor, an improvement of $\Theta_{\mu\nu}$ is necessary. Energy-momentum tensors are not unique and can be modified without changing the conserved charges $(H, \boldsymbol{P})$. A compelling reason for improvement is that $\Theta_{\mu\nu}$ appears in Einstein's field equations as the source of the gravitational field, and it is assumed that the tensor is covariantly conserved and symmetric [89]. The canonical tensor (19.64) obtained via the Noether theorem is covariantly conserved, but not symmetric. Moreover, it is not invariant under local gauge transformations (these are discussed in detail in Chap. 20)

$$A_\mu \mapsto A_\mu - \partial_\mu \lambda \,. \tag{19.66}$$

So it depends on the freely selectable gauge, i.e., it takes different values for physically equivalent gauge potentials and therefore cannot possibly be an observable quantity.

Finally, it is noted that $\mathcal{L}_{\mathrm{ED}}$ is independent of $\dot{A}_0$ and therefore the momentum field conjugate to $A_0$ vanishes. Systems with this property are called singular systems. Their singular behavior stems from the invariance of the Lagrange density under local gauge transformations (19.66). Singular systems are systems with constraints, the treatment of which is somewhat cumbersome in the canonical formulation. I refer to the extensive literature, for example the review [90].

### 19.4.2 Improvement of Noether Currents

Only for simple scalar fields is the canonical energy-momentum tensor symmetric. But one can improve a non-symmetric tensor with the help of *Belinfante's symmetrization* [91]. This allows us to construct a covariantly conserved and symmetric tensor that can be inserted into the Einstein field equations.

We again assume that the Lagrange density is invariant under a symmetry transformation $\phi \to \phi + \delta\phi$ up to a total derivative,

$$\delta\mathcal{L} = \partial_\mu V^\mu \,. \tag{19.67}$$

This is the case for all symmetries and in particular space-time symmetries, including a possible conformal symmetry (see Chap. 21) or a supersymmetry.

In (19.50) we had proven that the divergence-free Noether current density has the following form,

$$J^\mu = \frac{\partial \mathcal{L}}{\partial(\partial_\mu \phi)}\, \delta\phi - V^\mu \,. \tag{19.68}$$

To improve $J^\mu$, we note that $V^\mu$ in (19.67) is only determined up to a term of the form $\partial_\nu A^{\mu\nu}$ with anti-symmetric $A^{\mu\nu}$. Therefore, the Noether current is not unique,

$$J^\mu \longmapsto J^\mu - \partial_\rho A^{\mu\rho}, \quad A^{\mu\rho} = -A^{\rho\mu} \,. \tag{19.69}$$

For spatially localized fields, the Noether charge does not change with this modification,

$$J^0 \mapsto J^0 - \partial_i A^{0i}, \quad \text{so that} \quad Q \mapsto Q.$$

Such improvements have been extensively discussed in the literature [78, 83].

In electrodynamics, we can add the correction term $\partial_\rho (F^{\mu\rho} A^\nu)$ to the tensor $\Theta^{\mu\nu}$ in (19.64). The improved tensor is covariantly conserved, symmetric, and gauge invariant,

$$T_{\mu\nu} = F_\mu{}^\rho F_{\rho\nu} + \frac{1}{4}\eta_{\mu\nu} F^{\rho\sigma} F_{\rho\sigma}.　\tag{19.70}$$

The associated conserved charges are well known from electrodynamics.

**Conserved Energy and Conserved Momentum of $E$ and $B$**
An electromagnetic field carries the following energy and momentum:

$$H = \frac{1}{2}\int_{x^0} d\boldsymbol{x} \left(\boldsymbol{E}^2 + \boldsymbol{B}^2\right) \quad \text{and} \quad \boldsymbol{P} = \int_{x^0} d\boldsymbol{x}\, \boldsymbol{E} \wedge \boldsymbol{B}.　\tag{19.71}$$

The "correct" tensor (19.70) is also obtained if the Maxwell field is minimally coupled to the gravitational field, the corresponding action is varied with respect to the metric, and finally it is assumed that the metric is that of Minkowski space. This is carried out in Sect. 21.2.4.

### Example: Dirac Spinor Field

The free Dirac field has the Lagrange density (19.43), which is equivalent to

$$\mathcal{L}_{\mathrm{M}}(\psi) = \tfrac{i}{2}\left(\bar{\psi}\gamma^\mu \partial_\mu \psi - (\partial_\mu \bar{\psi})\gamma^\mu \psi\right) - m\bar{\psi}\psi.　\tag{19.72}$$

The associated Noether current density is the canonical energy-momentum tensor,

$$\Theta^{\mu\nu} = \frac{i}{2}\left(\bar{\psi}\gamma^\mu \partial^\nu \psi - (\partial^\nu \bar{\psi})\gamma^\mu \psi\right) - \eta^{\mu\nu}\mathcal{L}_{\mathrm{M}}(\psi).　\tag{19.73}$$

The energy stored in the field $H = P^0$ is equal to

$$H = \int d\boldsymbol{x}\, \psi^\dagger h \psi, \quad h = \alpha^i \partial_i + \beta m.　\tag{19.74}$$

◀

The canonical tensor $\Theta^{\mu\nu}$ is not symmetric and must be improved, similar to $\Theta^{\mu\nu}$ in electrodynamics. The starting point for the improvement are the relations

$$\{\gamma^{\mu\rho}, \gamma^{\nu}\} = -2\gamma^{\rho}\gamma^{\mu\nu} + 2\eta^{\mu\rho}\gamma^{\nu} - 2\eta^{\nu\rho}\gamma^{\mu} = -2\gamma^{\mu\nu}\gamma^{\rho} - 2\eta^{\mu\rho}\gamma^{\nu} + 2\eta^{\nu\rho}\gamma^{\mu},$$
$$(19.75)$$

which follow from $\{\gamma^{\mu}, \gamma^{\nu}\} = 2\eta^{\mu\nu}$. With the help of the Dirac equation for $\psi$ in (18.61) and that for $\bar{\psi}$ in (18.62), the tensor field $A^{\mu\rho\nu} = \frac{i}{8}\bar{\psi}\{\gamma^{\mu\rho}, \gamma^{\nu}\}\psi = -A^{\rho\mu\nu}$ now obeys the identity

$$\partial_{\rho}A^{\mu\rho\nu} = \frac{i}{4}\big((\partial^{\mu}\bar{\psi})\gamma^{\nu}\psi - (\partial^{\nu}\bar{\psi})\gamma^{\mu}\psi + \bar{\psi}\gamma^{\mu}\partial^{\nu}\psi - \bar{\psi}\gamma^{\nu}\partial^{\mu}\psi\big).\qquad(19.76)$$

If we subtract $\partial_{\rho}A^{\mu\rho\nu}$ from $\Theta^{\mu\nu}$ in (19.73), we find the improved tensor $T^{\mu\nu}$, which leads to the same conserved quantities as $\Theta^{\mu\nu}$.

---

**Improved Energy-Momentum Tensor for the Dirac Spinor Field**
A symmetric and covariantly conserved energy-momentum tensor of the Dirac theory is

$$T^{\mu\nu} = \frac{i}{4}\big(\bar{\psi}\gamma^{\mu}\partial^{\nu}\psi + \bar{\psi}\gamma^{\mu}\partial^{\nu}\psi - (\partial^{\nu}\bar{\psi})\gamma^{\mu}\psi$$
$$- (\partial^{\mu}\bar{\psi})\gamma^{\nu}\psi\big) - \eta^{\mu\nu}\mathcal{L}_{\mathrm{M}}(\psi),\qquad(19.77)$$

---

For solutions of the Dirac equation, $\mathcal{L}_{\mathrm{M}}(\psi) = 0$, and therefore in the literature one also finds the expressions (19.73) and (19.77) without the last term proportional to $\mathcal{L}_{\mathrm{M}}$. Then one ends up with an only seemingly different expression for the energy $H$.

---

**Task**

Show with the help of the Dirac equation and partial integration in $x$, that

$$H = i\int d\boldsymbol{x}\,\psi^{\dagger}\dot{\psi} = \frac{i}{2}\int d\boldsymbol{x}\big(\psi^{\dagger}\dot{\psi} - \dot{\psi}^{\dagger}\psi\big).\qquad(19.78)$$

---

## 19.5   Lorentz Transformations and Angular Momentum

After applying the Noether theorem to translations in space and time, which led to the components of $\Theta^{\mu\nu}$, we will now determine the Noether currents for the Lorentz symmetry. Since the latter includes rotations in space as a subgroup, the "spatial Noether charges" will be identified as components of the angular momentum.

In Sect. 19.1 we argued that in a relativistic field theory a matter field transforms under Poincaré-transformations according to $\phi'(x') = D(\Lambda)\phi(x)$. As earlier we assume that the target space $\mathcal{T}$ is linear, in which case $\Lambda \mapsto D(\Lambda)$ is a representation. The induced representation $\mathcal{D}$ (often denoted by $D_*$) of the Lorentz algebra $\mathfrak{so}(1, d-1)$ maps the base elements $M_{\mu\nu}$ in (14.77)) into a basis for the representation matrices,

$$\mathcal{D}(M_{\mu\nu}) = S_{\mu\nu}, \quad S_{\mu\nu} = -S_{\nu\mu}. \tag{19.79}$$

In particular, the $S_{\mu\nu}$ obey the same commutation rules (14.79) as the matrices $M_{\mu\nu}$. In this form, the result also applies to spinor fields which transform according to a representation $D$ of the spin group $\mathrm{Spin}(1, d-1)$, as detailed in Sect. 18.3.3. Then the spin matrices $S_{\mu\nu}$ are elements of an induced representation $\mathcal{D}$ of the Lie algebra $\mathfrak{spin}(1, d-1)$.

Now we proceed exactly as in Sect. 18.3.3, to determine the infinitesimal form of

$$\phi(x) \longrightarrow e^{\frac{i}{2}\omega^{\mu\nu}S_{\mu\nu}}\phi\left(e^{-\omega}x\right). \tag{19.80}$$

The small change $\delta_\omega\phi$ in $\phi \to \phi + \delta_\omega\phi$ is given by

$$\delta_\omega\phi = \frac{i}{2}\omega^{\mu\nu}\left(L_{\mu\nu} + S_{\mu\nu}\right), \quad L_{\mu\nu} = \frac{1}{i}(x_\mu\partial_\nu - x_\nu\partial_\mu). \tag{19.81}$$

We note this important result:

**Infinitesimal Lorentz Transformations**
The infinitesimal generators $J_{\mu\nu}$ of the Lorentz transformations in

$$\delta_\omega\phi = \frac{i}{2}\omega^{\mu\nu}J_{\mu\nu}\phi \equiv \frac{i}{2}(\omega, J)\phi \tag{19.82}$$

are the sum of an orbital and a spin part, $J_{\mu\nu} = L_{\mu\nu} + S_{\mu\nu}$.

Since $[L_{\mu\nu}, S_{\alpha\beta}] = 0$, the $J_{\mu\nu}$ satisfy the same Lorentz algebra as the $L_{\mu\nu}$ and $S_{\mu\nu}$. A scalar field describes particles without spin, and the spin part is indeed missing, $\delta_\omega\phi = \frac{i}{2}(\omega, L)\phi$.

---

**Task**

Confirm with an explicit calculation, that the differential operators $L_{\mu\nu}$ satisfy the commutation rules of the Lorentz algebra (14.79).

The results are valid in any dimension. In 4 dimensions, the $J_{ij}$, $L_{ij}$ and $S_{ij}$ with only spatial indices form the Lie subalgebra $\mathfrak{so}(3)$ of infinitesimal rotations in space. This becomes clear when one introduces the components

$$J_i = -\frac{1}{2}\epsilon_{ijk} J^{jk}, \quad \epsilon_{123} = 1, \tag{19.83}$$

which satisfy the known commutation relations (15.2).

## 19.5.1 Noether Charges for Lorentz Symmetry

As with translations, the variation of the Lagrange density for infinitesimal Lorentz transformations is a total divergence. Since $\mathcal{L}$ transforms like a scalar field, we have

$$\delta_\omega \mathcal{L} = \frac{i}{2}\omega^{\mu\nu} L_{\mu\nu}\mathcal{L} = \partial_\mu V_\omega^\mu, \quad \text{with} \quad V_\omega^\mu = -\omega^{\mu\rho} x_\rho \mathcal{L}. \tag{19.84}$$

If we substitute for the variation of the field $\delta_\omega \phi$ in the covariantly conserved Noether current

$$J_\omega^\mu = \frac{\partial \mathcal{L}}{\delta(\partial_\mu \phi)}\delta_\omega \phi - V_\omega^\mu \tag{19.85}$$

the result (19.82), then we find the simple form

$$\frac{\delta \mathcal{L}}{\partial(\partial_\mu \phi)}\delta_\omega \phi = \frac{i}{2}\frac{\partial \mathcal{L}}{\partial(\partial_\mu \phi)}\omega^{\rho\sigma} J_{\rho\sigma}\phi. \tag{19.86}$$

The subtraction of $V_\omega^\mu$ then leads to the Noether current

$$J_\omega^\mu = \frac{\omega^{\rho\sigma}}{2}\left(i\frac{\partial \mathcal{L}}{\partial(\partial_\mu \phi)}J_{\rho\sigma} + (\delta_\rho^\mu x_\sigma - \delta_\sigma^\mu x_\rho)\mathcal{L}\right). \tag{19.87}$$

Herein, the spin-independent terms can be expressed through the canonical energy-momentum tensor (19.51)—they are equal to $x_\rho \Theta^\mu{}_\sigma - x_\sigma \Theta^\mu{}_\rho$. For each real parameter $\omega^{\mu\nu}$ with $\mu > \nu$ we obtain a covariantly conserved current.

**Noether Currents of Lorentz Symmetry**
The Noether currents for Lorentz invariance are the $\frac{1}{2}d(d-1)$ currents $J^{\mu\rho\sigma} = -J^{\mu\sigma\rho}$ in

$$J_\omega^\mu = \frac{1}{2}\omega_{\rho\sigma} J^{\mu\rho\sigma}, \quad J^{\mu\rho\sigma} = x^\rho \Theta^{\mu\sigma} - x^\sigma \Theta^{\mu\rho} + i\frac{\partial \mathcal{L}}{\partial(\partial_\mu \phi)}S^{\rho\sigma}\phi.$$

$$\tag{19.88}$$

The corresponding conserved Noether charges are

$$Q^{\rho\sigma} = -Q^{\sigma\rho} = \int_t \mathrm{d}\boldsymbol{x}\, J^{0\rho\sigma}\,. \tag{19.89}$$

The space-time or space-space components have the explicit form

$$Q^{i0} = \int_{x^0} \mathrm{d}\boldsymbol{x}\, \left( x^i \mathcal{H} - \pi x^0 \partial^i \phi - \mathrm{i}\pi\, S^{0i}\phi \right),$$

$$Q^{ij} = \int_{x^0} \mathrm{d}\boldsymbol{x}\, \pi \left( x^i \partial^j \phi - x^j \partial^i \phi + \mathrm{i} S^{ij}\phi \right). \tag{19.90}$$

---

### Question

Why is $S^{\mu\nu} = M^{\mu\nu}$ for a vector field? What would $S^{\mu\nu}$ be for a tensor field $\phi_{\mu\nu}$?

A calculation shows (see Problem 19.7), that these charges generate those Lorentz transformations from which they originated,

$$\{\phi, Q^{\mu\nu}\} = \mathrm{i}\, J^{\mu\nu}\phi\,. \tag{19.91}$$

This is the general result (19.39), applied to Lorentz transformations.

---

### Example: Electrodynamics

In electrodynamics, $S^{\mu\nu} = M^{\mu\nu}$ and the Noether currents have the form

$$J^{\mu\rho\sigma} = \eta^{\mu\rho} x^\sigma \mathcal{L}_{\mathrm{ED}} + \frac{1}{2} F^{\mu\alpha} x^\sigma \partial^\rho A_\alpha + \frac{1}{2} F^{\mu\sigma} A^\rho - \left(\rho \leftrightarrow \sigma\right). \tag{19.92}$$

Similar to $\Theta^{\mu\nu}$, these are not gauge invariant and need to be improved. To do this, we replace the $J^{\mu\rho\sigma}$ with the improved currents

$$\mathcal{J}^{\mu\rho\sigma} = J^{\mu\rho\sigma} + \partial_\alpha A^{\mu\alpha\rho\sigma}, \qquad A^{\mu\alpha\rho\sigma} = \frac{1}{2} F^{\mu\alpha} \left( x^\rho A^\sigma - x^\sigma A^\rho \right) = -A^{\alpha\mu\rho\sigma}\,. \tag{19.93}$$

Because $A^{\mu\alpha\rho\sigma}$ is anti-symmetric in $(\mu, \alpha)$, the $\mathcal{J}^{\mu\rho\sigma}$ are also covariantly conserved, and because we added a total divergence, the conserved charges do not change. The improved currents can now be expressed through the improved and gauge invariant energy-momentum tensor in (19.71),

$$\mathcal{J}^{\mu\rho\sigma} = x^\rho T^{\mu\sigma} - x^\sigma T^{\mu\rho}\,. \tag{19.94}$$

◄

---



Prove the simple formula (19.94). The starting point could be the result (19.92).

---

## 19.6   Symmetries in Quantum Field Theories

For bosonic fields, which transform according to representations of the Lorentz group, the transition from a classical field theory to the corresponding quantum field theory (QFT) is done using the correspondence rule: according to this, the classical field $\phi(x)$ and its canonically conjugated momentum field $\pi(x)$ at the spatial point with coordinate $x$ are assigned the field operators $\hat{\phi}(x)$ and $\hat{\pi}(x)$. The Poisson bracket between classical fields is replaced by $-\,$i times the commutator of the corresponding field operators. In particular, the fundamental Poisson brackets (19.15) turn into the following commutators:

$$[\phi(x), \pi(y)] = i\delta(x - y) \quad , \quad [\phi(x), \phi(y)] = [\pi(x), \pi(y)]) = 0 \,. \qquad (19.95)$$

Here and in the following, we suppress the hat over the symbols for the operators to keep the notation simple.

In the Heisenberg picture, the evolution of the field operators is determined by the equations of motion

$$i\frac{d}{dt}\phi(x) = [\phi(x), H] \quad \text{and} \quad i\frac{d}{dt}\pi(x) = [\pi(x), H] \qquad (19.96)$$

The Hamilton operator of the QFT is obtained by replacing the fields in the Hamilton function $H(\phi, \pi)$ of the classical field theory with the corresponding field operators.[4] The solution of the equations of motion (with given initial conditions) determines the quantum field at all times, $\phi(x) = \phi(x^0, x)$. It acts on states in the Hilbert space $\mathcal{H}$ of the QFT. This can be a Fock space over a vacuum state $|0\rangle$ with minimal energy.

Wigner's Theorem 17.1 states that each symmetry acts as a linear unitary (or anti-linear antiunitary) transformation on the state vectors. For the Poincaré-symmetry this means

$$\phi(x) \mapsto D(\Lambda)\phi\big(\Lambda^{-1}(x - a)\big) = U(\Lambda, a)\phi(x)U^{-1}(\Lambda, a) \,, \qquad (19.97)$$

with unitary operator $U$, which depends on the shift $a$ and the Lorentz transformation $\Lambda$ (or spin transformation). For the translations, we obtain the unitary operator $U(a)$ by replacing in the classical expression (19.61), the Poisson bracket in (19.58)

---

[4] The mathematical problems that arise, such as divergent expressions, are not further discussed at this point.

by $-\mathrm{i}$ times the commutator,

$$\phi(x-a) = \mathrm{e}^{-\mathrm{i}\,\mathrm{ad}_{aP}}\phi, \quad \text{with} \quad \mathrm{ad}_F\, G = [F, G] \,. \tag{19.98}$$

If we expand the exponential function in powers of $a$, we obtain

$$\mathrm{e}^{-\mathrm{i}\,\mathrm{ad}_{aP}}\phi = \phi - \mathrm{i}[aP, \phi] - \frac{1}{2!}[aP, [aP, \phi]] + \frac{\mathrm{i}}{3!}[aP, [aP, [aP, \phi]]] + \dots$$

$$= \mathrm{e}^{-\mathrm{i}aP}\phi\,\mathrm{e}^{\mathrm{i}aP}, \quad \text{with} \quad aP = a^\mu P_\mu \,. \tag{19.99}$$

---

**Task**

Try to prove this identity.

The comparison of (19.98) with (19.99) reveals that

$$\phi(x-a) = U(a)\phi(x)U^{-1}(a), \quad \text{with} \quad U(a) \equiv U(\mathbb{1}, a) = \mathrm{e}^{-\mathrm{i}aP} \,. \tag{19.100}$$

The infinitesimal form of the transformation is obtained by comparing the orders $\mathcal{O}(a)$,

$$[P_\mu, \phi] = -\mathrm{i}\partial_\mu\phi \,. \tag{19.101}$$

A similar approach for the Lorentz transformations leads to the more general relations for Poincaré transformations in (19.97).

## 19.6.1 Covariance of the Correlation Functions

In a relativistic QFT, the state with the smallest energy—called the vacuum state and denoted by $|0\rangle$—cannot have negative energy. Since a negative energy can become arbitrarily negative after changing the inertial frame, this would contradict the stability of the vacuum. In fact, the vacuum state should have no energy, no momentum, and no angular momentum. This requirement is, as we will see below, equivalent to the requirement that $|0\rangle$ is invariant under Poincaré transformations,

$$U(\Lambda, a)|0\rangle = |0\rangle \iff U^{-1}(\Lambda, a)|0\rangle = |0\rangle \,. \tag{19.102}$$

In accordance with the equivalence principle, $|0\rangle$ looks the same in all inertial frames.

For the translations in space and time, we use (19.100) to connect the field operator at $x$ with the field operator at the origin:

$$\phi(0) = U(x)\phi(x)U^{-1}(x) \quad \text{or} \quad \phi(x) = \mathrm{e}^{\mathrm{i}xP}\phi(0)\,\mathrm{e}^{-\mathrm{i}xP} \,. \tag{19.103}$$

For the two-point function, the translation invariance of the vacuum (19.102) then yields

$$\langle 0|\phi(x)\phi(y)|0\rangle = \langle 0|\,e^{ixP}\phi(0)\,e^{-i(x-y)P}\phi(0)\,e^{-iyP}|0\rangle$$

$$= \langle 0|\phi(0)\,e^{-i(x-y)P}\phi(0)|0\rangle = \langle 0|\phi(x-y)\phi(0)|0\rangle = W(x-y)\,.$$

$$(19.104)$$

This expresses the homogeneity of spacetime.

The two-point function also transforms covariantly under Lorentz transformations. We use (19.97) in the form

$$U(\Lambda)\phi(x)U^{-1}(\Lambda) = D(\Lambda)\phi(\Lambda^{-1}x), \quad U(\Lambda) \equiv U(\Lambda, 0)\,. \tag{19.105}$$

For a scalar field with $D(\Lambda) = \mathbb{1}$, the Lorentz invariance of the vacuum yields the relation

$$W(x) = \langle 0|\phi(x)\phi(0)|0\rangle = \langle 0|\phi(\Lambda^{-1}x)\phi(0)|0\rangle = W(\Lambda^{-1}x)\,, \tag{19.106}$$

which expresses the invariance of the two-point function under Lorentz transformations. In contrast, the two-point function of a vector field $V_\mu(x)$ satisfies the following relation,

$$W_{\mu\nu}(x) \equiv \langle 0|V_\mu(x)V_\nu(0)|0\rangle = \Lambda_\mu{}^\rho \Lambda_\nu{}^\sigma W_{\rho\sigma}\left(\Lambda^{-1}x\right)\,. \tag{19.107}$$

These covariance conditions significantly restrict the possible forms of the two-point function. Similar restrictions also exist for higher correlation functions.

---

**Question**

What are the covariance conditions for the two-point function $S(x,y) = \langle 0|\psi(x)\bar{\psi}(y)|0\rangle$ of a Dirac spinor field?

---

So how do we find the infinitesimal generators of *internal symmetries* in a QFT? The starting point is again the classical result, this time in the form (19.39), together with the replacement of the Poisson bracket by the commutator,

$$\{\phi, Q_X\} = \delta_X\phi \longmapsto [\phi, Q_X] = i\delta_X\phi\,. \tag{19.108}$$

The last relation means that $Q_X$ generates the infinitesimal symmetries,

$$\delta_X\phi = i[Q_X, \phi]\,, \tag{19.109}$$

and the finite symmetry transformations are obtained by exponentiation,

$$\phi \longmapsto U\phi U^{-1}, \qquad U = \exp\left(iQ_X\right)\,. \tag{19.110}$$

> **Generators of Symmetries**
> The generators $Q_X$ of symmetries in a quantum theory are obtained by replacing the fields in the classical Noether charges with the corresponding field operators.

The more symmetries a relativistic QFT admits, the more its correlation functions are restricted. Important extensions of Lorentz invariance are the conformal symmetry or a supersymmetry or, even more, the conformal symmetry together with a supersymmetry in superconformal field theories. In extreme cases, these symmetries are so large that they significantly restrict or even determine the correlation functions. As a consequence, such theories have a restricted dynamics, which can even be incompatible with interactions. Well-known examples are two-dimensional conformal field theories, which are discussed in Chap. 21.

## 19.7  Exercises for Chap. 19

**Problem 19.1 (Delta-Distribution).**  Prove the identities

$$f(y)\delta'(x-y) = f(x)\delta'(x-y) + \partial_x f(x)\delta(x-y),$$

$$2\pi i\,\delta(\xi) = \frac{1}{\xi - i\epsilon} - \frac{1}{\xi + i\epsilon} \implies 2\pi i\,\frac{d^n}{d\xi^n}\delta(\xi)$$

$$= (-1)^n n!\left(\frac{1}{(\xi - i\epsilon)^{n+1}} - \frac{1}{(\xi + i\epsilon)^{n+1}}\right).$$

**Problem 19.2 (Variational Derivatives).**  Let $u(x)$ be a real function of the real variable $x$. The first and second variational derivative of a functional $F[u]$ are defined by

$$F[u + \delta u] = F[u] + \int dx\,\frac{\delta F}{\delta u(x)}\delta u(x)$$

$$+ \frac{1}{2}\int dxdy\,\frac{\delta^2 F}{\delta u(x)\delta u(y)}\delta u(x)\delta u(y) + O(\delta u^3).$$

Now let $F[u]$ be a functional of the form

$$F[u] = \int dx\, f(u, u_x, u_{xx}, \dots).$$

Here, $u_x$ is the first and $u_{xx}$ the second derivative of $u$ with respect to $x$. Prove that the first variational derivative of $F$ has the following form:

$$\frac{\delta F}{\delta u(x)} = \frac{\partial f}{\partial u(x)} - \frac{\mathrm{d}}{\mathrm{d}x}\frac{\partial f}{\partial u_x(x)} + \frac{\mathrm{d}^2}{\mathrm{d}x^2}\frac{\partial f}{\partial u_{xx}(x)} - \cdots$$

$$= \sum_{n=0,1,\dots} (-1)^n \left(\frac{\mathrm{d}}{\mathrm{d}x}\right)^n \frac{\partial f}{\partial u^{(n)}(x)}\,.$$

Now prove that the second variational derivative of $F$ is given by:

$$\frac{\delta^2 F[u]}{\delta u(x)\delta u(y)} = \delta(x-y)$$

$$\sum_{m,n=0,1,\dots} \left((-)^n \frac{\mathrm{d}^n}{\mathrm{d}x^n} f_{(nm)}(x)\frac{\mathrm{d}^m}{\mathrm{d}x^m} + (-)^m \frac{\mathrm{d}^m}{\mathrm{d}x^m} f_{(mn)}(x)\frac{\mathrm{d}^n}{\mathrm{d}x^n}\right)\,.$$

We used the notation

$$f_{(mn)}(x) = \frac{\partial^2 f}{\partial u^{(m)}(x)\partial u^{(n)}(x)}\,.$$

Remark: You may assume that the function $u(x)$ and all its derivatives fall off for $|x| \to \infty$ so quickly, that no surface terms appear.

**Problem 19.3 (Product Rule and Jacobi Identity for Poisson Bracket).** Points in the phase space of a classical field theory are field configurations $(\phi(x), \pi(x))$. Observables are functionals $F[\phi, \pi]$ and $G[\phi, \pi]$, i.e. functions from the phase space to the real numbers. Their (standard) Poisson bracket is

$$\{F, G\} = \int \mathrm{d}x \left(\frac{\delta F}{\delta\phi(x)}\frac{\delta G}{\delta\pi(x)} - \frac{\delta F}{\delta\pi(x)}\frac{\delta G}{\delta\phi(x)}\right)\,.$$

Prove the product rule

$$\{F, GH\} = G\{F, H\} + \{F, G\}H \implies \{FG, H\} = F\{G, H\} + \{F, H\}G\,.$$

Let $H[\phi, \pi]$ be another observable. Prove the Jacobi identity

$$\{F, \{G, H\}\} + \{H, \{F, G\}\} + \{G, \{H, F\}\} = 0\,.$$

Hint: The arguments are analogous to those in classical mechanics. The essential difference is that in a field theory the number of degrees of freedom is infinite.

**Problem 19.4 (An Integrable System).** Let $u(x)$ be a real function of the variable $x \in \mathbb{R}$ and $u_x$ the first, $u_{xx}$ the second, etc. derivative of $u$ with respect to $x$. Prove that the variational derivative of a functional $F$ with local density, $F[u] = \int dx\, f(u, u_x, u_{xx}, \dots)$, has the following form:

$$\frac{\delta F}{\delta u(x)} = \frac{\partial f}{\partial u(x)} - \frac{d}{dx}\frac{\partial f}{\partial u_x(x)} + \frac{d^2}{dx^2}\frac{\partial f}{\partial u_{xx}(x)} - \dots.$$

Now define the following bracket between two functionals,

$$\{F, G\} = \int dx\, \frac{\delta F}{\delta u(x)}\frac{d}{dx}\frac{\delta G}{\delta u(x)}.$$

Prove that this defines a Poisson bracket, i.e. it is antisymmetric, bilinear and satisfies the Jacobi identity. The Hamiltonian function is

$$H = \int dx \left( u^3 + \tfrac{1}{2}u_x^2 \right).$$

Determine the Hamilton equation of motion for $u(t, x)$ and investigate whether it is a known equation in mathematical physics.

**Problem 19.5 (Conserved Quantities).** A continuous symmetry implies a conserved quantity according to Noether's theorem. Determine the symmetries and the corresponding conserve quantities of the following systems:

- free particle in $d$ dimensions,
- motion of a planet in the Newtonian gravitational potential of a black hole.

**Problem 19.6 (Lorentz Invariance of Electrodynamics).** Prove that the Lagrange density $\mathcal{L}_{ED}$ of Maxwell's theory in (19.62) transforms under translations as follows: $\mathcal{L}'_{ED}(x') = \mathcal{L}_{ED}(x)$. What does this imply for the infinitesimal variation $\delta_a \mathcal{L}_{ED}$ and why does this imply the invariance of the action?

**Problem 19.7 (Infinitesimal Rotations).** Prove the relations (19.91), which state that the Noether charges $Q^{\mu\nu}$ in (19.88) generate the infinitesimal Lorentz transformations.

**Problem 19.8 (Energy-Momentum Tensor for Real Scalar Field I).** Here we determine the Poisson brackets of those components of $T_{\mu\nu} \equiv \Theta_{\mu\nu}$ in (19.53) (for a scalar field the tensor does not need to be improved), for which the result can be written in terms of the $T_{\mu\nu}$ without using the equations of motion (off-hell). One avoids errors if one considers the spatially smeared energy and momentum densities

$$H_f = \int d\boldsymbol{x}\, f(\boldsymbol{x})T_{00}(0, \boldsymbol{x}), \qquad P_a = \int d\boldsymbol{x}\, a^i(\boldsymbol{x})T_{0i}(0, \boldsymbol{x}), \tag{19.111}$$

where the test functions $f$ and $a^i$ should be sufficiently smooth and decaying so that partial integration is allowed. First confirm the following Poisson brackets:

$$\{H_f, H_{f'}\} = \int d\boldsymbol{x}\, T_{0i}\,(f\,\partial_i f' - f'\partial_i f)\,,$$

$$\{P_a, H_f\} = \int d\boldsymbol{x}\, \big(T_{00}\, a^i\,\partial_i f + T_{ij}\, f\partial^i a^j\big)\,,$$

$$\{P_a, P_{a'}\} = \int d\boldsymbol{x}\, T_{0i}(a^j\partial_j a'^i - a'^j\partial_j a^i)\,.$$

After eliminating the test functions, you should get the Poisson brackets

$$\{T_{00}(\boldsymbol{x}), T_{00}(\boldsymbol{y})\} = -\big(T_{0i}(\boldsymbol{x}) + T_{0i}(\boldsymbol{y})\big)\partial^i \delta(\boldsymbol{x} - \boldsymbol{y}),$$

$$\{T_{00}(\boldsymbol{x}), T_{0i}(\boldsymbol{y})\} = \big(T_{00}(\boldsymbol{y})\partial_i - T_{ij}(\boldsymbol{x})\partial^j\big)\delta(\boldsymbol{x} - \boldsymbol{y}), \qquad (19.112)$$

$$\{T_{0i}(\boldsymbol{x}), T_{0j}(\boldsymbol{y})\} = \big(T_{0i}(\boldsymbol{x})\partial_j + T_{0j}(\boldsymbol{y})\partial_i\big)\delta(\boldsymbol{x} - \boldsymbol{y})\,.$$

for energy and momentum densities.

**Problem 19.9 (Energy-Momentum Tensor for Real Scalar Field II).** We turn to the remaining commutators of $T_{\mu\nu}$. This task is relatively challenging and is explained in more detail for the brackets $\{T_{ij}(\boldsymbol{x}), T_{pq}(\boldsymbol{y})\}$. First we define the spatially smeared stress tensor,

$$T_b = \int d\boldsymbol{x}\, b^{ij}(\boldsymbol{x}) T_{ij}(0, \boldsymbol{x})\,,$$

and show

$$\{T_b, T_{b'}\} = \int d\boldsymbol{x}\, \left\{2(b\pi)\partial_i\big(b'^{ij}\partial_j\phi\big) - (b\nabla b')(\pi\nabla\phi)\right\} - \left\{b \leftrightarrow b'\right\}\,.$$

Here the trace $b = b^{ij}\delta_{ij}$ appears (and analogously for $b'$). Now justify that the first integral with the help of partial integrations can be written as follows,

$$\int d\boldsymbol{x}\, (b\pi)\partial_i\big(b'^{ij}\partial_j\phi\big) = -\int d\boldsymbol{x}\, \partial_i b\, b'^{ij}\pi\,\partial_j\phi - \int d\boldsymbol{x}\, b\, b'^{ij}\partial_i\pi\,\partial_j\phi\,.$$

The Poisson bracket can now almost be expressed by the components $T_{0i} = \pi\,\partial_i\phi$. Only the last integral with $\partial_i\pi\,\partial_j\phi$ does not appear in $T_{\mu\nu}$. Here we make use of the equations of motion (19.26). Now show in the third step that these equations imply

$$\dot{T}_{ij} = \partial_i\pi\,\partial_j\phi + \partial_i\phi\,\partial_j\pi + \delta_{ij}\big(\pi\Delta\phi - 2\pi V' - \nabla\phi\nabla\pi\big)\,.$$

Solve for $\partial_i \pi \partial_j \phi$ and insert the result into $\{T_b, T_{b'}\}$. You should obtain the following Poisson bracket:

$$\{T_b, T_{b'}\} = \int d\boldsymbol{x} \left\{ 2b^{ik}\partial_k b' - b\partial_k b')T_{0k} + b'b^{ij}\dot{T}_{ij} \right\} - \{b \leftrightarrow b'\}.$$

Now calculate the remaining Poisson brackets

$$\{H_f, T_b\} = \int d\boldsymbol{x} \left\{ 2f\pi\partial_i(b^{ij}\partial_j\phi) - f\pi\nabla(b\nabla\phi) - b\pi\nabla(f\nabla\phi) + 2fb\pi V' \right\}$$

$$= -\int d\boldsymbol{x} \left\{ 2b^{ij}\partial_i f T_{0j} + fb^{ij}\dot{T}_{ij} \right\}$$

$$\{P_a, T_b\} = \int d\boldsymbol{x} \left\{ 2a^i\partial_i\phi\,\partial_p(b^{pq}\partial_q\phi) - b\pi\partial_i(a^i\pi) - a^i\partial_i\phi\nabla(b\nabla\phi) + ba^i\partial_i V \right\}$$

After eliminating the test functions, you should obtain the following commutation relations:

$$\{T_{ij}(\boldsymbol{x}), T_{pq}(\boldsymbol{y})\} = 2\big(T_{0i}(\boldsymbol{x})\delta_{pq}\partial_j + T_{0p}(\boldsymbol{y})\delta_{ij}\partial_q\big)\delta(\boldsymbol{x}-\boldsymbol{y})$$

$$- \big(T_{0k}(\boldsymbol{x})\partial_k + T_{0k}(\boldsymbol{y})\partial_k\big)\delta(\boldsymbol{x}-\boldsymbol{y})$$

$$+ \big(\delta_{pq}\dot{T}_{ij} - \delta_{ij}\dot{T}_{pq}\big)\delta(\boldsymbol{x}-\boldsymbol{y}),$$

$$\{T_{00}(\boldsymbol{x}), T_{ij}(\boldsymbol{y})\} = \big(T_{0i}(\boldsymbol{y})\partial_j + T_{0j}(\boldsymbol{y})\partial_i\big)\delta(\boldsymbol{x}-\boldsymbol{y}) - \dot{T}_{ij}(\boldsymbol{x})\delta(\boldsymbol{x}-\boldsymbol{y})$$

**Problem 19.10 (Complex Scalar Field).** Let $\phi$ be a complex scalar field, which satisfies the KG equation. The action is

$$S_M[\phi] = \int d^4x \left( \partial_\mu\phi^*\partial^\mu\phi - m^2\phi^*\phi \right).$$

Find the momentum fields canonically conjugated to $\phi$ and $\phi^*$. Calculate the Heisenberg equations and show that these lead to the Klein-Gordon equation.

# Gauge Theories

**20**



All known fundamental interactions in nature are modeled by gauge theories: These are Maxwell's theory, the Weinberg-Salam model of electroweak interaction (which includes Maxwell's theory), Quantum Chromodynamics, which describes strong interaction, and Einstein's theory of gravity. The electromagnetic and gravitational forces are long-range, whereas the weak and strong forces are short-range. The standard model of particle physics models all (microscopic) interactions except gravitation. It describes the fields associated with elementary particles and their propagation in Minkowski space. In this chapter, we will see that gauge theories are determined by a few specifications—essentially the types of particles and symmetries.

First we discuss the gauge principle, with the help of which one can transform a global gauge symmetry into a local one. However, this is only possible if gauge potentials are introduced, which in the quantized theory describe the exchange particles of the interaction. In the standard model, these are the photon of the electromagnetic interaction, the $W^{\pm}$ bosons and the $Z$ boson of the electroweak interaction, and the gluons of the strong interaction. In this chapter, we again choose natural units, in which $c = \hbar = 1$.

Hermann Weyl had already introduced a variable length scale as a gauge factor in an extension of the general theory of relativity in 1919 [92]. In the subsequent work "Elektron und Gravitation" within the framework of quantum mechanics, he then formulated his gauge theory in the modern sense [93]. The simplest example of a gauge theory is electrodynamics with Abelian gauge group U(1). Gauge theories with non-Abelian gauge groups were introduced by C.N. Yang and R.L.

© The Author(s), under exclusive license to Springer-Verlag GmbH, DE, part of Springer Nature 2026
A. Wipf, *Symmetries in Physics*, https://doi.org/10.1007/978-3-662-72675-4_20

Mills in 1954 [94]. A good readable introduction to gauge theories can be found in the book [95], and a nice overview with references is contained in the article by O'Raifeartaigh and Straumann [96]. Detailed presentations of classical and quantized gauge theories can be found for example in the texts [81,97].

## 20.1   Gauge Transformations and Minimal Coupling

You first encountered the potentials in your lecture on electrodynamics. In solving the homogeneous Maxwell equations, one is naturally led to potentials, and their introduction seems to be just a mathematical trick. That this is not the case, we will clarify in the following.

We start with the homogeneous Maxwell equations for the electromagnetic field,

$$\nabla \cdot \boldsymbol{B} = 0 \quad , \quad \nabla \times \boldsymbol{E} + \frac{\partial \boldsymbol{B}}{\partial t} = 0 . \tag{20.1}$$

These are solved by introducing a vector potential $\boldsymbol{A}$ and a scalar potential $\varphi$,

$$\boldsymbol{B} = \nabla \times \boldsymbol{A} \quad , \quad \boldsymbol{E} = -\frac{\partial \boldsymbol{A}}{\partial t} - \nabla \varphi . \tag{20.2}$$

Now the apparent difficulty arises that different potentials give rise to identical electromagnetic fields. If one replaces $\varphi$ and $\boldsymbol{A}$ by

$$\varphi' = \varphi - \frac{\partial \lambda}{\partial t} \quad \text{and} \quad \boldsymbol{A}' = \boldsymbol{A} + \nabla \lambda \tag{20.3}$$

with a function $\lambda$ that depends on position and time, then one obtains the same fields $\boldsymbol{E}, \boldsymbol{B}$.

---

**Task**

Show that $\varphi'$, $\boldsymbol{A}'$ lead to the same electromagnetic fields as $\varphi$, $\boldsymbol{A}$.

---

The transformations (20.3) are called gauge transformations and $\lambda(x)$ is called the gauge function. Accordingly, the configurations $\phi$, $\boldsymbol{A}$ and $\phi'$, $\boldsymbol{A}'$ are called gauge equivalent. Together, the scalar and vector potential form the $4-$potential

$$(A^\mu) = \begin{pmatrix} \varphi \\ \boldsymbol{A} \end{pmatrix} = \begin{pmatrix} A^0 \\ \boldsymbol{A} \end{pmatrix} . \tag{20.4}$$

In relativistic notation, the gauge transformation (20.3) takes on a more elegant form,

$$A_\mu \longrightarrow A'_\mu = A_\mu - \partial_\mu \lambda . \tag{20.5}$$

> **Gauge Transformations**
> A transformation (20.5) with arbitrary gauge function $\lambda(x)$ is called a gauge transformation. The potentials $A_\mu$ and $A'_\mu$ are called gauge equivalent. They lead to the same gauge fields and are physically indistinguishable.

In the literature, one often finds the statement that the gauge potentials contain more information than the gauge fields. The Aharanov-Bohm effect is almost always cited as an argument. That this conclusion does not necessarily apply has been demonstrated, for example, in [98] (for non-Abelian gauge field theories, the situation is somewhat more complicated).

In the transition from $\boldsymbol{E}, \boldsymbol{B}$ to $A^\mu$, redundant degrees of freedom have been introduced, and therefore classes of potentials correspond to the same electromagnetic field. Thus, a 4−potential can be transformed as in (20.5), without "changing the physics". This freedom in the choice of the potential can be used to demand a so-called gauge condition for the potential. Such a condition should always be achievable with the help of a suitable gauge transformation (20.5), and should, given an electromagnetic field, determine the potential as much as possible. For example, the Lorenz gauge $\partial_\mu A^\mu = 0$ can always be satisfied, and in this gauge the inhomogeneous Maxwell equations simplify to

$$\Box A^\mu = j^\mu, \qquad (j^\mu) = \begin{pmatrix} \rho \\ j \end{pmatrix}. \tag{20.6}$$

Here, we are led to the electric 4−current density $j^\mu$. Since $j^\mu$ and $A^\mu$ are vector fields under Lorentz transformations, the field equation (20.6) has the same form in all inertial frames.

## 20.2 Gauge-Covariant Derivative

The Hamilton function for a charged point particle with charge $e$ in an external electromagnetic field contains the potentials $\varphi$ and $\boldsymbol{A}$,

$$H = \frac{1}{2m}\left(\boldsymbol{p} - e\boldsymbol{A}\right)^2 + e\varphi. \tag{20.7}$$

It leads to the well-known Lorentz equation of motion for the charged point particle and is obtained by making the following minimal coupling (also called minimal substitution) in the Hamilton function of the free particle,

$$E \longrightarrow E - e\varphi \quad, \quad \boldsymbol{p} \longrightarrow \boldsymbol{\pi} = \boldsymbol{p} - e\boldsymbol{A}, \tag{20.8}$$

see Problem 20.1. An external electromagnetic field is thus included by replacing energy and momentum according to the rules (20.8). Remembering that the energy $E$ and the spatial momentum $\boldsymbol{p}$ together form the components of the 4−momentum $p^\mu$, the minimal coupling (20.8) can be written as follows:

$$p_\mu \longrightarrow p_\mu - eA_\mu \,. \tag{20.9}$$

The transition from classical mechanics to quantum mechanics in position space is done using the correspondence rules $E \longrightarrow i\partial_t$ and $\boldsymbol{p} \longrightarrow -i\nabla$. Combining the minimal coupling with the correspondence rules, we obtain the coupling of matter fields to an electromagnetic field using the substitutions

$$p_\mu \longmapsto p_\mu - eA_\mu \longmapsto i\big(\partial_\mu + ieA_\mu\big) \equiv iD_\mu \,. \tag{20.10}$$

On the right-hand side is the so-called covariant derivative $D_\mu$.

**Covariant Derivative**

The coupling of a charged particle to the electromagnetic field is achieved by replacing the partial derivatives with the covariant derivatives in the corresponding field theory,

$$\partial_\mu \longmapsto D_\mu = \partial_\mu + ieA_\mu \,. \tag{20.11}$$

Why is the derivative called covariant? To answer this question, we check how it changes when we substitute $A'_\mu = A_\mu - \partial_\mu \lambda$ for $A_\mu$,

$$D_\mu(A') = \partial_\mu + ieA'_\mu = \partial_\mu + ie(A_\mu - \partial_\mu \lambda) = e^{ie\lambda}\big(\partial_\mu + ieA_\mu\big)e^{-ie\lambda} \,. \tag{20.12}$$

This means that the covariant derivative transforms homogeneously under gauge transformations,

$$D_\mu(A') = gD_\mu(A)g^{-1}, \quad \text{with} \quad g(x) = e^{ie\lambda(x)} \in U(1) \,. \tag{20.13}$$

For each point in spacetime, $g(x)$ is in the group U(1) and accordingly the function $x \mapsto g(x)$ is a U(1)-valued field.

In Sect. 19.3, we saw how a complex scalar field, which describes spinless charged particles of charge $e$, transforms under global gauge transformations,

$$\phi(x) \longmapsto \phi'(x) = g\phi(x), \quad g = e^{ie\lambda} \,. \tag{20.14}$$

Global here refers to the fact that $g$ takes the same value at all points in Minkowski space. The Lagrange density is invariant under global U(1) transformations

$$\mathcal{L}_{\mathrm{M}}(\phi) = \partial_\mu \phi'^* \partial^\mu \phi' - V(\phi'^* \phi') = \mathcal{L}_{\mathrm{M}}(\phi') \,, \tag{20.15}$$

and the corresponding covariant conserved Noether current $j^\mu$ is identified as the density of the electric current. We can now elevate the global gauge symmetry to a local one if we replace partial derivatives with covariant derivatives,

$$\mathcal{L}_{\mathrm{M}}(\phi, A) = (D_\mu \phi)^* D^\mu \phi - V(\phi^* \phi) \,. \tag{20.16}$$

If we now transform the scalar field and the gauge potential *simultaneously* with a space- and time-dependent (local) gauge transformation,

$$\phi'(x) = g(x)\phi(x), \quad A'_\mu(x) = A_\mu(x) - \partial_\mu \lambda(x), \qquad g(x) = \mathrm{e}^{\mathrm{i}e\lambda(x)} \,, \tag{20.17}$$

then $D_\mu \phi$ transforms just like $\phi$ under local gauge transformations,

$$D'_\mu \phi' = (g D_\mu g^{-1})(g\phi) = g(D_\mu \phi), \quad g = g(x) \,. \tag{20.18}$$

Here, $D'_\mu$ is the covariant derivative formed with the transformed potential $A'_\mu$.

**Gauge Invariance of $\mathcal{L}_{\mathrm{M}}$**
The Lagrange density (20.16), in which partial derivatives were replaced by covariant derivatives, is invariant under the local gauge transformations of $\phi$ and $A_\mu$ in (20.17).

So, a global symmetry can be made local if we introduce a new field, a vector potential $A_\mu$, also called gauge potential.[1]

In Problem 20.2, it is shown that the Euler-Lagrange equation for the gauge-invariant Lagrange density (20.16) with $V(\phi^* \phi) = \mu^2 \phi^* \phi$ is the covariant Klein-Gordon equation,

$$D_\mu D^\mu \phi + \mu^2 \phi = 0 \,. \tag{20.19}$$

It is obtained from the Klein-Gordon equation (19.6) for the free field, if we replace the ordinary derivatives with the covariant derivatives.

---

[1] There exist variations of this construction, for example for nonlinear sigma models, see e.g. [99].

> **Task**
>
> Convince yourself that if $(\phi, A_\mu)$ satisfy the Klein-Gordon equation (20.19), the gauge-transformed fields $(\phi', A'_\mu)$ do as well.

The newly introduced vector field $A_\mu$ will, like the original matter field, propagate. Its dynamics will be determined by a yet to be found Lagrange density. This should be invariant under local gauge transformations, since only the electromagnetic field—and not the potential—is observable.

To construct this Lagrange density, we first determine the commutator of two covariant derivatives. With the definition (20.11), we get

$$[D_\mu, D_\nu] = ie\big(\partial_\mu A_\nu - \partial_\nu A_\mu\big) \equiv ie F_{\mu\nu}\,, \qquad (20.20)$$

where the antisymmetric field strength tensor $F_{\mu\nu}$ introduced in Sect. 19.1 appears. Since the covariant derivative transforms homogeneously, see (20.13), the gauge equivalent potentials $A_\mu$ and $A'_\mu$ have the same field strength, $F_{\mu\nu}(A') = F_{\mu\nu}(A)$.

The action of electrodynamics is the spacetime integral over the Lorentz- and gauge-invariant Lagrange density $\mathcal{L}_{\mathrm{ED}}$ in (19.62)

$$S_{\mathrm{ED}}[A] = \int \mathrm{d}^4x\, \mathcal{L}_{\mathrm{ED}}(A)\,. \qquad (20.21)$$

A mass term of the form $\mu^2 A_\mu A^\mu$ is not allowed in $\mathcal{L}_{\mathrm{ED}}$, as it is not gauge invariant. Using the variation of the field strength tensor

$$\delta F_{\mu\nu} = \delta\big(\partial_\mu A_\nu - \partial_\nu A_\mu\big) = \partial_\mu \delta A_\nu - \partial_\nu \delta A_\mu \qquad (20.22)$$

we find after a partial integration for the variation of the action

$$\delta S_{\mathrm{ED}} = -\frac{1}{2}\int \mathrm{d}^4x\, \delta F_{\mu\nu}\, F^{\mu\nu} = \int \mathrm{d}^4x\, \delta A_\nu \partial_\mu F^{\mu\nu} \quad \text{that is} \quad \frac{\delta S_{\mathrm{ED}}}{\delta A_\nu} = \partial_\mu F^{\mu\nu}\,. \qquad (20.23)$$

The complete action for the system, consisting of scalar field and gauge potential, is

$$S[\phi, A] = S_{\mathrm{ED}}[A] + S_{\mathrm{M}}[\phi, A] = \int \mathrm{d}^4x\, \big(\mathcal{L}_{\mathrm{ED}}(A) + \mathcal{L}_{\mathrm{M}}(A, \phi)\big)\,, \qquad (20.24)$$

with Lagrange density $\mathcal{L}_{\mathrm{M}}$ in (20.16). Using the variational characterization of the Noether current in Problem 20.4, the Euler-Lagrange equations take the following simple form,

$$\partial_\mu F^{\mu\nu} = j^\nu \quad \text{and} \quad D_\mu D^\mu \phi + \mu^2 \phi = 0\,, \qquad (20.25)$$

where the Noether current for the (global) gauge transformations appears. On both sides of the equations are tensors, so that the field equations have the same form in all inertial frames. The field strength tensor and the Noether current are both gauge invariant, while $D_\mu D^\mu \phi$ transforms under gauge transformation just like $\phi$. This means that the field equations are gauge invariant or gauge covariant. Therefore, if a configuration $(\phi, A_\mu)$ solves the field equations, then so does every gauge-equivalent configuration $(\phi', A'_\mu)$.

### 20.2.1  Minimal Coupling of Charged Fermions

Dirac spinor fields describe charged fermions with spin $\frac{1}{2}$ and satisfy the Dirac equation—for free particles this is Eq. (18.61). It is the Euler-Lagrange equation for the density $\mathcal{L}_M(\psi) = \bar{\psi}(i\slashed{\partial} - m)\psi$ in (19.43). This density is invariant under global U(1) transformations of the spinor field $\psi \mapsto e^{ie\lambda}\psi$, which transforms solutions of the Dirac equation into solutions. According to Emmy Noether's theorem, this inner symmetry leads to the covariantly conserved current density $j^\mu$ in (19.45).

This "global gauge invariance" can again be extended to a local gauge invariance if we introduce a gauge potential and replace partial with covariant derivatives.

> **Dirac Equation**
>
> The Dirac equation for a charged electron in the electromagnetic field is
>
> $$ i\slashed{D}\psi = m\psi, \quad \text{with} \quad \slashed{D} = \gamma^\mu D_\mu, \quad D_\mu = \partial_\mu + ieA_\mu. \tag{20.26} $$
>
> It is the Euler-Lagrange equation for the Lorentz invariant action
>
> $$ S_M[\psi, A] = \int d^4x\, \mathcal{L}_M(\psi, A_\mu), \quad \mathcal{L}_M(\psi, A_\mu) = \bar{\psi}\left(i\slashed{D} - m\right)\psi. \tag{20.27} $$

The Lagrange density $\mathcal{L}_M$ is invariant under local gauge transformations of the fields

$$ \psi(x) \mapsto e^{ie\lambda(x)}\psi(x) \quad , \quad A_\mu(x) \mapsto A_\mu(x) - \partial_\mu\lambda(x). \tag{20.28} $$

The action of the entire system is the sum of the Maxwell and Dirac term in (20.21) and (20.27),

$$ S[\psi, A] = S_{\mathrm{ED}}(A_\mu) + S_M(\psi, A_\mu). \tag{20.29} $$

It is the starting point in the quantization of electrodynamics. The quantized electrodynamics, briefly QED (for Quantum Electrodynamics), is one of the most successful and best-tested physical theories. Here, the photons are the quanta of the $A_\mu$ field and the electrons and positrons are the quanta of the $\psi$ field.


## 20.3    Non-Abelian Gauge Theories

Non-Abelian gauge theories are theories with local gauge invariance, where the gauge transformations at each spacetime point are elements of a non-Abelian group $G$. In the standard model of particle physics only the non-Abelian gauge groups SU(2) and SU(3) occur. Let $g \in G$ be an element of the gauge group $G$ and $g \mapsto U(g)$ be an $n$-dimensional representation of $G$. Then the fermions in a multiplet (a representation) of $G$ transform under global gauge transformations,

$$\psi \longmapsto U(g)\psi, \quad \psi = \begin{pmatrix} \psi_1 \\ \vdots \\ \psi_n \end{pmatrix}. \tag{20.30}$$

Note that each $\psi_b$ with $b = 1,\ldots,n$ is a 4-component spinor. The gauge transformations do not act on the spin indices $\alpha$ of the $\psi_b = (\psi_{b\alpha})$, but only on the indices $b$. For the quarks of Quantum Chromodynamics there are three color indices $b = 1, 2, 3$ (in addition there are the flavor degrees of freedom, which are inert under gauge transformations). The Lagrange density for free fermions with equal masses is

$$\mathcal{L}_\mathrm{M}(\psi) = \bar{\psi}\left(\mathbb{1} \otimes (i\slashed{\partial} - m)\right)\psi, \tag{20.31}$$

with Dirac-conjugated spinor

$$\bar{\psi} = (\bar{\psi}_1, \ldots, \bar{\psi}_n), \quad \bar{\psi}_b = \psi_b^\dagger \gamma^0. \tag{20.32}$$

The unit matrix $\mathbb{1}$ in the tensor product

$$\mathbb{1} \otimes (i\slashed{\partial} - m) = \mathrm{diag}(i\slashed{\partial} - m, \ldots, i\slashed{\partial} - m)$$

acts in color space and $\slashed{\partial}$ acts in spin space. In color space, the free Dirac operator is proportional to the identity matrix and the Lagrange density $\mathcal{L}_\mathrm{M}$ is therefore invariant under *global gauge transformations*

$$\psi \longrightarrow U\psi, \quad \bar{\psi} \longrightarrow \bar{\psi}U^{-1} \quad \text{with} \quad U = U(g), \tag{20.33}$$

which only act in color space.

### 20.3.1 Local Gauge Invariance

As in the Abelian case, we now want to construct a theory that is invariant under local gauge transformations $g(x) \in G$. We consider again a color multiplet of Dirac spinors, which transforms under the defining representation $U(g) = g$ of a matrix group,

$$\psi(x) \longrightarrow g(x)\psi(x). \tag{20.34}$$

Again, local gauge invariance is achieved by minimal coupling to a gauge potential. Here we replace partial derivatives with covariant derivatives with vector potential $A_\mu$,

$$\partial_\mu\psi \longrightarrow D_\mu\psi = \left(\partial_\mu + ieA_\mu\right)\psi. \tag{20.35}$$

The covariant derivative should transform covariantly, i.e. $D_\mu\psi$ should transform like $\psi$. This is equivalent to the requirement

$$D_\mu(A') = gD_\mu(A)g^{-1} \implies ieA'_\mu = g\left(\partial_\mu + ieA_\mu\right)g^{-1}. \tag{20.36}$$

From this we read off the gauge transformation of the gauge potential,

$$A'_\mu = gA_\mu g^{-1} - \frac{i}{e}g\partial_\mu g^{-1}. \tag{20.37}$$

Because the last term lies in the Lie algebra $\mathfrak{g}$ of $G$, it is natural to consider $A_\mu$ as a vector field with values in this Lie algebra (more precisely $iA_\mu$ is in $\mathfrak{g}$). Then also the first term $gA_\mu g^{-1}$ is in $\mathfrak{g}$.

If the matter field does not transform according to the defining but an arbitrary representation $g \mapsto U(g)$, then its covariant derivative is

$$D_\mu\psi = \left(\partial_\mu + ieU_*(A_\mu)\right)\psi. \tag{20.38}$$

Here the gauge potential appears in the representation $U_*$ of the Lie algebra $\mathfrak{g}$ induced by $U$.

---

**Task**

Show that the covariant derivative (20.38) for a field $\psi$ in an arbitrary representation $U$ of the gauge group transforms under local gauge transformations as follows:

$$D_\mu(A') = U(g)D_\mu(A)U^{-1}(g). \tag{20.39}$$

From this it immediately follows the gauge invariance of the Lagrange density

$$\mathcal{L}_{\mathrm{M}}(\psi) = \bar{\psi}\left(\mathrm{i}\slashed{D} - m\right)\psi, \qquad \slashed{D} = \gamma^\mu D_\mu . \tag{20.40}$$

What is still missing is the non-Abelian generalization of $\mathcal{L}_{\mathrm{ED}}$ in (19.62). This must be invariant under local gauge transformations and coincide with the density $\mathcal{L}_{\mathrm{ED}}$ for the gauge group U(1). For its construction we consider, as in the Abelian case, the commutator of two covariant derivatives in the defining representation:

$$[D_\mu, D_\nu] = [\partial_\mu + \mathrm{i}e A_\mu, \partial_\nu + \mathrm{i}e A_\nu] = \mathrm{i}e\left(\partial_\mu A_\nu - \partial_\nu A_\mu + \mathrm{i}e[A_\mu, A_\nu]\right) \equiv \mathrm{i}e F_{\mu\nu} . \tag{20.41}$$

By construction, the components $F_{\mu\nu}$ of the antisymmetric field strength tensor lie in the Lie algebra $\mathfrak{g}$. From the transformation behavior of the covariant derivative in (20.39) one can read off how these components transform,

$$F_{\mu\nu}(A') = g F_{\mu\nu}(A) g^{-1} . \tag{20.42}$$

In contrast to electrodynamics, the field strength tensor is no longer gauge invariant and transforms according to the adjoint representation of the gauge group. The Lorentz- and gauge-invariant action for the gauge field $A_\mu$ now has the form

$$S_{\mathrm{YM}}[A] = \int \mathrm{d}^4 x \, \mathcal{L}_{\mathrm{YM}}(A_\mu), \quad \mathcal{L}_{\mathrm{YM}}(A_\mu) = -\frac{1}{4}\,\mathrm{Tr}\left(F^{\mu\nu} F_{\mu\nu}\right) . \tag{20.43}$$

It is gauge invariant because the trace of $F^{\mu\nu} F_{\mu\nu}$ is invariant under conjugations (20.42).

The action $S_{\mathrm{YM}}$ describes a pure Yang-Mills theory without matter fields and generalizes the action of Maxwell's theory. In the non-Abelian case, the field strength

$$F_{\mu\nu} = \partial_\mu A_\nu - \partial_\nu A_\mu + \mathrm{i}e[A_\mu, A_\nu] \tag{20.44}$$

is however non-linear in the gauge potential, as it contains the commutator of two potentials. The Lagrange density thus contains terms that are quadratic, cubic and quartic in the potential. The corresponding field equations—the Yang-Mills equations—are therefore non-linear partial differential equations. Thus, in the non-Abelian theory, the superposition principle of the linear Maxwell theory no longer applies.

The gauge potential and the field strength are $\mathfrak{g}$-valued and can be expanded according to a basis $\{X_a\}$ of $\mathfrak{g}$,

$$A_\mu = A_\mu^a X_a \quad \text{and} \quad F_{\mu\nu} = F_{\mu\nu}^a X_a . \tag{20.45}$$

The relations (20.44) for the components of the field strength tensor are

$$F_{\mu\nu}^{a} = \partial_{\mu} A_{\nu}^{a} - \partial_{\nu} A_{\nu}^{a} - e f_{bc}{}^{a} A_{\mu}^{b} A_{\nu}^{c} , \tag{20.46}$$

where the real structure constants of the Lie algebra appear.

**The Action of Gauge Theories with Fermions**

The equations of motion of a non-Abelian gauge theory for fermions are the Euler-Lagrange equations of the Lorentz- and gauge-invariant action

$$S[\psi, A] = S_{\mathrm{YM}}[A] + S_{\mathrm{M}}[\psi, A] = \int d^{4}x\, \mathcal{L}_{\mathrm{YM}}(A_{\mu}) + \int d^{4}x\, \mathcal{L}_{\mathrm{M}}(\psi, A_{\mu}) \tag{20.47}$$

with the Lagrange densities (20.43) and (20.40).

A gauge theory is fixed by specifying the gauge group $G$, the representations of the matter fields, the universal coupling constant $e$ in the covariant derivative and further parameters in the matter sector. In the Weinberg-Salam model of the electroweak interaction, these are the parameters in the Higgs potential, in the CKM matrix or the Yukawa couplings.

## 20.3.2  Infinitesimal Gauge Transformations

We now assume that the group element $g$ is close to the identity and expand

$$g = e^{ieX} \approx \mathbb{1} + ieX \implies g^{-1} \approx \mathbb{1} - ieX . \tag{20.48}$$

Except for the factor i, $X$ is in the Lie algebra $\mathfrak{g}$ of the gauge group, briefly called gauge algebra. Physicists have a preference for Hermitian matrices, and therefore we like to write an i in the expansion (20.48). We insert this expansion into (20.34) and (20.37) and obtain, up to linear order in $X$, the transformation rules

$$\phi' = g\phi \approx \phi + ieX\phi$$

$$A_{\mu}' = g A_{\mu} g^{-1} - \frac{i}{e} g\, \partial_{\mu} g^{-1} \approx A_{\mu} - ie[A_{\mu}, X] - \partial_{\mu} X , \tag{20.49}$$

from which we read off the infinitesimal gauge transformations of the fields,

$$\delta\phi = ieX\phi \quad \text{and} \quad \delta A_{\mu} = -\partial_{\mu} X - ie[A_{\mu}, X] \equiv -D_{\mu} X . \tag{20.50}$$

From (20.42) immediately follows the infinitesimal transformation of the field strength tensor,

$$\delta F_{\mu\nu} = ie[X, F_{\mu\nu}]. \tag{20.51}$$

In Abelian gauge theories all commutators vanish and $F_{\mu\nu}$ is gauge invariant.

---

**Task**

Show that the variation of the field strength has the simple form (20.51).

---

For unitary gauge groups, the $X$ are Hermitian matrices[2] for which $(X, Y) = \mathrm{Tr}(XY)$ defines a scalar product. We choose a basis $X_a$ with

$$(X_a, X_b) = \delta_{ab}. \tag{20.52}$$

If we expand the $\mathfrak{g}$-valued fields according to this basis and insert these expansions into the transformation formulas (20.50) and (20.51), we obtain the following infinitesimal transformations of the fields,

$$\delta A_\mu^a = -(D_\mu\lambda)^a = -\partial_\mu\lambda^a + ef^a_{\ bc}A_\mu^b\lambda^c \quad \text{and} \quad \delta F_{\mu\nu}^a = ef^a_{\ bc}F_{\mu\nu}^b\lambda^c. \tag{20.53}$$

The gauge invariance of the Lagrange density $\mathcal{L}_M$ leads to covariantly conserved current densities (also called gauge currents)

$$J_a^\mu \equiv -\frac{\partial\mathcal{L}_M}{\partial A_\mu^a}. \tag{20.54}$$

To prove this, we calculate the variation of the matter action under an arbitrary gauge transformation—this variation must vanish—and use the result (20.53) for the change of the potential together with the equation of motion $\delta S_M/\delta\phi = 0$,

$$\delta S_M = \int d^4x \left(\frac{\partial\mathcal{L}_M}{\partial A_\mu^a}\right)\delta A_\mu^a = -\int d^4x\, J_a^\mu\delta A_\mu^a$$

$$= \int d^4x\, J_a^\mu\left(\partial_\mu\lambda^a - ef^a_{\ bc}A_\mu^b\lambda^c\right) = 0.$$

We assume that $\lambda^a$ vanishes in the spatial and temporal infinity, so that we may integrate by parts. Then we get

$$(D_\mu J^\mu)_a = \partial_\mu J_a^\mu - ef_{ab}^{\ \ c}A_\mu^b J_c^\mu = 0. \tag{20.55}$$

---

[2] The infinitesimal generators are anti-unitary, but with the factor i in (20.48) the $X$ are Hermitian.

For $A_\mu = 0$, the gauge currents are equal to the conserved Noether currents associated with the global gauge transformations, $\partial_\mu (J_a^\mu|_{A=0}) = 0$.

---

**Example: Gauge Currents for Scalar Fields and Dirac Fields**

A Dirac field has the following Lagrange density and gauge currents:

$$\mathcal{L}_M(\psi, A_\mu) = \bar{\psi}(i\slashed{D} - m)\psi, \quad J_a^\mu(\psi) = e\bar{\psi}\gamma^\mu X_a \psi \,. \tag{20.56}$$

For scalar fields, the Lagrange density and gauge currents have the form

$$\mathcal{L}_M(\phi, A_\mu) = (D_\mu\phi)^\dagger D^\mu\phi + V(\phi^\dagger\phi), \quad J_a^\mu(\phi) = -ie(D^\mu\phi, X_a\phi) + h.c \tag{20.57}$$

◀

## 20.3.3 Field Equations

We consider a general gauge theory for scalar particles and Dirac fermions, and assume that the matter fields transform according to the defining representation of the gauge group. The straightforward generalization to arbitrary representations will be discussed afterwards. By appropriately choosing the gauge group and representations for the matter particles, one obtains the gauge theory of the electroweak or the strong interaction.

The Lagrange density contains three terms: the Yang-Mills term, the contribution of the scalar fields and that of the fermions:

$$\mathcal{L} = \mathcal{L}_{YM}(A_\mu) + \mathcal{L}_M(\phi, A_\mu) + \mathcal{L}_M(\psi, A_\mu), \quad \mathcal{L}_{YM}(A_\mu) = -\frac{1}{4}F_a^{\mu\nu}F_{a\mu\nu}\,, \tag{20.58}$$

where the densities $\mathcal{L}_M$ are given in (20.57) and (20.56). A possible Yukawa interaction between fermions and scalars has not yet been considered here. The Higgs potential in $\mathcal{L}_M(\phi, A_\mu)$ must be gauge invariant, $V(g\phi) = V(\phi)$.

---

**Task**

Prove the following variations,

$$\delta S_{YM}[A] = \int \delta A_\mu^a (D_\mu F^{\mu\nu})_a \,,$$

$$\delta S_M[\phi, A] = -\int \left(\delta\phi, D^2\phi + \frac{\partial V}{\partial\phi^\dagger}\right) - \int \delta A_\mu^a J_a^\mu(\phi)\,, \tag{20.59}$$

$$\delta S_M[\psi, A] = \int \delta\bar{\psi}(i\slashed{D} - m)\psi + h.c. - \int \delta A_\mu^a J_a^\mu(\psi)\,.$$

These contain the current densities $J_a^\mu$ of the scalar fields and Dirac spinor fields in (20.54).

**Field Equations of a Gauge Theory**

The field equations for the theory with Lagrange density (20.58) are the Yang-Mills (YM) equation, the covariant Klein-Gordon equation and the covariant Dirac equation:

$$(D_\mu F^{\mu\nu})_a = J_a^\nu(\phi) + J_a^\nu(\psi)\,,$$

$$D^\mu D_\mu \phi + \frac{\partial V}{\partial \phi^\dagger} = 0\,, \tag{20.60}$$

$$(i\slashed{D} - m)\psi = 0\,.$$

In a space without matter the Yang-Mills equations are $D_\mu F^{\mu\nu} = 0$. These are nonlinear partial differential equations, whose solutions are not all known.

## 20.4   Quantum Chromodynamics (QCD)

QCD describes the strong interaction and is a gauge theory with gauge group SU(3). Its elementary building blocks of matter are the quarks, from which mesons and baryons are composed. There are 6 types of quarks: up quarks, down quarks, strange quarks, charm quarks, bottom quarks and top quarks. The quarks are fermions with spin $\frac{1}{2}$. Their masses and some of their quantum numbers are given in the Table 20.1.

Note the mass hierarchy: Up, down and strange quarks are much lighter than charm, top and bottom quarks. If we only consider the three light flavors – the heavy ones are rarely found in nature—then we can set their masses equal in a first approximation, $m_u = m_d = m_s$. In this approximation, the Lagrange densities

**Table 20.1** The 6 quarks have very different masses. The electric charge $e$ and masses of the particles are given. Each quark in this table comes in 3 colors. The masses are taken from the Particle Data Group [100]

| Generation | Name | Symbol | $e$ | Mass |
|---|---|---|---|---|
| 1 | Up | u | $\frac{2}{3}$ | 2.16 MeV/c$^2$ |
| 1 | Down | d | $-\frac{1}{3}$ | 4.67 MeV/c$^2$ |
| 2 | Charm | c | $\frac{2}{3}$ | 1.27 GeV/c$^2$ |
| 2 | Strange | s | $-\frac{1}{3}$ | 93 MeV/c$^2$ |
| 3 | Top | t | $\frac{2}{3}$ | 172.76 GeV/c$^2$ |
| 3 | Bottom | b | $-\frac{1}{3}$ | 4.18 GeV/c$^2$ |

of the three flavors are identical,

$$\mathcal{L}_{\mathrm{M}}(\psi, A_\mu) = \sum_{f=u,d,s} \bar{\psi}_f \left(i\slashed{D} - m\right) \psi_f = \bar{\psi}\left(i\slashed{D} - m\right) \otimes \mathbb{1}_f\, \psi, \qquad \psi = \begin{pmatrix} u \\ d \\ s \end{pmatrix}. \tag{20.61}$$

The last notation makes clear that for equal quark masses the Lagrange density exhibits a global $U_V(3)$ flavor symmetry. The index $V$ indicates that it is a vectorial symmetry (this term will be explained after Eq. (20.68)).

The three light quarks transform according to the fundamental representation 3 of the $U_V(3)$:

$$\begin{pmatrix} u \\ d \\ s \end{pmatrix} \longmapsto U_V \begin{pmatrix} u \\ d \\ s \end{pmatrix}, \qquad U_V \in U_V(3), \tag{20.62}$$

while their antiparticles transform according to the complex conjugate representation $\bar{3}$. A bound state of a quark and an antiquark then transforms according to

$$3 \otimes \bar{3} = 8 \oplus 1 \quad \text{or} \quad \square \otimes \begin{matrix}\square\\\square\end{matrix} = \begin{matrix}\square\square\\\square\end{matrix} \oplus \begin{matrix}\square\\\square\end{matrix}, \tag{20.63}$$

and therefore octets and singlets of mesons should exist. A bound state of three quarks, on the other hand, transforms as $3 \otimes 3 \otimes 3 = 10 \otimes 8 \otimes 8 \otimes 1$, and therefore the baryons should appear as singlets, octets or decuplets. Since the particles in an irreducible multiplet are connected by the (approximate) flavor symmetry, they should have (approximately) the same mass. The hadrons can indeed be very well classified into this idealized three-flavor model. Figure 20.1 shows the meson octet and the baryon octet.

If we neglect the masses of the light quarks and set $m_u = m_d = m_s = 0$—this is referred to as the chiral limit – then an extended flavor symmetry exists. In this

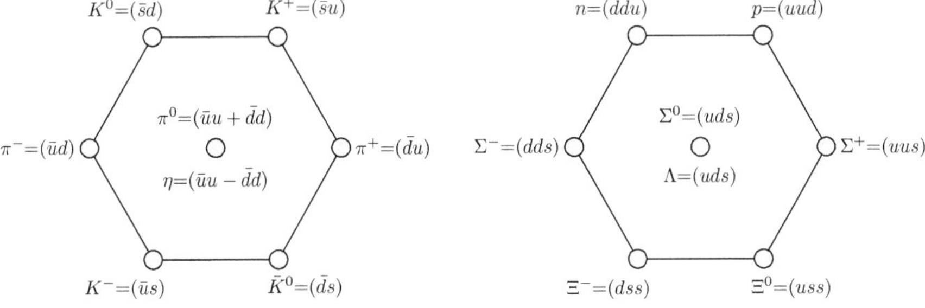


**Fig. 20.1** The octets of pseudoscalar mesons with $J^P = 0^-$ and baryons with $J^P = 1/2^+$. The flavor symmetry is only approximately realized, since the quarks have different masses. Therefore, the hadrons in a multiplet also have slightly different masses

case, $\mathcal{L}_\mathrm{M}(\psi, A_\mu)$ decomposes into two terms: a density for left-handed and one for right-handed quarks. To see this, we decompose the spinor field of the up quark,

$$u = u_R + u_L, \quad u_{R,L} = P_{R,L}u, \quad \text{with} \quad P_R = \tfrac{1}{2}(\mathbb{1} + \gamma_5), \quad P_L = \tfrac{1}{2}(\mathbb{1} - \gamma_5), \tag{20.64}$$

with orthogonal projectors $P_R, P_L$ with $P_R P_L = P_L P_R = 0$ and $P_R + P_L = \mathbb{1}$. These contain the Hermitian matrix $\gamma_5 = i\gamma^0\gamma^1\gamma^2\gamma^3$.

---

**Task**

Show the relations $P_R \slashed{D} P_R = P_L \slashed{D} P_L = 0$, as well as $\bar{u}_R = \overline{P_R u} = \bar{u}P_L$ and $\bar{u}_L = \overline{P_L u} = \bar{u}P_R$.

From these relations follows the decomposition

$$\bar{u}\slashed{D}u = \bar{u}_R \slashed{D} u_R + \bar{u}_L \slashed{D} u_L, \tag{20.65}$$

and with the same decompositions for the $d$ and $s$ quarks, for three massless quarks we get

$$\mathcal{L}_\mathrm{M}(\psi, A_\mu) = i\bar{\psi}_R(\slashed{D} \otimes \mathbb{1}_f)\psi_R + i\bar{\psi}_L(\slashed{D} \otimes \mathbb{1}_f)\psi_L, \quad \psi_{R,L} = \begin{pmatrix} u_{R,L} \\ d_{R,L} \\ s_{R,L} \end{pmatrix}. \tag{20.66}$$

In the chiral limit, $\mathcal{L}_\mathrm{M}$ is thus invariant under global $\mathrm{U}_R(3) \times \mathrm{U}_L(3)$ flavor transformations

$$\psi_R \longmapsto U_R\psi_R \quad \text{and} \quad \psi_L \longmapsto U_L\psi_L, \qquad U_R, U_L \in \mathrm{U}(3). \tag{20.67}$$

An equivalent parametrization of the symmetry is

$$\psi \mapsto \big(U_V P_R + U_V P_L\big)\psi, \quad \psi \mapsto \big(U_A P_R + U_A^{-1} P_L\big)\psi, \qquad U_V, U_A \in \mathrm{U}(3). \tag{20.68}$$

The transformation with $U_V$ is called vectorial, since it acts equally on right- and left-handed fermions. The transformation with $U_A$ is called axial because it rotates right-handed and left-handed fermions in opposite directions. Any unitary matrix $U$ can be decomposed according to $U = e^{i\lambda} V$ with $\det V = 1$. This decomposition leads to the identification of $\mathrm{U}(3)$ with $\mathrm{SU}(3) \times \mathrm{U}(1)/\mathbb{Z}_3$, see Problem 8.4. This means that in the chiral limit the flavor rotations form the symmetry group (modulo discrete subgroup)

$$\mathrm{SU}_R(3) \times \mathrm{U}_R(1) \times \mathrm{SU}_L(3) \times \mathrm{U}_L(1). \tag{20.69}$$

In general, not all global symmetries of a classical field theory are realized in the quantized theory—an example is the axial $U_A(1)$ in QCD. The so-called $U_A(1)$ anomaly is responsible for the fact that QCD only allows for a slightly reduced symmetry.

---

**QCD in the Chiral Limit**
In the chiral limit of three massless quarks, QCD has, in addition to local gauge invariance, the global flavor symmetry

$$SU_V(3) \times SU_A(3) \times U_V(1) . \tag{20.70}$$

The last symmetry factor leads to the conservation of baryon number.

---

However, this is not the whole story. If (20.70) were the symmetry group of QCD in the chiral limit, then the mesons and baryons would have to appear in multiplets of this group. In fact, they only appear in multiplets of the smaller group $SU_V(3)$. The axial $SU_A(3)$ is not visible in the mass spectrum of the bound states. In fact, it is not realized, because it is spontaneously broken. In QCD, a scalar condensate is formed, similar to the spontaneous magnetization in a ferromagnet, and this condensate completely breaks the axial $SU_A(3)$. Finally, we remind that the flavor symmetry in nature is only approximately realized. The quarks are neither massless nor of equal mass. More on this can be found, for example, in [101].

## 20.4.1   QCD Is an SU(3) Gauge Theory

Quarks are peculiar particles: their electric charges are non-integer multiples of the elementary charge and they have never been observed in nature. Hence, there are no free quarks as asymptotic particles that can be detected. They only appear in bound states, these are the mesons and baryons. This so-called *confinement of quarks* is explained by their mutual strong interaction.

QCD is a gauge theory with gauge group $SU_c(3)$ (not to be confused with the approximate flavor symmetry $SU_V(3)$), and the gluons—these are the particles described by the vector potential—are responsible for the confinement. Each of the six quarks comes in 3 colors, so, for example, there is a color multiplet of three up quarks or a color multiplet of three down quarks,

$$(u_a) = \begin{pmatrix} u_r \\ u_g \\ u_b \end{pmatrix}, \quad (d_a) = \begin{pmatrix} d_r \\ d_g \\ d_b \end{pmatrix}, \quad \dots . \tag{20.71}$$

An up quark can therefore be red, green or blue, or, according to the rules of quantum theory, any superposition of these three colors. The mixing of colors is done with an $SU_c(3)$ transformation

$$\psi \longrightarrow U_c\psi, \quad \psi = u, d, s, \ldots, \quad U_c \in SU_c(3). \tag{20.72}$$

In total, there are 6 triplets, so 18 different quarks.

The gauge symmetry is local and exact, in contrast to the approximate global flavor symmetry. All 6 flavors transform according to the fundamental representation 3 of $SU_c(3)$

$$\psi_{f,a}: \quad a \in \{1, 2, 3\}: \text{ color index}, \quad f \in \{u, d, c, s, t, b\}: \text{ flavor index}.$$

The $\mathfrak{su}(3)$-valued vector potential describes the 8 gluons,

$$A_\mu = A_\mu^a X_a, \qquad a = 1, \ldots 8. \tag{20.73}$$

Now one follows the rules in Sect. 20.3 to obtain the Lagrange density of QCD:

$$\mathcal{L}_{\text{QCD}} = -\frac{1}{4}\sum_{a=1}^{8}(F_{\mu\nu}^a)^2 + \sum_{f=1}^{n_f}\sum_{a,b=1}^{3}\bar{\psi}_{f,a}\left(i\slashed{\partial}\delta^{ab} - e\gamma^\mu A_\mu^c X_c^{ab} - \delta^{ab}m_j\right)\psi_{f,b}.$$

$$\tag{20.74}$$

By construction, this density is invariant under local $SU_c(3)$ gauge transformations. For equal quark masses, it is also invariant under flavor transformations.

## 20.4.2  Flavor Symmetries and Spontaneous Symmetry Breaking

From Table 20.1 we see that the up, down, and strange quarks have relatively small masses. Therefore, $\mathcal{L}_M(\psi, A_\mu)$ has, in addition to the local gauge symmetry, the approximate chiral symmetry (20.69). This (classical) symmetry includes 18 covariantly conserved Noether currents. If we denote the traceless and Hermitian generators of the global flavor-$SU_V(3)$ with $T_f$, $f = 1, \ldots, 8$, then they have the form

$$J^\mu = \bar{\psi}\gamma^\mu\psi, \quad J_f^\mu = \bar{\psi}\gamma^\mu T_f\psi, \quad J_5^\mu = \bar{\psi}\gamma^\mu\gamma_5\psi, \quad J_f^{5\mu} = \bar{\psi}\gamma^\mu\gamma_5 T_f\psi. \tag{20.75}$$

Note that the generators $T_f$ of the flavor group and the generators $X_a$ of the gauge group act in different spaces. The Noether charge for the $U_V(1)$ current $J^\mu$ is the baryon number. The charges for the $SU_V(3)$ currents $J_f^\mu$ generate the vector symmetry.

As we have seen, the (approximate) vector symmetry $\mathrm{SU}_V(3)$ is visible in the spectrum of hadrons. The $\mathrm{SU}_A(3)$, on the other hand, is not realized because it is spontaneously broken: in the chiral limit, the vacuum state of QCD is invariant under vectorial transformations, but not under axial ones. This is expressed by the fact that the associated Noether charges either annihilate the vacuum state (leave it invariant) or do not leave it invariant,

$$Q_f|0\rangle = 0 \quad \text{but} \quad Q_f^5|0\rangle \neq 0, \quad f = 1,\ldots,8. \tag{20.76}$$

According to the important Goldstone theorem, there then exist in the chiral limit 8 massless excitations, the so-called Goldstone bosons. For 3 massless flavors, these are the particles in the meson octet in Fig. 20.1. In reality, $\pi^\pm$, $\pi^0$, $K^\pm$, $K^0$, $\bar{K}^0$, $\eta$ are relatively light, but not massless, because the light quarks are not massless either. These mesons are also called pseudo-Goldstone bosons. In addition, the light quarks have different masses, which gives rise to a mass splitting within the octet. Due to the axial anomaly, the axial current $J_5^\mu$ is no longer covariantly conserved as in the classical theory. This anomaly is the prototype for many anomalies in quantum field theories.

## 20.5 Weinberg-Salam Model

The Weinberg-Salam model very successfully describes the electroweak interaction, which is responsible, for example, for the $\beta$-decay of the neutron. The weakly interacting matter particles are the leptons, the quarks, and the Higgs particle. Quarks and leptons come in three families, as shown in Table 20.2. The first family contains the up and down quark, the electron and the electron neutrino. Except for the mass, the families are identical and it is sufficient in the following to consider the members of the first family.

The gauge group is the four-dimensional semi-simple Lie group $\mathrm{SU}_L(2) \times \mathrm{U}_Y(1)$. Accordingly, there exist 4 gauge potentials

$$A_a^\mu : \ \mathrm{SU}_L(2) - \text{gauge fields}, \qquad B^\mu : \ \mathrm{U}_Y(1) - \text{gauge field}, \tag{20.77}$$

which describe the gauge bosons $W^\pm$, $Z^0$, $\gamma$ of the electroweak interaction. These mediate the force between the matter particles. Somewhat unusual is the fact that

**Table 20.2** The three families of matter (fermions), the gauge bosons and the Higgs particle. Except for the gluons $g$, all participate in the electroweak interaction

| I | II | II | | |
|---|----|----|---|---|
| $u$ | $c$ | $t$ | $\gamma$ | $H$ |
| $d$ | $s$ | $b$ | $g$ | |
| $\nu_e$ | $\nu_\mu$ | $\nu_\tau$ | $Z^0$ | |
| $e$ | $\mu$ | $\tau$ | $W^\pm$ | |

right- and left-handed fermions transform according to different representations of the gauge group.

---

**Leptons and Quarks of the First Family in the Weinberg-Salam Model**
The right-handed fermions

$$e_R, \, u_R, \, d'_R, \tag{20.78}$$

are SU(2) singlets, while the left-handed fermions are SU(2) doublets,

$$\begin{pmatrix} \nu_{eL} \\ e_L \end{pmatrix}, \, \begin{pmatrix} u_L \\ d'_L \end{pmatrix}. \tag{20.79}$$

Note that right-handed neutrinos do not occur in the standard model.

---

The down-like quark $d'$ is an eigenstate of the weak interaction. However, this is not the mass eigenstate $d$ measured by detectors. The same applies to the quarks $s$ and $b$. The transition between the mass eigenbasis $d, s, b$ and the eigenbasis of the weak interaction $d', s', b'$ is provided by the Cabibbo–Kobayashi–Maskawa matrix (CKM matrix).

The left-handed fermions transform according to the defining two-dimensional representation of the gauge group $SU_L(2)$. For example, the weak isospin, which is represented by the generator $\tau_3$ in the Lie algebra, acts on the up-quark as follows:

$$\tau_3 u_R = 0, \quad \tau_3 \begin{pmatrix} u_L \\ d'_L \end{pmatrix} = \frac{1}{2} \begin{pmatrix} +u_L \\ -d'_L \end{pmatrix}. \tag{20.80}$$

The members of each multiplet carry the same $U_Y(1)$-hypercharge $Y$ and the electric charge of a fermion is calculated according to the formula by Gell-Mann and Nishijima

$$Q = \tau_3 + \frac{1}{2}Y. \tag{20.81}$$

For fermions of the first family, the assignment of the quantum numbers $Y$ and $Q$ is shown in Table 20.3:

The coupling constant of the $U_Y(1)$ is $g'$ and that of the SU(2) is $g$. Then the covariant derivatives of the right- and left-handed fermions are given by

$$D_\mu \psi_R = \left( \partial_\mu + i \frac{g'}{2} y B_\mu \right) \psi_R \quad , \quad D_\mu \psi_L = \left( \partial_\mu + i \frac{g'}{2} y B_\mu + i \frac{g}{2} A^a_\mu \tau_a \right) \psi_L , \tag{20.82}$$

**Table 20.3** The fermions of the first family with their electric and hypercharges

|  | $\begin{pmatrix} \nu_{eL} \\ e_L \end{pmatrix}$ | $\begin{pmatrix} u_L \\ d'_L \end{pmatrix}$ | $e_R$ | $u_R$ | $d_R$ |
|---|---|---|---|---|---|
| Y | $-1$ | $\frac{1}{3}$ | $-2$ | $\frac{4}{3}$ | $-\frac{2}{3}$ |
| Q | $\begin{pmatrix} 0 \\ -1 \end{pmatrix}$ | $\frac{1}{3}\begin{pmatrix} 2 \\ -1 \end{pmatrix}$ | $-1$ | $\frac{2}{3}$ | $-\frac{1}{3}$ |

where the hypercharge $y$ of the respective multiplet appears. With this, we now form the gauge-invariant fermionic part of the Lagrange density,

$$\mathcal{L}_{\mathrm{M}}(\psi, A_\mu) = i\bar{\psi}_R \gamma^\mu D_\mu \psi_R + i\bar{\psi}_L \gamma^\mu D_\mu \psi_L \, ,$$

which includes a (not explicitly written) sum over the multiplets of the first family. An explicit mass term for the fermions does not appear here.

### 20.5.1 Higgs Mechanism

Fermions and three of the four gauge bosons receive their mass via the mechanism named after Peter Higgs. This requires a scalar field, which in the Weinberg-Salam model is a complex SU(2)-doublet with hypercharge $Y = 1$,

$$\phi = \frac{1}{\sqrt{2}} \begin{pmatrix} \phi_3 + i\phi_4 \\ \phi_1 + i\phi_2 \end{pmatrix}, \qquad Q \stackrel{Y=1}{=} \begin{pmatrix} 1 \\ 0 \end{pmatrix}. \tag{20.83}$$

The covariant derivative of the scalar field is accordingly

$$D_\mu \phi = \left( \partial_\mu + i\frac{g'}{2} B_\mu + i\frac{g}{2} A_\mu^a \tau_a \right)\phi \, . \tag{20.84}$$

The fermions and gauge bosons receive a mass through their interaction with the scalar particles, which condense at low temperatures, similar to the Cooper pairs in superconductors. The condensation originates from the self-interaction of the scalar particles, described by the gauge-invariant Lagrange density

$$\mathcal{L}_{\mathrm{M}}(\phi, A_\mu) = (D_\mu \phi)^\dagger D^\mu \phi - V(\phi) \, . \tag{20.85}$$

The quartic potential with a negative coefficient of the quadratic term,

$$V(\phi) = -c^2 \phi^\dagger \phi + \lambda (\phi^\dagger \phi)^2$$

has a local maximum at the origin. A corresponding "Mexican hat potential" for $\phi \in \mathbb{C}$ (and not $\mathbb{C}^2$, as in the Weinberg-Salam model) is shown in Fig. 20.2. All minima of the potential are possible classical ground states (vacuum states). We

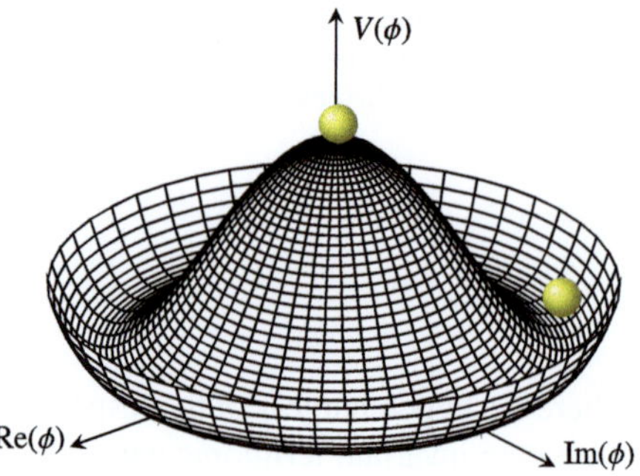


**Fig. 20.2**  A Mexican hat potential for a complex scalar field

choose an electrically neutral one, and the assignment of charges in (20.83) suggests the following choice,

$$\langle \phi \rangle = \frac{1}{\sqrt{2}} \begin{pmatrix} 0 \\ v \end{pmatrix} . \tag{20.86}$$

---

**Question**

Which subgroup of $SU_L(2) \times U_Y(1)$ leaves the ground state (20.86) invariant?

---

The vacuum expectation value (20.86) is no longer invariant under all gauge transformations, but only under gauge transformations in the stabilizer group of $\langle \phi \rangle$. This is a U(1)-subgroup of the gauge group. In this way, the original gauge symmetry is broken to the abelian gauge symmetry U(1),

$$SU_L(2) \times U_Y(1) \longrightarrow U_Q(1) . \tag{20.87}$$

At the minimum of the potential, the squared vacuum expectation value is equal to $v^2/2$.

## 20.5.2  Mass Generation

We decompose the scalar field into its condensate part $\langle \phi \rangle$ in (20.86) and the quantum fluctuations away from the condensate,

$$\phi \longmapsto \langle \phi \rangle + \phi , \tag{20.88}$$

and substitute this decomposition into the Lagrange density $\mathcal{L}_M$. If we expand up to the second order in the quantum fields, we can read off the masses of the particles they describe, at least in the semi-classical approximation. The square of the covariant derivative of the scalar field has the expansion

$$(D_\mu \phi)^\dagger (D^\mu \phi) = \partial_\mu \phi^\dagger \partial^\mu \phi + \frac{1}{2} A_\mu^A M_{AB}^2(v) A^{B\mu} + \dots , \tag{20.89}$$

from which we read off the masses of the vector bosons

$$A_\mu^A = (A_\mu^1, A_\mu^1, A_\mu^3, B_\mu) \tag{20.90}$$

To do this, we must diagonalize the resulting mass matrix

$$M^2(v) = \frac{v^2}{2} \begin{pmatrix} g^2 & 0 & 0 & 0 \\ 0 & g^2 & 0 & 0 \\ 0 & 0 & g^2 & -gg' \\ 0 & 0 & -gg' & g'^2 \end{pmatrix}$$

**Masses of the Gauge Bosons**
The mass matrix has the following eigenstates and eigenvalues:

$$W^\pm: \quad m^2 = \frac{1}{4}g^2 v^2 \,,$$

$$Z: \quad m^2 = \frac{1}{4}v^2(g^2 + g'^2) \,, \tag{20.91}$$

$$A: \quad m^2 = 0 \quad (\text{Photon})\,.$$

Three of the four gauge bosons receive a mass of the order $gv$, and one gauge boson remains massless. The latter is identified with the photon.

To read off the mass of the fermions, we expand the Yukawa interaction between the scalar field and fermion fields up to the second order,

$$\mathcal{L}_{\text{Yuk}}(\phi, \psi) = \gamma_e \begin{pmatrix} v_{eL} \\ e_L \end{pmatrix} \langle \phi \rangle \, e_R + \cdots \propto (\gamma_e v)\, e_L e_R + \ldots \,. \tag{20.92}$$

It follows that the fermions receive a mass proportional to the product of the vacuum expectation value $v$ and the Yukawa coupling $\gamma$ via the Higgs mechanism. Note that the neutrinos remain massless, as the first component of the condensate $\langle \phi \rangle$ vanishes.

## 20.6   Exercises for Chap. 20

**Problem 20.1 (Hamilton's Equations of Motion for a Charged Point Particle)**
Show that Hamilton's equations of motion for a point particle with Hamilton function (20.7) lead to Lorentz's equations of motion for the particle's position.

**Problem 20.2 (Klein-Gordon Equation in an External Electromagnetic Field)**
Show that the Euler-Lagrange equation for the Lagrange density (20.16) with
$V(\phi^*\phi) = \mu^2\phi^*\phi$ has the following form,

$$D_\mu D^\mu \phi + \mu^2 \phi = 0 .$$

Here, $A_\mu$ is an external potential and $\phi$ is varied.

**Problem 20.3 (Bianchi Identity)**  Prove the Bianchi identity

$$D_\mu F^a_{\nu\rho} + D_\nu F^a_{\rho\mu} + D_\rho F^a_{\mu\nu} = 0 .$$

Convince yourself that for a U(1) gauge field, the Bianchi identity corresponds to
the two Maxwell equations $\operatorname{div} \boldsymbol{B} = 0$ and $\operatorname{rot} \boldsymbol{E} = -\partial_t \boldsymbol{B}$.

**Problem 20.4 (Noether Current for a Minimally Coupled Scalar Field)** We
consider a minimally coupled complex scalar field with Lagrange density (20.16).

- Verify that the gauge current has the following form:

$$J^\mu = -\mathrm{i}e\left((D_\mu\phi)^\dagger \phi - \phi^\dagger D_\mu\phi\right) .$$

- The Lagrange density is gauge invariant, $\mathcal{L}_\mathrm{M}(\mathrm{e}^{\mathrm{i}\lambda}\phi, A_\mu - \mathrm{i}\partial_\mu\lambda) = \mathcal{L}_\mathrm{M}(\phi, A_\mu)$.
  Show now that the following alternative formula for the Noether current follows:

$$J^\mu(x) = -\frac{\partial \mathcal{L}_\mathrm{M}(\phi, A_\mu)}{\partial A_\mu}\bigg|_{A=0} = -\frac{\delta S_\mathrm{M}[\phi, A]}{\delta A_\mu(x)}\bigg|_{A=0} .$$

**Problem 20.5 (Mass Term for Gauge Bosons)** Show that a mass term
$\frac{1}{2}m^2 A^a_\mu A^{a\mu}$ for the non-Abelian gauge potential is not invariant under SU($N$)
gauge transformations.

**Problem 20.6 (Chern-Simons Theory)** Consider a real vector field $B_\mu$ in a three-
dimensional spacetime with action

$$S = \int d^3x \left( -\frac{1}{4}G_{\mu\nu}G^{\mu\nu} + g\epsilon^{\mu\nu\rho} B_\mu \partial_\nu B_\rho \right) ,$$

with totally antisymmetric tensor $\epsilon^{\mu\nu\rho}$ ($\epsilon^{012} = 1$), real coupling constant $g$ and field
strength tensor $G_{\mu\nu} = \partial_\mu B_\nu - \partial_\nu B_\mu$.

1. Is the action invariant under gauge transformations $B_\mu \mapsto B'_\mu = B_\mu - \partial_\mu\alpha$?
2. What are the equations of motion?
3. Express these through the dual field strength $\tilde{G}^\mu = \frac{1}{2}\epsilon^{\mu\nu\rho}G_{\nu\rho}$. Does the result
   look familiar to you?

# Conformal-Invariant Field Theories (CFT) 21



The group of Poincaré-transformations was thoroughly examined in Sects. 5.5 and 14.6. Now we turn to an important extension of the Poincaré-group—the conformal group. This includes all angle-preserving mappings in Minkowski space. In more than two dimensions, the group is finite-dimensional and contains the Lorentz transformations, translations, dilations, special-conformal transformations and inversion. In two spacetime dimensions, the group is infinite-dimensional. Most of the time, we ignore the difficulties that can arise when defining the composition of two such (only locally defined) transformations.

Conformal invariant field theories, briefly called conformal field theories, are field theories with the conformal group as symmetry group. They are very special relativistic field theories. Most physical models possess a characteristic length scale—whose inverse is identified as the smallest mass in the system—and are therefore not invariant under scale transformations (dilations) and thus also not conformally invariant. Conversely, theories with only massless particles are good candidates for scale-invariant theories. For example, classical electrodynamics is conformally invariant. Scale invariance is also observed in systems near a continuous phase transition.

We start with the discussion of conformal symmetry and then characterize conformal field theories. We pay special attention to theories in $1 + 1$ dimensions, as these play a central role in recent developments. Conformal field theories in $1 + 1$ dimensions have the Virasoro algebra as an infinite-dimensional symmetry algebra [102]. Some allow an even larger symmetry algebra, for example a Kac-Moody algebra [103].


© The Author(s), under exclusive license to Springer-Verlag GmbH, DE,      477
part of Springer Nature 2026
A. Wipf, *Symmetries in Physics*, https://doi.org/10.1007/978-3-662-72675-4_21

Detailed presentations of conformal field theories and their symmetry algebras can be found in [64, 104, 105]. These theories find application in statistical physics and especially the theory of critical phenomena, two-dimensional quantum field theories as well as in particle physics and string theory. Structural and mathematical properties of conformal field theories are the subject of the monograph [106]. The historical development of conformal transformations and symmetries is outlined by a pioneer of the research field in [107].

## 21.1   Conformal Mappings

In Euclidean space, lengths and angles have the properties well-known from Euclidean geometry. As in Euclidean space, in Minkowski spacetime, the "scalar product" of two vector $\xi$ and $\zeta$ is defined according to $(\xi, \zeta) = \eta_{\mu\nu}\xi^\mu\zeta^\nu$. In analogy to Euclidean space, we call a *linear mapping* $\xi \mapsto J\xi$ in Minkowski spacetime $M$ "angle-preserving", if for two arbitrary $\xi, \zeta$ we have

$$(J\xi, J\zeta) = \lambda^2(\xi, \zeta) \implies J^T\eta J = \lambda^2\eta. \tag{21.1}$$

**Question**

In fact, it suffices to require the first condition for $\zeta = \xi$. How can the product $(\xi, \zeta)$ be obtained from the knowledge of all $(\xi, \xi)$?
Hint: The polarization formula may be helpful here.

For the scaling factor $\lambda = 1$, $J$ is a Lorentz transformation. In the following, we use the terms length and angle in the described sense.

**Example: Linear Conformal Transformation in $2d$ Minkowski Space**

In two-dimensional Minkowski space, the matrix of a linear conformal transformation has the form

$$J = \lambda \begin{pmatrix} \cosh\alpha & \sinh\alpha \\ \sinh\alpha & \cosh\alpha \end{pmatrix}.$$

In the Fig. 21.1, the images of the vectors $(\cos n\pi/6, \sin n\pi/6)$ with $n = 0, 1, \ldots, 11$ under the conformal transformation $\xi \mapsto J\xi$ are shown for the parameter values $\lambda = 2$ and $\alpha = 0.5$.

◄

We begin with the geometric definition of a (in general nonlinear) conformal transformations in d-dimensional Minkowski space. A transformation

$$x^\mu \mapsto y^\mu = y^\mu(x^\nu) \tag{21.2}$$

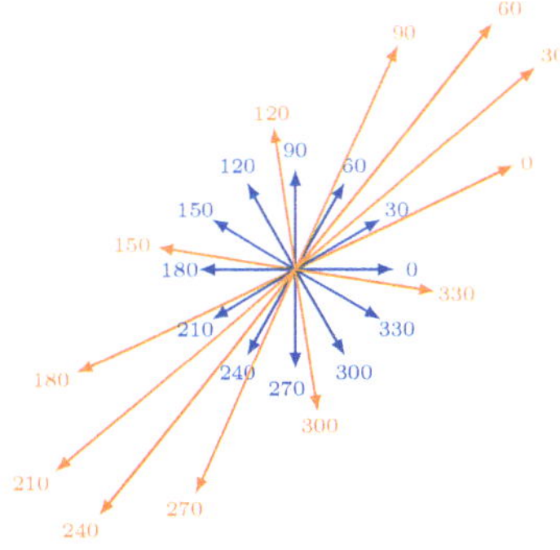

**Fig. 21.1** The linear conformal transformation $J$ maps the blue to the orange vectors

is called conformal if its (linear) tangential transformations are angle-preserving. In other words: if $\xi(x)$ and $\zeta(x)$ are the tangential vectors of two curves intersecting at $x$, then a conformal transformation (21.2) preserves the angle between these vectors.[1] This is equivalent to

$$(J\xi, J\xi) = e^{2\sigma(x)}(\xi, \xi), \quad \text{where} \quad J^\mu_\alpha = \frac{\partial y^\mu}{\partial x^\alpha} \tag{21.3}$$

is the Jacobian matrix of the mapping (21.2) at the point $x$. This matrix $J$ is, up to a scale factor $e^\sigma$, the matrix of a generally space- and time-dependent Lorentz transformation, $J(x) = e^{\sigma(x)}\Lambda(x)$.

**Conformal Mapping**

A mapping is conformal if the Lorentz-invariant length element changes only by a (generally $x$-dependent) factor,

$$\eta_{\mu\nu}\, \mathrm{d}y^\mu \mathrm{d}y^\nu = \eta_{\mu\nu} \frac{\partial y^\mu}{\partial x^\alpha} \frac{\partial y^\nu}{\partial x^\beta}\, \mathrm{d}x^\alpha \mathrm{d}x^\beta = e^{2\sigma(x)} \eta_{\alpha\beta}\, \mathrm{d}x^\alpha \mathrm{d}x^\beta . \tag{21.4}$$

Under conformal mappings in Minkowski space, light-like vectors are mapped into light-like vectors or light cones into light cones.

**Task**

Show that for the inverse mapping $y^\mu \mapsto x^\mu(y^\alpha)$ we have

$$\eta_{\alpha\beta} \frac{\partial x^\alpha}{\partial y^\mu} \frac{\partial x^\beta}{\partial y^\nu} = e^{-2\sigma(y)} \eta_{\mu\nu} . \tag{21.5}$$

This means that it is also angle-preserving with conformal factor $e^{-\sigma}$.

---

[1] We use the same notation as in Sect. 5.5.

Finite conformal mappings $x^\mu \mapsto y^\mu(x)$ can map points in $M$ to infinity or spacelike separated into timelike separated points. Lüscher and Mack [108] were able to show that the covering space $\tilde{M} = S^{d-1} \times \mathbb{R}$ of $M$ has a global causal structure and the universal covering group of the conformal group operates on it respecting the causal structure.[2] The problem with causality thus arises from the fact that an attempt has been made to identify the different sheets in $\tilde{M}$.

For infinitesimal conformal mappings generated by a vector field $X(x)$ near the identity

$$y^\mu \approx x^\mu + X^\mu(x) \implies \frac{\partial y^\mu}{\partial x^\alpha} \approx \delta^\mu_{\ \alpha} + \frac{\partial X^\mu}{\partial x^\alpha} \equiv \delta^\mu_{\ \alpha} + X^\mu_{\ ,\alpha}, \tag{21.6}$$

such problems do not occur. For the identity mapping, $\sigma$ in (21.3) is zero, and therefore for infinitesimal conformal mappings $e^{2\sigma} \approx 1 + 2\sigma$. Now we expand the second relation in (21.4) to first order and obtain

$$\eta_{\alpha\beta} + X_{\alpha,\beta} + X_{\beta,\alpha} = \eta_{\alpha\beta} + 2\sigma\,\eta_{\alpha\beta}. \tag{21.7}$$

Here we see that the symmetrized derivative of $X$ must be proportional to $\eta_{\alpha\beta}$. Taking the trace of this relation, we find

$$\sigma = \frac{1}{d}\,\partial_\rho X^\rho \implies X_{\alpha,\beta} + X_{\beta,\alpha} = \frac{2}{d}\,\eta_{\alpha\beta}\partial_\rho X^\rho. \tag{21.8}$$

The last relation is the conformal Killing equation.

**Conformal Killing Vector Fields**
Conformal mappings are generated by conformal Killing vector fields $X^\mu$. These satisfy the conformal Killing equation

$$X_{\mu,\nu} + X_{\nu,\mu} = \frac{2}{d}\,\eta_{\mu\nu}(\partial_\rho X^\rho). \tag{21.9}$$

If the source term on the right-hand side vanishes, then $X_\mu$ is called a Killing vector field. For these, the function $\sigma$ vanishes according to (21.8) and the mapping (21.2) is length-preserving.

---

[2] Coverings and covering groups were discussed in Sect. 8.4.2.

### 21.1.1 Conformal Mappings in $d \geq 3$ Dimensions

In $d > 2$ dimensions, there exist only a finite number of independent solutions of the conformal Killing equation. To prove this, we rewrite the second derivatives of $\mathrm{div}X = \partial_\rho X^\rho$:

$$\partial_\mu \partial_\nu (\partial^\rho X_\rho) = \partial_\mu \partial^\rho (\partial_\nu X_\rho + \partial_\rho X_\nu - \partial_\rho X_\nu) \overset{(21.9)}{=} \partial_\mu \partial^\rho \left( \frac{2}{d} \eta_{\nu\rho} \mathrm{div}X - \partial_\rho X_\nu \right)$$

$$= \frac{2}{d} \partial_\mu \partial_\nu \mathrm{div}X - \Box(\partial_\mu X_\nu) \,.$$

Since the expression is symmetric in the indices $\mu$ and $\nu$, we also have

$$\left( \frac{2}{d} - 1 \right) \partial_\mu \partial_\nu (\mathrm{div}X) = \frac{1}{2} \Box(\partial_\mu X_\nu + \partial_\nu X_\mu) \overset{(21.9)}{=} \frac{1}{d} \eta_{\mu\nu} \Box(\mathrm{div}X) \,. \qquad (21.10)$$

Taking the trace of this matrix equation, we obtain

$$(d - 1)\Box(\mathrm{div}X) = 0 \,. \qquad (21.11)$$

This means that in $d \neq 1$ dimensions $\mathrm{div}X$ is a harmonic function.[3] Due to (21.10), all second derivatives $\partial_\mu \partial_\nu (\mathrm{div}X)$ then vanish in $d > 2$ dimensions. Therefore, $\mathrm{div}X$ is a linear function and accordingly $X^\mu$ a quadratic function of the coordinates,

$$X^\mu = a^\mu + b^\mu_{\ \alpha} x^\alpha + \frac{1}{2} c^\mu_{\ \alpha\beta} x^\alpha x^\beta, \quad \text{with} \quad c^\mu_{\ \alpha\beta} = c^\mu_{\ \beta\alpha} \,. \qquad (21.12)$$

If we insert this $X^\mu$ into the conformal Killing equation (21.9), we find

$$b_{\mu\nu} + b_{\nu\mu} + (c_{\mu\nu\alpha} + c_{\nu\mu\alpha}) x^\alpha = \frac{2}{d} \eta_{\mu\nu} (b^\rho_{\ \rho} + c^\rho_{\ \rho\alpha} x^\alpha) \,.$$

A comparison of coefficients makes it clear that the antisymmetric part $\omega_{\mu\nu}$ of $b_{\mu\nu}$ is undetermined and the symmetric part must be proportional to the metric tensor, $b_{\mu\nu} = \omega_{\mu\nu} + \lambda \eta_{\mu\nu}$. Similarly, one decomposes $c_{\mu\nu\alpha}$ into a symmetric and antisymmetric part in the first two indices and concludes

$$c_{\mu\nu\alpha} = 2\eta_{\mu\nu} c_\alpha + \omega_{\mu\nu\alpha} \,, \quad \text{with} \quad \omega_{\mu\nu\alpha} = -\omega_{\nu\mu\alpha} \,,$$

and constant vector $c^\mu$. Since $c_{\mu\nu\alpha}$ must be symmetric in the last two indices according to (21.12), we finally obtain

$$c_{\mu\nu\alpha} = 2(\eta_{\mu\nu} c_\alpha + \eta_{\mu\alpha} c_\nu - \eta_{\nu\alpha} c_\mu) \,.$$

---

[3] Similar to the terms length and angle, we adopt the term "harmonic" from Euclidean and call a solution of $\Box\phi = 0$ a harmonic function.

**Table 21.1** The conformal mappings in $d > 2$ dimensions. Listed are the conformal Killing vector fields $X^\mu$, the corresponding finite conformal mappings $y(x)$ and their conformal factor $e^\sigma$. We use the abbreviation $N = 1 - 2(c, x) + c^2 x^2$

| Mapping | $X^\mu(x)$ | $y(x)$ | $e^{\sigma(x)}$ |
|---|---|---|---|
| Translations | $a^\mu$ | $x^\mu + a^\mu$ | 1 |
| Lorentz transformations | $\omega^\mu{}_\alpha x^\alpha$ | $\Lambda^\mu{}_\alpha x^\alpha, \ \Lambda = e^\omega$ | 1 |
| Dilatation | $\lambda x^\mu$ | $e^\lambda x^\mu$ | $e^\lambda$ |
| Special conformal transformations | $2(c \cdot x)x^\mu - x^2 c^\mu$ | $\frac{1}{N}(x^\mu - x^2 c^\mu)$ | $\frac{1}{N}$ |

**Conformal Killing Vector Field in $d \geq 3$ Dimensions**

In $d \geq 3$ dimensions, a conformal Killing vector field has the form

$$X^\mu = a^\mu + \omega^\mu{}_\nu x^\nu + \lambda x^\mu + \left\{2(c \cdot x)x^\mu - x^2 c^\mu\right\}, \quad \text{with} \quad \omega_{\mu\nu} = -\omega_{\nu\mu}. \tag{21.13}$$

It is determined by the values of the parameters $(a^\mu, \omega_{\mu\nu}, \lambda, c^\mu)$ and therefore the number of independent vector fields is

$$d + \frac{d(d-1)}{2} + 1 + d = \frac{(d+2)(d+1)}{2}. \tag{21.14}$$

Each conformal Killing vector field generates a finite conformal mapping $y(x)$ as listed in Table 21.1: The inversion at the Lorentzian unit sphere in Minkowski space, consisting of the spacelike hyperboloids $t^2 - x^2 = 1$ and the timelike hyperboloid $t^2 - x^2 = -1$, has the form

$$x^\mu \mapsto \frac{x^\mu}{(x, x)}, \tag{21.15}$$

and is also a conformal mapping. But it reverses the orientation and therefore is not in the connected component of the identity mapping.

**Example: Inversion in Two-Dimensional Minkowski Space**

Under an inversion at the Lorentzian unit sphere, the coordinate lines $(t, x_0)$ and $(t_0, x)$ are mapped to the curves

$$\frac{1}{t^2 - x_0^2}(t, x_0) \quad \text{and} \quad \frac{1}{t_0^2 - x^2}(t_0, x).$$

In the Fig. 21.2, the world lines $(t, x_0)$ of stationary objects with $x_0$ in $\{0.4, 0.65, 1, 2\}$ and their image curves are shown. Points in the forward and

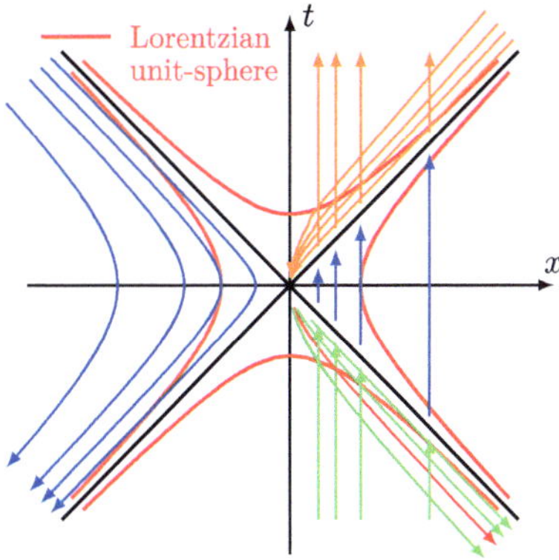

**Fig. 21.2** The worldlines of 4 stationary observers and their images under an inversion at the unit hypersphere in $2d$ Minkowski space



backward light cone are reflected at the Lorentzian unit sphere. Points that are spacelike relative to the origin are additionally reflected at the origin due to $(x, x) < 0.$ ◀

With the inversion one can obtain the special conformal transformations. For this, one considers an inversion, followed by a translation with subsequent inversion,

$$x^\mu \mapsto \frac{x^\mu}{(x, x)} \mapsto \frac{x^\mu}{(x, x)} - c^\mu \mapsto \frac{x^\mu - x^2 c^\mu}{N}. \tag{21.16}$$

The inversion is singular on the light cone. The problem can be solved by extending the domain by a null cone, which is added at infinity (in Euclidean space one adds a point at infinity).

## 21.1.2 Conformal Mappings in $d = 2$ Dimensions

In two spacetime dimensions, $\mathrm{div}\,X$ is still a harmonic function, but from (21.10) we can no longer infer that the second derivatives of $\mathrm{div}\,X$ vanish. Therefore, the set of conformal mappings contains significantly more elements than in higher dimensions. The structure and properties of conformal mappings in two dimensions become clearer after introducing light cone coordinates,

$$x^\pm = x^0 \pm x^1, \qquad \partial_\pm = \frac{1}{2}(\partial_0 \pm \partial_1), \qquad \eta_{+-} = \eta_{-+} = \frac{1}{2}. \tag{21.17}$$

In these coordinates, the conformal Killing equation (21.9) simplifies to

$$\partial_- X^+ = \partial_+ X^- = 0, \tag{21.18}$$

and therefore every vector field with light cone components $\{X^+, X^-\} = \{X^+(x^+), X^-(x^-)\}$ is a conformal Killing vector field.

**Task**

Show that in $1 + 1$ dimensions the conformal Killing equation is equivalent to (21.18).

As a basis for the conformal Killing vector fields, one chooses the monomials,

$$X_n^+(x^+) = (x^+)^n, \quad X_n^-(x^-) = (x^-)^n, \qquad n = 0, 1, 2, 3, \ldots \tag{21.19}$$

A look at Table 21.1 reveals, that the translations, Lorentz transformations, dilations and special conformal transformations occurring in all spacetime dimensions are generated by those $X^\mu$ which are constant, linear and quadratic in $x^\mu$. In 2 dimensions, these 6 conformal Killing vector fields generate the Lie group $SO(2, 2) \cong SL(2, \mathbb{R}) \times SL(2, \mathbb{R})$ of Möbius transformations

$$x^\pm \longrightarrow y^\pm = \frac{a^\pm x^\pm + b^\pm}{c^\pm x^\pm + d^\pm}, \qquad \begin{pmatrix} a^\pm & b^\pm \\ c^\pm & d^\pm \end{pmatrix} \in SL(2, \mathbb{R}). \tag{21.20}$$

As functions of the light cone coordinates, they are listed in Table 21.2.
A general finite conformal transformation generated by the conformal Killing vector fields $X^+(x^+)$ and $X^-(x^-)$ has the form

$$x^+ \mapsto y^+(x^+) \quad \text{and} \quad x^- \mapsto y^-(x^-). \tag{21.21}$$

These transformations define the infinite-dimensional group of conformal transformations in $1 + 1$ dimensions. It contains the Möbius transformations as a subgroup. We will investigate the Lie algebra of the conformal group in Sect. 21.2, when we study its action on fields of a relativistic field theory. There we will also explicitly calculate the commutation relations of the conformal algebra.

**Table 21.2** The conformal transformations in $SL(2, \mathbb{R}) \times SL(2, \mathbb{R})$ (last column) with associated conformal Killing vector fields (second column)

| Transformations | $X^\pm(x)$ | $y^\pm(x)$ |
|---|---|---|
| Translations | $a^\pm$ | $x^\pm + a^\pm$ |
| Lorentz transformations | $\pm \omega x^\pm$ | $e^{\pm\omega} x^\pm$ |
| Dilations | $\lambda x^\pm$ | $e^\lambda x^\pm$ |
| Special conformal transformations | $4d^\pm (x^\pm)^2$ | $x^\pm (1 - 4d^\pm x^\pm)^{-1}$ |

## 21.2  Conformally Invariant Field Theories (CFT)

If one scales all lengths by a factor $e^\lambda$ in a physical system without intrinsic length scale—one speaks of a scale-invariant system—its properties do not change. It turns out that many scale-invariant theories are even invariant under conformal transformations. This is particularly true for unitary, relativistic and scale-invariant theories (with discrete spectrum of scale dimension) in 2 spacetime dimensions [108, 109].

### 21.2.1  Weyl Invariant Field Theories

There exists a constructive characterization of conformal field theories (see [110]). One can show, that classical field theories, which can be covariantly and Weyl-invariantly coupled to the gravitational field—encoded in the metric tensor $g_{\mu\nu}$ of curved spacetime—define conformally invariant theories in Minkowski space. For example, a mass term would destroy the Weyl invariance. We begin with the Weyl invariance of generally covariant field theories.

---

**Weyl-Invariant Field Theories**

A field theory is called Weyl-invariant if its field can be coupled to the gravitational field in such a way that the action is invariant under Weyl transformations of the metric and the field

$$g_{\mu\nu}(y) \mapsto e^{2\sigma(y)} g_{\mu\nu}(y) \quad \text{and} \quad \phi(y) \mapsto e^{-w_\phi \sigma(y)} \phi(y), \tag{21.22}$$

with Weyl weight $w_\phi$,

$$S\big(e^{2\sigma(y)} g_{\mu\nu}(y),\, e^{-w_\phi \sigma(y)} \phi(y)\big) = S\big(g_{\mu\nu}(y), \phi(y)\big). \tag{21.23}$$

If the matter is described by several fields, these generally have different Weyl weights.

---

A matter field can be covariantly coupled to the gravitational field by replacing partial derivatives with covariant derivatives and contracting tensors with $g_{\mu\nu}$ or $g^{\mu\nu}$ instead of $\eta_{\mu\nu}$ or $\eta^{\mu\nu}$. Furthermore, the invariant volume element is $\sqrt{-g}\, d^d x$, where $g = \det(g_{\mu\nu})$.

### Example: Electrodynamics in 4 Dimensions

If we apply these rules to $S$ in (20.21), this leads to the action of Maxwell's theory invariant under general coordinate transformations,

$$S[g_{\mu\nu}, A_\mu] = -\frac{1}{4} \int d^4x \sqrt{-g}\, g^{\mu\nu} g^{\alpha\beta} F_{\mu\alpha} F_{\nu\beta}, \quad F_{\mu\nu} = \nabla_\mu A_\nu - \nabla_\nu A_\mu .$$

$$(21.24)$$

In the expression for the antisymmetric field strength tensor $F_{\mu\nu}$, the covariant derivatives can be replaced by ordinary derivatives. The action (21.24) is Weyl-invariant if we assign the Weyl weight 0 to the vector potential (and field strength tensor),

$$S[e^{2\sigma} g_{\mu\nu}, A_\mu] = -\frac{1}{4} \int d^4x \, (e^{4\sigma} \sqrt{-g})(e^{-2\sigma} g^{\mu\nu})(e^{-2\sigma} g^{\alpha\beta}) F_{\mu\alpha} F_{\nu\beta}$$

$$= S[g_{\mu\nu}, A_\mu] . \qquad (21.25)$$

We have used that the determinant $\det (g_{\mu\nu})$ has the Weyl weight 8. The classical Yang-Mills theories introduced in Chap. 20 are also Weyl-invariant in $d = 4$ dimensions. ◄

### Question

Massless scalar fields and Dirac spinor fields can be covariantly and Weyl-invariantly coupled to the gravitational field in any dimension. What are their Weyl weights?

Now we come to the diffeomorphisms, which have the form (21.2) in local coordinates. Under a diffeomorphism, the covariant components of a tensor field transform according to

$$\phi_{\mu\nu\ldots}(x) = \frac{\partial y^\alpha}{\partial x^\mu} \frac{\partial y^\beta}{\partial x^\nu} \cdots \phi_{\alpha\beta\ldots}(y) . \qquad (21.26)$$

The conformal invariance in Minkowski space can now be justified as follows [110]:

1. In the first step, we choose curvilinear coordinates in Minkowski space. Here, $\eta_{\alpha\beta}$ and $\phi_{\alpha\beta}$ transform as tensor fields,

$$\left\{ x^\alpha, \eta_{\alpha\beta}, \phi_{\alpha\beta\ldots}(x) \right\} \longmapsto \left\{ y^\alpha, \frac{dy^\mu}{dx^\alpha} \frac{dy^\nu}{dx^\beta} \eta_{\mu\nu}, \frac{dy^\mu}{dx^\alpha} \frac{dy^\nu}{dx^\beta} \cdots \phi_{\mu\nu\ldots}(y) \right\} .$$

$$(21.27)$$

For a generally covariant theory, the action $S[g_{\mu\nu}, \phi]$ is invariant under this coordinate transformation. If $x \mapsto y$ is a conformal mapping, then according to (21.4)

$$\frac{\mathrm{d}y^\mu}{\mathrm{d}x^\alpha} \frac{\mathrm{d}y^\nu}{\mathrm{d}x^\beta} \eta_{\mu\nu} = \mathrm{e}^{2\sigma(x)} \eta_{\alpha\beta}, \tag{21.28}$$

so that the transformed metric in (21.27) is equal to the Minkowski metric up to a conformal factor $\mathrm{e}^{2\sigma}$.

2. In the second step, we bring the metric back to Minkowski form with a compensating Weyl transformation (21.22) with conformal factor $\mathrm{e}^{-2\sigma}$,

$$\left\{ y^\mu, \ \mathrm{e}^{2\sigma} \eta_{\alpha\beta}, \ \frac{\mathrm{d}y^\mu}{\mathrm{d}x^\alpha} \frac{\mathrm{d}y^\nu}{\mathrm{d}x^\beta} \cdots \phi_{\mu\nu\ldots}(y) \right\}$$

$$\longmapsto \left\{ y^\mu, \ \eta_{\mu\nu}, \ \mathrm{e}^{w_\phi \sigma(x)} \frac{\mathrm{d}y^\mu}{\mathrm{d}x^\alpha} \frac{\mathrm{d}y^\nu}{\mathrm{d}x^\beta} \cdots \phi_{\mu\nu\ldots}(y) \right\}. \tag{21.29}$$

For a Weyl-invariant field theory, the action is then also invariant under the transformation (21.29), which leads back to the original Minkowski metric.

So if we first perform the coordinate transformation (21.27) and then the compensating Weyl transformation (21.29), we find the following symmetry in Minkowski space:

$$\left\{ x^\mu, \eta_{\mu\nu}, \phi_{\alpha\beta\ldots}(x) \right\} \longmapsto \left\{ y^\mu(x), \eta_{\mu\nu}, \mathrm{e}^{w_\phi \sigma(x)} \frac{\partial y^\mu}{\partial x^\alpha} \frac{\partial y^\nu}{\partial x^\beta} \phi_{\mu\nu\ldots}\big(y(x)\big) \right\}, \tag{21.30}$$

where $\sigma$ is determined by the relation in (21.28).

> **Weyl Invariance Implies Conformal Invariance**
> A relativistic field theory that can be coupled covariantly and Weyl-invariantly to the gravitational field $g_{\mu\nu}$ is invariant under conformal mappings in Minkowski space.

## 21.2.2 Consequences of Conformal Invariance

We have shown that under conformal mappings in Minkowski space a matter field $\phi_{\mu\nu\ldots}$ transforms according to (21.30). Often it is desirable, for example in the derivation of Ward identities, to know the infinitesimal symmetry transformations.

To this end, we consider, similar to (21.6), conformal mappings $x^\mu \mapsto y^\mu(x^\nu)$ close to the identity. For these, $y^\mu \approx x^\mu + X^\mu(x)$ and according to (21.8) $\sigma = \frac{1}{d}\partial_\mu X^\mu$.

---

**Infinitesimal Conformal Transformation of a Tensor Field**

Under an infinitesimal conformal mapping with conformal Killing vector field $X$, a tensor field transforms according to

$$\delta_X \phi_{\alpha\beta\dots} = (L_X\phi)_{\alpha\beta\dots} + \frac{w_\phi}{d}\,\mathrm{div}(X)\,\phi_{\alpha\beta\dots}\,. \tag{21.31}$$

Here, $L_X$ is the Lie derivative in the direction of $X$,

$$(L_X\phi)_{\alpha\beta\dots} = X^\mu \partial_\mu \phi_{\alpha\beta\dots} + \partial_\alpha X^\mu \phi_{\mu\beta\dots} + \partial_\beta X^\mu \phi_{\alpha\mu\dots} + \dots . \tag{21.32}$$

For dilations, $X^\mu = \lambda x^\mu$ and accordingly

$$(\delta_\lambda\phi)_{\alpha\beta\dots} = \lambda\big(x^\mu \partial_\mu + \Delta_\phi\big)\phi_{\alpha\beta\dots}, \qquad \Delta_\phi = s_\phi + w_\phi\,. \tag{21.33}$$

The conformal weight $\Delta_\phi$ is the sum of the mixed rank $s_\phi$ and the Weyl weight $w_\phi$ of $\phi_{\alpha\beta\dots}$. The mixed rank is the number of covariant minus the number of contravariant indices.

---

**Question**

Why does a classical scalar field have $\Delta_\phi = \frac{1}{2}(d-2)$ and a Dirac spinor field $\Delta_\psi = \frac{1}{2}(d-1)$?

The action of a classical conformal field theory (QFT) in Minkowski space is invariant under conformal mappings. This does not imply that the corresponding quantum field theory inherits these symmetries. As we saw in Chap. 20, a symmetry of the classical theory can be spontaneously or anomalously broken in the quantum theory. Quantization requires regularization and renormalization, and these can lead to anomalies. If the conformal symmetry exhibits no anomalies, quantum theory should be conformally invariant—similar to how relativistic quantum field theories are Poincaré-invariant. For example, the 2-point correlation function of two matter fields[4] with Weyl weights $\Delta_1$ and $\Delta_2$,

$$\langle 0|\phi^{(1)}_{\mu\nu\dots}(x_1)\phi^{(2)}_{\alpha\beta\dots}(x_2)|0\rangle \equiv G^{(2)}_{\mu\nu\dots,\alpha\beta\dots}(x_1,x_2)\,, \tag{21.34}$$

---

[4] It is always assumed, $x_1$ and $x_2$ are chronologically ordered, i.e. $x_1^0 \geq x_2^0$.

should transform covariantly under conformal mappings $x \mapsto y(x)$,

$$G^{(2)}_{\mu\nu\ldots,\alpha\beta\ldots}(x_1, x_2)$$

$$= e^{w_1\sigma(x_1)}\, e^{w_2\sigma(x_2)}\, \frac{\partial y_1^\sigma}{\partial x_1^\mu}\frac{\partial y_1^\rho}{\partial x_1^\nu}\cdots \frac{\partial y_2^\gamma}{\partial x_2^\alpha}\frac{\partial y_2^\delta}{\partial x_2^\beta}\cdots G^{(2)}_{\sigma\rho\ldots,\gamma\delta\ldots}(y_1, y_2)\,. \qquad (21.35)$$

**Ward Identities of Conformal Symmetry**

The infinitesimal form of (21.35) is a differential equation for the 2-point function,

$$\sum_{i=1}^{2}\left[L_{X(x_i)} + \frac{w_i}{d}(\operatorname{div}X)(x_i)\right]G^{(2)}_{\mu\nu\ldots,\alpha\beta\ldots}(x_1, x_2) = 0\,, \qquad (21.36)$$

and is called a Ward identity of conformal symmetry.

This Ward identity uniquely determines the 2-point correlation function. We demonstrate this for two scalar fields, for which the Lie derivative is equal to the derivative in the $X$ direction, and the Ward identity simplifies:

$$\sum_{i=1}^{2}\left[X^\mu(x_i)\frac{\partial}{\partial x_i^\mu} + \frac{\Delta_i}{d}(\operatorname{div}X)(x_i)\right]G^{(2)}(x_1, x_2) = 0 \qquad (\Delta_i = w_i)\,. \qquad (21.37)$$

For translations and Lorentz transformations, $X$ is source-free and (as argued in Sect. 19.6.1) $G^{(2)}$ only depends on $\xi^\mu\xi_\mu$, where $\xi = x_2 - x_1$ denotes the difference vector. The Ward identity for dilations with $X = \lambda x$ then simplifies to

$$\left(\xi\partial_\xi + \Delta_1 + \Delta_2\right)G^{(2)}(\xi) = 0\,, \qquad (21.38)$$

with the solution

$$G^{(2)}(x_1, x_2) = \frac{C_{12}}{|x_2 - x_1 - i\epsilon|^{\Delta_1+\Delta_2}}\,. \qquad (21.39)$$

The Feynman $i\epsilon$-prescription was used to handle the poles of $G^{(2)}$.

If one inserts the Killing vector field belonging to the special conformal transformations into (21.37), one finds

$$(\Delta_1 - \Delta_2)\, G^{(2)}(x_1, x_2) = 0\,. \qquad (21.40)$$

The 2-point function of two scalar fields is thus only non-zero when they have the same conformal weights (see Problem 21.2). After a linear transformation of the fields, which diagonalizes the 2-point function, one finds

$$
\langle 0|\phi^{(1)}(x_1)\phi^{(2)}(x_2)|0\rangle = \begin{cases} 0 & \text{if } \Delta_1 \neq \Delta_2 \\ C_{12}|x_2 - x_1 - i\epsilon|^{-2\Delta} & \text{if } \Delta_1 = \Delta_2 \equiv \Delta\,. \end{cases} \tag{21.41}
$$

For conformally invariant theories without intrinsic length scale, the correlations are long-range and have a typical power-law behavior.

If one repeats the same calculation for the 3-point function of scalar fields, then one finds with similar arguments

$$
\langle 0|\phi^{(1)}(x_1)\phi^{(2)}(x_2)\phi^{(3}(x_3)|0\rangle = \frac{C_{123}}{\xi_{12}^{\Delta_1+\Delta_2-\Delta_3}\,\xi_{23}^{\Delta_2+\Delta_3-\Delta_1}\,\xi_{31}^{\Delta_3+\Delta_1-\Delta_2}}\,, \tag{21.42}
$$

with $\xi_{ij} = |x_i - x_j|$. In contrast to the 2-point and 3-point functions, the higher correlation functions are not determined by the $SO(d, 2)$-Ward identities, but only restricted. In 2 dimensions, however, the conformal group is infinite-dimensional and also determines the higher correlation functions (at least for a large class of conformally invariant models).

### 21.2.3 Conformal Algebra in $d$ Dimensions

In (21.31) we determined the infinitesimal conformal transformations of tensor fields. In particular, for a scalar field

$$
\delta_X\phi = \left(X^\mu\partial_\mu + \frac{\Delta_\phi}{d}\,\text{div}\,X\right)\phi \qquad (\Delta_\phi = w_\phi)\,. \tag{21.43}
$$

If we write the general conformal Killing vector field $X^\mu$ as in (21.13) as a linear combination of infinitesimal displacements, Lorentz transformations, dilations and special conformal transformations, then with the help of $\text{div}\,X = d\lambda + 2d(c \cdot x)$ we find

$$
i\delta_X\phi = \left(a^\mu P_\mu + \tfrac{1}{2}\omega^{\mu\nu}L_{\mu\nu} + \lambda D + c^\mu K_\mu\right)\phi\,. \tag{21.44}
$$

Here, the (Hermitian) generators are the momenta $P_\mu$ as infinitesimal translations, the generalized angular momenta $L_{\mu\nu}$ as infinitesimal Lorentz transformations, the scale operator $D$ as infinitesimal dilations and the $K_\mu$ as infinitesimal special conformal transformations,

$$
P_\mu = i\partial_\mu\,, \qquad\qquad L_{\mu\nu} = \frac{1}{i}(x_\mu\partial_\nu - x_\nu\partial_\mu)\,,
$$

$$
D = i(x^\mu\partial_\mu + \Delta_\phi)\,, \quad K_\mu = i(2x_\mu x^\alpha\partial_\alpha - x^2\partial_\mu + 2\Delta_\phi x_\mu)\,. \tag{21.45}
$$

The Lie algebra generated by these generators is the conformal algebra—an extension of the Poincaré algebra (14.88). The commutation relations for the generators $\{P_\mu, L_{\mu\nu}\}$ are well-known, and what remains is the calculation of the remaining commutators. Thus, due to $[D, x^\mu] = \mathrm{i}x^\mu$ and $[D, \partial_\mu] = -\mathrm{i}\partial_\mu$ (up to a factor i), the scale operator $D$ simply determines the dimension of an operator in natural units. In particular, $D$ commutes with the dimensionless components $L_{\mu\nu}$. Furthermore, the $K_\mu$ form a vector operator and therefore have the same commutation relations with $L_{\mu\nu}$ as the momenta $P_\mu$. With the remaining commutators, which are easy to calculate, we find the following commutation relations:

---

**Conformal Algebra**

$$[P_\mu, P_\nu] = [K_\mu, K_\nu] = [D, D] = 0\,,$$

$$[L_{\mu\nu}, P_\rho] = \mathrm{i}(\eta_{\rho\mu}P_\nu - \eta_{\rho\nu}P_\mu), \quad [L_{\mu\nu}, K_\rho] = \mathrm{i}(\eta_{\rho\mu}K_\nu - \eta_{\rho\nu}K_\mu)\,,$$

$$[L_{\mu\nu}, L_{\rho\sigma}] = \mathrm{i}(\eta_{\mu\rho}L_{\nu\sigma} + \eta_{\nu\sigma}L_{\mu\rho} - \eta_{\mu\sigma}L_{\nu\rho} - \eta_{\nu\rho}L_{\mu\sigma})\,, \tag{21.46}$$

$$[D, P_\mu] = -\mathrm{i}P_\mu, \quad [D, K_\mu] = \mathrm{i}K_\mu, \quad [D, L_{\mu\nu}] = 0\,,$$

$$[P_\mu, K_\nu] = 2\mathrm{i}\eta_{\mu\nu}D + 2\mathrm{i}L_{\mu\nu}\,.$$

---

**Task**

Confirm the commutation relations in the last line of (21.46).

---

These are the commutation relations of the Lie algebra $\mathfrak{so}(2, d)$ of the pseudo-orthogonal group $\mathrm{SO}(2, d)$, also called the AdS (Anti-de Sitter) group (see Problem 8.9). To see this, we introduce the following coordinates in the Anti-de Sitter space $\mathrm{AdS}_{d+1} \cong \mathbb{R}^{2,d}$ with signature $(2, d)$:

$$\xi_m = (\xi_{-1}, x_\mu, \xi_d)\,. \tag{21.47}$$

As a length element, we choose

$$\mathrm{d}\xi^2 = \mathrm{d}\xi_{-1}^2 + \eta_{\mu\nu}\mathrm{d}x^\mu\mathrm{d}x^\nu - \mathrm{d}\xi_d^2 = \eta_{mn}\mathrm{d}\xi^m\mathrm{d}\xi^n\,. \tag{21.48}$$

We consider infinitesimal transformations in $\mathrm{AdS}_{d+1}$ and decompose the generators according to

$$(L_{mn}) = \frac{1}{2}\begin{pmatrix} 0 & P_\mu & D \\ -P_\mu & L_{\mu\nu} & P_\mu \\ -D & -P_\mu & 0 \end{pmatrix} + \frac{1}{2}\begin{pmatrix} 0 & K_\mu & D \\ -K_\mu & L_{\mu\nu} & -K_\mu \\ -D & K_\mu & 0 \end{pmatrix}\,. \tag{21.49}$$

Then the commutation rules (21.46) are written as follows

$$[L_{mn}, L_{pq}] = \mathrm{i}\big(\eta_{mp} L_{nq} + \eta_{nq} L_{mp} - \eta_{mq} L_{np} - \eta_{np} L_{mq}\big).$$ (21.50)

These are exactly the commutation relations of the Lie algebra $\mathfrak{so}(2, d)$ with dimension

$$\dim(\mathfrak{so}(2, d)) = \frac{1}{2}(d + 2)(d + 1),$$ (21.51)

which is exactly the number of parameters of the conformal mappings in (21.14).

## 21.2.4 Energy-Momentum Tensor and Bessel-Hagen Currents

First, we convince ourselves that the trace of the energy-momentum tensor of a Weyl-invariant theory vanishes. We prove this for the metric tensor $T_{\mu\nu}$, which appears as the source of the gravitational field in Einstein's theory of gravitation, see the remarks in Sect. 19.4.2. This improved tensor is obtained when the matter field is covariantly coupled to the gravitational field and the corresponding invariant action is varied with respect to the metric,

$$T_{\mu\nu} = \frac{2}{\sqrt{-g}} \frac{\delta}{\delta g^{\mu\nu}} S[g_{\mu\nu}, \phi].$$ (21.52)

This tensor is symmetric and covariantly conserved, $\nabla_\mu T^{\mu\nu} = 0$. The symmetry is manifest and the covariant conservation follows from the diffeomorphism invariance of $S$ [111]. In the limit $g_{\mu\nu} \mapsto \eta_{\mu\nu}$ one obtains an improved energy-momentum tensor in Minkowski space.

**Example: Metric Energy-Momentum Tensor of Electrodynamics**

To vary the generally covariant action of the electromagnetic field in (21.24) with respect to the metric, we need the variation

$$\delta \det(g_{\mu\nu}) = g\, g^{\mu\nu} \delta g_{\mu\nu} \Rightarrow \delta\sqrt{-g} = \tfrac{1}{2}\sqrt{-g}\, g^{\mu\nu} \delta g_{\mu\nu} = -\tfrac{1}{2}\sqrt{-g}\, g_{\mu\nu} \delta g^{\mu\nu}.$$ (21.53)

With this, one finds the variation formula

$$\delta S = \tfrac{1}{4} \int \sqrt{-g}\, \delta g^{\mu\nu} \big(\tfrac{1}{2} g_{\mu\nu} F^{\alpha\beta} F_{\alpha\beta} - 2 F_{\mu\alpha} F_\nu{}^\alpha\big),$$ (21.54)

and with (21.52) the metric energy-momentum tensor

$$T_{\mu\nu} = F_{\mu\alpha} F^\alpha{}_\nu + \tfrac{1}{4} g_{\mu\nu} F^{\alpha\beta} F_{\alpha\beta}.$$ (21.55)

In the limit $g_{\mu\nu} \rightarrow \eta_{\mu\nu}$ we recover the improved energy-momentum tensor (19.70). ◄

In a Weyl-invariant field theory, the variation of the left-hand side of (21.23) with respect to the field $\sigma$ vanishes, and it follows

$$
0 = \frac{\delta S}{\delta \sigma}\Big|_{\sigma=0} = \left( \frac{\delta S}{\delta g^{\mu\nu}} \frac{\delta g^{\mu\nu}}{\delta \sigma} + \frac{\delta S}{\delta \phi} \frac{\delta \phi}{\delta \sigma} \right)_{\sigma=0} = -\sqrt{-g}\, T_{\mu\nu} g^{\mu\nu} - \frac{\delta S}{\delta \phi} w_\phi \phi \, .
\tag{21.56}
$$

---

**Trace of the Energy-Momentum Tensor**
If the Weyl weights $w_\phi$ of all matter fields vanish, then $T^\mu_\mu = 0$ for arbitrary fields. One also says that the trace vanishes off-shell. If $w_\phi \neq 0$, then $T^\mu_\mu$ only vanishes when $\phi$ is a solution of the Euler-Lagrange equation $\delta S/\delta\phi = 0$. One says that the trace vanishes on-shell.

---

For example, in electrodynamics (in 4 dimensions), the Weyl weight vanishes and $T_{\mu\nu}$ in (21.55) is off-shell trace-free. On the other hand, a massless Dirac spinor field in $d \neq 1$ dimensions has $w_\psi \neq 0$ and $T_{\mu\nu}$ in (19.77) is only on-shell trace-free.

With the trace-free, symmetric, and covariantly preserved $T^{\mu\nu}$ of a conformal field theory, one can easily obtain the $\frac{1}{2}(d+1)(d+2)$ covariantly conserved Noether current densities: These are the current densities named after Erich Bessel-Hagen.

---

**Bessel-Hagen Current Densities**
For each conformal Killing vector field $X$, the Bessel-Hagen current density

$$
j_X^\mu = T^{\mu\nu} X_\nu
\tag{21.57}
$$

is covariantly conserved.

---

This follows from the properties of $T^{\mu\nu}$ and the conformal Killing equation for $X_\mu$,

$$
\partial_\mu j_X^\mu = T^{\mu\nu} X_{\nu,\mu} = \frac{1}{2} T^{\mu\nu}(X_{\nu,\mu} + X_{\mu,\nu}) = \frac{1}{d} \eta_{\mu\nu} T^{\mu\nu} \partial X = 0 \, .
\tag{21.58}
$$

For example, the Noether current and the Noether charge for the dilations take the form

$$
j_D^\mu = T^{\mu\nu} x_\nu \quad \text{and} \quad Q_D = \int d\boldsymbol{x}\, T^{0\nu} x_\nu \, .
\tag{21.59}
$$

## 21.3   Conformal Field Theories in $1+1$ Dimensions

In two dimensions, a CFT has, in addition to the 6 charges $P_\mu, L_{01}, D$ and $K_\mu$, infinitely many other charges, because the conformal group is infinitely dimensional. The infinitely dimensional symmetry algebra generated by the charges is the Virasoro algebra. We start with the symmetry algebra of a classical conformal field theory. Its central extension is the Virasoro algebra.

### 21.3.1  Witt Algebra

A classical CFT in 2 dimensions has the algebra named after Ernst Witt as its symmetry algebra. In the following, it is convenient to use the light cone coordinates $x^\pm$ in (21.17) again. Although the spatial coordinate $x^1$ lies in $\mathbb{R}$, we denote it—as in higher dimensions—with $x$, to distinguish it from the spacetime coordinate $x = (x^0, x^1) \equiv (t, x)$.

In light cone coordinates,

$$T^\mu_{\ \mu} = \eta^{\mu\nu} T_{\mu\nu} = 4\, T_{+-} = 0 \,, \tag{21.60}$$

and the covariant conservation of the energy-momentum tensor, $\partial_+ T^{+\pm} + \partial_- T^{-\pm} = 0$ implies $\partial_\pm T^{\pm\pm} = 0$, which means that the remaining components of $T_{++}$ and $T_{--}$ are chiral fields:

$$T_{++} = T_{++}(x^+) \quad \text{and} \quad T_{--} = T_{--}(x^-) \,. \tag{21.61}$$

According to (21.18), the components of conformal Killing vector fields are chiral, so that the components of the Bessel-Hagen current densities (21.57) are also chiral,

$$j_{+X} = T_{++} X^+ \quad \text{and} \quad j_{-X} = T_{--} X^- \,. \tag{21.62}$$

The chiral fields $X^+$ and $X^-$ can be chosen independently, and therefore the charges associated with the two current densities are separately conserved. If one chooses for $X^\pm$ the monomials in (21.19), then one obtains the conserved charges

$$Q_n^+ = -\int dx\, (x^+)^{n+1} T_{++}(x^+) = -\int dx\, x^{n+1} T_{++}(x) \,,$$

$$Q_n^- = -\int dx\, (x^-)^{n+1} T_{--}(x^-) = -\int dx\, x^{n+1} T_{--}(x) \,. \tag{21.63}$$

Since these are time-independent, we were allowed to choose the time $t = 0$ under the integrals on the right. The charges $Q_n^{\pm}$ generate two commuting infinite-dimensional Witt algebras.

---

**Example: Witt Algebra for Classical Scalar Field**

We consider a real scalar field $\phi$ with Lagrangian density (19.53), where we set $V = 0$. In $1 + 1$ dimensions, the components of the energy-momentum tensor are

$$T_{00} = T_{11} = \frac{1}{2}\left(\pi^2 + \phi'^2\right), \quad T_{01} = T_{10} = \pi\phi'. \tag{21.64}$$

These have the Poisson brackets

$$\{T_{00}(x), T_{00}(y)\} = \{T_{01}(x), T_{01}(y)\} = \left(T_{01}(x) + T_{01}(y)\right)\delta'(x - y),$$

$$\{T_{00}(x), T_{01}(y)\} = \{T_{01}(x), T_{00}(y)\} = \left(T_{00}(x) + T_{00}(y)\right)\delta'(x - y),$$

see also Problem 19.8. Thus, $T_{++} = \frac{1}{2}(T_{00} + T_{01})$ has the following Poisson bracket:

$$\{T_{++}(x), T_{++}(y)\} = \left(T_{++}(x) + T_{++}(y)\right)\delta'(x - y). \tag{21.65}$$

With the help of the derivation rule, one finds

$$\{Q_m^+, Q_n^+\} = \int \mathrm{d}x\mathrm{d}y\, x^{m+1}y^{n+1}\left(T_{++}(y)\partial_x - T_{++}(x)\partial_y\right)\delta(x - y)$$

$$= (m - n)\, Q_{m+n}^+.$$

In Problem 21.3 it is further proven that the charges $Q_m^+$ commute with the charges $Q_n^-$ and the $Q_m^-$ have identical Poisson brackets as the $Q_m^+$. This proves that the massless real field in $1 + 1$ dimensions has two copies of the Witt algebra as (classical) symmetry algebra:

$$\{Q_m^{\pm}, Q_n^{\pm}\} = (m - n)\, Q_{m+n}^{\pm}, \quad \{Q_m^+, Q_n^-\} = 0. \tag{21.66}$$

We will see that the Virasoro algebra is a central extension of the Witt algebra.
◀

---

In Problem 21.4 we convince ourselves that also for an interacting scalar field the charges $Q_n^{\pm}$ in (21.63) fulfill the Witt algebra (21.66).

Under an infinitesimal mapping $x \mapsto x + X(x)$ ($x$ and $X$ stand for $x^{\pm}$ and $X^{\pm}$) a scalar field $\phi$ with $w_\phi = 0$ transforms according to $\delta_X \phi(x) = X\phi'(x)$. Show that for the basis $X_n(x) = -x^{n+1}$ of conformal Killing vector fields the corresponding Lie derivatives

$$L_n = -x^{n+1}\partial_x \tag{21.67}$$

fulfill the same commutation rules as the $Q_m^{\pm}$, i.e. $[L_m, L_n] = (m-n)L_{m+n}$.


## 21.3.2 Virasoro Algebra

The previous arguments and results were predominantly of a classical nature and the question arises to what extent they still apply in a conformal quantum field theory (CQFT). This question was answered early on by Lüscher and Mack [112]. The starting point is a Wightman theory with a Hilbert space $\mathcal{H}$ with vacuum $|0\rangle$, which carries a unitary representation $U(\Lambda, a)$ of the Poincaré group. Furthermore, the spectrum properties $P_0 \geq 0$ and $P^2 \geq 0$ should hold. In addition to these usual axioms, it is assumed that dilations are also unitarily implemented,

$$V(\lambda)\phi(x)V(\lambda)^{-1} = \lambda^{\Delta}\phi(\lambda x), \tag{21.68}$$

and a symmetric and covariantly conserved energy-momentum tensor with weight $\Delta = 2$ exists, which generates the translations

$$\int \mathrm{d}\boldsymbol{y}\left[T_{0\mu}(y), \phi(x)\right] = -\mathrm{i}\partial_\mu\phi(x). \tag{21.69}$$

Using similar arguments as in classical field theory, one can show [112] that the properties (21.60) and (21.61) hold as operator identities for $T_{\mu\nu}$, i.e., that the operator $T_{\mu\nu}$ in QFT is traceless and the remaining components are chiral.

We calculate the commutator of $T_{++}$ at two different spacetime points. Because $T_{++}$ depends only on $x^+$, the commutator of $T_{++}(x^+)$ and $T_{++}(y^+)$ must vanish for $x^+ \neq y^+$. Otherwise, there would be two spacelike separated points with non-commuting $T_{\mu\nu}$, which would contradict Einstein causality. The following considerations are for $T_{++}(x^+)$. To keep the notation simple, we temporarily set $T_{++} \equiv T$ and $x^+ \equiv x$.

We can therefore assume

$$[T(x), T(y)] = \sum A_k(x)\, \delta^{(k)}(x-y),$$

$$A_k(x) = \frac{1}{k!}\int \mathrm{d}\boldsymbol{y}\,(y-x)^k\,[T(x), T(y)]. \tag{21.70}$$

Because $T_{\mu\nu}$ has the conformal weight $\Delta = 2$, the coefficient $A_k(x)$ has the weight $3 - k$ and

$$V(\lambda)A_k(x)V(\lambda) = \lambda^{3-k}A_k(\lambda x) . \tag{21.71}$$

Due to (21.41), this determines the 2-point correlation function,

$$\langle 0|A_k(x)A_k(y)|0\rangle = \alpha_k(x - y - i\epsilon)^{2k-6} . \tag{21.72}$$

This result implies that the expectation values

$$\langle 0|A_k(\bar{f})A_k(f)|0\rangle = \|A_k(f)|0\rangle\|^2, \qquad A_k(f) = \int dx\, f(x)A_k(x) , \tag{21.73}$$

for $k > 3$ would not be positive, which is not allowed in a Hilbert space. Therefore, all $A_k$ and thus all $\alpha_k$, with $k > 3$ must vanish.

---

**Question**

Why could the squared norm in (21.73) become negative for $\alpha_{k>3} \neq 0$?

Of the remaining coefficients, $A_3$ is constant, as it has weight 0. Due to locality, it commutes with all fields and is therefore a multiple of the identity,

$$A_3 = -\frac{ic}{24\pi} . \tag{21.74}$$

The spatial integral

$$\int dx\, T(x) = \frac{1}{2}\int dx\left(T_{00}(x) + T_{01}(x)\right) \equiv P \tag{21.75}$$

generates translations in the direction of the light cone coordinate $x$, so that

$$A_0 = \int dy\, [T(x), T(y)] = [T(x), P] = i\partial_x T(x) . \tag{21.76}$$

This leads to the intermediate result

$$[T(x), T(y)] = -\frac{ic}{24\pi}\delta'''(x-y) + iA_1(x)\,\delta'(x-y) + i\partial_x T(x)\,\delta(x-y) . \tag{21.77}$$

To determine $A_1$, one uses the fact that the right-hand side must be antisymmetric in $x$ and $y$. Using the first relation in Problem 19.1, one then finds

$$\partial_x A_1 = 2\partial_x T . \tag{21.78}$$

The integration constant vanishes, as $A_1$ has weight $\Delta = 2$. For the other non-vanishing component $T_{--}(x^-)$, one finds exactly the same result.

**Virasoro Algebra for Energy-Momentum Tensor**

The non-vanishing light cone components $T_{++}(x^+)$ and $T_{--}(x^-)$ of the energy-momentum tensor of a two-dimensional CQFT commute with each other and each satisfy the commutation rules of the Virasoro algebra,

$$[T(x), T(y)] = i\partial_x T(x)\delta(x - y) + 2iT(x)\delta'(x - y)$$

$$- \frac{i}{24\pi} c\, \delta'''(x - y)\,. \tag{21.79}$$

The Virasoro algebra is thus the symmetry algebra of every two-dimensional CQFT. Note that the commutator in (21.79) only closes up to a multiple of the identity. The corresponding coefficient $c$ is the central charge and characterizes the CQFT. We will later give a physical interpretation of $c$.

**Question**

How do the Witt algebra (21.65) of the classical CFT and the Virasoro algebra (21.79) of the quantized CFT differ?

We will return to the connection between the Witt and Virasoro algebra.

**Lemma 2 (Central Charge).** *In a CQFT, the central charge $c$ is always positive.*

**Proof.** Due to the translation and dilation invariance, $\langle 0|T|0\rangle = 0$ holds, and therefore

$$\langle 0|[T(x), T(y)]|0\rangle = -\frac{ic}{24\pi}\delta'''(\xi), \qquad \xi = x - y\,. \tag{21.80}$$

$T(x)$ has conformal weight 2 and according to (21.41) a 2-point correlation function $\alpha(\xi - i\epsilon)^{-4}$. If we substitute for $\delta'''$ the representation given in Problem 19.1, then $\alpha = c/8\pi^2$ follows, or that the correlation function in momentum space looks as follows,

$$\langle 0|T(x)T(y)|0\rangle = \frac{c}{48\pi^2} \int_0^\infty dp\, p^3\, e^{-ip(x-y)}\,. \tag{21.81}$$

Then the expectation value of the square of the smeared Hermitian operator $T_f = \int dx\, f(x)T(x)$ is equal to

$$\langle 0|T_f^2|0\rangle = \big\| T_f|0\rangle \big\|^2 = \frac{c}{48\pi^2} \int_0^\infty \mathrm{d}p\, p^3 |\tilde{f}(p)|^2 \,.$$

In a Hilbert space, the left-hand side must be positive.

**Sign of $c$**

A conformally invariant quantum field theory in a Hilbert space has $c > 0$.

**Example: Central Charge for a Real Scalar Field**

The operators $T_{++}$ and $T_{--}$ must be regularized because the product of quantum fields at the same location is not defined. It is sufficient here, to normally order the occurring products of the field operator. Using the known 2-point correlation function of the scalar field in 2 dimensions,

$$\langle 0|\phi(x^+)\phi(y^+)|0\rangle = -\frac{1}{4\pi} \log \mu(x^+ - y^+ - i\epsilon)\,, \tag{21.82}$$

and the Wick rule, one finds for $T(x) = \partial_+\phi\partial_+\phi$

$$T(x)T(y) =: T(x)T(y): -\frac{1}{\pi}\frac{1}{(x - y - i\epsilon)^2} : \partial_+\phi(x)\partial_+\phi(y)$$

$$: +\frac{1}{8\pi^2}\frac{1}{(x - y - i\epsilon)^4}\,.$$

If one subtracts the corresponding expression with $x$ and $y$ interchanged, then one obtains with the last relation in Problem 19.1 the commutator

$$[T(x), T(y)] = i\,T'(x)\delta(x - y) + 2i\,T(x)\delta' - \frac{i}{24\pi}\delta(x - y)''' \tag{21.83}$$

$$= i\big(T(y)\partial_x - T(x)\partial_y\big)\delta(x - y) - \frac{i}{24\pi}\delta'''(x - y)\,. \tag{21.84}$$

A comparison with (21.79) shows that $c = 1$. ◄

### 21.3.3  On the Significance of the Central Charge

The difference between the Witt and Virasoro algebra is the term proportional to the central charge $c$, and this is due to the necessary regularization of the UV divergences in quantum field theories. There exist alternative and more direct characterizations of $c$. A geometric characterization starts with the Bessel-Hagen charges, which generate the conformal transformations (21.31) of tensor fields. For a conformal Killing vector field $X(x)$ with light cone coordinate $x = x^+$ we have

$$\delta_X \phi(x) = \mathrm{i} \int \mathrm{d}y\, X(y)\, [T(y), \phi(x)] = (L_X \phi)(x) + \frac{w_\phi}{2} X'(x)\, \phi(x)\,, \qquad (21.85)$$

where $X'$ is the derivative of $X$ with respect to the light cone coordinate $x^+$. A field transforming in this way is referred to as a primary field with conformal weight $\Delta_\phi = s_\phi + w_\phi$. For the energy-momentum tensor, $w_T = 0$ and $s_T = 2$. If $T_{++}$ were primary, we would conclude

$$\delta_X T = L_X T = X T' + 2 X' T\,. \qquad (21.86)$$

A comparison with (21.79) shows, however, that $T$ is not a primary field,

$$\delta_X T = \mathrm{i} \int \mathrm{d}y\, X(y)[T(y), T] = X T' + 2 X' T - \frac{c}{24\pi} X'''\,. \qquad (21.87)$$

The anomalous term proportional to $X'''$ vanishes only for constant, linear and quadratic conformal Killing vector fields. So exactly for conformal mappings in the six-dimensional $\mathrm{SL}(2, \mathbb{R}) \times \mathrm{SL}(2, \mathbb{R})$ subgroup of Möbius transformations (21.20). Because $c$ is positive, the anomalous term does not vanish. It is responsible for the fact that in the quantized theory $T$ does not transform like a tensor field under conformal mappings.

## Weyl or Trace Anomaly

The anomalous term in the commutation relation (21.87) can be attributed to the trace anomaly. To better understand this, we couple the matter field to gravity as before and choose a conformally flat metric and isothermal light-cone coordinates,

$$g_{\mu\nu} = \mathrm{e}^{2\sigma(x)} \eta_{\mu\nu}, \quad \mathrm{d}s^2 = g_{\mu\nu} \mathrm{d}x^\mu \mathrm{d}x^\nu = \frac{1}{2}\, \mathrm{e}^{2\sigma} (\mathrm{d}x^+ \mathrm{d}x^- + \mathrm{d}x^- \mathrm{d}x^+)\,, \qquad (21.88)$$

with non-vanishing Christoffel symbols

$$\Gamma^+_{++} = g^{+-} \partial_+ (g_{-+}) = 2\partial_+ \sigma, \qquad \Gamma^-_{--} = g^{-+} \partial_- (g_{+-}) = 2\partial_- \sigma\,. \qquad (21.89)$$

When quantizing a CFT in an external gravitational field, a non-vanishing trace of the tensor $T_{\mu\nu}$ remains after regularization and possibly renormalization. This trace anomaly must have the mass dimension 2 like $T_{\mu\nu}$, be a local scalar, and vanish in flat space. The only scalar with this property is the Ricci scalar $R$, so that

$$g^{\mu\nu} T_{\mu\nu}[g, x] = 4\,\mathrm{e}^{-2\sigma} T_{+-} = \frac{c}{24\pi} R, \qquad R = -8\,\mathrm{e}^{-2\sigma} \partial_+ \partial_- \sigma\,, \qquad (21.90)$$

with an undetermined constant $c \in \mathbb{R}$. More precisely, $T_{+-} = -c/12\pi \cdot \partial_+\partial_-\sigma$, multiplied by the identity in the Hilbert space of the CQFT.

If we now write the covariant conservation $\nabla_\mu T^{\mu\nu} = 0$ in light cone coordinates (21.88) and use (21.90), we obtain

$$\partial_\pm\left(T_{\mp\mp}[g,x] + \frac{c}{12\pi}\, e^\sigma \partial_\mp^2 \, e^{-\sigma}\right) = 0\,. \tag{21.91}$$

**Task**

Prove this useful result.

The integration of these differential equations leads to

$$T_{\pm\pm}[g,x] = -\frac{c}{12\pi}\, e^\sigma \partial_\pm^2 \, e^{-\sigma} + T_{\pm\pm}[\eta, x^\pm]\,, \tag{21.92}$$

with a $\sigma$-independent and generally operator-valued integration constant $T_{\pm\pm}(\eta, x^\pm)$. These are the components of the energy-momentum tensor in Minkowski space with metric $\eta_{\mu\nu}$.

Under angle-preserving coordinate transformations $x^\pm \longrightarrow y^\pm(x^\pm)$, the conformally flat metric tensor transforms according to

$$ds^2 = e^{2\sigma(x)}dx^+dx^- = e^{2\sigma(x)}\frac{dx^+}{dy^+}\frac{dx^-}{dy^-}dy^+dy^- = e^{2\tilde\sigma(y)}dy^+dy^-\,, \tag{21.93}$$

and the tensor $T[g,x]$ transforms covariantly,

$$T_{++}[e^{2\sigma(x)}\eta, x^+] = \left(\frac{dy^+}{dx^+}\right)^2 \tilde T_{++}[e^{2\tilde\sigma(y)}\eta, y^+]\,. \tag{21.94}$$

With (21.92), we then find for $T_{\pm\pm}[\eta, x^\pm]$ the unusual transformation law

$$\tilde T_{++}[\eta, y^+] = \left(\frac{dx^+}{dy^+}\right)^2 T_{++}[\eta, x^+] - \frac{c}{24\pi}\{x^+, y^+\},$$

$$\{x, y\} \equiv \frac{x_{yyy}}{x_y} - \frac{3}{2}\left(\frac{x_{yy}}{x_y}\right)^2\,, \tag{21.95}$$

under conformal mappings $(\eta_{\mu\nu}, x^\pm) \mapsto (\eta_{\mu\nu}, y^\pm)$. The other non-vanishing component $T_{--}[\eta, x^-]$ transforms just like $T_{++}[\eta, x^+]$.

The inhomogeneous term $\{x, y\}$ in the transformation formula (21.95) contains the first, second, and third derivative of $x$ with respect to $y$. It is the Schwarz derivative of the mapping $y \mapsto x(y)$ named after Hermann Schwarz. It vanishes exactly for the Möbius transformations (21.20), see Problem 21.6. As before, we conclude that only for conformal mappings in the Möbius subgroup SL$(2, \mathbb{R}) \times$

SL$(2, \mathbb{R})$ does $T_{\mu\nu}$ transform like a tensor. For the other conformal mappings, $T_{\mu\nu}$ transforms anomalously. For infinitesimal mappings $y^+ \mapsto x^+ + X^+$, the (linearized) Schwarz derivative is equal to the third derivative of the conformal Killing field $X^+$ with respect to the light cone coordinate. Accordingly, (21.95) simplifies to the known infinitesimal transformation (21.87) and we conclude:

> **Meaning of $c$**
>
> The constant $c$ in the trace anomaly (21.90) is equal to the positive central charge in the Virasoro algebra (21.79).

### 21.3.4  Virasoro Algebra for Expansion Coefficients

Since the Möbius transformations

$$x \mapsto y = \frac{ax + b}{cx + d} \quad \text{with} \quad \frac{dy}{dx} = \frac{ad - bc}{(cx + d)^2} = \frac{1}{(cx + d)^2} \tag{21.96}$$

are not anomalous, the correlation functions of the CQFT fulfill the classical Ward identities of the Möbius mappings in SL$(2, \mathbb{R})$—for the 2-point correlation function, the identities (21.35). In particular, the $n$-point correlation functions of $T_{++}(x^+) \equiv T(x)$ transform as follows:

$$G_n(y_1, \ldots, y_n) = \prod_{j=1}^{n} (cx_j + b)^{-4} G_n(x_1, \ldots, x_n) . \tag{21.97}$$

The corresponding infinitesimal Ward identities with constant, linear, and quadratic conformal Killing field $x^k$, $k = 0, 1, 2$, are

$$\sum_{j=1}^{n} \left( x_j^k \partial_{x_j} + 2k x_j^{k-1} \right) G_n(x_1, .., x_n) = 0, \qquad k = 0, 1, 2 . \tag{21.98}$$

The Möbius mappings are singular for certain $x \in \mathbb{R}$. To regularize the singularities, one compactifies $\mathbb{R}$ to $\mathbb{R} \cup \{\infty\} \cong S^1$ using the stereographic projection

$$\mathbb{R} \ni x \longmapsto z \in S^1, \quad \text{where} \quad z = -i\,e^{i\tau}, \ x = \tan \frac{\tau}{2} . \tag{21.99}$$

It can now be shown [108] that the energy-momentum tensor can be "lifted" to an operator-valued distribution on $S^1$,

$$T(x) \longrightarrow \tilde{T}(z) = \left( \cos \tfrac{1}{2}\tau \right)^{-4} T\left( \tan \tfrac{1}{2}\tau \right), \tag{21.100}$$

so that its $n$-point correlation functions

$$\tilde{G}_n(z_1, \ldots, z_n) = \langle 0 | \tilde{T}(z_1) \cdots \tilde{T}(z_n) | 0 \rangle \qquad (21.101)$$

transform covariantly with respect to the now regularly acting Möbius group $SL(2, \mathbb{R})$. The Virasoro algebra of the $T_{\pm\pm}$ in (21.79) implies the following commutation relations for the $\tilde{T}(z)$:

$$[\tilde{T}(z), \tilde{T}(w)] = \frac{2}{3\pi} ic \left\{ \delta'''(\tau - \sigma) + \delta'(\tau - \sigma) \right\}$$
$$+ 4i \left\{ \tilde{T}(w)\partial_\tau - \partial_\sigma \tilde{T}(w) \right\} \delta(\tau - \sigma) . \qquad (21.102)$$

The coefficients in the Laurent series for $\tilde{T}(z)$ are proportional to

$$L_n = \frac{i}{4} \oint_{S^1} dz \, z^{-n-1} \tilde{T}(z), \qquad L_n^\dagger = L_{-n} , \qquad (21.103)$$

and in Problem 21.7 we calculate, starting from the relations (21.102), their commutators.

---

**Virasoro Algebra for Expansion Coefficients $L_n$**
The operators $L_n$ in the Laurent series for $\tilde{T}_{++}$ satisfy the commutation relations

$$[L_m, L_n] = (m-n)L_{m+n} + \frac{c}{12}(m^3 - m)\delta_{m+n}, \qquad L_n^\dagger = L_{-n} .. \qquad (21.104)$$

The analogously defined operators $\bar{L}_n$ in the Laurent series of $\tilde{T}_{--}$ satisfy identical commutation relations and the $L_n$ commute with the $\bar{L}_m$.

---

The infinite-dimensional Lie algebra (21.104) was introduced by Miguel Virasoro in 1970 in the context of string theory [113]. For the generators $\{L_0, L_1, L_{-1}\} \in \mathfrak{sl}(2, \mathbb{R})$ and $\{\bar{L}_0, \bar{L}_1, \bar{L}_{-1}\} \in \mathfrak{sl}(2, \mathbb{R})$, which together generate the Poincaré transformations, dilations and special conformal transformations, the term proportional to the central charge vanishes. Again, we see that the two $\mathfrak{sl}(2, \mathbb{R})$ Lie subalgebras of the two Virasoro algebras do not experience quantum corrections.

The Virasoro algebra contains, in addition to the generators $\{L_n | n \in \mathbb{Z}\}$, the identity operator, which lies in its center. The last term in (21.104) proportional to the $c$ contains the identity in Hilbert space and is called *central extension*. In this sense, the Virasoro algebra is the central extension of the classical Witt algebra. In Appendix 21.4 we prove that the central extension of the Witt algebra (up to equivalence) is unique. We then speak of the universal central extension.

Two extensions are equivalent, if they belong to two different bases $\{L_n\}$ and $L'_n = L_n + \alpha_n$.


### 21.3.5  Representations of the Virasoro Algebra

In the analysis of representations of the conformal algebra, it is advantageous to choose the Virasoro algebra in the form (21.104) as a starting point. Because

$$T_{00}(x) = T_{++}(x^+) + T_{--}(x^-) \quad \text{and} \quad T_{01}(x) = T_{++}(x^+) - T_{--}(x^-)$$

$$(21.105)$$

the energy and momentum of the CQFT are given by

$$H = L_0 + \bar{L}_0 + \text{const} \quad \text{and} \quad P = L_0 - \bar{L}_0 \,. \tag{21.106}$$

Because the generators $\{L_n\}$ and $\{\bar{L}_n\}$ define identical and commuting Virasoro Lie algebras, it is sufficient to study the action of the $L_n$ in the Hilbert space of the CQFT in the following.

Let $|\Delta\rangle$ be an eigenvector of $L_0$ with weight $\Delta$. Then $L_n|\Delta\rangle$ is also an eigenvector, but with weight $\Delta - n$. In the proof, we make use of the relation $[L_0, L_n] = -nL_n$:

$$L_0|\Delta\rangle = \Delta|\Delta\rangle \implies L_0 L_n|\Delta\rangle = (\Delta - n)L_n|\Delta\rangle \,. \tag{21.107}$$

The $L_n$ with positive $n$ lower the weight and the $L_n$ with negative $n$ raise it. In a Fock space representation with energy bounded from below, there exists a state $|\Delta\rangle$ with the lowest $L_0$ weight $\Delta$:

$$L_0|\Delta\rangle = \Delta|\Delta\rangle, \qquad L_n|\Delta\rangle = 0, \quad n > 0 \,. \tag{21.108}$$

The Fock space is then spanned by the vectors

$$\psi_N = L_{-1}^{n_1} L_{-2}^{n_2} \cdots |\Delta\rangle \quad \text{with} \quad L_0 \psi_N = (\Delta + N)\psi_N, \quad N = \sum j\, n_j \,.$$

$$(21.109)$$

The construction of the Fock space representation is modeled after the construction of representations of simple Lie algebras with highest weight, see Chap. 16. The state $|\Delta\rangle$ with minimal $L_0$ weight $\Delta$ corresponds to the state with highest weight $\lambda$.

The norm of a vector $\neq 0$ in the Hilbert space of the CQFT must be positive. This requirement restricts the possible values of $c$ and $\Delta$. For example, using (21.104) and (21.108), we find for the squared norm of the states $L_{-n}|\Delta\rangle$

$$\|L_{-n}|\Delta\rangle\|^2 = \langle\Delta|L_n L_{-n}|\Delta\rangle = \langle\Delta|[L_n, L_{-n}]|\Delta\rangle$$

$$= \left(2n\Delta + \frac{c}{12}n(n^2 - 1)\right)\||\Delta\rangle\|^2 . \tag{21.110}$$

From the positivity for $n = 1$ it follows $\Delta \geq 0$ and from that for $n \to \infty$ it follows $c \geq 0$, which was already implied by the positivity of the 2-point correlation function of the energy-momentum tensor. The investigation of positivity in the entire Fock space is simplified if one takes into account that $\psi_N$ and $\psi_M$ are orthogonal for $N \neq M$. Therefore, the (infinite) norm matrix $\langle\psi_N|\psi_M\rangle$ decomposes into finite norm matrices for the finite-dimensional subspaces with fixed $L_0$ eigenvalue $\Delta+N$. The positivity of the scalar product in these subspaces was thoroughly analyzed by Friedan et al. [114] with the result that either $c \geq 1$, or, if $0 < c < 1$, $c$ must take one of the following rational values,

$$c = 1 - \frac{6}{n(n + 1)}, \qquad n = 2, 3, 4, \ldots . \tag{21.111}$$

For a given $n$, only the following highest weights $\Delta$ are allowed:

$$\Delta = \frac{[(n + 1)p - nq]^2 - 1}{4n(n + 1)}, \qquad p = 1, 2, \ldots, n - 1, \quad q = 1, 2, \ldots, p . \tag{21.112}$$

For a proof, I refer to the detailed presentation of CFT in [104].

The theories with $c < 1$ are called minimal conformal models or simply minimal models. These can be identified with lattice models at the critical point, e.g., the Ising model, the tricritical Ising model, Potts models, $O(n)$ models or models with Majorana or Dirac fermions [104].

## 21.4  Appendix A: Central Extension of the Witt Algebra

The question arises about the uniqueness of the central extension in the Virasoro algebra. Any extension has the form

$$[L_m, L_n] = (m - n)L_{m+n} + c_{m,n}, \qquad c_{m,n} = -c_{n,m} , \tag{A.1}$$

where the $c_{m,n}$ lie in the center of the Lie algebra. The Jacobi identity for $L_m, L_n, L_p$ restricts the choice of the central extension,

$$(n - p)c_{m,n+p} + (m - n)c_{p,m+n} + (p - m)c_{n,p+m} = 0 . \tag{A.2}$$

If we set $p = 0$ in this relation, then we find with $c_{m,n} = -c_{n,m}$ the relationship

$$(m + n)c_{m,n} = -(m - n)c_{0,m+n} \,. \tag{A.3}$$

For $m + n \neq 0$ we can solve for $c_{m,n}$ and obtain

$$c_{m,n} = -\frac{m - n}{m + n} c_{0,m+n}, \quad m + n \neq 0 \,. \tag{A.4}$$

In the other case $m + n = 0$, (A.3) leads to the condition $c_{0,0} = 0$, which however holds for $c_{m,n} = -c_{n,m}$ anyway. The $c_{n,-n}$ with $n \neq 0$ are not restricted by (A.3). If we set in (A.4) $c_{0,m+n} = -(m + n)g_{m+n}$, which we may do for $m + n \neq 0$ without restricting $c_{0,m+n}$, then we can summarize the intermediate result as follows:

$$c_{m,n} = f_m \delta_{0,m+n} + (m - n)(m + n)g_{m+n}, \quad f_{-m} = -f_m \,. \tag{A.5}$$

To further restrict the $f_m$, we insert this form into the condition (A.2) with $n = 1$ and $p = -m - 1$. This leads to the recursion relation

$$(m - 1)f_{m+1} - (m + 2)f_m + (2m + 1)f_1 = 0 \,, \tag{A.6}$$

which has the general solution $f_m = am^3 + bm$. Therefore, a solution of (A.2) has the form

$$c_{m,n} = \left(\tfrac{c}{12}m^3 + bm\right)\delta_{0,m+n} + (m - n)(m + n)g_{m+n} \,, \tag{A.7}$$

with undetermined constants $c, b$ and undetermined function $g_k$.

---

**Task**

    We have only found necessary conditions on the $c_{m,n}$. Check whether the $c_{m,n}$ in (A.7) actually satisfy the conditions (A.2).

We have obtained the general solution of (A.2). But many of these solutions are equivalent. To justify this, we change the basis of the Virasoro algebra by adding a multiple of the identity (the central element) to the $L_n$. The $L'_n = L_n + \alpha_n$ again form a Virasoro algebra, but with shifted central charge

$$c'_{m,n} = c_{m,n} - (m - n)\alpha_{m+n} \,. \tag{A.8}$$

Of course, the $c'_{m,n}$ also satisfy the relations (A.3). If one now chooses

$$\alpha_k = kg_k + \frac{1}{2}\left(b - \tfrac{c}{12}\right)\delta_{0,k} \,, \tag{A.9}$$

then one finds the previously used form

$$c'_{m,n} = \frac{c}{12}(m^3 + m)\delta_{0,m+n} \, . \tag{A.10}$$

If we identify equivalent extensions with $c_{mn}$ and $c_{m,n} - (m-n)\alpha_{m+n}$, then the central extension of the Witt algebra is unique. For example, one can transform away the term linear in $m$ in (A.10) by a basis change. In this form, one finds the Virasoro algebra in the literature.

## 21.5  Exercises for Chap. 21

**Problem 21.1 (Conformal Killing Vector Fields).** Prove that the vector field (21.13) satisfies the conformal Killing equation (21.9).

**Problem 21.2 (Conformal Ward Identity).** Show that the Ward identity for the special conformal transformations with conformal Killing vector field (see Table 21.1)

$$X^\mu = 2(c \cdot x)x^\mu - x^2 c^\mu$$

leads to the condition (21.40).

**Problem 21.3 ( Witt Algebra I).** Verify that the charges $Q_n^-$ for a real massless scalar field in $1+1$ dimensions have identical Poisson brackets as the charges $Q_n^+$ in (21.66). Also show that $\{Q_m^+, Q_n^-\} = 0$.

**Problem 21.4 ( Witt Algebra II).** An interacting scalar field has in 2 dimensions a traceless classical energy-momentum tensor (19.53) with $T_{++} = \frac{1}{2}T_{01}$. Using the commutation relations for $\{T_{01}(x), T_{01}(y)\}$ in (19.112), show that the charges $Q_n^+$ satisfy the Witt algebra.

**Problem 21.5 (Representation of the Conformal Algebra on Vector Fields).** How does the form of the generators (21.45) change when they act on the electromagnetic potential in 4 dimensions? Calculate the commutation relations of the generators.

**Problem 21.6 (Schwarz Derivative).** The Schwarz derivative of a complex function $f(z)$ is

$$(\mathcal{S}f)(z) = \frac{f'''}{f'} - \frac{3}{2}\left(\frac{f''}{f'}\right)^2$$

Prove the following statements:

1. The Schwarz derivative of every Möbius transformation

$$m(z) = \frac{az + b}{cz + d}, \quad ad - bc \neq 0,$$

   vanishes.
2. If $f$ and $g$ are non-constant meromorphic functions, then $\mathcal{S}f = \mathcal{S}g$ if and only if there is a Möbius transformation $m$ such that $m \circ f = g$.

**Problem 21.7 (Virasoro Algebra).** After compactification of the light cone coordinates, the Virasoro algebra in position space takes the form (21.102). Now show that the Laurent coefficients defined in (21.103) satisfy the Virasoro algebra (21.104) in "momentum space".

# References


1. Hermann Weyl. *The theory of groups and quantum mechanics*. Courier Corporation, 1950.
2. Eugene Wigner. *Group theory: and its application to the quantum mechanics of atomic spectra*, volume 5. Elsevier, 2012.
3. J. F. Cornwell. *Group Theory in Physics*. Academic Press, Amsterdam, Boston, 1984.
4. Michael Artin. *Algebra*. Pearson Education Limited, second edition, 2013.
5. Asim Orhan Barut and Ryszard Raczka. *Theory of Group Representations and Applications*. World Scientific, Singapur, 1986.
6. S. Sternberg. *Group Theory and Physics*. Cambridge University Press, Cambridge, 1995.
7. Robert Gilmore. *Lie Groups, Physics, and Geometry - An Introduction for Physicists, Engineers and Chemists*. Cambridge University Press, Cambridge, 2008.
8. Morton Hamermesh. *Group Theory and Its Application to Physical Problems*. Courier Corporation, New York, 2012.
9. Wolfgang Ludwig and Claus Falter. *Symmetries in Physics - Group Theory Applied to Physical Problems*. Springer Science & Business Media, Berlin, Heidelberg, 2012.
10. M. I. Petrashen and J. L. Trifonov. *Applications of Group Theory in Quantum Mechanics*. Courier Corporation, New York, 2013.
11. Wu-Ki Tung. *Group Theory in Physics*. World Scientific, Singapur, 1985.
12. Mildred S. Dresselhaus, Gene Dresselhaus, and Ado Jorio. *Group Theory - Application to the Physics of Condensed Matter*. Springer Science & Business Media, Berlin, Heidelberg, 2007.
13. Howard Georgi. *Lie Algebras In Particle Physics - from Isospin To Unified Theories*. CRC Press, Boca Raton, Fla, 2018.
14. H.F Jones. *Groups, Representations and Physics*. Taylor & Francis, New York, 1998.
15. Jean Pierre Serre. *Linear Representations of Finite Groups*. Springer, Berlin, Heidelberg, New York, graduate texts in mathematics 42 edition, 1977.
16. Max Wagner. *Gruppentheoretische Methoden in der Physik - Ein Lehr- und Nachschlagewerk*. Springer, Berlin, Heidelberg, 2001.
17. Alexey P. Isaev and Valery A. Rubakov. *Theory of Groups and Symmetries: Representations of Groups and Lie Algebras, Applications*. World Scientific Publishing Company Pte Limited, Singapur, 2020.
18. Jialun Ping, Fan Wang, and Jin-quan Chen. *Group Representation Theory For Physicists (2nd Edition)*. World Scientific Publishing Company, Singapore, 2002.
19. Donald E. Knuth. *The Art of Computer Programming - Volume 3: Sorting and Searching*. Addison-Wesley Professional, Boston, 1998.
20. R. Pauncz. *The Symmetric Group in Quantum Chemistry*. CRC Press, Boca Raton, Fla, 2018.
21. Dennis Morris. *Finite Groups - a Simple Introduction*. Abane & Right, Brotton, 2016.
22. John Rose. *A Course on Group Theory*. Dover Books on Advanced Mathematics, Mineola, NY, 1994.
23. B. Steinberg. *Representation Theory of Finite Groups: An Introductory Approach*. Springer, New York, Dordrecht, Heidelberg, London, 2011.



A. Wipf, *Symmetries in Physics*, https://doi.org/10.1007/978-3-662-72675-4

24. John D. Dixon and Brian Mortimer. *Permutation Groups*. Springer Science & Business Media, Berlin, Heidelberg, 2012.

25. E.A. O'Brien H.U. Besche, B. Eick. The groups of order at most 2000. *Res. Announc. Amer. Math. Soc.*, 7:1–4, 2001.

26. Wolfgang Rindler. *Relativity: Special, General, and Cosmological*. Oxford University Press, Oxford, second edition, 2006.

27. Domenico Giulini. *Special Relativity: A First Encounter - 100 years since Einstein*. OUP Oxford, New York, London, 2011.

28. Reinhard Meinel. *Spezielle und allgemeine Relativitätstheorie für Bachelorstudenten*. Springer, Berlin, Heidelberg, New York, third edition, 2025.

29. Christopher Bradley and Arthur Cracknell. *The Mathematical Theory of Symmetry in Solids - Representation Theory for Point Groups and Space Groups*. Oxford University Press, New York, 2010.

30. Richard Liboff. *Primer for Point and Space Groups*. Springer Science & Business Media, Berlin, Heidelberg, 2012.

31. David M. Bishop. *Group Theory and Chemistry*. Courier Corporation, New York, 2012.

32. Alan Vincent. *Molecular Symmetry and Group Theory - A Programmed Introduction to Chemical Applications*. John Wiley & Sons, New York, 2013.

33. M.M. Julian, C. Carla Slebodnick, and F.T. Julian. *Foundations of Crystallography with Computer Applications*. CRC Press, Boca Raton, third edition, 2024.

34. Mark A. Armstrong. *Groups and Symmetry*. Springer Science & Business Media, Berlin, Heidelberg, 2013.

35. P.J. Morandi. Symmetry groups: The classification of wallpaper patterns. *Mathematics*, 482/526, 2007. New Mexico State University.

36. Frank A. Farris. *Creating Symmetry - The Artful Mathematics of Wallpaper Patterns*. Princeton University Press, Princeton, 2015.

37. Kristopher Tapp. *Matrix Groups for Undergraduates*. American Mathematical Soc., Heidelberg, second edition, 2016.

38. Mikio Nakahara. *Geometry, Topology and Physics*. Taylor & Francis, Boca Raton, 2003.

39. Rainer Oloff. *The Geometry of Spacetime: A Mathematical Introduction to Relativity Theory*. Springer Nature, 2018.

40. Claude Chevalley. *Theory of Lie Groups*. Dover Publications Inc., 2018.

41. Adolf Hurwitz. Ueber die Erzeugung der Invarianten durch Integration. *Nachr. Akad. Wiss. Göttingen*, pages 71–89, 1897.

42. Hermann Weyl. *The Classical Groups - Their Invariants and Representations*. Princeton University Press, Princeton, 2016.

43. J. von Neumann. Die Einführung Analytischer Parameter in Topologischen Gruppen. *Ann. of Math.*, 34:170–190, 1933.

44. A. Haar. Der Massbegriff in der Theorie der kontinuierlichen Gruppen. *Ann. of Math.*, 34:147–169, 1933.

45. Martin Ammon and Johanna Erdmenger. *Gauge/Gravity Duality*. Cambridge University Press, Cambridge, 2015.

46. J. Balog, L. O'Raifeartaigh, P. Forgacs, and A. Wipf. Consistency of String Propagation on Curved Space-Times: An SU(1,1) Based Counterexample. *Nucl. Phys. B*, 325:225, 1989.

47. Charles W. Curtis. *Pioneers of Representation Theory: Frobenius, Burnside, Schur, and Brauer*. American Mathematical Soc., 1999.

48. Gordon James and Martin Liebeck. *Representations and Characters of Groups*. Cambridge University Press, Cambridge, 2001.

49. Victor E Hill. *Groups and Characters*. CRC Press, Boca Raton, Fla, 2018.

50. J. Sutherland Frame, G. de B. Robinson, and Robert M. Thrall. The hook graphs of the symmetric group. *Canadian Journal of Mathematics*, 6:316–324, 1954.

51. William Fulton and Joe Harris. *Representation Theory - A First Course*. Springer Science & Business Media, Berlin, Heidelberg, 2013.

52. William Fulton. *Young Tableaux - With Applications to Representation Theory and Geometry.* Cambridge University Press, Cambridge, 1997.

53. Dudley E. Littlewood. *The Theory of Group Characters and Matrix Representations of Groups.* AMS Chelsea Publishing, Province, second edition, 2006.

54. Hossein Abbaspour and Martin A. Moskowitz. *Basic Lie Theory.* World Scientific, Singapur, 2007.

55. S. Uhlmann, R. Meinel, and A. Wipf. Ward identities for invariant group integrals. *J. Phys. A*, 40:4367–4390, 2007.

56. Hans Samelson. *Notes on Lie Algebras.* Springer Science & Business Media, Berlin, Heidelberg, 2012.

57. J.E. Humphreys. *Introduction to Lie Algebras and Representation Theory.* Springer Science & Business Media, Berlin, Heidelberg, 2012.

58. Nathan Jacobson. *Lie Algebras.* Courier Corporation, New York, 1979.

59. Robert N. Cahn. *Semi-Simple Lie Algebras and Their Representations.* Dover Publications Inc., 2006.

60. T. Bröcker and T.tom Dieck. *Representations of Compact Lie Groups.* Springer Science & Business Media, Berlin, Heidelberg, 2013.

61. Matthias Bartelmann, Björn Feuerbacher, Timm Krüger, Dieter Lüst, Anton Rebhan, and Andreas Wipf. *Theoretische Physik.* Springer, Berlin, Heidelberg, New York, 2014.

62. Brian Hall. *Lie Groups, Lie Algebras, and Representations - An Elementary Introduction.* Springer, Berlin, Heidelberg, 2015.

63. R. W. Carter. Conjugacy classes in the weyl group. *Compositio Mathematica*, 25(1):1–59, 1972.

64. Jürgen Fuchs and Christoph Schweigert. *Symmetries, Lie Algebras and Representations - A Graduate Course for Physicists.* Cambridge University Press, Cambridge, 2003.

65. Anatoliy U. Klimyk and Jirí Patera. Antisymmetric orbit functions. *Symmetry Integrability and Geometry-methods and Applications*, 3:023, 2007.

66. Hans Freudenthal. Zur Berechnung der Charaktere der halbeinfachen Lieschen Gruppen, I. *Indag Math.*, page 369, 1954.

67. B. Kostant. A formula for the multiplicity of a weight. *Trans. Amer. Math. Soc*, 16:53, 1959.

68. Kurt Gottfried and Tung-Mow Yan. *Quantum Mechanics: Fundamentals.* Springer Science and Business Media, Berlin, Heidelberg, 2013.

69. Alberto Galindo and Pedro Pascual. *Quantum Mechanics I.* Springer Science and Business Media, Berlin, Heidelberg, 2012.

70. Steven Weinberg. *Lectures on Quantum Mechanics.* Cambridge University Press, Cambridge, 2015.

71. E. U. Condon and G. H. Shortley. *The Theory of Atomic Spectra.* Cambridge University Press, Cambridge, 1935.

72. H. Goldstein. Prehistory of the runge–lenz vector. *Am. J. Phys.*, 43:737, 1975.

73. H. Goldstein. More on the prehistory of the laplace or runge–lenz vector. *Am. J. Phys.*, 44:1123, 1976.

74. A. Kirchberg, J. D. Lange, P. A. G. Pisani, and A. Wipf. Algebraic solution of the supersymmetric hydrogen atom in d-dimensions. *Annals Phys.*, 303:359–388, 2003.

75. J. J. Sakurai. *Advanced Quantum Mechanics.* Pearson Education, Incorporated, Old Tappan, 2006.

76. Armin Wachter. *Relativistic Quantum Mechanics.* Springer, Berlin, Heidelberg, 2011.

77. Florian Scheck. *Quantum Physics.* Springer, Berlin, Heidelberg, Dordrecht, London, second edition, 2013.

78. Steven Weinberg. *The Quantum Theory of Fields.* Cambridge University Press, Cambridge, 1995.

79. Joel Scherk. Extended Supersymmetry and extended Supergravity theories. *NATO Sci. Ser. B*, 44:0479, 1979.

80. Lev Davidovic Landau and Evgenij M. Lifsic. *Klassische Feldtheorie.* Deutsch, Frankfurt am Main, 1997.

81. Florian Scheck. *Classical Field Theory: On Electrodynamics, Non-Abelian Gauge Theories and Gravitation.* Springer, Berlin, Heidelberg, second edition, 2018.

82. Davison E. Soper. *Classical Field Theory.* Courier Dover Publications, Mineola, New York, 2008.

83. Sidney Coleman. *Quantum Field Theory - Lectures of Sidney Coleman.* World Scientific, Singapur, 2019.

84. Michel E. Peskin and Daniel V. Schroeder. *An Introduction To Quantum Field Theory.* CRC Press, Taylor & Francis Group, Boca Raton, London, New York, 2019.

85. Gordon Walter Semenoff. *Quantum Field Theory: An Introduction.* Springer Nature, Singapore, 2018.

86. Matthew D. Schartz. *Quantum Field Theory and the Standard Model: With 191 Exercises.* Cambridge Universit Press, Cambridge, 2014.

87. A. Pais and G. E. Uhlenbeck. On Field theories with nonlocalized action. *Phys. Rev.*, 79:145–165, 1950.

88. Ph. Blanchard and E. Bruning. *Variational Methods in Mathematical Physics: A Unified Approach.* Springer, Berlin, Heidelberg, 2014.

89. Steven Weinberg. *Gravitation and Cosmology - Principles and Applications of the General Theory of Relativity.* Wiley, New York, 1972.

90. Andreas W. Wipf. Hamilton's formalism for systems with constraints. *Lect. Notes Phys.*, 434:22–58, 1994.

91. F. J. Belinfante. On the current and the density of the electric charge, the energy, the linear momentum and the angular momentum of arbitrary fields. *Physica*, 7(5):449–474, May 1940.

92. H. Weyl. A New Extension of Relativity Theory. *Annalen Phys.*, 59:101–133, 1919.

93. H. Weyl. Electron and Gravitation. 1. (In German). *Z. Phys.*, 56:330–352, 1929.

94. Chen-Ning Yang and Robert L. Mills. Conservation of Isotopic Spin and Isotopic Gauge Invariance. *Phys. Rev.*, 96:191–195, 1954.

95. L. O'Raifeartaigh. *Group Structure of Gauge Theories.* Cambridge University Press, Cambridge, 1986.

96. L. O'Raifeartaigh and N. Straumann. Gauge theory: Historical origins and some modern developments. *Rev. Mod. Phys.*, 72:1–23, 2000.

97. Stefan Pokorski. *Gauge Field Theories.* Cambridge University Press, Cambridge, 2000.

98. L. O'Raifeartaigh, N. Straumann, and A. Wipf. On the origin of the Aharonov-Bohm effect. *Comments Nucl. Part. Phys.*, 20(1+2):15–22, 1991.

99. V. A. Novikov, Mikhail A. Shifman, A. I. Vainshtein, and Valentin I. Zakharov. Two-Dimensional Sigma Models: Modeling Nonperturbative Effects of Quantum Chromodynamics. *Phys. Rept.*, 116:103, 1984.

100. P. A. Zyla et al. Review of Particle Physics. *PTEP*, 2020(8):083C01, 2020.

101. Steven Weinberg. *The Quantum Theory of Fields: Volume 3, Supersymmetry.* Cambridge University Press, Cambridge, 2005.

102. A. A. Belavin, Alexander M. Polyakov, and A. B. Zamolodchikov. Infinite Conformal Symmetry in Two-Dimensional Quantum Field Theory. *Nucl. Phys. B*, 241:333–380, 1984.

103. P. Goddard, A. Kent, and David I. Olive. Unitary Representations of the Virasoro and Supervirasoro Algebras. *Commun. Math. Phys.*, 103:105–119, 1986.

104. Philippe Francesco, Pierre Mathieu, and David Senechal. *Conformal Field Theory.* Springer, Berlin, Heidelberg, New York, 2011.

105. Ralph Blumenhagen and Erik Plauschinn. *Introduction to Conformal Field Theory - With Applications to String Theory.* Springer Science, Berlin, Heidelberg, 2009.

106. Martin Schottenloher. *A Mathematical Introduction to Conformal Field Theory.* Springer Science & Business Media, Berlin, Heidelberg, 2008.

107. H. A. Kastrup. On the Advancements of Conformal Transformations and their Associated Symmetries in Geometry and Theoretical Physics. *Annalen Phys.*, 17:631–690, 2008.

108. M. Luscher and G. Mack. Global Conformal Invariance in Quantum Field Theory. *Commun. Math. Phys.*, 41:203–234, 1975.

109. Joseph Polchinski. Scale and Conformal Invariance in Quantum Field Theory. *Nucl. Phys. B*, 303:226–236, 1988.

110. Arne Dettki and Andreas W. Wipf. Finite size effects from general covariance and Weyl anomaly. *Nucl. Phys. B*, 377:252–280, 1992.

111. Norbert Straumann. *General Relativity*. Springer Netherlands, Dordrecht Heidelberg London New York, 2014.

112. M. Luscher and G. Mack. The energy momentum tensor of critical quantum field theories in 1+1 dimensions. *Hamburg, unpublished*, 1976.

113. M. A. Virasoro. Subsidiary conditions and ghosts in dual-resonance models. *Phys. Rev. D*, 1:2933–2936, May 1970.

114. Daniel Friedan, Zong-an Qiu, and Stephen H. Shenker. Conformal Invariance, Unitarity and Two-Dimensional Critical Exponents. *Phys. Rev. Lett.*, 52:1575–1578, 1984.

# Index